THIS BOOK READY FOR ISSUE   MAY   1 1985

MSU LIBRARIES

RETURNING MATERIALS:
Place in book drop to
remove this checkout from
your record. FINES will
be charged if book

Precious Metals
in the Later Medieval
and Early Modern Worlds

Precious Metals in the Later Medieval and Early Modern Worlds

Edited by J. F. Richards

Carolina Academic Press
Durham, North Carolina

International Standard Book Number: 0-89089-224-5
Library of Congress Catalog Card Number: 82-73059

© 1983 by J.F. Richards. All Rights Reserved
Printed in the United States of America

Carolina Academic Press
PO Box 8795 Forest Hills Station
Durham, North Carolina 27707

Contents

	page
Preface	vii
Introduction	3

Part 1: Medieval Monetary Flows

1 The Italian Gold Revolution of 1252: Shifting Currents in the Pan-Mediterranean Flow of Gold — 29
Thomas Walker

2 Monetary Flows—Venice 1150 to 1400 — 53
Louise Buenger Robbert

3 Money and Money Movements in France and England at the end of the Middle Ages — 79
Harry A. Miskimin

4 Bullion Flows and Monetary Contraction in late-Medieval England and the Low Countries — 97
John Munro

5 Monetary Movements in Medieval Egypt, 1171–1517 — 159
Jere L. Bacharach

6 Outflows of Precious Metals from Early Islamic India — 183
J.F. Richards

7 The China Connection: Problems of Silver Supply in Medieval Bengal — 207
John Deyell

Part 2: Monetary Expansion and Intensified Demand for Metals

8 Africa and the Wider Monetary World, 1250–1850 — 231
Philip D. Curtin

9 The Role of International Monetary and Metal Movements in Ottoman Monetary History 1300–1750 — 269
Halil Sahillioğlu

10 Gold and Silver Exchanges Between Egypt and Sudan, 16-18th Centuries — 305
Terence Walz

11 Silver Mines and Sung Coins—A Monetary History of Medieval and Modern Japan in International Perspective — 329
Kozo Yamamura and Tetsuo Kamiki

12 Vietnam and the Monetary Flow of Eastern Asia, Thirteenth to Eighteenth Centuries — 363
John K. Whitmore

Part 3: New World Metals

13 South American Bullion Production and Export 1550-1750 — 397
Harry E. Cross

14 New World Silver, Castile and the Philippines 1590–1800   425
    John J. TePaske
15 The Exports of Precious Metal from Europe to Asia by the
      Dutch East India Company, 1602-1795   447
    F.S. Gaastra
16 Silver in Seventeenth-Century Surat: Monetary Circulation
      and the Price Revolution in Mughal India   477
    Joseph J. Brennig

Index   497

Preface

I first began considering world monetary history six years ago in a series of conversations with Maureen Mazzaoui, my colleague and fellow-member of the Comparative World History Program at the University of Wisconsin, Madison. My interest in the medieval and early modern economic history of India and the Middle East converged with her interest in the early-modern economic history of Italy and the Mediterranean. Stimulated by a fresh reading of Andrew Watson's well-known essay "Back to Gold and Silver," *The Economic History Review* (1967), we began to realize the need for a more comparative, global view of monetary systems and their interaction before the period of European colonial domination. A preliminary visit and lecture by Andrew Watson, who supported and encouraged this view, led to our decision to organize and hold a conference on this topic at Madison. Further discussions, notably with our colleague Domenico Sella, reduced the scope of the conference to a reasonably manageable and important theme: the inter-regional flow of precious metals in the period 1200 to 1750 A.D.

Setting the date of the proposed conference for early September 1977, we allowed a full year for formulating and writing a coherent proposal. In the course of that year we also gradually identified and contacted a group of more than twenty historians, numismatists and other scholars who were actively interested in the monetary history of one of the major world regions in this period. The remaining months before the conference demanded a complicated fundraising effort and the actual organization of the conference. Finally, after this extended preparation we did convene the conference as planned. Participants giving papers and additional commentators assembled at the Elm Drive Conference Center on the lakeshore at Madison for five full days of discussions.

To my knowledge this was a first attempt to convene a meeting of scholars to consider both medieval and early modern monetary history on a world scale. We tried, reasonably successfully, to have at least one paper devoted to each of the conventionally defined-world regions (e.g., Southeast Asia, Middle East). We tried to balance European specialists with specialists concerned with Asia, Africa or Latin America. We tried to balance the conference chronologically, not quite successfully, by including equivalent numbers of medievalists and early modernists. In the end we brought together an international group of scholars of varied experience, age and reputation, and distinctive area and disciplinary specialities. In other words this was not, as is sometimes the case, a cozy group of specialists assembling once again to share recent research in a well-defined field of mutual interest. Instead, for all concerned the conference was a challenging and stimulating exposure to new materials and new colleagues.

Consideration of unfamiliar material contained in each paper meant that each conferee had to review his or her own research and historical understanding to make comparative judgements and to enter the ongoing discussion. This juxtaposition of topics engendered a comparative assessment by the group which was extremely valuable. I think that it is safe to say that each conferee departed with possible new answers to old questions; with a better grasp of inter-regional monetary connections; and with new issues and questions to consider. For their willingness to engage in this debate, and for their unfailing energy, warmth and civility—which added immeasurably to the value of the conference—Maureen Mazzaoui and I must thank all the participants.

Sixteen of the original conference papers appear in this volume. All have been revised by their authors subsequent to the Madison conference. All address the question of inter-regional fine metal flows during the period from 1200 to 1800 A.D. Common themes and shared motifs do emerge from these essays as I have tried to suggest in the introduction. But the collection as a whole does not and could not possibly present a definitive or systematic global portrait of changing precious metal transfers prior to the 19th century. Instead, together they offer a sampling, a beginning point for what we hope will be a furthering of this aspect of monetary history not explored. Until more monographic and systematic research is completed the state of the art rests close to this point in this important branch of world economic history.

I cannot emphasize how much the organization and convening of the 1977 Madison conference depended upon the enthusiasm, scholarship and support of Maureen Mazzaoui. Without her full partnership the conference could not have been held. Although she has been unable to assist in the editing of this volume, her work is very much a part of the volume as well as the conference.

Because of previous commitments, several of the original members of the Wisconsin conference were unable to publish their revised papers in this volume. Despite this loss, the contributions of Artur Attman, William Atwell, K. N. Chaudhuri, Marie Martin, and Arthur Seltman to the shaping of this volume are tangible and much appreciated. The discussants, Alan Heston, Andrew Watson, and John Sharpless, added invaluable disciplinary and research insights to the 1977 meeting. I am also grateful for advice and support from Philip Curtin, Kamal Karpat, Steven Feierman, Domenico Sella and others of my colleagues on the Comparative World History Committee at the University of Wisconsin. Many other persons around the country helped in identifying participants as well.

Preface

   Maureen Mazzaoui and I are indebted to the National Endowment for the Humanities for a grant toward the expenses of the conference. Dean E. David Cronon supplied approximately half the funding required for the conference from sources with the College of Arts and Science at the University of Wisconsin. The latter assistance was indispensable. Finally, I should like to acknowledge the support of the editor and publisher of Carolina Academic Press, Keith Sipe, for his energy and vision in bringing this volume to publication.

J. F. Richards
Durham, NC

Precious Metals in the Later Medieval and Early Modern Worlds

Introduction

I

Monetary history flourishes today as a sub-field of economic history. Especially for the medieval and pre-modern periods in European and Mediterranean history, the study of coinage, currency systems, minting policies and practices and allied topics is a rigorous field of inquiry which has made a considerable contribution to our understanding of the economy and culture of Europe and North America. We also can cite impressive gains in our understanding of the evolution of monetary systems in other areas of the world. Increasingly South Asian, Japanese, Russian historians—well-trained in economic theory, in statistics, as well as in the fast-developing techniques of the numismatist—have been studying the characteristics and development of the monetary systems of their own and other societies. Since the 1950's the issues raised by Karl Polanyi and his followers regarding prevailing assumptions of micro-economic theory in regard to exchange, to value, to wealth and toward the theory of markets has further encouraged serious research in this branch of economic history and theory.

But despite these gains our understanding of changing medieval and early modern monetary systems is still far from satisfactory. We have some difficulty in defining the limits of discrete monetary systems or in distinguishing levels or enclosures within larger systems of exchange. An especially weak area lies in the determination of ties between various economies and regions. For the medieval and early modern periods (defined as approximately 1200 to 1800 A.D. for the purposes of this volume) wide usage of metallic currencies in the precious metals (gold and silver) was characteristic of most monetary systems in Europe, Asia, and an increasingly large portion of Africa. The movement of fine metals, either as coin or bullion, was one of the most important linkages between the economies of each region and sub-region.

The premise of this volume and the organization of the conference which preceded it is that to fully understand monetary systems in the period 1200 to 1800 A.D. we must study inter-regional flows of gold and silver. Insofar as possible, we need to know the direction, magnitude, fluctuation and duration of gold and silver transfers between major economic regions (e.g., western

Europe) as well as between subregions (e.g., northern Italy). Contemporary theorists and men of affairs in those centuries agreed that the presence or absence of supplies of gold or silver adequate for coinage was important. They also knew that, by and large, the movement of these metals tended to be predictable over fairly long periods of time, although the level of supply could not always be predicted nor controlled. We may not wish to give to these movements the same importance attributed to them by contemporaries; nonetheless, the flow of fine metals was always an important variable in the operation of any medieval or premodern monetary economy. The force of demand for monetary purposes (as well as for display or other uses) can be seen in the ongoing hunt for new areas of production in the medieval world.

A further assumption shaping this volume is that more self-consciously comparative research and discussion is essential. Comparison is necessary first to discover these multiple links established by the transfer of precious metals and secondly to describe the characteristics of these flows. Until quite recently a major obstacle to truly comparative work has been a continuing concentration by monetary historians and economic historians upon the New World supplies of precious metals which surged into Europe and the Mediterranean world in the mid-sixteenth century. Obviously American treasure imports were the most important new source of precious metal supplies since antiquity, but over-concentration on this development should not reduce the complexities of world monetary history to a single change—no matter how dramatic. Thus, even medieval historians look at economic conditions in the 15th century (or earlier) with an eye to questions of price inflation as opposed to growth in productivity in the following century. The scale of production by the New World mines too readily suggests that other sources and other monetary flows were inconsequential. Over concentration upon imports of silver through Seville tends also to distort our perception of the monetary systems and economies of Asia in the pre-modern period. These societies often are seen as merely passive recipients of the silver and gold drained from Europe, which having somehow lost its dynamic quality, comes to rest in hoarded caches in India or China.

Our fascination with the New World treasure continues the fascination and puzzlement of 16th century Europe with this issue. After the Discoveries, within a remarkably short time, European societies bothered by inadequate supplies (often bordering on famines) of silver and gold supplies were suddenly wealthy—through no discernible effort on their part. It is thus only slightly an exaggeration to assert that the American treasure imports initiated by the mid-sixteenth century, a lively, broad-ranging pan-European debate and inquiry into monetary theory, or in Foucault's terms "the analysis of wealth."[1] For over two centuries men of affairs and scholars grappled with the monetary problems and issues continually raised by the influx of American treasure and its movement through the economies of Europe and adjacent

---

1. Michel Foucalt, *The Order of Things*, English translation (New York, 1971), p. 168.

Introduction 5

regions. From writers as early as Bodin in the 1560's to Law and Locke in the mid-eighteenth century discussion of the true role of newly produced gold and silver supplies within the monetary order of Europe remained a central issue.

Foucault postulates an important alteration in the conception of precious metals as money. This reflected the general shift from "Renaissance" to "Classical" thought:

> For the Renaissance "economists," ... the ability of money to measure commodities, as well as its exchangeability, rested upon its intrinsic value.... [Gold and silver] possessed, both in the natural scale of things and in themselves, an absolute and fundamental price, higher than any other, to which the value of any and every commodity could be referred.... Fine metal ... had a *price;* for this reason too ... it was a *measure* of all prices; and for this reason, finally, one could *exchange* it for anything else that had a price.[2]

Foucault argues that this view underwent a dramatic reversal in the early years of the 17th century:

> In the 17th century, these three properties [price, measure and exchange] are still attributed to money, but they are all three made to rest, not on the first (possession of price), but on the last (substitution for that which possesses price). Whereas the Renaissance based the *functions* of coinage (measure and substitution) on the double nature of its intrinsic *character* (the fact that it was precious), the 17th century turns the analysis upside down; it is the exchanging function that serves as a foundation for the other two characters (its ability to measure and its capacity to receive a price) thus appearing as *qualities* deriving from that *function*.[3]

If Foucault's analysis is correct, we might suggest (although he does not do so directly) that the sudden abundance of American treasure in the late sixteenth—early 17th century, as well as the perceived threat of continuing hemorrhage to the East, must have encouraged reflection upon the dynamic of money—and especially upon the role of precious metals. Throughout the premodern period scholarly and popular preoccupation, if not obsession, with the American stocks of bullion and coin continued. Even Adam Smith, engaged in establishing a new, radical definition of wealth (based upon labour) in the late 18th century, felt obliged to discuss in a "Digression" the dynamic impact of New World silver upon prices.[4] Perhaps not so coincidentally mercantilist economic analysis and the great period of New World treasure each ended by the last years of the 18th century.

Not surprisingly, modern scholarly interest in the systematic comparative study of precious metal transfers gained great impetus from data published in the 1920's and 1930's by E. J. Hamilton on the magnitude and trend of

---
2. *Ibid.*, p. 174.
3. *Ibid.*
4. Adam Smith, *An Inquiry Into the Nature and Causes of The Wealth of Nations*, ed. by Edwin Cannan (Modern Library edition, 1937), pp. 176-242. "Digression concerning the Variations in the Value of Silver during the course of the Four last centuries."

Spanish treasure imports into Seville.[5] Thereafter, in the immediate post-World War II period, Fernand Braudel, Lucien Fèbvre and other leading figures of the French *Annales* school have fostered medieval and early modern monetary history (in contrast to purely numismatic studies) as a fully-developed area of historical research. Braudel, his students, and the dozens of historians whom he has influenced have treated large-scale gold and silver movements as a major variable to be subjected to systematic description and analysis. Not necessarily returning to the fullblown mercantilist position, historians within this school (in the broadest definition) have pursued the notion of an intimate connection between price levels, productivity, and monetary supply in the medieval and early modern periods. These historians, perceiving the dynamic potential of precious metal flows—as seen most notably in the case of the Spanish treasure in the Mediterranean—have gradually moved toward consideration of this question in the regions beyond Europe.

Fernand Braudel, in the course of his assessment of the Mediterranean economies of the 16th century, comments that "whatever their origin, precious metals once absorbed into Mediterranean life were fed into the stream that continually flowed eastward."[6] In Braudel's view the evidence for this movement in the 16th and 17th centuries was direct, unambiguous and overwhelming. Ubiquitous circulation of coin minted in Italy or New Spain throughout the far reaches of Asia was but one fact among many confirming this massive drain of metals.

> Away to the east flowed these currencies, out of the Mediterranean circuit into which it had often required so much patience to introduce them. The Mediterranean as a whole operated as a machine for accumulating precious metals, of which, be it said, it could never have enough. It hoarded them only to lose them all to India, China, and the East Indies. The great discoveries may have revolutionized routes and prices, but they did not alter this fundamental situation, no doubt because it was still a major advantage to westerners to have access to the precious merchandise of the East, ...[7]

In a later passage, Braudel makes this point even more emphatically. Commenting upon the increasing values of silver (expressed in its ratio to gold) as it travelled from Italy to the Levant and on toward Asia in the sixteenth century he contrasts the demand for silver in Europe to that of China. The former hovered at a gold-silver ratio of 1:12 to the latter's ratio of 1:4. Thus the uninterrupted eastward movement of silver is scarcely surprising:

> This Italy-China axis, beginning in America and running right round the world either through the Mediterranean or round the Cape of Good Hope, can be considered

---

5. Earl J. Hamilton, *American Treasure and the Price Revolution in Spain, 1501-1650* (Cambridge, Mass., 1934).

6. Fernand Braudel, *The Mediterranean*, 2 vols. (New York, 1966), I, 464.

7. *Ibid.*

Introduction 7

a structure, a permanent and outstanding feature of the world economy which remained undisturbed until the twentieth century.[8]

Thus, Braudel carefully set into his paradigm of the Mediterranean world economy a northwest to southeast flow of precious metals. This was rapidly augmented, but not begun, by the import of New World silver into Seville in the 16th century.

As they have for numerous of Braudel's other assertions about premodern Mediterranean economic and social history a generation of French historians have devoted meticulous attention to studies ". . . in depth and in detail, with a documentation and methods neither of which were available to Braudel in the late forties when *La Mediterranee* was brought to birth."[9] Certainly, the post World War II rigor and sophistication of detailed monetary history owes much to Braudel. We can see this clearly in the deepening of Braudel's insights by some of his colleagues and students. Equally significant is the extension of monetary research. Thus, Pierre Chaunu's grand work on the New World-Atlantic-Seville linkage amplifies E. J. Hamilton's seminal study of Spanish silver exports.[10] Ömer Lufti Barkan has led the efforts of Turkish scholars to extend the Braudelian paradigm to the Ottoman empire—an approach which has been most fruitful for Ottoman monetary and economic history.[11] But the single work which offers the closest approximation to The Mediteranean is *L'économie de l'empire portugais aux xve et xvie siècles* by V. Magalhães-Godinho, a Portuguese scholar directly influenced by Braudel.[11] Godinho has created a model for the economy of the Indian Ocean and its linkages with Portugal via both the Atlantic and Mediterranean which is worthy of Braudel. A central theme is of course the movement of New World metals into the Indian ocean world economy in the pre-modern period. Beyond this, Godinho details the earlier and ongoing exploitation by the Portuguese of African gold (both west and east coast) and gold and silver in Southeast Asia. He describes in detail the complex interaction among the varied regional monetary systems within the Indian Ocean. Despite its underutilization by Anglo-American, South Asian and other scholars without access to French, Godinho's study remains the standard for all monetary historians venturing beyond the Mediterranean.

More recently, Pierre Vilar has set out a synthesis of these and other materials in a major comparative study. Titled *A History of Gold and Money*,

---

8. *Ibid.*, pp. 499-500.

9. J. H. Hexter, "Fernand Braudel and the *Monde Braudellien* . . . ," *Journal of Modern History*, v. 44 (1972): 480-539, esp. 529.

10. Pierre Chaunu, *Seville et l'Atlantique (1504-1650)*, Paris, 3 volumes, 1959 and other articles and essays.

11. See, for example, Ömer Lufti Barkan, "La 'Mediterranee' de Fernand Braudel vue d'Istamboul," *Annales*, E.S.C., IX, 1954; and Barkan, "The Price Revolution of the Sixteenth Century: A Turning Point in the Economic History of the Near East," *International Journal of Middle East Studies*, VI (1975): 3-28.

1450-1920, this work originated in a course of lectures first delivered at the Sorbonne. In his discussion of the pre-modern period Vilar begins with the Portuguese exploitation of African gold, but moves rapidly on to a consideration of the central problem: the accelerated inflow of newly-produced silver and gold from the New World and the subsequent reaction of the European economy to that development. Here Vilar is not concerned solely with the price rise of the 16th century, but assesses the rise of new monetary and commercial institutions as part of the dynamic rise of Europe's economy in the period between 1500 and 1800 A.D. Precious metal flows figure largely in Vilar's discussion, although they are not the principal object of his attention. The text and maps establish an overall pattern to these flows of precious metals in the early modern period. Nonetheless, despite its many virtues the work is not, as Vilar claims, an adequate treatment of the entire world. He does describe briefly monetary relationships between Europe and monsoon Asia, mediated by the Portuguese in the 16th century, and the movement of New World metals. But once again his focus is Europe—the only center for dynamic monetary and economic change in the pre-modern world.[12]

The above sketch fails to mention dozens of scholars and historians who have made other important contributions to world medieval and pre-modern monetary history and the narrower subject of monetary links within and beyond Europe—not all of whom are directly associated with or necessarily influenced by the *Annales* program. For example, Artur Attman (one of the participants in the Madison conference) treats Baltic, Eastern European, and Russian trade and monetary flows in two studies recently published in English translation.[13] William Atwell, another member of the conference, has begun what will be a series of detailed studies of the role of silver imports for the economy of Ming China.[14] Eliahyu Ashtor's assessment of Mid-Eastern Islamic economic and social history during the medieval and pre-modern periods adds considerable depth to our understanding of the circulation and flow of silver and gold within the classical locus of the Islamic world: Egypt, Syria, Iraq, and Iran.[15]

Generally speaking, however, correction of the European bias in monetary history demands that comparative juxtaposition which was so effective in the presentation of papers in Madison in 1977 and which is in part the object of this volume. That is, by presenting a range of case studies which examine important long-distance monetary flows from various world regions, we

---

12. Pierre Vilar, *A History of Gold and Money, 1450-1920* translated by Judith White (London, 1976) from the 1974 French edition. Frank Perlin has published a substantial review which argues this point very effectively. See Frank Perlin, "A History of Money in Asian Perspective," *The Journal of Peasant Studies*, VII (1980): 235-44.

13. Artur Attman, *The Russian and Polish Markets in International Trade, 1500-1650* (Goteborg, 1973); and *The Struggle for Baltic Markets, Powers in Conflict, 1558-1618* (Goteborg, 1979).

14. William S. Atwell, "Notes on Silver, Foreign Trade and the Late Ming Economy," *Ch'ingshih wen-t'i*, III: 1-33.

15. Eliahyu Ashtor, *A Social and Economic History of the Middle East* (Berkeley, 1976).

Introduction 9

begin to understand the extent to which these systems were interdependent. And, we might add, these systems were converging or growing in their degree of interdependence between the start our of period in the early 13th century and the closing years of the 18th century.

To discover the configuration of precious metal flows between localities, regions and continents at any given period, or to detect significant changes over time, an elevated supra-regional comparative perspective and an elongated temporal view are both essential. Obviously, these larger movements were formed by myriad individual transactions of varying size and direction carried out by traders, officials, and other contemporaries. The algebraic sum of these transactions determined the overall drift of gold and silver in the medieval and early modern periods. The process of attaining this perspective may be likened to that undertaken by archaeologists who use aerial photographs to detect the pattern of long disused irrigation systems whose outline is invisible at ground level.

Though it is not usually the practice to do so, we may, for the medieval and early-modern periods tentatively identify certain types of large-scale precious metal flows. Perhaps most easily defined and bounded was the journey of newly mined and refined metal from point of production to first use as money or for industrial purposes. (Within this category we can include the frequent practice of converting newly-mined metal into coin in mints established just adjacent to the mining region.) Whether in the form of bullion or just-minted coin the outward flow of gold or silver toward its first utilization can be viewed as a commodity movement. That is, the continuing output of alluvial workings, or mine, smelter and, occasionally, mint was an extractive process set in motion by demand expressed for those metals for monetary and exchange purposes (among others). The series of transfers and transactions which pulled new metal supplies from the zone of production in a consistent flow to satisfy expressed demand moved a commodity—despite occasional use of these supplies for exchange purposes en route.

A second long-term, large-scale monetary flow is a bit more complex and somewhat harder to delineate. In contrast to the commodity flow of new metal, relatively consistent linear movements of trade currencies provided another mechanism for inter-regional transfer of precious metals along the primary trade arteries of the medieval and pre-modern civilizations. Those currencies possessing the widest distribution and acceptance in long-distance trade—the hyperperon, the ducat, the dinar, the real—were the most effective vehicles for this form of movement. Certainly after emission from the mints of the issuing authority these coins did move within or "circulate" among the normal trade channels dominated by that state. But in addition to what might be seen as a reciprocal or circular movement of these coins, a series of local, sub-regional and regional exchange mechanisms drew off part of this flow in a unilinear fashion toward one or more ultimate destinations or repositories. On the whole this linear movement was determined by balance of payment mechanisms necessary for trade between disparate economies of varying productivity. Reflecting as they did basic economic relationships, these movements were known to contemporaries and often persisted over decades or even centuries.

The third category of precious metal flows in this period derives its impetus more narrowly from political rather than market or economic demands. That is, forceful imposition by one regime upon another for various types of tribute, ransom or plunder could cause large-scale movements of precious metals. To this list we might add remittances sent by individuals profiting from participation in some form of colonial or dominant alien power demands upon a subject population. While it is often the case that these transfers were merely episodes, more enduring political relationships—such as the decades of tribute in gold extracted by the Ottomans from Egypt—could sustain a long-term flow of precious metals. (Of course problems of definition do arise, if over time, economic integration accompanying intensifying policies of direct administration, altered the character of monetary movements within a more unified polity. Political exchanges presumably assumed the circulatory patterns of a coherent monetary system.) Nonetheless, political transfers of gold and silver must be reckoned as having a significant economic impact upon regional economies and upon the shape of world monetary interaction.

II

Walker, in "The Italian Gold Revolution of 1252" analyzes the interregional monetary system of the thirteenth-century Mediterranean World. The breakup of the Byzantine empire by 1204 A.D. dislodged the Byzantine gold hyperperon as a stable gold coin accepted for use in inter-regional trade. At the same time the Egyptian dinar and its several Crusader state imitations and counterfeits also faltered by the beginning of the 13th century. Faced with a serious impediment and added cost to their burgeoning trade, the North Italian city states began to coin their own high quality gold currencies which were of a fineness and quality sufficient for long-distance trading acceptance and circulation.

But this monetary initiative demanded commodity supplies of gold for the Italian mints. Early medieval Genoa and Florence quickly turned to African gold as the most plausible source for a reliable supply of gold. Reaching across the Mediterranean, the merchant oligarchies of the European city states obtained gold from the Hafsid rulers of Tunis. The latter port-city had become the northernmost entrepôt for the exchange of Sub-Saharan African gold brought overland by caravan for European goods and services. From Tunis the stream of gold moved across the Mediterrean to the Florentine mints via Sicily and Pisa. Or in an increasingly favored route African gold was shipped direct from Tunis to Genoa or Florence. Other, lesser amounts of gold (usually gold dust carried in bags) found its way through the complex trade network dominated by the Italian trade colonies scattered around the

Mediterranean. The heightened economic strength of Northern Italy and its trade dominance created an equivalent African demand. This in turn attracted and perhaps accelerated the extended African commodity flow of gold from its point of production in the interior overland to Tunis and thence through the Mediterranean trading networks to the Italians mints.

In her discussion of medieval Venetian monetary movements, Robbert also describes the inflow of new commodity gold to Venice for minting purposes. Throughout the 13th and 14th centuries the Venetians relied upon both their own and foreign coinage; the 2.2 gram Venetian silver grosso for domestic circulation and limited export, and the gold Byzantine hyperperson or Arab gold dinar for trans-Mediterranean trade. Demand for silver induced German merchants to import silver overland from the rapidly developing mining regions of Bohemia, Hungary, Silesia and Saxony. Even after output from these mines began to decline after 1350 Venetian moneyers obtained a near-monopoly of the silver produced in the newer mines of Serbia and Bosnia. Ragusan merchants served as entrepreneurs and middlemen to deliver these supplies to Venice. One stream of silver in commodity form moved overland to the Venetian city-state from eastern European mining regions via Germany. Another stream moved by sea from the highly productive Bosnian mines (after 1295 A.D.) via Ragusa on the Dalmatian Coast to Venice.

In 1284 A.D. Venice uttered the first gold ducats of 3.559 grams weight at 24 carats fineness. Soon supplanting Florentine issues, the gold ducat of nearly unvarying fineness and weight served the Mediterranean world and beyond for over five hundred years as a stable medium of trade. Commodity supplies of metal for the ducat included gold from the mines of Hungary and Bohemia which arrived in Venice on the overland route in exchange for Venetian luxury imports. By the first quarter of the 14th century Venetian minting demands for the ducat also diverted increasing supplies of West African gold from Northern Italy. Obtained at Tunis by Venetian merchants this African gold bullion followed the familiar cross-Mediterranean sea route. Finally, Robbert notes that Venice did obtain considerable supplies of gold by sea for minting from its trading stations on the Black Sea. Since no substantial gold mining areas existed in Anatolia, Russia or Central Asia in this period, it is unlikely that this route relied upon newly produced gold as a source of supply. Instead gold bullion obtained as payment from Mamluk Egypt or Muslim India for Turkish slaves or horses funneled through the lands controlled by the Golden Horde to eventually arrive at Venice.

The great economic strength of Venice, derived from trade, industrial production, and fiscal services, was reflected in the monetary dominance achieved by the Venetian currency. By the 14th century both gold ducat and silver grosso were widely accepted and used in trade throughout the Mediterranean and its adjacent regions. The pattern of circulation was shaped in large measures by Venetian trading stations and colonies around the Mediterranean. Nevertheless, as Robbert emphasizes, both ducat and grosso were subject to the southeastern pull of the Indian Ocean economies. The net effect of this pronounced linear flow, which continued for several centuries, was to drain Venetian and other Italian trading currencies toward India. By the 13th and 14th centuries, if not well before, western European monetary systems were also directly involved in the long-distance flow of precious metals. Miskimin, in "Money and Money Movements in France and England at the End of the Middle Ages" points out that western Europe after 1360, in common with the Mediterranean countries, suffered an increasing shortage of silver for coin manufacture. Rejecting earlier analyses based upon the influence of a more favorable gold to silver ratio in western Europe, Miskimin argues that lessened economic productivity in the wake of the Black Death weakened the balance of payments position of England and France in relation to Italy. In other words intensifying medieval trade with the Levant, increasingly dominated by the Italian city-states, was the mechanism for draining precious metals—and especially silver—from the north. Lessened productivity in western Europe exacerbated this effect.

In his revised and amplified essay Munro has constructed a new extended set of mint production statistics for England (1235-1500) and the Low Countries (1335-1500). Using this comprehensive data series, he reviews the more recent literature devoted to the various questions regarding late medieval European monetary supplies. Munro addresses the intricate questions of a possible monetary contration; of its timing; of a possible bullion "famine" and of the relationship to these developments to more general economic trends within the 14th and 15th century European economy. For the past forty years these have been among the most intensely studied and debated areas in world monetary history. His argument essentially supports the Miskimin and Day theses on late-medieval monetary constriction caused, in large measure, by continuing outflows to the East. Munro also argues, however, that beginning about the late 1360's this tendency was exacerbated by a European mining slump, by coin wear, by hoarding and by industrial use of precious metals. He considers the role of European pro-gold mint-ratios in causing the outflow of silver to the Levant in this period (as suggested by Watson). The mint-ratios were unlikely "in themselves able to instigate international bullion flows in the late-medieval world." On the other hand the ratios cannot be ignored for they could determine which metal predominated in the flows outward from Europe. Merchants employed what Munro terms "indirect arbitrage" in both goods and metal. Thus, the so-called silver famine of the 14th century was probably a result of pro-gold western European monetary policies, as Watson argues. Yet in the 15th century mint outputs of gold declined sharply as the price of gold rose and we can see a general deflation in England and the Burgundian low countries during the middle decades of the 15th century.

Bacharach's description of monetary movements in pre-Ottoman medieval Egypt reveals the monetary difficulties of a regime presiding over an active Mediterranean economy which lacked indigenous sources of precious

Introduction

metals. He concludes that between 1171 and 1399 A.D. successive Muslim rulers of Egypt managed, despite occasional crises, to induce commodity inflows of gold and silver sufficient for state minting needs. Gold brought into Egypt by the West African overland caravan route seems to have arrived in quantities which allowed the minting of large numbers of high quality dinars in this period. Silver supplies were somewhat more uncertain. Egypt's silver supply did not come directly from an area of new production. Instead the Ayyubid and Mamluk rulers apparently depended upon control of Syria where European and Central Asian silver was far more abundant (presumably due to trading surpluses and subsidies for the Crusader states). The mechanism for inducing silver inflows to the Egyptian mints was therefore largely political: the flow of tribute, taxation and state-forced trade. In the absence of this political connection with Syria, Egyptian silver currencies suffered accordingly.

In spite of Egypt's reasonably successful efforts to supply and issue gold and silver coinages, as an entrepôt economy, situated between the Mediterranean and the Indian Ocean trading zones, it was subject to the steady drain of precious metals south and east. Before 1399 it is probable that, over the longterm, Egyptian losses in precious metals to the Indian Ocean were balanced off by profits gained from its own industrial and agricultural exports, as well as by payments for "invisible" trade and fiscal services. That is, precious metals originating in Europe or elsewhere may have passed through Egypt, but not created a deficit. After 1399, according to Bacharach, Egypt suffered increasing monetary difficulties. Egypt's trade balance with Europe shifted in favor of the latter—especially in industrial goods (glassware, metalware, textiles, etc.). In the early 15th century Egypt's ability to draw substantial commodity inflows of African gold into its mints declined relative to the Italian demand. By this time the Venetian ducat had overtaken the dinar as the leading trade coinage of the Mediterranean. Weaker ties with Syria meant that silver supplied from that region were much reduced. For at least two decades at the beginning of the 15th century Mamluk Egypt suffered a true silver famine before recovery slowly began. Silver dirhams issued by the Mamluk Sultans underwent continuous debasement as the drain of precious metal to the east continued. Throughout the century in Bacharach's words Egypt experienced "political, economic and monetary crisis and chaos."

My own essay, "Outflows of Precious Metals from Early Islamic India," first addresses the question of the long-standing power of India to attract continuing imports of precious metals. From antiquity this has been a prominent feature of the pre-modern world trading pattern. In an argument to be further developed I insist that a simplistic assumption that an excessive propensity of hoarding (i.e., some form of cultural greed) offers an inadequate explanation for the steady movement of gold and silver toward India. Instead we must look to the true economic position of the subcontinent. In the medieval Islamic period (13th to 15th centuries) North and Central India displayed a high rate of industrial production, export-oriented cash crop agriculture, well-developed transport and fiscal services, and stable political systems. Political conquest and growing centralized state power exerted from Lahore, Delhi and Laknauti encouraged a resurgence of urbanized society in the north. In general, one might argue for a largely self-sufficient economy. Therefore a favorable balance of payments and steady inflows of precious metals paid for India's goods which were much in demand in other regions of the world.

We oversimplify if we assume that medieval India, both North and South, only passively absorbed precious metals. Several distinct types of reverse mechanisms moved precious metals out of India. Recurring plunder and tribute extracted by Muslim rulers from Hindu or Buddhist monarchs pushed large amounts of gold and silver into Central Asia, Persia and beyond. Gradually as Muslim rule expanded in India these trans-subcontinental commodity transfers gave way to a steadier flow of remittances to the same destinations by Muslim officials, soldiers and traders. Largess and pious donations, royal and private, tapped the wealth of India for the shrines of the Islamic heartlands. Also extremely important was the movement of coin which paid for the import of military horses overland from Central Asia and Tibet and by sea from the Persian Gulf. It is doubtful if, in total, these outflows exceeded the long-term favorable monetary balance of payments enjoyed by medieval India. Yet the size and extent of these outflows suggest that they may well have had significant impact upon other regions.

Deyell in "The China Connection: Problems of Silver Supply in Medieval Bengal" underscores the difficulties of new Muslim rulers and their moneyers in implementing and sustaining silver and gold coinages. Between 1200 and 1500 A.D. we know of no meaningful sources of domestic mining for silver or gold upon the subcontinent. Deyell argues that inland the 14th-century Sultans of Delhi relied upon released stocks of precious metals from the south for a supply of both gold and silver. On the other hand the independent Sultans of Bengal probably obtained their supplies of silver from a source and a route which has been little recognized: southwest China and eastern Burma by means of overland trade routes using pack animals. Deyell points to the overland route as an important alternative to the well-known sea route to India from southeast Asia.[16] Although Deyell does not emphasize this point it seems likely that the new Turkish and Afghan dynasties of Muslim-held Bengal generated a new intensity to Bengal's demand for silver in contrast to the monetary policies of their Hindu predecessors. The inflows of commodity silver destined for minting purposes over both the overland and sea routes from southeast Asia to India may have been greatly amplified after 1200 A.D.

---

16. I should add that his view of precious metal imports coming overland from the northeast, contradicts my assumption that Indian metals flowed northeast to pay for imports of horses over the same routes. This contradiction obviously will need further study.

III

Between the mid-fifteenth and mid-sixteenth century—the generally accepted divide between "medieval" and "early modern" times—global monetary systems and the patterns of precious metal flows changed dramatically. The most significant of these changes was of course the American discoveries and the exploitation of new, abundant sources of New World gold and silver. The extent to which cheap American silver affected world economic history has long been studied and debated. But emphasis upon American silver imports has somewhat overshadowed a number of other important changes in the world monetary pattern. During this transitional century newly-formed, larger, more centralized states or economic enterprises acted decisively to meet their supply needs for precious metals from older and new sources of production *before* the full onset of imports from the New World. The following set of essays examines these changes as well as the impact of American treasure in synoptic studies for four areas: Africa; the growing Ottoman domains centered upon Anatolia and the Balkans; Japan; and Vietnam.

Curtin, in "Africa and the Wider Monetary World, 1250-1850," discusses the integration of sub-Saharan Africa as a commodity supplier of gold bullion to Eurasian (and North African) monetary systems. From the 12th through the 15th centuries total demand and total exports of sub-Saharan gold apparently moved steadily upward. In addition to overland trans-Saharan exports (estimated by Curtin at as much as 1500 k.g. per year) gold bullion from Ethiopian placer deposits moved via Red Sea ports to Egypt. Gold from the eastern African interior moved to Sofala and other East African ports and thence as part of the merchandise of Arab traders across the Arabian Sea to the Persian Gulf.

The 15th century saw a rising demand for African gold in Europe and Mediterranean economies. The post-1450 maritime ventures of the Portuguese around the coast of West Africa established a new European-West African sea link. The new route permitted the Portuguese to obtain gold in the West African ports. Apparently this new expression of heightened European demand for newly-produced gold did not divert the bulk of the northward trans-Saharan traffic in that metal. In fact, if anything, overland traffic enjoyed a resurgence with firm Moroccan political control in the latter years of the 16th century. The probable response to intensified market demand (expressed in rising prices) was heightened production. As Curtin points out, Dutch and English interloping ventures within the Portuguese monopoly further added demand pressure for West African gold stocks in the 16th century.

In East Africa by the early 16th century, Portuguese intrusion into the Indian Ocean caused a shift in the outward commodity flow of East African gold away from the Persian Gulf. Moved largely by Portuguese official and unofficial traders, Zambezi gold bullion traveled directly east across the

Arabian Sea from Mozambique and Sofala to "Golden Goa" and its South Indian markets.

The pattern of pre-modern African gold movements set in the early 16th century—two direct maritime links (West Africa-Europe; East Africa-India) and the overland Saharan route to North Africa—remained intact for two centuries. As Curtin notes, however, an important change in this model occurred in the 18th century. Shortly after the turn of the century, the rising price of west African gold induced a flow of Brazilian bullion and minted cruzados, carried by Portuguese slavers, to Africa. Dutch traders re-exported part of this to Europe. The entry of New World precious metals into the African economy had begun. Several decades later (by 1760) steep rises in the price of gold in the Gold Coast and Senegal virtually halted the export of African gold to Europe. Increasing political consolidation in states such as Akwamu and Dahomey seems to have placed a steadily rising value upon gold for store of value and state transaction purposes.

The trade and monetary patterns of the Mediterranean economy underwent a distinct shift during the 15th century as a result of the rising power and territorial expansion of the Ottoman empire. Ottoman seizure of Constantinople in 1453 and establishment of the dynasty's seat in the former Byzantine capital achieved final closure of the Balkan and Anatolian territories of the empire. Before this date Ottoman moneyers had faced dwindling supplies of precious metals from war plunder and from the mines of Anaolia (the usual sources). By mid-century, however, Ottoman advances into Serbia and Bosnia brought the most productive gold and silver mining regions of Europe into Turkish hands. Ottoman authorities established direct state monopolies over the mines and established or authorized mints adjacent to them. Both mines and mints leased on farming terms to German or other private entrepreneurs. As a result a substantial share of the precious metal ouput of eastern Europe was diverted eastward from its former destinations, Ragusa and Venice, toward the Ottoman treasuries in Constantinople. These new commodity supplies of metal encouraged Sultan Muhammad to introduce a larger silver akce (about ten grams) in 1470 and a gold sultani (about 3.5 grams) in 1749. The latter was intended to pass as an equivalent to the Venetian ducat.

Nevertheless, despite accession of output from these mines, the Ottoman empire did not succeed in establishing its own international trade currency, nor indeed, its own dominant domestic currency. Unlike its contemporary Muslim empire, that of the Mughals in India, the Ottoman authorities did not insist upon the reminting of foreign coinage at its ports of entry. Nor, apparently, did it insist, as did the Mughals, that coin entering its treasuries for tax payments be that of its own issue. Consequently, the Venetian ducat remained the most numerous and widely accepted coin for inter-regional trade purposes within the empire. We can see this clearly in Sahillioğlu's analysis of the contents of Sultan Muhammad's treasury, for example.

After 1517, Ottoman annexation of Egypt and Syria from the Mamluks

Introduction

began direct transfer of gold coin and bullion as tribute from those regions to Constantinople. Ottoman fiscal records reveal that over a million gold sultanis minted in Alexandria and Damascus moved as tribute each year to the Ottoman central treasury. Much of this politically directed flow of treasure was undoubtedly originally African gold (see Walz's essay discussed below). That all of this transferred tribute remained in the coffers of the Ottoman capital is unlikely. It is more likely that a large share moved by means of official disbursements into trade channels and eventually into the eastward flow of treasure.

Between 1470 and 1580 Ottoman fiscal authorities had somewhat better success in controlling their silver currencies than their gold coinage. Commodity silver came from the Balkan mines in relatively larger quantities and value than gold in this period. The Ottoman state also prohibited, with some degree of success, the export of silver coin or bullion toward the east by private traders. Within the empire three reasonably well-defined silver currency zones emerged: in Anatolia and the Balkans, that of the akce; in parts of Iran and Iraq the local Persian shahi (issued by Ottoman mints); and in Egypt, the small Mamluk silver coin, the pare. None of these issues fully dominated domestic or interregional trade and commerce but each retained a respectable degree of acceptance and circulation.

By the 1580's however a wave of American silver engulfed the Ottoman monetary system. Spanish reales (termed gurus) by the Ottomans surged into Constantinople and the other trade centers of the empire. The influx of cheap silver meant that the Ottomans lost control of their silver currency. By the end of the century silver American coins had become the chief circulating medium in the empire. New World silver imports travelling overland from Western Europe through the Balkans caused the cessation of mining and minting for all save the most productive mines in Serbia and Bosnia. Thus, the eastward commodity movement of newly mined and minted silver coins from the Balkans was largely replaced by the flow of American coin. (That this had serious consequences for Ottoman balance of payments adjustments is not at all surprising.)

By the end of the century the eastward flow of tribute in gold coin from Egypt and Syria also dried up as gold became more and more expensive in relation to silver and as European gold imports to the Levant diminished sharply. The former transfer of gold to Istanbul shifted to a silver transfer of the same nominal value. Finally, the inflow of American silver created a growing disparity between the value of silver in the western Ottoman provinces of the empire (Rumelia, Anatolia, North Africa) and Iraq and Iran. High silver demand in Iran and in the lands to the east intensified the drainage of silver reales (often reminted as shahis). The official barriers erected by the Ottomans to stop this failed to do so.

As Sahillioğlu concludes, throughout the 17th and much of the 18th centuries the Ottoman lands served as a transit zone for American silver which was valued at successively higher rates in its progression from Europe, to Istanbul or Cairo, to Safavid Iran or Mughal India. Paradoxically the inflow of cheap silver coin steadily moving east virtually wrecked the monetary machinery of the Ottoman state. The ready availability of Spanish coin discouraged Ottoman mining. The 17th century also saw the simultaneous cessation of territorial expansion and a steady series of treasury imbalances for the fiscally hard-pressed regime.

Under Ottoman rule during the 16th and 17th centuries the demand for gold in Egypt was probably intensified by the recurring need to transfer a large monetary surplus each year (as much as 600,000 gold coins) from Cairo to Constantinople. This pressure resulted in an apparent increase in the commodity flow of gold to Egypt from the Western Sudan (often referred to as Takrur, actually the land of the Niger Bend). After reviewing the scattered documentary evidence Walz calculates that gold dust carried with the annual Haj pilgrim caravans has very likely been greatly overestimated. On the other hand, data from estate records of merchants which he has used from the Cairo archives underscore the previously underestimated role of private trade in carrying Egyptian cloth to the gold region and Sudanese gold to Egypt in the early modern period. He stresses the pivotal importance of the development of the Timbuktu to Cairo route in this movement of gold as new sources and new market towns appeared. The increasingly powerful ruler of Agades, a city and kingdom astride this route, fostered close commercial ties with Ottoman Cairo. Further testimony to the demand exerted by Cairo for Takrur gold can be found in the emergence of an important alternative movement of new metal across Libya, not north to Tripoli, but northeastward to Cairo in exchange for Egyptian goods. Ottoman conquest of Sawakin and Massawa, the Red Sea ports, in the mid-sixteenth century may also have fostered regular direct inflows of newly washed or mined gold from Ethiopia to Egypt. To summarize, the Ottoman annexation apparently strengthened rather than weakened Egypt's monetary position. In contrast to the shortages characteristic of fifteenth-century Mamluk Egypt, the rulers of Ottoman Egypt were able to attract enlarged imports of African gold.

Egypt, along with the rest of the Mediterranean world, received new inflows of American silver from Europe in the early modern period. Much of this inflow, as we might anticipate, continued to move into the Indian Ocean track as it had for centuries. However, Walz has drawn our attention to the emergence of a new movement of silver and copper from Egypt south into the Eastern Sudan. By the late 17th century the rulers of Sinnar, whose economy had developed more intimate ties with Ottoman Egypt, had turned away from the monetized cloth, iron and other artifacts of the region's monetary system toward the usual Islamic system. Gold, silver and copper coin, both foreign and (increasingly) indigenous coins, had replaced the older system as monetary integration of the region into the Mediterranean and Indian Ocean economies proceeded.

Often overlooked in the ongoing debate over the possible effects of New World commodity flows of silver in the pre-modern period is the sudden emergence of Japan as a major producer and exporter of silver in the 16th century. As Yamamura and Kamiki emphasize, the monetary system of medieval Japan (prior to 1469) hinged on Japanese use of imported Chinese cash or strings of copper coin for use in the island society's monetizing economy. For not fully understood reasons, Japan's kingdoms had ceased minting their own coins since the 10th century A.D. Between that point and the late 15th century the principal monetary commodity inflow into Japan consisted of huge numbers of Chinese copper coins. Some Chinese silver, probably in the form of silver bar coins, also moved to Japan in this early period. Japan, in return, exported gold derived largely from domestic alluvial production. By and large throughout the medieval centuries this pattern of metallic exchanges did not alter in its essential features. During the late 13th to the late 14th centuries A.D., however, the Mongol conquest of Sung China and attempted invasion of Japan did cause a sharp diminution in bilateral island-mainland trade and monetary exchange.

Paradoxically, after 1460, intense pressure for state funds felt by the warring states of Japan in the civil war period encouraged investment in precious metal mining throughout the country. New tunnelling techniques which came into use very quickly improved output. In the 1520's a major breakthrough occurred when immigrant Korean miners diffused knowledge of an improved Chinese silver-smelting technique. The "blowing ash" method used molten lead to extract pure silver from its ore. Similar technical improvements occurred in gold-mining. Both developments were reflected in a surge of precious metal output in Japan. By 1588, the Tokugawa Bakafu broke with longstanding tradition to strike its own coin in large issues. After 1520, large quantities of Japanese silver began to move into the export trade. Yamamura and Kamiki estimate that during the eighty years between 1560 and 1640, Portuguese trading vessels carried between 7.35 million to 9.45 million kilograms of silver from Japan to Macao, to Malacca and to points beyond those entrepôt. Japanese imports in this period consisted mainly of Chinese raw silk and relatively small amounts of New World gold bullion.

This relatively sudden intrusion of Japanese precious metal exports into the world economy ended nearly as abruptly. The "seclusion policy" adopted by the Tokugawa in the 1630's directed against Iberian missionaries and Portuguese traders began to constrict exports of silver. Only a few Dutch merchants confined to a small number of trading stations and some Chinese merchants continued to transport Japanese silver exports. Until all silver exports were competely prohibited in 1668, these traders carried a sharply reduced cargo estimated at just over one hundred thousand kilograms average per year.

"Vietnam and the Monetary Flow of Eastern Asia, 13th to 18th Centuries" by J. K. Whitmore underscores the role of Vietnam in commodity production of both silver and gold. By the 13th century alluvial workings had given way to numerous gold mines concentrated in the hilly zone north of the Red River delta. Similarly numerous silver mines were in active production in the same mountainous zone. Like those sub-Saharan African regimes which exploited that continent's gold resources the Vietnamese kingdoms did not use either gold or silver for monetary purposes. Until the mid-nineteenth century, the Vietnamese currency system relied upon Chinese style copper cash. The latter either was imported from China or manufactured domestically. We can thus assume that a large proportion of Vietnamese production of silver and gold was available for export. Whitmore's data, though scattered, suggest that Vietnamese precious metal exports were probably as great in magnitude as those of southwest China and Burma. Medieval exports of gold and silver flowed from Vietnam in two major streams: as tributary payments to China (and for one period under the Ming occupation as direct taxation) and as commodity exports to Java and later to Malacca—the principal maritime entrepôt in Southeast Asia.

During the 16th century we find in Southeast Asia, as elsewhere, a decisive shift in the configuration of precious metal movements. In part this shift reflects the familiar arrival of New World and Japanese silver, but in part it also derives from political consolidation coupled with tighter integration of Vietnam into the growing sphere of inter-regional trade. Whitmore argues that creation of a new entrepôt, the port of Hoi-an (also Faifo) on the south-central coast of Vietnam by the recently victorious Nguyen regime marked a critical shift in Vietnam's economic role. Under Nguyen protection and encouragement Hoi-an became a focus for north-eastern maritime trade with China and Japan. By the 1580's Hoi-an received a heavy influx of Japanese silver carried by Japanese or European traders. Another significant, as yet unmeasured, inflow of American silver arrived at the port from the Philippines. Apparently the greater portion of these new supplies paid for imported silk, porcelain, books, copper and other Chinese products in high demand. As a result of this exchange silver moved steadily to Canton from Hoi-an. Another part of the silver import paid for Southeast Asian spices sold at the new port. Thus, some incoming silver travelled to Malacca now under Portuguese control. Under the onslaught of cheap foreign silver Vietnamese maritime exports of the metal were probably ended. However production continued at the mines.

Throughout the early modern period substantial amounts of Vietnamese gold and silver continued to travel overland northward to China in the form of gifts of worked gold and silver in what was a politically inspired transfer of precious metals. Through tribute and trade, carried both overland and by sea, China remained a primary recipient for Vietnamese precious metal production.

IV

Cross's contribution "South American Bullion Production and Export 1550-1750 A.D." is an analysis of the trends and volume of gold and silver output in the Spanish mines of the New World. We are reminded that large-scale silver

Introduction 21

exports from the Spanish colonies did not really begin until after the spectacular strikes of the late 1540's (especially at Potosí in present-day Bolivia). By far the greatest rise in silver output came after adoption, in 1573, of the mercury amalgamation process for refining of low-grade argentiferous ore. (The latter incidentally allows us to carry out calculations for production based upon the amounts of mercury imported for use in the mines.) In the 1570's the quantities of New World silver exported soared dramatically. Mints located adjacent to the mines manufactured Spanish silver pesos (or reales) of a standard weight and fineness for export. For the next two centuries truly massive amounts of silver, largely in the form of coin, moved steadily by galleon from the mines of Bolivia, Columbia and Mexico to Seville, and by the Pacific route to Manila.

Basing his estimates upon various data series already developed, Cross concludes that between 1571 and 1700 A.D. total silver production in the viceroyalty of Peru (excluding the lesser Mexican production) reached an average annual figure of 214,500 kilograms of pure silver. The total for the entire one hundred twenty-nine year period is thus 27.7 million kilograms of pure silver. Peruvian production alone constituted over two-thirds of estimated total world output in the 17th century. Moreover, Cross argues that E. J. Hamilton's figures for Spanish silver production, which are based upon import data series from surviving records in Seville "do not accurately reflect the levels of American production and . . . far understate the amounts of bullion flowing to Europe." Aggregated production figures for the Viceroyalty of Peru in the 17th century do not reveal the steep drop in production which Hamilton postulated from the official Spanish import figures.

Spanish gold output from lower and central Columbia (within the Viceroyalty of Peru) beginning in the first half of the 16th century, while far from as impressive as silver was of considerable value. Those estimates which have been made point to annual average production ranging from over 1700 kilograms in the 16th to over 3100 kilograms in the 18th centuries. Thus Columbian gold combined with limited workings in lower Peru and Chile, contributed from one fifth to perhaps one-fourth of total world output in those centuries. Most of this Spanish gold bullion found its way into European or Asian maritime export channels without delay.

But the most impressive gain in New World gold production came much later—in Brazil under Portuguese colonial rule. Using Lisbon fleet registration figures adjusted upward for unregistered bullion, Cross calculates annual average production figure at nearly eleven thousand kilograms of refined gold per year. This was exported annually between 1712 and 1755 from Brazil to Portugal. These conservative output figures underscore the impact of Brazilian gold production upon world stocks in the first half of the 18th century. As is well known most of that exported bullion found its way from Lisbon to England as a result of the new commercial dependency of Portugal formalized in the terms of the Methuen treaty of 1703.

Using data compiled from manuscript records of the Spanish royal treasuries in the New World, TePaske's essay, "New World Silver and the Philippines, 1590-1800" also addresses the question of treasure exports. Focusing his attention upon official shipments of silver, TePaske finds that official accounts from the Vice-royalties of Peru and Mexico support E. J. Hamilton's conclusions regarding the trend of official shipments of treasure. Shipment of silver bullion direct from the colonies to Seville peaked early in the 16th century and gradually declined to 1660 (the last year of Hamilton's figures). After 1660 they dropped more steeply to rise only gradually in the first half of the 18th century. Nevertheless this conclusion does not necessarily contradict Cross's estimates of sustained silver production in the 17th century (given above). Officially recorded treasure shipments were probably one-fourth of private remittances in the 16th and 17th centuries.

TePaske also argues that rising proportion of Crown revenues from silver remained in the New World in order to pay for increasing defense, administrative and other needs. By 1680 the Viceroy of Peru was sending on average no more than 8 percent (4346 kilograms of silver) of the total official revenue of the colony to Seville each year. Official remissions of silver to Spain from the Viceroyalty of Mexico on the other hand which remained relatively constant, outstripped those from Peru in the latter decades of the 17th century. Official consignments of Mexican silver sent on the Manila galleons paid for much of the costs of Spanish control of the islands in this period as well. However, these steady shipments of Mexican silver, averaging about 5100 kilograms (200,000 silver pesos) annually were dwarfed by the flow of coined silver through private commercial channels. Manila and Acapulco served as the maritime terminii for a massive exchange of New World silver for Asian goods. Continued powerful demand forces on each end encouraged continuing illegal trade and the outward smuggling of New World silver to Manila—the gateway to Asia. Estimating absolute magnitude is difficult but the drain of silver toward Asia across the Pacific may well have absorbed by far the greater portion of New World silver output in the 16th, 17th, and 18th centuries.

By 1600 A.D. the Dutch and English East Indian companies, holding national monopolies for direct trade to Asia, had opened a conduit for the flow of New World precious metals eastward. The East Indiamen of northern Europe sailed from the Atlantic coasts around Africa to Cape Town and thence direct across the Arabian Sea to West India and the Straits of Malacca. New transport speed and efficiency encouraged the opening of a new market in Europe for manufactured Indian textiles in bulk in addition to spices and other long-familiar Asian commodities. Throughout the 17th and 18th centuries European demand for Asian goods steadily expanded to fashion a major economic linkage between these two distant world regions. Asian demand for European goods, however, although increasing, was never fully equivalent. Instead, as is well known, precious metals drawn largely from New World coin and bullion provided a means of payment for the commodities of Asia.

Gaastra, in his contribution to this volume, describes the mechanisms

Introduction

employed by the Dutch East Indian Company to assemble silver and gold in large quantities (coined or bullion) and estimates the annual quantities shipped on the basis of the Company's own records. According to the annual decisions of the Company's governing body, the Heren XVII, combined gold and silver shipments to Asia which began at an average value of just over 500,000 guilders a year in 1602-10 had reached the one million plus mark in 1660-1670. Between 1680-90 the decennial average value (at just under two million guilders) began a much steeper ascent during the next fifty years to peak at 6.6 million guilders annual average value for treasure in the decade of the 1720's. Although a precise breakdown of content for treasure shipments is not possible, Gaastra estimates that the greater proportion shipped from Holland by the Company between 1602-1695 took the form of Spanish reales (Mexican preferred), bar silver, and bar gold from the New World. A somewhat lesser amount consisted of various types of Dutch silver and gold coins. From one perspective, at least, the Dutch East Indian simply acted as a European way station for the flow of New World silver and pumped this out to its trading stations in the east as a commodity. More often than not reales valued by weight remained in their original boxes packed at the Mexican or Peruvian mints set up adjacent to the mines. Only after arrival at Batavia or any of the other Dutch trading stations did the reales pass into circulation or bullion enter local mints. In other words, at least part of the copious New World treasure flow was a direct transfer from the point of production and working to its eventual, far distant, point of monetary circulation in Asia.

That Dutch East India Company bullion shipments rose in value until the 1720's and only gradually fell off thereafter is clear from the aggregate figures derived from Heren XVII resolutions. What of actual quantities of silver and gold? How do these quantities compare with Cross's production estimates?

Decades	Peruvian production	Dutch VOC silver shipments
	kilograms of pure silver	kilograms of pure silver
1601-10 (rounded)	1,300,000	54,000
1630-40	2,500,000	89,000
1670-80	1,200,000	108,000
1690-1700	1,200,000	260,000

Abbreviated though it is, this sampling of Peruvian output and Dutch silver shipment figures does encourage the assumption that direct Dutch commodity transshipments of New World silver probably did not exceed ten percent of total production until the last years of the 17th century, if then. This is certainly true if we refine the calculation by adding the output of Mexican mines in the 17th century and deducting indeterminate amounts of silver collected by the Company from intra-European channels and reminted into Dutch coin. Nevertheless the proportion of silver directly transferred to

Asia undoubtedly rose in the 18th century, when, as Gaastra's figures reveal, the decennial averages soon were triple that of the previous century. Rapidly declining Peruvian silver production and export (discussed by TePaske) only partially balanced by sustained Mexican exports, very likely increased the ratio of Dutch company shipments to New World production as well. A full calculation of the monetary and economic impact of New World precious metals upon 17th and 18th-century Europe must consider the mode of transmittal, the destination, and ultimate use of much of this treasure.

Any estimation of direct Europe to Asia precious metal movements must consider the exports of other East India companies. The chief rival of the Dutch, the English East India company, also channelled large sums in silver and gold coin and bullion eastward to finance its trade. Quantitative data recently published by K. N. Chaudhuri offers a basis for tentative comparison of the magnitude of shipments made by each company. Thus for the hundred years between 1660 and 1760 A.D. (the period in which Gaastra's and Chaudhuri's figures coterminous) the total shipments of each company are as follows:

Dutch East India Company		English East India Company	
Silver	Gold	Silver	Gold
3,126,785 kilograms	21,874 kilograms	3,749,000	20,384
Average Annual Shipments:			
31,268	218	37,490	203[17]

Over the long term, the two companies clearly exported treasure on a roughly equivalent basis. Silver shipments made by the Dutch were in total somewhat less (about 83%) than those made by the English. Dutch gold shipments, which continued throughout the period, were slightly more—largely because the English company stopped sending gold entirely after 1715 A.D. Together, the two Companies unquestionably transferred over one-tenth of New World silver production directly to the Indian Ocean.

Gaastra also distinguishes between destinations for this steady flow of treasure from Amsterdam. Primary recipient was the Company's treasury at Batavia (Jakarta), capital of the East India Company territorial possessions. From Batavia the Dutch merchants sent either silver or gold in various forms to meet local monetary requirements. Rapidly rising trade with Bengal encouraged steep increases in silver inflows to that region after the 1660's. The new abundance of silver contrasts with the situation described by Deyell for Bengal in the medieval period (see above).

The Dutch company ships also served as carriers for existing intra-Asian

---

17. K. N. Chaudhuri, *The Trading World of Asia and the English East India Company, 1660-1760* (Cambridge, 1978), Table A.7, "The Quantity of silver and gold exported by the company 1600-1760," p. 177. Dr. Chaudhuri, a participant in the 1977 Madison Conference, read a paper embodying these results.

Introduction 25

flows of precious metals. For example after 1643 the Dutch factory at the Persian port of Gombroon in the Persian gulf sent substantial treasure in the form of Persian silver abbasis, reales, and gold ducats to Surat, the leading port of Gujarat in Western India. This of course was the final link in the European-Levantine-Persian drain of precious metals.

Between 1630 and 1685 the Dutch augmented European supplies of (primarily) New World silver and gold for the India trade with newly accessible Japanese silver and Taiwanese gold. After loading Japanese silver Dutch Indiamen bypassed Batavia in a direct haul to eastern India where Japanese silver was put into circulation. Alternatively some ships stopped at Taiwan to exchange Japanese silver for Taiwanese gold before sailing to India. As Gaastra shows in an appended table, direct Japanese treasure exports usually exceeeded those sent from Europe in the decades between 1640 and 1660. Thus, access to Japanese output partially explains relatively low European exports prior to 1670 and the sudden surge in New World exports thereafter.

The concluding essay in this volume by Joseph Brennig offers an analysis of the precious metal imports into the Indian port of Surat during the 17th century. Until the 1720's Surat retained its long-standing role as the leading seaport for Gujarat and the entire western Indian seaboard. Well before the arrival of the Portuguese, Surat and its companion ports Broach, and Cambay, had been the base for the wide-ranging Indian Ocean trading activities of the vigorous, Gujarati trading communities. In the seventeenth century Surat was the premier terminus for the inflow of gold and silver to India. Unfortunately, due to the destruction of many Mughal records in the Maratha wars of the 18th century, and gaps in European trading company records forwarded to London or Amsterdam, completion of extended data series for treasure imports into Surat (as well as for other Indian ports) requires a laborious process of historical reconstitution and interpolation. Records of ships and their cargos arriving at Surat for a sample year (1643-44) abstracted from accounts kept in the Dutch trading station in that port are revealing. Cargos carried by a total of seventeen Asian and European ships comprised the following:

	Number	*Value in Reales*	*Kilograms*
Silver Reales	649,768	649,768	17,414
Silver Abbasis	289,000	94,514	3,533
Dutch coin	55,000	53,117	1,424
Silver Bars (*tael*)	158,000	187,718	5,031
Total		985,117	26,401

The Dutch count shows another 16,284 gold ducats imported on these ships, which, if reckoned as reales, equalled an additional 1663 kilograms of silver. Thus, the total precious metal imports into Surat port listed by the Dutch factors for the single year 1643-44 (which may well not be a complete listing) is more than three times as great as the *total* Dutch average annual exports of treasure shipped from Amsterdam (9046 kilograms) in the decade of the 1640's. Of the total Surat imports listed in 1643-44, as Brennig emphasizes, Dutch imports direct from Amsterdam in 1644 were scarcely five percent (1,424 kilograms) of that observed from all sources. As Brennig comments: "the Companies . . . moved only a minor portion of the precious metal passing from Europe to India before the 18th century . . ." (p. 2). Prior to the late 17th century upsurge of northern European trading company silver and gold shipments, the far older trans-Mediterranean, overland drain of metals through the Red Sea and the Persian Gulf to western India seems to have kept intact its central importance in the world's long-distance monetary transfers.

At Surat, as in other Mughal controlled ports of entry for the subcontinent, most of the incoming metal went directly to the imperial mints for melting and reissue as silver rupees or gold mohurs. Unlike the Ottoman empire, the fiscal administrators of the Mughal empire did not permit large inflows of foreign coin to circulate in an uncontrolled fashion. Various regulations requiring payment of the land tax and other taxes in imperial coin, as well as discounts on coin of older issue (presumably for wear) encouraged virtually immediate presentation at the mints. Mughal monetary policy was also aided by the commodity exchange or balance of trade advantage which Indian goods possessed. That is, in contrast to the Levantine provinces of the Ottoman empire, India did not serve as a conduit for the further eastward movement of precious metals. Instead two great streams of metal converged upon the subcontinent.

Part I

Medieval monetary flows

1

The Italian gold revolution of 1252: Shifting currents in the pan-Mediterranean flow of gold

THOMAS WALKER

The 13th century brought to Western Europe a new political and economic maturity. This new situation encouraged more than one government to attempt significant monetary reform.[1] Of these, contemporaries and later generations have been most impressed with the return to the minting of gold in 1252. Historians have not yet come to full agreement on all the parameters of this development, but three questions have emerged. First, why did European states feel that a return to gold coinage was desirable? Second, why

---

1. The impressive revival of western commerce first of all created the need for a large coin to replace the debased silver pennies circulating since Carolingian days. For the importance of this change, see Roberto S. Lopez, "Prima del ritorno all'oro nell' Occidente duecentesco: primi denari grossi d'argento," *Rivista storica italiana*, Vol. LXXIX (1967), pp. 174-187; Gino Luzzatto, *An Economic History of Italy*, trans. by Philip Jones (New York, 1961), p. 123; and Philip Grierson, "The Origins of the Grosso and of Gold Coinage in Italy," *Numismaticky Sbornik*, Vol. XII (1973), pp. 33-48. While the economic demand for such a step seems to have existed by the 12th century, Marc Bloch has argued that a certain degree of political sophistication and consolidation was first necessary to thrust aside feudal or customary restrictions upon such novelties as the reassertion of regal or state coinage monopolies. Bloch's ideas can be found in his *Esquisse d'une histoire monétaire de l'Europe* (Paris, 1954), p. 35. This process is most evident in France, where a long series of suppressions of local mints preceded the key reforms of 1266 by Louis IX. Consult, for example, Étienne Fournial, *Histoire monétaire de l'Occident médiéval* (Paris, 1970), p. 83 and Thomas N. Bisson, "Coinages and Royal Monetary Policy in Languedoc During the Reign of Saint Louis," *Speculum*, Vol. XXXII (1957), pp. 452-466.

Venice issued the first large silver coin in the West, the grosso. Doge Enrico Dandolo struck this coin as part of his program "to strengthen the commercial and political prestige" of the Venetian state. For the circumstances of its appearance, see Louise Buenger Robbert, "Reorganization of the Venetian Coinage by Doge Enrico Dandolo," *Speculum*, Vol. XLIX (1974), pp. 48-60. Florence may have issued her grosso in 1237 as part of that commune's resistance to the increasingly anachronistic efforts of Frederick II to keep Tuscany in subjection to the Empire. For more information, confer Ferdinand Schevill, *Medieval and Renaissance Florence* (New York 1961), pp. 290-295. Even the new golden *Augustalis* of Frederick II was a symbolic expression of that emperor's resumption of ancient Roman imperial prerogatives of coinage. For a review of this interpretation, see Robert S. Lopez, "Back to Gold, 1252," *Economic History Review*, Series 2, Vol. IX (1956-1957), p. 227. These kinds of developments are another aspect of what R. W. Southern has viewed as the progressive secularization and rationalization of society in the 12th and 13th centuries. For an elaboration, consult his *The Making of the Middle Ages* (New Haven, 1953), pp. 41-49 and 96-98. This, in effect, was the *sine qua non* of coinage reform.

29

did that resumption occur in Italy and, more precisely, in Genoa and Florence? Finally, why did it occur in 1252?

In tracing the vicissitudes of gold coinage, a broad approach which examines African, Islamic, Byzantine and European developments seems necessary. The important economic changes of the thirteenth century occurred in an *oikoumene* centered in the entire Mediterranean basin and stretching into even more distant economic spheres, which were tied together by the Mediterranean axis. Political or economic changes in any one of these spheres could and did influence others.

Historians have not yet attempted such a Braudelian study. Undoubtedly, scholars have shied away from the difficulties of doing research in several extremely disparate areas, where language barriers force them to rely on secondary evidence. The variability of surviving sources also makes it extremely difficult for an historian to amass comparable data.[2] Despite such admitted problems, the very existence of the medieval Mediterranean *oikoumene* seems to justify this attempt to discover a working hypothesis to explain the occurrence and success of different gold coinages there in the high middle ages.

Throughout the medieval period, a stable unit of international coinage was of critical importance for the development and prosperity of the Mediterranean economy. Merchants in international trade demanded recognizable and trustworthy coins from foreigners who came to do business in their markets. Only a few coins performed such a stabilizing role. The most important examples are the Byzantine solidus and later the Venetian ducat.[3] This kind of coinage could be trusted and used by all traders within the Mediterranean economic system.[4] The need for coinage acceptability was so great that medieval societies often imitated the currency of more mature economic systems in whose markets they hoped to trade.[5] Thus there existed

---

2. It is an unenviable task for the historian of this period to have to weigh and compare evidence of such a diverse nature. For example, the Latin West provides a relative abundance of commercial data. Similar records originating in the Near East or Africa are almost non-existent. Even within each area, extremely variable information is available from individual places and different time periods. Thus in the West, for example, the historian of Genoese commerce has far greater resources to work with than his colleague who works on Venetian materials. In the Islamic world, the Cairo archives continue to offer specialists unpublished materials on Egypt. On the other hand, historians of Northwest and West Africa have more recently relied to a far greater extent on evidence gathered by archeology, anthropology, and similar auxiliary fields.

3. On the role of certain coinages as units of international currency, use Robert S. Lopez, "The Dollar of the Middle Ages," *Journal of Economic History*, Vol. XI (1951), pp. 209-212; Carlo Cipolla, *Money, Prices, and Civilization in the Mediterranean World* (Princeton, 1956), pp. 15-20; and Philip Grierson, "La moneta veneziana nell'economia mediterranea del '300 e '400," *La Civiltà veneziana del Quattrocentro* (Florence, 1957), p. 78.

4. As Lopez, "Dollar," pp. 220-225 was careful to point out, however, such an argument by no means assumes that an absolutely stable coinage, which did not even allow for controlled economic growth, was of ultimate benefit to the civilization which struck it.

5. There is a wide variety of examples of this phenomenon. This paper will deal at some length with the Syrian besant of the crusader states, the Guiscardian tari in southern Italy, and

a constant demand for metals for quality coinage to meet such trade requirements. Its focus shifted widely as trade patterns evolved.

On the other hand, bullion supply to meet that demand was equally critical. The available flow of precious metals in this period varied considerably. It depended on changing political, economic, and technological circumstances. Yet it seems to be true that gold was in fact the metal of choice for international coinage in the Mediterranean. Although this is certainly not axiomatic world-wide, any economically developed Mediterranean state which has had access to significant gold supplies indigenously or through favorable competition in the international gold market, has attempted at least a bimetallic currency. The tradition of Roman imperial coinage in the West and Byzantium, as well as the classical conceptions of gold coinage incorporated into Islamic ideology, helped to perpetuate that tendency.[6]

Therefore, this essay will attempt to demonstrate that available evidence leads to the conclusion that the basic cause for changes from silver to gold currency and *vice versa* may well have been shifts in supply and demand for gold on a Mediterranean scale. This essay will try to demonstrate how at least through the mid-13th century such shifts seem traceable to access to and control of the changing gold supply.

In the early 1960s, Philip Grierson demonstrated the effect of different gold-silver ratios in the early middle ages in the three major medieval civilizations of western Eurasia. Grierson argued that the different values placed upon those metals in each civilization caused their inexorable flow to where they would bring the highest price.[7] Without challenging these assumptions, Cipolla asked the significant question: why did the ratios change? Cipolla was the first historian to emphasize demand as the critical factor here. He argued that the ratios changed because gold demand fell in the West as it abandoned the gold standard.[8]

---

the milarès of southern France. Other examples can be added. For instance, S. M. Stern, "Tari," *Studi medievali*, serie terza, Vol. XI (1970), pp. 187-88 describes how the first south Italian taris, especially from Amalfi, imitated Islamic coinage of Sicily. Bartolomeo Lagumina, "Studi sulla Numismatica Arabo-Normanna di Sicilia," *Archivio storico siciliano*, nuova serie, Vol. XVI (1891), p. 1 indicates that the first Norman coins struck in Palermo were exact reproductions of Islamic coins previously struck there. Paul Balog and Jacques Yvon, "Monnaies a légendes arabes de l'Orient latin," *Revue numismatique*, Sixième série, Vol. I (1958), p. 137 refer to similar coinage in 11th century Barcelona and 12th century Castile.

    6. Bloch, *Esquisse*, pp. 14-15. Many historians have also alluded to the nearly mystical respect for gold among the barbarian successor states in the West. Philip Grierson, "Muslim Coins in Thirteenth Century England," in *Near Eastern Numismatics, Iconography, Epigraphy and History*, ed. by Dickran K. Kouymjian (Beriut, 1974), p. 37 gives an excellent 13th century example in which Henry III purchased large Muslim gold coins to serve as alms for great religious festivals.

    7. Philip Grierson, "The Monetary Reforms of 'Abd al-Malik," *Journal of the Economic and Social History of the Orient*, Vol. III (1960), pp. 241-264.

    8. Carlo Cipolla, "Sans Mahomet, Charlemagne est inconcevable," *Annales—Économies • Sociétés • Civilisations*, Vol. XVII (1962), pp. 130-136.

Nevertheless, it seems necessary to understand demand in a wider context. As western participation in international commerce dropped significantly, the demand for gold to participate in it declined as well. In other words, the West simply did not have the economic means or desire to compete in the international gold market. As a result, demand fell until the gold-silver ratio reached its low point in the West. Only then did specie flow follow the observed directions and ratios.

Following up Grierson's insights, Andrew Watson in 1967 masterfully combined previously uncorrelated data from most major medieval states, with which he established the "general movement of gold-silver ratios."[9] But Cipolla's question for Grierson seems equally applicable to Watson's data. This essay will try to formulate a possible explanation for that gold-silver ratio shift in the context of the changing supply of and demand for gold throughout the Mediterranean basin.

The center of gold production for the Mediterranean lay outside it in West Africa. It seems to be critically important to trace changes in the intensity and direction of the gold flow north in order to understand the vicissitudes of Mediterranean gold coinage. Interest in West African gold production among modern European historians began in 1935, when E. F. Gautier called attention to its reputation in Islamic tradition.[10] But the credit for focusing historians upon West Africa itself as a key to the understanding of so much of Mediterranean economic change belongs to Maurice Lombard, who examined in depth the sources of gold in classical Islam.[11] More recent research by historians of Africa has begun to illuminate the complexity of that gold trade.[12] Thus changes in the supply of West African gold now seem intimately related to the history of Mediterranean gold coinage.

The West African gold trade only became possible after profound changes occurred in the trans-Saharan trade between the classical age and the middle ages. These changes included the spread of the camel as a beast of burden capable of reliable desert transportation; the diffusion of Berber tribes along both shores of the Sahara to act as a human bridge; the spread of Islam as a

---

9. Andrew M. Watson, "Back to Gold—and Silver," *The Economic History Review*, Second Series, Vol. XX (1967), pp. 1-34.

10. E. F. Gautier, "L'or du Soudan dans l'histoire," *Annales d'histoire économique et sociale*, Vol. VII (1936), pp. 113-123.

11. Maurice Lombard, "Les bases monétaires d'une suprématie économique: l'or musulman du VIIe au XIe siècle," *Annales—Économies • Sociétés • Civilisations*, Vol. II (1947), pp. 143-160. In the early period of the unified caliphate, the dethesaurisation of Christian church treasures in Egypt and Syria made available large gold stocks for monetary use. The same phenomenon occurred in Persia, where a silver coinage had allowed the accumulation of a massive golden treasure. The pharaonic tombs were also systematically plundered. Yet all this gold contributed only to a bimetallic currency, before West African gold began to flow to the Mediterranean.

12. An excellent example is the article by Jean Devisse, "Routes de commerce et échanges en Afrique occidentale en relation avec la Méditerranée," *Revue d'histoire économique et sociale*, Vol. L (1972), pp. 42-73 and 356-397.

common religion among Arab, Berber, and Sudanese traders; and the trade-inspired creation of large empires in the western sudan.[13]

This gold trade began sporadically quite early in response to demand in North Africa. The earliest mention of Ghana as a land of gold occurs in the writings of al-Fazari of Baghdad in the 8th century.[14] Ghana emerged as a commercial center at which salt from desert mines could be exchanged for gold from regions further south.[15] As in later medieval sudanese empires, gold played practically no role in internal commerce. It was a luxury item for the court, or it was used to secure articles of foreign exchange.[16] Therefore, these empires were undoubtedly eager to expand their gold commerce.

The earliest exploited West African gold came from the Bambuk fields along the upper Senegal.[17] As the peacemaker in the region, Ghana controlled its export north. As early as the 10th century, a rare list of Arab gold mines written by a native of far-off South Arabia described the mines of the sudan in this way: "The most productive gold mine in the world is the mine of Ghanah in the country of the Maghrib."[18]

By this time, the gold-salt exchange had become regularized. Gold flowed north in increasing quantities.[19] Many of the successor states to the caliphate were changing from a bimetallic to a monometallic gold standard. Lombard attributed this switch to a massive infusion of gold from West Africa to a Mediterranean basin then dominated by Islam.[20] Such a significant influx into an already thriving economy probably caused gold to rise in value as demand accelerated. This is the converse of what happened in the West in the early middle ages. It may also be the ultimate explanation for the widened

---

13. Nehemia Levtzion, *Ancient Ghana and Mali* (London, 1973), p. 124 and Vitorino Magalhães-Godinho, *L'économie de l'empire portugais aux XVe et XVIe siècles* (Paris, 1969), pp. 99-101.

14. Levtzion, *Ancient Ghana*, p. 3.

15. Both commodities were highly valued in this trade, since salt was not available in the western sudan. For a discussion of this trade, see J. D. Fage, "Ancient Ghana: A Review of the Evidence," *Transactions of the Historical Society of Ghana*, Vol. III (1957), pp. 82-83.

16. Marion Malowist, "Le commerce d'or et d'esclaves au Soudan Occidental," *Africana Bulletin*, Number 4 (1966), p. 51.

17. For an excellent discussion of changing centers of West African gold production and the interrelated shifting trade routes, consult Raymond Mauny, *Tableau géographique de l'Ouest africaine au Moyen Age* (Amsterdam, 1967), pp. 426-439.

18. The translation is from D. M. Dunlop, "Sources of Gold and Silver in Islam According to al-Hamdānī (10th Century A.D.)," *Studia Islamica*, Vol. VIII (1957), pp. 29-39. Dunlop also points out that four centuries later, Ibn Khaldun considered the sudan to be the only source of Islamic gold. Westerners would later be just as enthusiastic in their efforts to reach the sources of this fabulous gold flow. When Spain seized Tripoli in 1511, Fernand Braudel, "Monnaies et civilisations: De l'or du Soudan à l'argent d'Amérique," *Annales—Économies • Sociétés • Civilisations*, Vol. I (1946), p. 12 claims that the Spaniards assumed they were conquering a mountain of gold.

19. Devisse, "Routes de commerce," p. 49.

20. Lombard, "Les bases," pp. 150-151.

gold-silver ratios which Watson documented during the silver famine in Islam from the tenth century.[21]

A good illustration of the importance of the control of this West African gold supply occurs in the states of the Maghrib. The history of North African gold coinage in the high middle ages demonstrates the vital correlation of gold with sovereignty in Islam. The gold ports of the northern sahel had long been in the control of the Kharijites.[22] In 909, the Fatimids proclaimed an independent caliphate. The Kharijites bitterly opposed this move. They revolted against Fatimid attempts to rule the entire Maghrib. The Kharijite revolution cut off the Fatimid gold supply, which had come across the central Sahara. Because they needed gold for their ambitious political objectives, the Fatimids moved west and seized Sijilmasa in 951. Here they issued gold coins on their own authority. For the first time, the Fatimids helped to forge the great trans-Saharan trade route of the far west for regular caravan traffic. The trade route running from Ghana through Awdaghost to Sijilmasa became the great port of entry for gold to the Maghrib and beyond.[23]

North Africa in the 10th century was also the scene of a bitter imperial rivalry between the Fatimids of Ifriqiya and the Umayyads of Spain. Abd al-Rahman III seized Ceuta in 928, which allowed the Umayyads to initiate a gold coinage. In the following year, he proclaimed himself caliph.[24] This fierce struggle increased demand for West African gold and most likely contributed to the evolution of Ghana as its supplier.

The Fatimids controlled Sijilmasa through the dynamic reign of al-Muizz in Ifriqiya. Under him, the Fatimids conquered Egypt in 969. Thereupon, the Umayyads took Sijilmasa, which they controlled with only a short interruption to the fall of their dynasty in the early 11th century. Thus a strong

21. Watson, "Back to Gold," pp. 2-7 and chart on p. 27. The same process of accelerated demand in a booming economy may have been responsible for the rise in the gold-silver ratio in the West after 1252.

22. For a good examination of the role played by the Kharijite states, especially the Rustemites at Tahert, see Tadeusz Lewicki, "L'État nord-africain de Tāhert et ses relations avec le Soudan occidental à la fin du VIIIe et au IXe siècle," *Cahiers d'études africaines*, Vol. VIII (1962), pp. 513-535.

23. For a survey of the commercial role of Sijilmasa, consult J.-M. Lessard, "Sijilmassa—la ville et ses relations commerciales au XIe siècle d'après El Bekri," *Hespéris-Tamuda*, Vol. XI, pp. 5-36. The relationship of these political changes to the gold trade is discussed in Jean Devisse, "La question d'Audaghost," in Denise Robert, Serge Robert, and Jean Devisse, *Tegdaoust I. Recherches sur Aoudaghost* (Paris, 1970), pp. 54-63 and Devisse, "Routes de commerce," p. 53. Archeological excavations at Awdaghost reveal that all recognizable gold coinage discovered there is of Fatimid origin. Ibn Hawkal celebrated the economic vitality of this route. He claimed to have visited Awdaghost in 951/952. There he allegedly saw a check written for 42,000 dinars to be cashed at a partner's business house in Sijilmasa. For a discussion of this incident, see Nehemia Levtzion, "Ibn-Hawqal, the Cheque, and Awdaghost," *Journal of African History*, Vol. IX (1968), pp. 223-233.

24. Braudel, "Monnaies," p. 11 was the first to demonstrate that Spain began to mint gold in the 10th century only when the Umayyads there had seized Ceuta and the gold route to the interior.

gold coinage appeared for the first time in the far west at the very moment that the Umayyads exercised conscious control of a major source of world gold supply.[25] By the year 1000, trade was flowing across the cultural boundary in northern Spain, and Spanish texts began to allude to the famous *mancusi*.[26] The return to gold in the West had begun.

The minting of gold in Andalusia fell dramatically with Umayyad political fragmentation in 1031 and failure to control the African side of Gibraltar. However, this trend markedly reversed itself with the Almoravid conquest. By 1086, this dynasty controlled the entire west from the Ebro to the Senegal, with its center at Marakesh. For the next half a century, they issued a stable dinar of almost pure gold. Known as Marobitins, they circulated as a major unit of international coinage in the European west and to some extent even in the eastern Mediterranean.[27]

The Umayyads had minted gold at least partially to symbolize their independence. The Almoravids, however, struck the names of the Abbasid caliphs on their coins. Therefore, Almoravid gold could fulfill a wider economic function. The vast majority of the coins which were struck in their Spanish mints were issued in ports of the Mediterranean from which the Pisans carried them far afield as early as 1100. Their African mints lay in the cities of the sahel, close to the source of Almoravid gold, the western sudan.[28]

After the Almoravids, the Almohads forged a new empire in the mid-12th century. Their empire was the only native dynasty to rule the entire Maghrib. Until their defeat at Las Navas de Tolosa in 1212, the Almohads firmly controlled all the northern trans-Saharan trade termini.[29] As ideological restorers of Islamic tradition, they took their minting responsibilities quite

---

25. Devisse, "La question," p. 70 and G. C. Miles, *The Coinage of the Umayyads of Spain* (New York, 1950), p. 26. Umayyad mint output reached a peak in the 990s when they firmly controlled both sides of Gibraltar.

26. J. Gautier-Dalché, "Monnaie et économie dans l'Espagne du nord et du centre," *Hespéris-Tamuda*, Vol. III (1962), p. 65. For a review of the literature on the mancus, see J. Duplessy, "La circulation des monnaies arabes en Europe occidentale du VIIIe au XIIIe siècle," *Revue numismatique*, Cinquième série, Vol. XVIII (1956), pp. 108-116; Philip Grierson, "Carolingian Europe and the Arabs: The Myth of the Mancus," *Revue belge de philologie et d'histoire*, Vol. XXXII (1954), pp. 1059-1074; and especially the discussion in *L'occidente e l'islam nell'alto Medioevo*, Vol. I, pp. 408-512 in *Settimane di studio del Centro italiano di studi sull'alto Medioevo*, Vol. XII (Spoleto, 1965).

27. The most convincing exposition of Almoravid currency as an international coinage of repute is the article by Ronald A. Messier, "The Almoravids. West African Gold and the Gold Currency of the Mediterranean Basin," *Journal of the Economic and Social History of the Orient*, Vol. XVII (1974), pp. 31-41. Duplessy, "La circulation," pp. 112-119 and 138-141 also gives some representative examples of the Marobitin circulating beyond Spain.

28. Devisse, "Routes de commerce," pp. 65-68 and Harry W. Hazard, *The Numismatic History of Late Medieval North Africa* (New York, 1952), pp. 96-112.

29. A good recent overview of the Almohad state can be found in Roger Le Tourneau, *The Almohad Movement in North Africa in the Twelfth and Thirteenth Centuries* (Princeton, 1969). For a general survey of all these political changes in the Maghrib, see Charles-André Julien, *History of North Africa*, trans. by John Petrie (London, 1970), pp. 100-121.

seriously. They reformed the coinage weights to conform to classical tradition and zealously maintained the fineness of their gold coinage.[30] As long as their power in North Africa remained unchallenged, their control of the gold trade assured them a strong gold coinage. But when their state began to break up in the early 13th century, silver coinage became much more prevalent there in commercial transactions. The turbulent political and economic situation of the disintegrating Almohad state may have helped to cause a temporary break-down of the long-distance gold trade.[31] The resulting reliance on silver may also explain the proliferation of imitations of silver Islamic coins in southern France in this period.[32]

Thus it would seem clear that control over the trans-Saharan gold trade was a consistent political objective of Maghribi states. The Fatimids had organized the trade routes of the far western Sahara in the 10th century. In the 11th century, the Almoravids restored Berber control over the entire route. They conquered Sijilmasa and Awdaghost. In 1076, they even took Ghana itself for a time and began to Islamicize the region.[33]

However, there were unexpected, adverse, long-term effects of this Almoravid expansion so far south upon the western trade routes. It dispersed the Soninke power base of Ghana.[34] As gold from Boure in modern Guinea began to enter international markets in the late 11th century, the southern termini of the trade routes began to shift slowly to the south and east.[35] Gold

---

30. Hazard, *Numismatic History*, pp. 66-67 and Robert Brunschvig, "Esquisse d'histoire monétaire Almohado-Hafside," in *Mélanges William Marçais* (Paris, 1950), p. 64.

31. A very important aspect of the fragmentation of the Almohad state may have been its inability to channel the gold trade to one or another center in its North African domain. This political deterioration coincides with the transitional period between the fall of Ghana and the rise of Mali. Therefore, it is hardly surprising that this period seems to have seen an increased emphasis upon silver coinage in the Almohad state. In his doctoral dissertation at the University of Wisconsin in 1968 entitled *Genoese Trade and the Flow of Gold, 1154-1253*, Henry L. Misbach has documented that specie flow in the early 13th century in each direction between Genoa and North Africa increasingly comprised silver. Information on the Almohad dirhem is available in Hazard, *Numismatic History*, p. 45 and Alfred Bel, "Contribution a l'étude des dîrhems de l'époque almohade," *Hespéris*, Vol. XVI (1933), pp. 1-68. Bel cites evidence that 13th century Almohad rulers struck false dinars made of copper. He also points out that Almohad coin hoards of this period consist of nearly exclusively silver coins.

32. The most thorough treatment of this coinage remains the work of Louis Blancard, *Le millarès* (Marseilles, 1876). Unfortunately, this work is unavailable in the United States. A summary can be read in a review by Adrien de Longpérier, "Le millarès, étude sur une monnaie du XIIIe siècle imitée de l'arabe par les chrétiens pour les besoins de leur commerce en pays maure, par Louis Blancard," *Journal des savants* (1876), pp. 428-441. Consult also M. Chaudruc de Crazannes, "De la monnaie arabe frappée dans le moyen âge par les évêques de Maguelone," *Revue archéologique*, 1. série, 5. année (1848), II, pp. 400-404.

33. Fage, "Ancient Ghana," p. 80 and Levtzion, *Ancient Ghana*, pp. 29-51.

34. Devisse, "La question," p. 109. Nevertheless, the process of economic decline in Awdaghost and Ghana was slow. Some prosperity lasted through the 12th century. This may have been related to the continued presence of a strong Almohad state to the north.

35. Mauny, *Tableau*, p. 316.

supplies reached the Sahara from the south at several points as Ghana lost the political power to funnel the entire gold supply. For a time by the early 13th century, therefore, political fragmentation in both North and West Africa may well have slowed the gold flow north.

Eventually, new political groupings in the sudan emerged, this time based on the Malinke peoples. The vestigial state of Ghana fell by 1224. In its place, the new and much more powerful kingdom of Mali arose in the area of the Niger bend about the middle third of the 13th century. Such political consolidation enormously increased the peace of the region and strengthened commerce. These changes in the sudan seem to have contributed to a marked increase in gold exports to the Maghrib and beyond. Commercial prosperity was a state goal. Revenue came from the taxation of merchants.[36] It is possible to get a vivid impression of the extent of Mali's prosperity from the ostentatious show of wealth by Mansa Musa on his 1324 pilgrimage, which greatly impressed the cosmopolitan Cairenes.[37] Finally, hostile nomads increasingly threatened the trans-Saharan trade routes of the far west.[38] Therefore, by the mid-13th century, these trade routes had shifted permanently toward the east, now linking primarily the Niger bend with Tunisia. This situation seems to be one major factor responsible for the flurry of European commerce with Tunisia around 1250.

Thus a series of powerful dynasties controlled the entire trading area from Spain to the western sudan for nearly three hundred years. This meant that almost all West African gold then entering the Mediterranean region probably came originally through northwest Africa. Consequently, as commerce became revitalized in the West and a demand for gold arose there, western maritime centers inevitably turned their attention to the Maghrib.[39]

To understand the changes in Mediterranean gold coinage, one must try to discover what happened to West African gold after it reached the Mediterranean basin. The history of Tunisian gold coinage is an excellent case study. There had been a long history of high-quality gold coinage in Ifriqiya under the Aghlabids and Fatimids until the Umayyad conquest of Sijilmasa.

After the Fatimids departed for Egypt, their Zirid puppets briefly retook Sijilmasa in the late 10th century, during which Umayyad gold output dropped.[40] Traditionally, historians have blamed the Hilali invasion launched against Ifriqiya by its Egyptian Fatimid overlords for the cessation of Zirid

36. Levtzion, *Ancient Ghana*, pp. 53-75 and 109; Fage, "Ancient Ghana," p. 86; and Devisse, "Routes de commerce," pp. 71-72. Mali soon expanded to control all the great sudanese trading centers, such as Walata, Timbuktu, and Gao. By 1260-1277, Mansa Uli became the first king of Mali to make the pilgrimage to Mecca.
37. For an account of his gold-laden camels, see E. W. Bovill, *The Golden Trade of the Moors* (New York, 1968), pp. 86-89.
38. Mauny, *Tableau*, pp. 430-436.
39. See below, pp. 21-23.
40. Devisse, "La question," p. 70.

gold coinage. Recently, however, some revisionists have begun to downplay the dramatic impact of the Hilali nomads.[41] A pre-Hilalian crisis in Ifriqiya may well be related in some way to the rise of the western Saharan trade routes. Zirid gold coinage ceased altogether shortly after the Almoravid conquest of Sijilmasa.[42] In any case, the Hilali did not destroy all political and social order in Ifriqiya and the nearby areas, but fragmented the traditional political state. Coastal cities became independent, each with some demand for luxury goods, including gold. Bougie especially developed as the seat of the Hamdannid dynasty. It continued to draw a portion of the gold trade, especially through Tlemcen. Djerba and Tripoli flourished as Kharijite markets and desert ports, at least until captured by the Normans in 1135 and 1146 respectively.[43]

Therefore, Ifriqiya continued to have some centers of surviving prosperity, which became targets for western economic expansion from the late 11th century. It seems quite likely that there was enough demand in these centers for gold to move east across North Africa to some extent. Until the Almohad conquest, not enough gold existed in any one center for a regular gold issue. *In toto,* however, this exportable gold may well have been enough to supply Sicilian mints. With the recreation of a strong Ifriqiyan state in the 13th century by the Hafsids, demand for gold there expanded dramatically. The trade routes which supplied it shifted east to meet that demand.[44] As a symptom of that change, the Hafsids began to issue a strong gold coinage.[45]

This problem of gold reaching Tunisia leads one to ask whether gold flowed across North Africa to Egypt throughout this period. Goitein has shown in his writings on the Geniza documents that a flourishing trade once existed between Egypt and Ifriqiya.[46] The Hilali invasion did end this phenomenon to a great degree.[47] By the mid-12th century, it seems to have been

---

41. Consult especially J. Poncet, "Le mythe de la 'catastrophe' hilalienne," *Annales— Économies · Sociétés · Civilisations,* Vol. XXII (1967), pp. 1099-1120 and Claude Cahen, "Quelques mots sur les Hilaliens et le nomadisme," *Journal of the Economic and Social History of the Orient,* Vol. XI (1968), pp. 130-133. For a vigorous rejection of this revision, see Hady R. Idris, "L'invasion hilālienne et ses conséquences," *Cahiers de civilisation médiévale,* Vol. XI (1968), pp. 353-369. The data in Messier, "The Almoravids," also seems to support the older historical tradition, which Ibn Khaldun himself originated.

42. Devisse, "Routes de commerce," pp. 68-69

43. Michael Brett, "Ifrīqiya as a Market for Saharan Trade from the Tenth to the Twelfth Century A.D.," *Journal of African History,* Vol. X (1969), pp. 347-364.

44. Levtzion, *Ancient Ghana,* p. 130.

45. Hazard, *Numismatic History,* p. 72 and Brunschvig, "Esquisse," pp. 69-74.

46. See, for example, S. D. Goitein, *A Mediterranean Society,* Vol. I (Berkeley, 1967), pp. 235-236.

47. Although it is still debatable whether the Hilali invasions themselves destroyed a thriving economy in Tunisia, historians are unanimous in blaming them for the end of east-west commerce across North Africa. For an examination of this question, see Eliyahu Ashtor, *Les métaux précieux et la balance des payements du Proche-Orient à la basse époque* (Paris, 1971), p. 17. There is some slight evidence that an east-west sudanese route across the Sahara linked the western sudan directly with Egypt in the 10th and 11th centuries. However, all indications are

extremely difficult even to sail between them.[48] Not surprisingly, a recent radio-chemical analysis of Almoravid and Fatimid gold indicates that Sudanese gold made up an ever-decreasing percentage of Egyptian Fatimid dinars after 1047.[49]

This depleted supply of gold proved to be an eventual disaster for Egyptian gold coinage. Various other causes also contributed to the exhaustion of Egyptian gold stocks. Egyptian gold mines in al-Allaqi in lower Nubia seem to have become unproductive by the 12th century. When Saladin rose to power, Fatimid supporters fled up the Nile and cut off even this diminished supply. The exploitation of pharaonic gold as a state monopoly had long depleted this resource as well. Finally, the influence of the crusades cannot be overemphasized.[50]

Due to the unstable political situation in the eastern Mediterranean, Egyptian exports fell dramatically and its balance of payments may have become unfavorable.[51] Egypt also had to pay a series of ransoms and tributes to the Kingdom of Jerusalem. A very large number of dinars seems to have passed in this way to the crusaders. In addition, continual military operations in Syria were a severe drain.[52]

But perhaps the most important influence which the crusader states had upon Egyptian coinage lay in their own imitation dinars called *besancii saracenati*. In a typical Mediterranean pattern, the less commercially sophisticated crusader states forged their own coinage on the model of the prestigious and widely circulating coinage of their neighbors, the Fatimids. This enabled the Christians to trade successfully in Muslim markets, which were accustomed to demand gold in commercial transactions. With the capture of the last Syrian mint at Tyre in 1122, this crusader gold coinage began to predominate in international trade between the Christian west and the Near East.[53]

---

that this route became abandoned until the 14th century owing to its high dangers. For a discussion of this problem, consult Nehemia Levtzion, "The Early States of the Western Sudan to 1500," in *History of West Africa*, ed. by J. F. A. Ajayi and Michael Crowder (New York, 1972), pp. 152-153.

48. S. D. Goitein, "Sicily and Southern Italy in the Cairo Geniza Documents," *Archivio storico per la Sicilia orientale*, Vol. LXVII (1971), p. 16.

49. Messier, "The Almoravids," pp. 36-40.

50. Hassanein Rabie, *The Financial System of Egypt, A.H. 564-741/A.D. 1169-1341* (London, 1972), pp. 169-170. Some Egyptian gold may have come through Southeast Africa via the ports of Sofala and Kilwa. Archeological evidence indicates peak production in Zimbabwe c. 900-1200, but the evidence for regular export to Egypt is not strong. Confer Roger Summers, *Ancient Mining in Rhodesia and Adjacent Areas* (Salisbury, 1969), pp. 189-210.

51. Subhi Y. Labib, "Handelsgeschichte Ägyptens im Spätmittelalter (1171-1517)," *Vierteljahrschrift für Sozial- und Wirtschaftsgeschichte*, Vol. XLVI (1965), p. 264.

52. Michel de Boüard, "Sur l'évolution monétaire de l'Egypte médiévale," *l'Egypte contemporaine*, Vol. XXX (1939), pp. 446-451 and Rabie, *Financial System*, pp. 169-170. For example, Shawar paid a yearly tribute of 100,000 dinars to Amalric in 1167. The following year, he agreed to pay one million dinars.

53. The first modern scholar to realize that Christians had begun to imitate Arab coinage during the crusades for trade with their Muslim neighbors was Henri Lavoix, *Monnaies à*

But the long range effects seem to have been traumatic. To ensure its acceptance, the crusader coinage was an exact reproduction of the Fatimid dinars, right down to traditional Kufic inscriptions. Intrinsically, however, they were quite debased. Following the principles of Gresham's Law, the bad crusader coins chased the good Fatimid ones out of circulation. Egyptians began to hoard gold.[54]

Due to this shortage of gold, Egypt ceased to mint high quality dinars of fixed weight already by 1130. Then Saladin debased the coinage and placed Egypt on a silver standard. Gold became a commodity with a floating value determined in the market.[55] Even though al-Kamil restored the fineness of Egyptian dinars in the 13th century, it is unclear whether the Ayyubids were ever able to strike a large quantity of gold. The circumstances which caused the 12th century gold crisis in Egypt continued into the 13th century. Historians have been unable to link Egypt with any fresh sources of gold until the 14th century, when the route to "Takrur", or West Africa, opened.[56] As

---

légendes arabes frappées en Syrie par les croisés (Paris, 1877). In this work, p. 28, Lavoix described how the crusaders issued both their own local feudal coinage, as well as these larger imitations meant to circulate much more widely. On this point, see also Balog and Yvon, "Monnaies," pp. 135-138. Louis Blancard, "Le besant d'or sarrazinas pendant les croisades," *Mémoires de l'académie des sciences, lettres et beaux-arts de Marseille* (1879-1880), pp. 151-180 proved that this coinage continued to be issued from crusader mints through the mid-13th century. These crusader coins were struck at Acre, Tripoli, Tyre, and Antioch. For the latest catalogue of this coinage, see Balog and Yvon, "Monnaies," pp. 145-168. A recent survey of crusader imitations of dirhems has been published by Michael L. Bates, "Thirteenth Century Crusader Imitations of Ayyūbid Silver Coinage: A Preliminary Survey," in *Near Eastern Numismatics, Iconography, Epigraphy and History*, ed. by Dickran K. Kouymjian (Beirut, 1974), pp. 393-409. This phenomenon may be an indication that silver had become increasingly important in Near Eastern commerce in this period, which may have been caused by the growing unreliability of gold coinage in the area.

54. Andrew S. Ehrenkreutz, "Arabic Dīnārs Struck by the Crusaders," *Journal of the Economic and Social History of the Orient*, Vol. VII (1964), pp. 177-179 and Rabie, *Financial System*, p. 171.

55. On these developments, consult especially Andrew S. Ehrenkreutz, "The Crisis of the Dīnār in the Egypt of Saladin," *Journal of the American Oriental Society*, Vol. LXXVI (1956), pp. 178-184; De Boüard, "Sur l'évolution," pp. 446-461; and Rabie, *Financial System*, pp. 174-178. This debased Egyptian coinage did not escape the critical eyes of foreign merchants. In a document of October, 1139, a Venetian charter from Damietta refers to *biçancii saracenati novi*. It is published in *Comitato per la pubblicazione delle fonti relative alla storia di Venezia: S. Giorgio Maggiore*, Vol. II, 1968, ed. by Luigi Lanfranchi, pp. 405-406.

56. The restored military power of Egypt, however, did create the foundations for political and economic revitalization, which the Mamluks would consolidate. Any significant improvement of the gold coinage seems not to have come until after 1260. For a brief summary of these changes, consult Paul Balog, *The Coinage of the Mamluk Sultans of Egypt and Syria* (New York, 1964), p. 16 and Labib, "Handelsgeschichte," p. 265. It may be that here, as in 13th century Italy, the emergence of a strong state would focus the demand for gold bullion. This could have helped to open new trade routes from the western sudan and affect gold prices throughout the entire Mediterranean basin in the later middle ages. The first clear proof of new interior caravan routes to Egypt from the western sudan does not occur until Mansa Musa's pilgrimage in 1324. This seems to be related to the Arab conquest of the Upper Nile and the opening of the Darfur

proof of a deepening gold crisis, Misbach cites figures from Genoese cartularies to show that practically no gold flowed from Alexandria to Genoa from 1213 to 1253. By way of contrast, the level of gold flow from Syria to Genoa remained fairly constant from 1154 to 1253.[57] However, his figures may only reflect payments of account. These could really involve shipment of commodities rather than specie or perhaps credit for further investment in the East.[58] In any case, the standard clearly remained silver. The permanence of this standard change in Egypt seems evidence enough that the crisis in gold in the 12th century was both real and long-lasting.

Similarly in this period, there were serious problems with the other major prestige unit of Mediterranean gold coinage, the hyperperon. Historians have never satisfactorily explained the sources of Byzantine gold. A considerable amount seems to have been once available from mines in the Balkans and Anatolia. But Byzantium permanently lost control of these areas by the late 12th and early 13th centuries.[59] Some gold in the Comneni era probably originated ultimately from North Africa via Italian trade.[60]

Until fairly recently, most historians had argued that the stable, high-quality Byzantine gold coinage began its inexorable decline in the mid-11th century.[61] However, it is clear from Italian commercial documents that the

---

route to Lake Chad and the Niger. By 1352, Ibn Battuta could speak of Egyptian colonies in Mali. According to Ibn Khaldun, 1200 caravans per year crossed between Egypt and Mali in the fourteenth century.

Nevertheless, for a period of about three hundred years from the mid-11th to the early 14th century, Egypt seems to have had no significant access to the West African supply of new gold into the Mediterranean. On these developments, see Ashtor, *Les métaux*, p. 18; Mauny, *Tableau*, p. 437; Rabie, *Financial System*, pp. 188-193; and Godinho, *L'économie*, pp. 120-123.

57. Henry L. Misbach, "Genoese Commerce and the Alleged Flow of Gold to the East, 1154-1253" *Revue internationale d'histoire de la banque*, Vol. III (1970), pp 77-78.

58. David Abulafia, *The Two Italies: Economic Relations Between the Norman Kingdom of Sicily and the Northern Communes* (Cambridge, 1977), p. 269.

59. Speros Vryonis, "The Question of the Byzantine Mines," *Speculum*, Vol. XXXVII (1962), pp. 7-14.

60. Michael F. Hendy, *Coinage and Money in the Byzantine Empire, 1081-1261* (Washington, D.C., 1969), p. 313 gives the provisions of the Byzantine-Pisan treaty of 1111 in which gold and silver imports into the empire were exempted from taxation. This may indicate that Byzantium in this period consciously encouraged gold imports and that Pisa was in an excellent position to perform a middle-man role. It is precisely the failure of Byzantine demand in the 13th century which no doubt discouraged this process and contributed to the decline of their gold coinage.

61. See, for example, Philip Grierson, "Coinage and Money in the Byzantine Empire," *Moneta e scambi nell'alto Medioevo. Settimane di studio del Centro italiano di studi sull'alto Medioevo*, Vol. VIII (Spoleto, 1961), p. 418. Some historians had noticed an increased fineness in some 12th century Byzantine coins, but they had been unable to comprehend the overall coinage system of the state in that period. Consequently, they attributed these coins to erratic minting policies as a manifestation of an already permanently damaged Byzantine economy. For examples of such views, see Georg Ostrogorsky, *History of the Byzantine State*, trans. by Joan Hussey (New Brunswick, 1969), p. 489 and G. I. Brătianu, "L'hyperpère byzantin et la monnaie d'or des républiques italiennes au XIIIe siècle," in *Mélanges Charles Diehl*, Vol. I (Paris, 1930) p. 42.

hyerperon was a widely circulating, highly respected coin throughout the 12th century.[62] Michael Hendy solved this enigma by explaining the Alexian monetary reform of 1092. This reform raised the declining fineness of Byzantine gold coinage from a low of 8 to a fairly stable 20.5 carats, where it remained with a few exceptions to the disaster of 1204.[63]

In that year, the crusaders broke up the empire and several successor states emerged. This event probably doomed any potential role for the once proud Byzantine coinage in international commerce. Only the empire of Nicea definitely continued to issue gold coins. But even there, the fineness of the hyperperon was only 16 carats in the early 13th century and later slipped even further. Moreover, the number of coins struck seems to have been limited. By 1250, nearly all references to hyperpera in commercial documents specifically describe them as coins of account.[64]

Throughout this period, the Venetians tried with some success to cope with this new situation. Until 1219 and perhaps even longer, the Venetians probably continued to issue some Comneni coins from the mints in Constantinople.[65] But since none of these coins have survived, it may well be the case that the Venetians relied on old 12th century coins and increasingly on their new grossi for commerce in the old Byzantine trading sphere. Owing to Venetian widespread political and economic influence, by 1220 the grosso seems to have become a basic unit for commerce in the Aegean and Balkan areas. The limited success of their coinage here as well as their sensitive political relationships with the Latin Empire and Nicea no doubt continued to dissuade the Venetians from such a radical step as minting their own gold

62. Some examples are outlined in Marc Bloch, "Le problème de l'or au moyen âge," *Annales d'histoire économique et sociale*, Vol. V (1933), p. 15. For a Byzantine scholar who has studied late Byzantine coinage and has come to the same conclusions, consult D. A. Zakythinos, "Crise monétaire et crise économique à Byzance du XIIIe au XVe siècle," *L'Hellenisme contemporaine*, Deuxième série, Vol. I (1947), pp. 167-168.
63. Hendy, *Coinage and Money*, pp. 3-14.
64. Hendy, *Coinage and Money*, pp. 224 and 246-248; Tommaso Bertelè, "Moneta veneziana e moneta bizantina," in *Venezia e il Levante fino al secolo XV*, ed. by Agostino Pertusi (Florence, 1973), Vol. I, pp. 106-108; and Zakythinos, "Crise monétaire," pp. 120-123. Despite its 13th century drop in fineness, the 20.5 carat standard of the hyperperon still fascinated Europeans. Hendy, pp. 15-16, claims that Frederick II may well have modeled his 20.5 carat *Augustalis* on it, while giving it a weight which equalled the Almohad-Hafsid double dinar. This would have given the coin wide repute in two trading areas of the Mediterranean. Such a move would have assured a wide acceptability, if Frederick had been able to expand his kingdom politically and economically. Such a move also made sense in the context of a declining Byzantine fineness, since an international standard was still needed in such a vital trading area.

Frederick probably did not realize that he was likely to have been following the example of Robert Guiscard. This Norman imperialist changed the fineness of the tari from about 22-23 carats down to 16.5 to match the contemporary Byzantine standard. Perhaps this anticipated Guiscard's conquests there. For the exact dating of these changes, consult Philip Grierson and W. A. Oddy, "Le titre du tari sicilien du milieu du XIe siècle à 1278," *Revue numismatique*, Sixième série, Vol. XVI (1974), pp. 127-128. G. I. Brătianu, "L'hyperpère," clearly connected the default of the Byzantine gold coinage with the gold revolution in Italy.
65. Bertelè, "Moneta veneziana," pp. 14-17 and Hendy, *Coinage and Money*, pp. 206-209.

until well beyond 1261. After that, the wide success of the florin probably mandated such a change.[66]

Thus both the hyperperon and the Egyptian dinar had faltered by the opening years of the 13th century. Only the crusader states continued to issue a regular gold coinage for use in Mediterranean commerce. Yet at this very moment, Levantine trade was becoming increasingly critical to the economies of the North Italian commercial powers.[67] With such a significant Italian commitment in the east Mediterranean, maintenance of a viable unit of international currency must have been a vital concern to those traders.

In 1250, however, a papal legate forbade the blasphemous practice of Christians issuing coins in the name of Muhammed. Pope Innocent IV confirmed his legate's censure in 1253. This ban forced the crusaders to issue only coins with Christian legends, although their Kufic script testified to their intended use in commerce with Muslims. Nevertheless, such coins could never have circulated widely, since they were now overtly coins of states with virtually no political or economic clout. Consequently, their minting ceased completely in 1258.[68] Since Pisa and Genoa concentrated their eastern commerce in Syria and Egypt after 1204, it is hardly surprising that a major initiative for a new gold coinage would come from the great commercial cities of Tuscany and Liguria.

Therefore, there was now an overwhelming need in the West to devise a new monetary unit for Mediterranean commerce. Western gold at first clearly did not come from the exploitation of European mines. The first major gold mine there, in Serbia, opened only in 1252.[69] Therefore, European gold coinage must have depended at least initially on imported gold. Some of it

66. Grierson, "La moneta veneziana," pp. 80-83 and Gino Luzzatto, "L'oro e l'argento nella politica monetaria veneziana dei secoli XIII e XIV," *Rivista storica italiana*, Serie V, Vol. II (1937), pp. 22-23. The ducat's charter of origin specifically cites the florin as its model. The rapid success of the florin in the Levant can be seen from a gold hoard buried at Acre in 1291. Of 630 coins discovered, over 600 were florins and only 21 were ducats. Such a rapid success indicates the preference for a gold coinage in the Levant.

67. Hilmar C. Krueger, "Genoese Trade with Northwest Africa in the Twelfth Century," *Speculum*, Vol. VIII (1933), pp. 385-389. Small shipments dominated the North African trade, whereas Levant transports were large; they demanded major investments and reaped large profits. Confer also Claude Cahen, "Notes sur l'histoire des Croisades et de l'Orient latin," *Bulletin de la faculté des lettres de Strasbourg*, Vol. XXIX (1950), p. 342. E. H. Byrne, "The Genoese Colonies in Syria," in *The Crusades and Other Historical Essays* (New York, 1928), pp. 139-182 has shown how the Genoese specifically reorganized their colonies in Syria in the 13th century to secure more control by and a greater profit for Genoa.

68. For a full discussion of this coinage, consult Philip Grierson, "A Rare Crusader Bezant with the *Christus Vincit* Legend," *American Numismatic Society Museum Notes*, Vol. VI (1954), pp. 169-178. The papal letter can be found in Lavoix, *Monnaies*, pp. 52-53.

69. Desanka Kovacevic, "Dans la Serbie et la Bosnie médiévales: Les mines d'or et d'argent," *Annales—Économies • Sociétés • Civilisations*, Vol. XV (1960), pp. 248-249 and Lopez, "Back to Gold," p. 233. Balint Homan, "La circolazione delle monete d'oro in Ungheria dal X al XIV secolo e la crisi europea dell'oro nel secolo XIV," *Rivista italiana di numismatica*, Vol. XXXV (1922), pp. 126-131 has ascribed the rapid expansion of gold mining in the Balkans in the late 13th century to the surging demand for gold in the West.

came from the Near East. Gold specie was among the treasures sent home by the crusaders.[70] But West Africa had indirectly provided gold to Italy for many years. Italians even seem to have used it for international commerce with increased regularity. Their trade with the Maghrib and its traditional politico-economic satellites of Spain and Sicily provided the single most important source of gold for the West.[71]

Historians have long debated the extent of this European gold trade with Africa. On one extreme, Lopez maintained that the Genoese viewed their African penetration as a commercial secret well worth guarding.[72] On the other hand, Bautier, for example, has argued that because the Genoese abandoned Morocco after 1235, they must not have been relying heavily on African gold.[73] In reality, the situation was more complex. It can only be understood in terms of the changing trans-Saharan trade routes, which have been described above.

Until 1212, Spain was one route by which gold entered Europe. Marobitins circulated fairly extensively in the West, as did to a lesser extent the mazmodina a dinar of the Almohads in Spain. One famous example dates from 1142. In that year, Abbot Peter the Venerable met with Alphonso VIII of Castile at Salamanca to re-establish the annual Hispanic stipend to Cluny at 2000 dinars.[74]

As in other frontier areas between Christianity and Islam, the attractiveness of commerce encouraged coin imitations to facilitate it. In 1173, the Castilians began to strike *morabiti alphonsi*. The Portuguese followed in the

70. For some examples of huge Islamic payments of gold to the conquering crusaders, consult Blancard, *Le besant d'or*, pp. 151-161. E. H. Byrne, "Genoese Trade with Syria in the Twelfth Century," *American Historical Review*, Vol. XXV (1919), p. 216 even points to a possible changing balance of trade with the Near East by the early 13th century, as European cloth exports increased.

71. North Italian commerce with the Maghrib goes back to the obscurity of the early 11th century, as Tyrrhenian fleets chased Arab raiders back to their ports. Soon there began that long series of commercial treaties allowing free access, reduced tariffs, and the establishment of fonduks. This culminated only in the 19th century with full colonialism. For detailed information on these treaties and their texts, consult Louis de Mas-Latrie, *Traités de paix et de commerce et documents divers concernant les relations des chrétiens avec les Arabes de l'Afrique septentrionale au moyen âge* (Paris, 1866).

72. Roberto S. Lopez, *Studi sull'economia genovese nel medio evo* (Turin, 1936), pp. 5-8 and 31-32. Lopez also pointed to the frequent passage of gold-paying Arabs on Genoese ships.

73. R.-H. Bautier, "Les relations commerciales entre l'Europe et l'Afrique du Nord et l'equilibre économique méditeranéen du XIIe au XIVe siècle," *Bulletin philologique et historique du comité des travaux historiques et scientifiques* (1953-1954), pp. 400-406. In his dissertation, pp. 81-88, Misbach has likewise argued that the decline of African gold exports was a permanent thing.

74. See above, p. 7. For the details of the Cluny stipend, consult Charles Julian Bishko, "The Spanish and Portuguese Reconquest, 1095-1492," in *A History of the Crusades*, ed. by K. M. Setton (Madison, 1975), Vol. III, p. 408. According to Jaime Vincens Vives, *An Economic History of Spain*, trans. by Frances M. López-Morillas (Princeton, 1969), p. 275, Ferdinand III in 1251 conceded special trading rights to the Genoese in Seville, which may have accorded them an opportunity to export gold.

The Italian gold revolution of 1252

late 12th century. After the Almohad debacle in 1212, they issued their own gold coins, now based on the Almohad dobla. Yet only Castile among European states continued to maintain an Islamic system of coinage until the reforms of Ferdinand and Isabella in the late 15th century. Perhaps this indicates the extremely slow integration of Castile into the expanding European commercial world.[75]

In Aragon, it was not until well into the 14th century that a regular gold coinage began. Probably the final disappearance of a strong Islamic bridge at Gibraltar in the early 13th century limited the direct flow of gold across the straits. Independent issues of gold in the Iberian peninsula would have to await penetration by their merchants to the trade routes now reaching central North Africa. The first known Catalan consulates and fonduks in North Africa date from 1253 at Tunis, the heart of the Hafsid state, and later at Bougie in 1259.[76] It is unlikely, however, that demand for gold in the Iberian peninsula was as great as in northern Italy, where merchants were deeply involved in Levantine trade. Nor does it seem probable that the Catalans in the 13th century had as yet as many profitable luxury exports to sell there as did the North Italians.

It is extremely difficult to get any real appreciation of the amount of gold which the North Italians could and did acquire in the western Mediterranean. For example, the Genoese cartularies often record units of account. In those cases, it is impossible to ascertain what coinage was actually involved. But much more importantly, the cartularies refer only to trade which passed through Genoa. The pattern of Italian trade in the Mediterranean, however, was far more complex. With Italian trading colonies all over the Mediterranean, trade often passed directly from one port to another without ever being registered by an Italian notary at home. A classic example is the voyage of Ibn Jubayr from Spain to Egypt in a Genoese ship which never touched Italy.[77]

---

75. Devisse, "Routes de commerce," pp. 376-377; Vives, *Economic History*, pp. 225-227 and 278-279; and Godinho, *L'économie*, p. 131.

76. Charles-Emmanuel Dufourcq, *L'espagne catalane et le maghrib aux XIIIe et XIVe siècles* (Paris, 1966), pp. 98-101.

77. By way of contrast, E. Lévi-Provençal and E. García Gómez, "La politica africana de 'Abd al-Rahmān III," *Al-Andalus*, Vol. XI (1946), p. 376 tell of a vessel of Islamic Spain which sailed directly to Egypt in 935 for trade. S. D. Goitein, "Medieval Tunisia. The Hub of the Mediterranean," in *Studies in Islamic History and Institutions* (Leiden, 1966), p. 309 also refers to regular Islamic commerce from Spain to the Levant in the 11th century. But by the 12th century, Christian commercial dominance was complete. According to A. Gateau, 'Quelques observations sur l'intérêt du voyage d'Ibn Jubayr pour l'histoire de la navigation en Mediterranée au XIIe siècle," *Hespéris*, Vol. XXXVI (1949), pp. 289-312, Ibn Jubayr's voyage illustrates not only regular Genoese routes from Ceuta to Alexandria and Acre to Messina, but it seems to demonstrate that Muslim commerce was restricted to cabotage along the coast of North Africa. Gateau reasonably assumed that the cause for this change was the new Christian dominance over the Mediterranean islands and the ports of the Levant, the keys to long-distance Mediterranean trade. Another indication of this state of affairs occurs in a quote from a Sicilian Muslim, hesitant to travel to Islamic Spain: "The sea belongs to the Romans, and our ships circulate there only by exposing themselves to great risks; only the mainland belongs to the Arabs." The quote and its

Thus Italians had the ability to move gold across the Mediterranean without bringing it directly to Europe, which was still exclusively on a silver standard.

It seems likely that Italians did use the profits of commerce from one area of the Mediterranean to trade successfully in another. Once again, the direction and intensity of the gold flow from West Africa seems to have directed to some degree European commercial attempts to secure gold there for further investment elsewhere. Almoravid and Almohad Spain was a favorite trading partner in the mid-twelfth century. But trade with Spain was hardly the most convenient way to acquire gold to trade in the Levant. Therefore, a major point of entry for West African gold through most of the 12th century may well have been Norman Sicily.

Sicily possessed a gold currency unparalled in Europe in the 12th century.[78] Numismatic evidence confirms the stability of the tari as long as it was struck.[79] Moreover, Norman monarchs were clearly able to amass and spend extremely large amounts of gold for political purposes.[80]

The source of Sicilian gold was undoubtedly North Africa. Sicilian relations there extended back at least to the time of Count Roger, whose friendship for the Zirid ruler Temin probably induced him to renounce the Pisan-Genoese expedition against al-Mahdiyya in 1087.[81] The African conquests of Roger II brought a significant amount of gold to the Norman treasury. In addition to booty, commercial taxes imposed in African ports also seem to have been very lucrative. Norman control of the entire central Mediterranean

---

circumstances can be found in J. K. Hyde, *Society and Politics in Medieval Italy* (London, 1973), pp. 30-31. Such a drastic change in the domination of Mediterranean commerce, which passed from nearly total Islamic to Christian control in a short time, demonstrates the need for scholarly research on medieval merchant marines.

78. For evidence regarding Sicilian gold coinage, see especially Domenico Spinelli, *Monete cufiche battute dai principi Longobardi, Normanni e Svevi nel regno delle Due Sicilie* (Naples, 1844); Arthur Engel, *Recherches sur la numismatique et le sigillographie des Normands de Sicile et d'Italie* (Paris, 1882); Lagumina, "Studi sulla numismatica," pp. 1-32; C. A. Garufi, "Moneta e Conii nella Storia del diritto siculo dagli Arabi ai Martini," *Archivio storico siciliano*, Nuova serie, Vol. XXIII (1898), pp. 1-171; Giulio Sambon, *Repertorio generale delle monete coniate in Italia (476-1266)* (Paris, 1912); Luigi dell'Erba, "La monetazione normanna nell'Italia meridionale e nella Sicilia," *Bollettino del circolo numismatico napoletano*, 1927, fasc. II; and *Corpus Nummorum Italicorum*, Vol. XVIII, *Italia meridionale continentale* (Rome, 1939).

79. H. Kowalski, "Zur Metrologie und zu den Beizeichen der Augustalen, Realen und Tari," *Revue belge de numismatique*, Vol. CXVII (1971), pp. 133-160 and Grierson and Oddy, "Le titre," pp. 123-134.

80. For example, the regent Margaret sent 700 ounces of gold to Pope Alexander III in 1168 to persuade him to disallow the election of Walter of the Mill as Archbishop of Palermo. King Tancred was able to indemnify Queen Joanna for the loss of her dowry with one million taris. He also gave 20,000 ounces of gold both to Joanna and King Richard for English support in the wars of his accession. See John J. Norwich, *The Kingdom in the Sun, 1130-1194* (London, 1970), pp. 297 and 367-370.

81. John J. Norwich, *The Normans in the South, 1016-1130* (London, 1967), p. 277. His source is Ibn al-Atir.

probably forced Italian shipping from the Tyrrhenian Sea to stop by necessity at some Norman port on the way east. This, of course, brought customs profits to the crown. Therefore, the Almohad conquest of the eastern Maghrib by 1160 was at least a partial economic disaster for Sicily.[82]

Yet despite the territorial loss of the North African possessions, Norman commerce with North Africa continued to be important. Wheat was the primary Sicilian export. Its sale in North Africa was systematically organized for the benefit of the crown, which held many wheat-producing lands in the royal demesne. Royal agents were present in al-Mahdiyya from 1117 to coordinate the sale of wheat there.[83] The Hilali invasion had nearly destroyed the rich agricultural base of the Tunisian hinterland. So anxious were Tunisian merchants to buy Sicilian grain that in 1141 some of them actually agreed to exchange pure gold for Sicilian gold at equal weight. Since the tari contained only 16.5 carats of gold, the crown netted quite a profit.[84] The rulers of Ifriqiya even promised an annual gold subsidy to the Normans from at least 1180 in exchange for the right to buy grain in Sicily.[85]

This Sicilian gold seems to have become quite attractive to North Italian merchants. Many years ago, Heyd recognized the importance of Sicily as a mid-point of Mediterranean trade.[86] A series of commercial treaties in the 12th century gave Italian maritime states extensive commercial rights in the Regno.[87] The North Italians exported manufactured products to Sicily, whose industry had become decadent in the 12th century.[88] The net balance

---

82. Abulafia, *The Two Italies*, p. 85. For an overview of Sicilian involvement in North Africa, consult G. La Mantia, "La Sicilia ed il suo dominio nell'Africa settentrionale dal secolo XI al XVI," *Archivio storico siciliano*, Nuova serie, Vol. XLIV (1922), pp. 154-255. More recently, see Hélène Wieruszowski, "The Norman Kingdom of Sicily and the Crusades," in *A History of the Crusades*, ed. by K. M. Setton (Madison, 1969), Vol. II, pp. 3-44.

83. Abulafia, *The Two Italies*, pp. 36-40.

84. Hady R. Idris, *La Berbérie orientale sous les Zīrīdes, X-XIIe siècles*, Vol. II (Paris, 1962), p. 665.

85. For a discussion of the provisions in these treaties, consult Wieruszowski, "The Norman Kingdom," pp. 18-33 and Mas-Latrie, *Traités*, pp. 52 and 123. Since the Hafsids were on friendly terms with the Hohenstaufen, al-Mustansir firmly opposed the Angevin take-over in Sicily and halted the traditional subsidy. Therefore, historians have pointed to this situation as at least one cause for the destination of the 1270 crusade of Louis IX. For a discussion of the controversial relationship of gold to this crusade, confer Robert S. Lopez, "Le facteur économique dans la politique africaine des papes," *Revue historique*, Vol. CXCVIII (1947), p. 119; Julien, *History*, pp. 143-144; and Joseph R. Strayer, "The Crusades of Louis IX," in *A History of the Crusades*, ed. by K. M. Setton, Vol. II (Madison, 1969), pp. 509-518.

86. Wilhelm von Heyd, *Histoire du commerce du Levant au moyen-âge*, trans. by Furcy Raynaud (Leipzig, 1885), Vol. I, pp. 133-145; Goitein, "Sicily and Southern Italy," pp. 9-33; and Abulafia, *The Two Italies*, pp. 79, 98, and 156.

87. For references to these treaties, consult Abulafia, *The Two Italies*.

88. Michele Amari, *Storia dei musulmani di Sicilia*, ed. by C. A. Nallino, Vol. III (Catania, 1937), pp. 802-805; F. Ciccaglione, "La vita economica siciliana nel periodo normanno-svevo," *Archivio storico per la Sicilia orientale*, Vol. X (1913), pp. 341-343; and James M. Powell, "Medieval Monarchy and Trade: The Economic Policy of Frederick II in the Kingdom of Sicily," *Studi medievali*, Serie terza, Vol. III (1962), p. 492.

of trade probably produced a surplus of gold for North Italian traders. This was then available for further investment elsewhere, most likely in the East where gold was essential for commerce.[89]

In addition, Sicily clearly seems to have served as a place where Italian silver could be exchanged for Sicilian gold.[90] This gold provided easier access to markets in the gold-oriented Levant. The tari seems to have played a pan-Mediterranean role. Fully 25% of all gold transactions in Genoese cartularies up to 1253 were recorded in taris.[91] For a while in the late 12th century, Sicily may have even served as a central bank for Mediterranean commerce. There, merchants may have come to rely on a sophisticated, large-scale transfer of accounts without requiring similarly large shipments of bullion, for which little evidence has survived.[92]

Yet for various reasons, Sicily was not destined to remain at the heart of Mediterranean commerce. For one thing, the chief termini of the gold flow in the 12th century lay in the far west. By the early 13th century, even Sicily had difficulty sustaining a gold coinage. In 1222, a major coinage reform of Frederick II seems to have issued exclusively silver. Frederick even outlawed the payment of internal debts in old gold to secure its sale to the state mint.[93] He issued his famous *Augustalis* in 1231 only after a new treaty with Tunisia had renewed the traditional gold subsidy. Thus, Frederick's apparent difficulties in securing gold for this issue probably doomed the *Augustalis* to fulfill only an ideological function.[94]

---

89. Henry L. Misbach, "Genoese Trade and the Role of Sicilian Coinage in the Mediterranean Economy, 1154-1253," *Revue internationale d'histoire de la banque*, Vol. V (1972), p. 306.

90. An excellent example can be found in the cartulary of Giovanni Scriba from October, 1158. Elia in Genoa made a sea-loan of £155 18s 6d Genoese to be repaid in Palermo for 81 ounces of taris and 66 ounces of paiola gold. Confer *Il cartolare di Giovanni Scriba*, ed. by M. Chiaudano and M. Moresco (Rome, 1935), Vol. II, p. 310, no. XVIII. For commentary, see Abulafia, *The Two Italies*, p. 98.

91. Misbach, "Genoese Trade," p. 305.

92. Abulafia, *The Two Italies*, pp. 226-232 and 255-280.

93. There is no clear agreement among historians on the question of whether taris continued to be issued throughout the reign of Frederick and, if so, in what quantity. The only contemporary source, the Chronicle of Riccardo di San Germano, master of the royal mint at Brindisi, is ambiguous. The chronicle can be found in the *Monumenta Germaniae Historica, Scriptores*, Vol. XIX, ed. by Georg Pertz (Hannover, 1866), pp. 340-343. C. A. Garufi, "Moneta e conii," pp. 104-127 has set the traditional interpretation that taris continued to be minted. More recently, however, Powell, "Medieval Monarchy," pp. 506-509 and Eugenio Travaglini, "La zecca di Brindisi nei documenti e scritti di epoca sveva," *Brundisii Res*, Vol. IV (1972), pp. 123-127 have argued that only silver coins were struck in this first coinage reform. Similarly, Oddy and Grierson, "Le titre," p. 133 demonstrate a fineness drop in the tari under Frederick, which may be symptomatic of gold flow problems. The numismatic evidence presented by Sambon, *Repertorio generale*, pp. 186-196 and in the *Corpus Nummorum Italicorum*, Vol. XVIII, pp. 194-211 seems to indicate a new profusion of silver denarii after 1220. But a lack of clear dating makes it impossible to rule out the continued emission of taris between 1222 and 1250.

94. Powell, "Medieval Monarchy," p. 511. The treaty appears in the *Monumenta Germaniae Historica, Constitutiones et Acta*, Vol. II, p. 187. There is, however, some problem with the dating of this treaty. According to Robert Brunschvig, "Note sur un traité conclu entre Tunis et

Secondly, the continued development of larger ships in the 12th and 13th centuries for Mediterranean commerce began to enable Italian merchants to bypass Sicily. Historians of the Genoese archives have noted the decreased frequency of stops in Sicily on voyages to Syria.[95] The Genoese of the 13th century were more likely to stop at Tunis and decide there if the prospects for a successful commercial venture to Syria were propitious.[96] Thus it may be the case that for a while in the 13th century, Tunisia served the same monetary function as 12th century Sicily. This could be related to the redirection of the gold routes there in mid-century. Further archival research would be necessary to determine the exact chronology of changes in sailing patterns to the east. It may also be true that this ability to sail directly to Syria created a new demand for gold coinage at home in Italy to use in Levantine trade.

Finally, one must consider the economic and political policies of Frederick II. Genoese commercial privileges in Sicily during Frederick's minority were so extensive that they threatened to turn the island into a Genoese colony. But the Diet of Capua in 1220 restored the *status quo ante* 1189, which *de facto* greatly reduced Genoese commercial freedom in Sicily.[97] Moreover, Frederick's pan-Italian policies pushed Genoa fully into the Guelf military alliance by 1230. Consequently, Genoese cartularies reveal no trade with Sicily from then until Frederick's death in 1250.[98]

Perhaps these difficulties in securing gold indirectly in Sicily had led Genoa to concentrate even more on commerce further west in Morocco, where the western trans-Saharan trade routes first reached the Mediterrane-

---

l'empereur Frédéric II," *Revue tunisienne*, Vol. III (1932), pp. 153-160, the date should in fact be August, 1221. His arguments are persuasive, although they do not seem to have been widely circulated. In any case, the treaty still points to Sicilian efforts to secure a regular gold supply, perhaps as early as the 1220s.

For a discussion of the ideological reasons for Frederick's new coin, see Garufi, "Moneta e conii," pp. 108-130. On the other hand, Lopez, "Back to Gold," p. 227 has argued that the fineness of the *Augustalis* reflected the natural alloy of Senegalese gold and silver and therefore indicated the origin of the metal. Arthur J. Sambon, "Monnayage de Charles I d'Anjou dans l'Italie méridionale," *Annuaire de la société française de numismatique*, Vol. XV (1890), pp. 55-56 has even been able to discover in the Angevin registers a clear reference that *aurum de paiola* was the actual source of the new gold issue. The Angevin registers also make it clear that Tunisian gold payments to the Sicilian crown were customary and traditional in 1272 and 1273. They specify that such payments had been made at least since the time of Frederick II. For the actual texts, consult Camillo Minieri-Riccio, *Saggio di codice diplomatico formato sulle antiche scritture dell'Archivio di stato di Napoli*, Vol. I (Naples, 1878), pp. 95-96 and 104. Also, see above, footnote 64.

95. Mario Chiaudano, "Genova e i Normanni," *Archivio storico pugliese*, Vol. XII (1959), p. 76 and Goitein, "Sicily and Southern Italy," p. 26.

96. Eugene H. Byrne, *Genoese Shipping in the Twelfth and Thirteenth Centuries* (Cambridge, 1930), pp. 36-47.

97. Powell, "Medieval Monarchy," pp. 444-469.

98. James M. Powell, "Genoese Policy and the Kingdom of Sicily, 1220-1240," *Medieval Studies*, Vol. XXVIII (1966), pp. 352-353.

an.[99] In 1161, Genoa and the then powerful Almohad state concluded a commercial treaty of great significance.[100] Before that year, there had been no recorded Genoese trade west of Bougie; after that treaty, trade with Ceuta predominated in North Africa.[101] Within a year, a Ceuta commenda mentioned Paiola, the land of gold. In 1184, the term *aurum de paiola* began to appear in Genoese documents for North Africa.[102] When their position in Sicily deteriorated in the 13th century, the Genoese even sent a massive naval expedition to aid the besieged Almohad caliph in Ceuta in 1235.[103] Clearly, the Genoese had by now attached great importance to their commerce with the Almohads. In vain, they tried desperately to protect their investment and privileges there.

Unfortunately, Genoa had backed a loser. In 1236, the Almohad governors of Tlemcen and Tunisia declared their independence. The Hafsids of Tunis were the new star over the Maghrib. In the 1230s alone, they signed trade treaties with Pisa, Venice, Genoa, and Sicily. By the 1250s, commercial embassies arrived in Tunis from as far away as Norway and Bornu. At the end of that decade, the sharif of Mecca recognized the Hafsid ruler as Caliph of Islam.[104] Tunisia had become the center of orthodoxy and commerce in North Africa. At first, the Genoese position in Tunis was the weakest of the major European commercial states. They received a weak treaty there in 1236 only after their show of force at Ceuta. They were unable to secure better provisions until after the death of Frederick II in 1250. This time the treaty included a specific provision for the purchase of gold at Tunis.[105]

There is even some evidence to indicate how acutely Genoa felt the loss of her privileged position in Morocco. The Genoese had begun themselves to try

---

99. Good accounts of Genoese trade with the Maghrib are available in Hilmar C. Krueger, "Genoese Trade," pp. 377-395 and "The Wares of Exchange in the Genoese-African Traffic of the Twelfth Century," *Speculum*, Vol. XII (1937), pp. 57-71. Consult also Lopez, *Studi*, pp. 5-45.

100. Mas-Latrie, *Traités*, p. 47.

101. Krueger, "Genoese Trade," pp. 379-382.

102. Lopez, *Studi*, p. 42.

103. The Almohad caliph was isolated in his great Mediterranean port of Ceuta by a rebellion led by his governor of Seville. According to Charles E. Dufourcq, "La question de Ceuta au XIIIᵉ siècle," *Hespéris*, Vol. XLII (1955), pp. 87-107, most areas of the Almohad empire were in full revolt against the legitimate ruler. For a thorough examination of this final full-scale Genoese naval expedition in the western Mediterrannean before the 15th century, the best work is Raffaele di Tucci, "Documenti inediti sulla spedizione e sulla mahona dei Genovesi a Ceuta, 1234-1237," in *Atti della Società ligure di storia patria*, Vol. LXIV (1935), pp. 271-340. Di Tucci concludes that the Genoese took the initiative in their own self-interest. This is hardly surprising, given the prominence of Morocco in Genoese trade until then.

104. Julien, *History*, pp. 121-143. Braudel, "Monnaies," p. 11 believed that the decline of large and powerful North African empires allowed the emergence of more city-states, where Christians were welcome. For a comparison of these treaties, consult Mas-Latrie, *Traités*, pp. 82-123.

105. Robert Brunschvig, *La Berbérie orientale sous les Ḥafsides*, Vol. I (Paris, 1940), pp. 22-43 and Mas-Latrie, *Traités*, p. 129.

The Italian gold revolution of 1252 51

to penetrate to the interior to meet caravans coming north. In the late 1240s, Pope Innocent IV, a Genoese, supported their efforts to sail down the Atlantic coast of Morocco. In 1253, the Genoese banking family of Leccacorve invested in a commenda to Safi in Morocco in the name of three nephews of the pope. The following year, Innocent requested the Almohads to cede fortresses along the Atlantic coast. It may be possible to see here a leading role for the Genoese pope. His approval of the discontinuance of the Crusader forged coins may have been connected to business opportunities for his family in Genoa. But it seems more reasonable that Innocent's efforts in Morocco reflect the sudden increased demand for gold in the West.[106]

Unlike the Genoese, the Pisan commercial position in Sicily was more consistent. Like their fellow North Italians, the Pisans had profited from concessions granted in the Hohenstaufen take-over.[107] But Pisa remained steadfastly Ghibelline. Consequently, Pisan commerce tended to predominate in Sicily right down to the Angevin conquest. Genoa's losses were Pisa's gains. Therefore, Pisa probably continued to draw off some gold in that trade through the 1260s.[108]

Moreover, Pisa was also deeply involved in Tunisia. A Pisan document of the early 13th century claimed that one-third of her trade was with Sicily and another third with North Africa.[109] The Pisans seem to have held the most favored position in Tunis in the 12th and 13th centuries.[110] They were therefore in the best position to reap the rewards of the shifting gold routes in North Africa. But since Pisa continued to support the Hohenstaufen to the bitter end, it would have been unthinkable for her to strike gold in the 1250s. The minting of gold was an imperial act, which would have been repugnant to Ghibelline Pisa, regardless of the economic advantages of such a move.

Nevertheless, it may be quite reasonable to conjecture that Guelf Florence, by contrast, may have been emboldened by the death of Frederick II to

---

106. Lopez, "Le facteur," p. 187 and Roberto S. Lopez, *La prima crisi della banca di Genova (1250-1259)* (Milan, 1956), pp. 75-79. The papacy had become involved quite early in Italian commercial designs. In 1067, Gregory VII backed the Pierleoni of Rome in their attempt to gain an important position at Bougie, seat of the Hamdannids and an important gold terminus. In 1087, the legate of Victor III led three hundred Genoese and Pisan ships against al-Mahdia, which was spared only upon payment of 100,000 dinars. For references, consult Harry W. Hazard, "Moslem North Africa, 1049-1394," in *A History of the Crusades*, ed. by K. M. Setton, Vol. III (Madison, 1975), p. 464 and Krueger, "Genoese Trade," pp. 377-388. For a general survey of medieval Atlantic commerce, see Raymond Mauny, *Les navigations médiévales sur les côtes sahariennes antérieures à la découverte portugaise, 1434* (Lisbon, 1960).

107. Abulafia, *The Two Italies*, pp. 172-173.

108. David Herlihy, *Pisa in the Early Renaissance* (New Haven, 1958), pp. 162-163. For an excellent overview of North Italian commercial and naval relations with the Kingdom of Sicily, consult Pietro Nardone, *Genova e Pisa nei loro rapporti col Mezzogiorno d'Italia fra la fine del secolo XII a gl'inizi del XIII* (Prato, 1923), pp. 3-73. The best account of Pisan support for the Hohenstaufen is the work of Steven Runciman, *The Sicilian Vespers* (Cambridge, 1958).

109. Herlihy, *Pisa*, p. 163.

110. André-E. Sayous, *Le commerce des Européens à Tunis depuis le XIIe siècle jusqu'à la fin du XVIe* (Paris, 1929), pp. 49-51.

make such a move. Already in the late 12th century, a series of Pisan-Florentine treaties make it clear that Florence was using the port of Pisa to export manufactured goods.[111] This could mean that over the long run, a good deal of the gold funneled through Pisa may ultimately have wound up in Florence. Pisa even assisted the Florentines in setting-up banking operations in Tunis, beginning in 1250.[112]

By 1252, therefore, Europeans dominated the commerce of the major entry points for sudanese gold into the Mediterranean basin. These merchants had already become paramount in Mediterranean commerce and were beginning to open up additional markets all over Europe. It seems reasonable for historians to see here an important correlation in Italy between the demand for a new reliable unit of international commerce in the Mediterranean and access to the supply of gold bullion. To the Italians of Genoa and Florence must go the credit for the monetary technology to make it work. Certainly, local conditions in Genoa and Florence must have played a role in the precise timing of the gold revolution.[113] But examination of that equally important question is beyond the scope of this paper.

---

111. Nardone, *Genova e Pisa*, pp. 62-65 and Gino Arias, *I trattati commerciali della repubblica fiorentina*, Vol. I (Florence, 1901), pp. 11-12, 18-19, 31-32, and 46-52.

112. Brunschvig, *La Berbérie*, pp. 38-45. The Sicily of Frederick II was also in a strong position. So profitable was the post of Sicilian consul at Tunis that Pietro Capuano of Amalfi paid 3000 taris for it in 1242.

113. For a review of the pertinent literature, see the extensive bibliographical footnotes in Roberto S. Lopez, "Settecento anni fa: il ritorno all'oro nell'occidente duecentesco," *Rivista storica italiana*, Vol. LXV (1953), pp. 19-55 and 161-198 and more recently, Corrado Astengo, "L'inizio della coniazione dell'oro a Genova ed una pubblicazione del Prof. R. S. Lopez della Yale University," *Rivista italiana di numismatica* (1961), pp. 13-57.

2

Monetary flows—Venice 1150 to 1400

LOUISE BUENGER ROBBERT

The Republic of Venice contributed substantially through her money to international trade in the high and later middle ages. Beginning in the late twelfth century, her merchant aristocracy directed her mints to produce a continuous series of sound, full-weight coins: first the Ziani penny, 1172-1192; then the large silver grosso, 1194-1356; and finally the gold ducat, 1284-1797. Each coin in its era remained full weight and reliably pure. With the grosso and ducat the Venetian merchants had a reliable instrument of exchange for their foreign commerce. The moneys of account based on these coins provided flexibility before the variations of supply and demand. Venice's neighbors and foreign hosts began to realize how dependable was the money of the Republic, and, by 1400, nearly everyone in the Venetian trading zone was using Venetian coins. Venice was the middleman between the European and the Near Eastern trading zones, and in the late fourteenth century her money functioned as the principal medium of international exchange.

After 1150, evidence of her commercial activity accumulates, and Venice appears with a money system based on silver. Venice received silver from the north to mint her own silver coins. These and other Italian silver coins, silver bullion, and money of account based on silver coins were used by Venetians to pay private debts, public taxes or bequests, in Venice itself and in Italy. Until about 1200 Venetian payments in Romania (the Greek East), the Crusader states and Muslim Egypt were made in the gold money of these areas. Since the Venetian merchants often referred to these gold coins in their commercial documents and in intergovernmental contracts, they were perfectly familiar with the use of gold coins. But the gold coins were not used for payments in Venice itself until late in the thirteenth century. Gold coins were acceptable in the eastern Mediterranean; silver coins were acceptable in Italy.

In the thirteenth century, these generalizations change. Venetian silver coins and silver bullion was shipped in increasing quantities into the eastern Mediterranean as the Venetian mints continued to receive silver from Ger-

---

The author gratefully acknowledges the stimulus of Dr. Maureen Mazzaoui and Dr. John Richards who requested this paper for the Workshop in Pre-Modern Monetary History, A.D. 1200-1750. It was originally presented at this Workshop, held at the University of Wisconsin, Madison, in September, 1977, in an earlier form which considered the data from A.D. 1150 only to 1300. After the conference, the author, at their request, continued the study to A.D. 1400. The conclusions are here presented publicly for the first time.

53

man merchants and then from Serbian mines. At mid-century Italian gold coins began to take the place of the Byzantine gold coinage in international trade. Throughout the fourteenth century Venice continued to export silver coins. Venetian gold ducats were not exported in volume until the mid-fourteenth century. By 1400 the number of silver and gold coins exported from Venice had become very great indeed.

What were these Venetian coins? The penny (denaro, denarius) was the basic silver coin in Venice, as in most medieval European principalities.[1] When the Venetian mints began issuing the penny again in 1172, after about a fifty year pause, it weighed 0.37 grams and was 230/1000 fine silver. Although these coins presented Venetian political independence by recognizing no outside sovereign, they were not sufficiently numerous to satisfy the Venetian demand for a local fractional currency. On the local markets, the Venetians also accepted other Italian pennies as well as silver pieces from across the Alps.

When, at the end of the twelfth century, the Venetians sensed the need for a larger coin, Doge Enrico Dandolo in 1194 created the first grosso.[2] This silver coin (also called the silver ducat or matapan) was much larger than the penny, weighing 2.2 grams and also more fine, at 965/1000 pure silver. Originally, probably supposed to be worth 24 pennies, the reality of its weight and silver content soon caused it to circulate outside Venice at slightly over 26 pennies. Concurrently Doge Dandolo discontinued the striking of his slightly debased penny, but the old pennies were not removed from circulation. These and other fractional Venetian coins served the local Venetian retail needs, while the grosso gradually became important as a medium of exchange for larger transactions and for the Levantine markets.

The official standard of value in Venice in the twelfth century was a money of account, the pound (libra, lira) equivalent to 240 silver pennies. The medieval pound was not a coin, but an officially recognized standard of value based on a penny. The Venetians used several pounds.[3] In the first half of the twelfth century the Venetian pound was called the libra denariorum nostre monete (the pound in pennies of our money) which was equivalent to 23.85 grams pure silver.[4] But in the middle of the twelfth century, the Venetian mints were not issuing any pennies, and after mid-century the money standard used most frequently by the Venetians in Venice was not the Venetian but the Veronese pound (libra denariorum veronensium) equivalent

---

1. This study will ignore the smaller Venetian coins, the quarter-penny and the half (or third-) penny, because they did not circulate outside Venice.

2. Louise Buenger Robbert, "Reorganization of the Venetian Coinage by Doge Enrico Dandolo," *Speculum*, Vol. 49, No. 1 (January 1974), p. 51.

3. A detailed analysis of the evolution of the several Venetian pounds from 1170 to 1400 has recently appeared, Frederic C. Lane, "The First Infidelities of the Venetian Lire," in Harry A. Miskimin, David Herlihy, and A. L. Udovitch, eds., *The Medieval City* (New Haven, 1977), pp. 43-63.

4. Louise Buenger Robbert, "The Venetian Money Market, 1150-1229" *Studi Veneziani* Vol. 13 (1971), p. 54.

to 240 pennies of Verona. Two Veronese pounds equalled one Venetian pound until 1150. These two moneys of account continued to have the same purchasing power in the free markets outside the control of either the Veronese or the Venetian state from 1172 to 1194.[5] After 1172 the Venetians began to use a money of account based on a penny from their own mints, the libra denariorum venetialium.

Also about 1172, Venice began to produce the first of her good silver coins. The new doge, Sebastiano Ziani, initiated a new Venetian penny with precisely the silver content of 1/240 the old Venetian pound, the libra denariorum nostre monete.[6] Concurrently, a new Venetian money of account, the libra denariorum venetialium (later called the lira di piccoli) represented the same purchasing power in silver bullion as the earlier Venetian money of account in the market controlled by the Venetian state. But in foreign markets it was worth 11% less in the first decade of the thirteenth century. This reflects the coinage of the first grosso by Doge Enrico Dandolo in 1194.[7] The grosso continued unchanged in weight and fineness until 1356. The grosso seemed a better coin because it was larger and much more fine than the silver penny; but 10 grossi contained 11% less silver than 240 of the pennies issued by the doges from 1172 to 1193. Sometime in the first fifty years of its existence, the grosso became legally equivalent in Venice to 26 pennies. In these same decades, the Venetians began to use another money of account known as the libra grossorum (or lira di grossi) equivalent to 240 grossi coins.

Thus in the early thirteenth century the Venetian money system was solidly based in silver. The best coin was the grosso, but Venetian pennies, half-pennies, and quarter pennies also circulated alongside foreign coins. Accounts were kept in pounds—the pound of 240 pennies (libra denariorum venetialium or lira di piccoli) and the pound of 240 grossi (libra grossorum or lira di grossi).

The use of these pounds, which were multiples of the actual coins, and their complex variations, gave the Venetian money market a legal instrument of flexibility and, perhaps, elasticity. In 1254 the Venetian state recognized that 1 lira di piccoli = 9 grossi and 5 pennies. This gave official recognition to the discount that Venetian merchants were giving to payments in grossi rather than in the fractional coinage. At this rate, this lira de grossi = 26 lira di piccoli = 239 grossi coins (rather than 240 grossi coins). Frederic Lane has recognized this lira di grossi as the lira di grossi manca (or lacking) because it lacked one grosso to becoming complete (equal to 240 grossi).[8]

---

5. i.e., in Padua and Monselice. Ibid., pp. 31-34.

6. The Veronese pennies began to depreciate about this date also. Ottorino Murari, "Denari Veronesi di un Ripostiglio del Sec. XII" *Rivista Numismatica* (1951-1952), pp. 17-18; and his "I denari Veronesi del periodo Comunale ed il ripostiglio di Vicenza" *Annuario Numismatica "Rinaldi"*, (1950).

7. Robbert, "Reorganization," pp. 55-57 and "Venetian Money Market," pp. 54-55.

8. Lane, "Infidelities," pp. 50-51 and also his "Le vecchie monete di conto veneziane ed il ritorno all'oro" *Atti dell'Istituto Veneto di scienze, lettere ed arti* (Classe scienze morali e lettere), Vol. 107 (1958-59), pp. 56-57.

Table 1. Ratio of Grosso to Penny[10]

Date	Ratio
1254	1:26–1/9
1265	1:27
1267	1:27
1269	1:28
1274	1:28
1275	1:30
1276	1:27
1290	1:32
1291	1:32
1298	1:32

In the sixth decade of the century several evidences of monetary strain appeared in Venice. The need for a fractional coinage became so great that the Venetian mints once more began to coin the penny. This time, in 1269, its weight was reduced from the 0.37 gram penny last issued by Doge Enrico Dandolo, to one of 0.289 grams; but the fineness remained the same as Dandalo's, 250/1000. Venetians hereafter used the lira di piccoli at this rate. Shortly after, in 1282, another new Venetian money of account was introduced, the lira a grossi (the lira tied to the grosso) which was the old lira di piccolo at the new rate of 1 lira = 9 grossi and 5 pennies. The older standard of value was now tied to the stable grosso coin, in contrast to the new lira di piccoli which was equivalent to 240 depreciating, old, and worn-out pennies. This lira a grossi became the most commonly used money of account in Venetian bookkeeping until the mid-fourteenth century.[9]

The purchasing power of the penny continued to decline as Table 1 illustrates. At the same time, the grosso remained a reliable coin, of the same weight and fineness as when it was first issued. However, in the 1280s and 1290s, grossi coins were becoming hard to get in Venice.[11] Not only were the grossi being clipped, but also false and lighter grossi became more common. Since the Venetian government did not want to debase its stable, successful grosso, they copied the Florentines and the Genoese and issued their first gold coin, the ducat, in 1284. Possibly the ducat was issued for the purpose of support-

---

9. Lane, "Infidelities," p. 48; and his "Vecchie Monete," p. 57; Roberto Cessi, *Problemi Monetari Veneziani (fino a tutto il sec. xiv)* (R. Accademia dei Lincei, Documenti finanziari della Repubblica di Venezia) (Padua, 1937), pp. xxxii-iii; and Gino Luzzatto, *Storia Economica di Venezia dall'XI al XVI Secolo* (Venice, 1961), p. 94.

10. Robbert, "Venetian Money Market," p. 49; Cessi, *Prob. Mon. Ven.*, p. xxxii.

11. Luzzatto, *Storia Econ. di Ven.*, p. 96; Cessi, *Prob. Mon. Ven.*, Docs. 56, 57, 60, 67, 68, 69; and Lane, "Infidelities," p. 55.

ing the grosso.[12] The ducat would support the grosso when the grosso was freely exchangeable with the ducat at a given rate. Silver remained the basis of the Venetian moneys of account, even after the first gold ducat appeared. Gold ducats, by law, were evaluated in terms of the silver based lira a grossi.[13]

* * *

What about gold and the Venetians? The gold ducat of 1284 was the first gold coin struck by Venice. However, the Venetians had been using gold coins for centuries in their Levantine business. Venetians and other Italian merchants had found that to do business in Byzantium, the Crusader states and Muslim Egypt, they had to use the money acceptable to the inhabitants of these regions. In twelfth century Romania, this meant the Byzantine hyperperon, which was the standard coin of the Byzantine empire from the coinage reforms of Alexius I Comnenus. In 1092 he had reformed the Byzantine coinage and returned to the standard of the solidus of Constantine, which he called the hyperperon.[14] These fine gold coins were struck by the Byzantine mints throughout the twelfth century, but were somewhat debased during the troubled reigns of the Angeli emperors from 1185 to the Fourth

---

12. Gino Luzzatto explained it, "L'oro e l'argento nella politica monetaria veneziana dei sec. xiii-xiv," in his *Studi di Storia Economica Veneziana* (Padua, 1954), p. 264. "The coining of the grosso of Constantinople and of other grossi in other Balkan countries, with legal parity with the Venetian grosso, but at a reduced intrinsic value, constituted a threat for the silver money of Venice, against which she reacted with the coining of the ducat of fine gold at a fixed rate of exchange with the grosso. In this way the Venetian grosso, always exchangeable with the gold ducat at an admittedly rather favorable rate, should be preferred to all other silver money circulating on the market of Constantinople. It remains, however, to be documented whether or not the foreign payments in Constantinople were also made by preference in gold or in silver." (my translation)

13. In 1284 the law stated that the mint would pay 18 grossi for a ducat, or 39 soldi a grossi = 1 ducat. The shilling (solidus) was another mythical multiple of an actual coin. It represented 1/20 of a pound or 12 pennies. A few months later, the government agreed to accept payment of taxes in ducats at the rate of 40 sol. a grossi = 1 ducat (a little less than 18-1/2 grossi). But the market rate fluctuated. Frederic Lane has called my attention to the fact that Venetian goldsmiths statutes and papal tax collectors considered 1 ducat worth 22, 24, or even 28 grossi. In practice, ducats were accepted as worth 24 grossi for about twenty years before their value as legal tender was raised from 18-1/2 to 24 grossi. After 1328, the ratio fell. Cessi, *Prob. Mon. Ven.* docs. 36, 40, and pp. xl-lxiii; Luzzatto, "L'oro e l'argento," p. 265; and Lane, "Infidelities," p. 53.

14. Tommaso Bertelè, "Moneta veneziana e moneta bizantina" in A. Pertusi, ed., *Venezia e il Levante fino al Secolo XV*, Vol. 1, No. 1 (Florence, 1973), p. 3 and App. 5 p. 65, gives two issues: one of 4.25 grams at 868/1000 fine gold, 98/1000 silver and 34/1000 copper, thus being almost 20-3/4 carats pure. Michael F. Hendy, *Coinage and Money in the Byzantine Empire, 1081-1261* (Dumbarton Oaks, Washington, D.C., 1969) listed a gold hyperperon of Alexius Comnenus at 820/1000 fine weighing 4.27 grams. The second hyperperon issued by Alexius' son John II Comnenus was 4.4 grams at 884/1000 fine gold, 88/1000 silver, and 28/1000 copper; or about 21 1/4 carats pure gold. These compare to Constantine's solidus of 4.55 grams at 980/1000 fine gold.

Crusade. Most of the Venetian business in Romania was conducted in terms of these gold hyperperon of the old weight (perperi auri veteres pensantes.) From 1152 to 1228, 1 hyperperon of gold of the old weight = 40 shillings Venetian = 480 Venetian pennies. After 1194 it equalled 20 grossi.[15] Each of these gold hyperperon of the old weight contained 3.5 to 3.8 grams pure gold.

The problems of Byzantine gold coinage become complex after 1204. The Latin Empire was established by the Venetian and French crusaders after their capture of Constantinople on the Fourth Crusade. French princes, loyal to the Latin West, reigned in Constantinople for fifty-six years with the help of the Venetians; and Latins ruled over much of mainland Greece, the Peloponnese (known as the Morea) and many Greek islands. But Greeks established Greek successor states in Asia Minor (Nicaea and Trebizond), in Epirus, and later also in Saloniki. The rulers of Saloniki struck silver coins, and the Greeks of Nicaea struck both gold and silver coins. I do not believe, as some scholars suggest, that the Venetians produced *Venetian* coins in the mints which they operated in Constantinople in the thirteenth century.

The Venetians in Constantinople probably did strike copper as well as gold coins in the *Byzantine* style for the Latin Empire of Constantinople from 1205 to 1261.[16] Bertelè strongly argued, and the Byzantinst Paul Lemerle suggested, that the mints continued to operate under the Latin emperors. No coins have survived. The Venetian merchants and the Venetian government continued to specify payments in good gold hyperperon in Romania in these years. The Byzantine chronicler, Nicetas Choniates stated that the Constantinopolitan mints did function under the crusader emperors. Pegolotti in *La Pratica della Mercatura*, written about 1340, referred to a Latin hyperperon of gold of 16-1/2 carats.[17] The Latin rulers of Constantinople apparently continued as many Greek customs and ceremonials as possible, and this probably included coinage. Why none of these Latin coins from Constantinople have survived to the twentieth century, must be conjectured. Possibly because the Latin gold hyperperon continued to be of fine gold and full weight, the commercial and monetary dislocations of the thirteenth and fourteen centuries, following Gresham's law, drove these coins into the melting pots of Europe.

During the middle years of the thirteenth century, as the gold hyperperon were probably issuing less frequently if at all from the Frankish-Venetian

15. Robbert, *Venetian Money Market*, pp. 61-66, 92-93.

16. Scholars do not agree on this point. For the negative opinion: G. Schlumberger, *Numismatique de l'Orient Latin* (Paris, 1878), pp. 274-277; Philip Grierson, "La Moneta Veneziana nell' Economia Mediterraneo del '300 e '400" in *La Civiltà Veneziana del Quattrocento*, Centro di Cultura e Civiltà della Fondazione Giorgio Cini (Florence, 1957), p. 81. For the positive side, Bertelè, *Moneta Ven.* pp. 14-17 and his Appendix I, Tables I, II.

17. F. Balduccio Pegolotti, *La pratica della mercatura*, ed. A. Evans (Cambridge, Mass., 1936), p. 288. Evans in his editorial footnote referred to another such Latin hyperperon at 15 1/2 carats in the 1442 work by Giov. di Ant. da Uzzano, "Pratica della Mercatura" ed. Pagnini, *Della Decima* Vol. 4, (Lisbon and Lucca, 1766), p. 88.

mints in Constantinople and from the Greek mints of Nicaea, the Italian gold coins first appeared. In 1231 the Emperor Frederick II introduced his short-lived gold augustale in Sicily.[18] In 1252 the Genoese initiated the genovino and Florence, the florin, at the same weight and fineness: 3.5 grams of 24 carat gold. The florin ultimately became the more important Italian gold coin and circulated throughout western Europe as well as on the Mediterranean littoral. The Venetian mints in these decades continued to strike their fine silver grossi, which served the Venetians well at home and abroad.

After the collapse of the Latin Empire in Constantinople and the return of the Greeks under Emperor Michael VIII Palaeologus in 1261, the Greeks in Constantinople once more struck the gold hyperperon. At first their gold coins were nearly as fine as the Comneni hyperperon. The second Palaeologan emperor, Andronicus II, issued, in 1282, a poorer quality hyperperon and subsequent issues of gold coins became increasingly poor. Silver coins began to be struck, including a Byzantine silver grosso. Fourteenth century Palaeologan rulers in Constantinople issued a hierarchy of Byzantine silver coins closely tied in weight to the Venetian grosso and the Chiarenza gros tournois. By 1400, the Byzantine silver hyperperon was payable in Byzantine silver coins, and equalled one-half the Venetian gold ducat.[19] The last Greek emperor to strike gold hyperperon was John V Palaeologus who reigned from 1347 to 1376.

This Byzantine experience bears out the thesis propounded by Andrew Watson that the Greek speaking east was a transition zone between European silver coinage and Muslim gold coinage up to 1270.[20] During the period of the Latin Empire, 1205-1261, the mints of Venice and Saloniki issued silver coins, and the mints of Nicaea issued both gold and silver coins. Probably Constantinople under the Latins issued gold and copper coins. Venetian commercial documents specified, for payments in the Greek east, Venetian pounds and also gold hyperperon.

Only in 1284, two years after the Palaeologi issued their poorer hyperperon, did Venice issue the gold ducat at 24 carats fine gold, weighing 3.559 grams. This approximated the gold hyperperon of the old weight and repeated the weight and fineness of the florin.[21] Although the Italian coins were slightly more pure than the good gold hyperperon of the old weight, and weighed

---

18. Weighing 5.3 grams at 854/1000 fine. P. Spufford, "Appendix, Coinage and Currency," in M. M. Postan and H. S. Habakkuk, eds., *Cambridge Economic History of Europe*, Vol. III (Cambridge, 1965), p. 601.

19. Speech by Tommaso Bertelè, "Il Libro dei Conti di Giacomo Badoer ed il problema dell'Iperpero Bizantino nella prima metà del Quattrocento," delivered to the XII Convegno "Volta" promosso dalla classe di Scienze Morali, Storiche e Filologiche, on the theme, "Oriente e Occidente nel Medioevo," *Accademia Nazionale dei Lincei*, (Fondazione "Alessandro Volta," Istituto dalle Società Edison di Milano) (Rome, 1957), pp. 255-258; see also his "Moneta veneziana," pp. 17-19.

20. A. M. Watson, "Back to Gold—and Silver," *Economic History Review* Second Series, Vol. 20, No. 1 (1967), 1-34, esp. pp. 7-10, 28-29.

21. Niccolò Papadopoli, *Le Monete di Venezia* Vol. 1 (Venice, 1893), pp. 137-138.

slightly less than the hyperperon, the Italian coins contained as much pure gold as the Byzantine coin. Several reasons for the Venetian ducat's appearing first in 1284 can be brought forward. First, a need existed for good gold coins for overseas trade, and the good hyperperon were no longer supplied in sufficient quantity nor quality by the mints of Constantinople.[22] Secondly, Venetian pride desired her own gold coin and would not be satisfied with using one of Florence. Thirdly, it was pointed out above that the Venetian silver coinage was faltering and the Venetian state issued the gold coin to bolster the silver grosso. Finally, it was possible that new supplies of gold became available to the Venetians. The gold ducat continued to be issued by the Venetian Republic for over five hundred years with only two minor 1% debasements in the sixteenth century. Napoleon's conquest of Venice in 1797 finally closed the independent Venetian mints, but Vienna continued the ducat for several decades.[23]

For the first fifty years following the introduction of the gold ducat, Venice attempted to maintain a bimetallic standard, with the gold ducat and the silver grosso and a fixed ratio between them. (Ultimately they accepted a gold standard based on the ducat.) Silver as well as gold coins issued from the Venetian mints, while the government tried to enforce the ratio. The principal money of account continued to be the lira a grossi, supplemented by several other moneys of account. Frederic Lane has analyzed these moneys of account with great subtlety.[24] In the third decade of the fourteenth century a monetary crisis forced government action. Table 2 illustrates these problems.

The gold and silver ratios for all of Europe also suggest a monetary crisis. The generally accepted gold and silver ratio for Italy in 1284 was 1:11.3. But an abundance of silver from Bohemia began to cheapen silver so that the gold and silver ratio in Venice between 1305 and 1310 reached over 1:14. The general European gold and silver ratio in 1311 was 1:13. In 1326-1328 silver became less plentiful (or more in demand in the Levant) and rose in value. By 1326 the ratio in Venice had returned to 1:10.[26]

The merchants controlling the Venetian state needed to make some adjustments in the 1320s. First the government tried to solve the problem by requiring that customs and other taxes had to be paid in good grossi and that all grossi that were underweight by wear or clipping should be destroyed. Although this was tempered by orders that the mint should produce more

22. Luzzatto, *Storia Econ. di Ven.* pp. 95-97; and Cessi, *Prob. Mon. Ven.*, pp. xxxviii-xlii.
23. Grierson, *Moneta Veneziana*, pp. 78-79.
24. For a more detailed discussion of changes in the Venetian money of account, Lane, "Vecchie monete," pp. 58-72 and also his "Infidelities," pp. 43-64.
25. Luzzatto, *Storia Econ. di Ven.*, pp. 97-99; Frederic C. Lane, *Venice, a Maritime Republic* (Baltimore, 1973), pp. 148-150.
26. Lane, *Venice, Mar. Rep.* pp. 148-149; his "Infidelities," p. 53; and above n. 13; Josef Janáček, "L'argent tchèque et le Méditerranée (XIV et XV siècles)," in *Mélanges en l'honneur de Fernand Braudel*, Vol. 1, *Histoire èconomique du monde méditerranéen 1450-1650* (Paris, 1973), pp. 247-249; and Cessi, *Prob. Mon. Ven.* pp. li-lxv.

Monetary flows—Venice 1150 to 1400   61

Table 2. Exchange rates between Ducat and Grosso coins[25]

Date	Rate	Type of rate
1284	1:18	official rate
1285	1:18–1/2	official rate
1305–1325	1:24	market rate
1328	1:20	market rate
	1:22	market rate
1328	1:24	official rate

pennies and that government agencies could accept pledges if grossi were not available, the result of these policies and of the increasing scarcity of silver was deflation. Grossi coins could be sold by weight abroad, but not in Venice, and the grossi continued to disappear.[27]

In 1228, in response to the decreasing supply of silver, the merchant politicians of Venice decided not to debase either the ducat or the grosso, but to change the official ratio between these two coins. One ducat should now equal 24 grossi coins, instead of 18-1/2 grossi. They also expressed this new relationship in terms of the money of account; 1 ducat = 2 solidi di grossi; or more simply still, 10 ducats = 1 lira de grossi.[28] Thus this money of account, which took its name and its significance from the principal silver coin, was now tied to the purchasing power of the gold coin. This simplified relationship between this Venetian money of account and the gold ducat proved to be very convenient and the lira di grossi was now used to express most financial data. Now payments could be made legally in the plentiful gold ducat, not with the grosso. Lane believes this marks the end of the Venetian attempt to have a bimetallic standard and the beginning of the gold standard.[29] This new ratio undervalued silver, so that grossi coins were hoarded or sent to the east where they bought gold.

Since the adjustments of 1328 did not solve the problem and since Venice would neither reissue the grosso nor withdraw it from circulation, it was not coined at all from 1356 to 1379.[30] The Venetian state continued its series of large silver coins by issuing two new coins in place of the grosso, which expressed a reduction of 10% in the silver content of the grosso. These new

27. The government policy-makers were called "hard-money, budget-balancing protectionists" by Lane, "Infidelities," p. 56; see also Cessi, *Prob. Mon. Ven.* pp. li-liii. Compare Lane, "Infidelities," pp. 56-59.

28. To remind the reader, the lira di grossi = the lira of 240 grossi coins; the lira a grossi = the lira of 240 pennies at the rate of 1 lira = 9 grossi and 5 pennies; and 1 lira di grossi = 26 lire a grossi.

29. Lane "Vecchie monete," pp. 72-75; and his "Infidelities," pp. 56-63. Luzzatto disagreed, placing this development in the fifteenth century, in his *Storia Econ. di Ven.*, pp. 213-214.

30. Luzzatto, *Storia econ. di Ven.* pp. 97-98; and Cessi, *Prob. Mon. Ven.*, pp. lxxxii-lxxxvii.

coins, the mezzanino and soldino first appeared about 1332. The mezzanino, weighing 1.242 grams at 780/1000 fine, represented a half grosso or 16 pennies, since on the domestic market 1 grosso = 32 pennies. The soldino, weighing 0.957 grams at 670/1000 fine, represented a third of a grosso or 12 pennies or 1/20 of the pound, since abroad 1 grosso = 36 pennies. After the mid-fourteenth century, the mezzanino was no longer issued; and the soldino, now struck at the same purity as the grosso (965/1000 fine), was made lighter in weight from time to time to keep pace with the changing price of silver. The Venetian mints at home and in her colonies struck another small silver and copper piece, the tornesello, beginning about 1354, but this coin only circulated in Romania, not in Venice.[31] The soldino soon became the most common silver coin in Venice, replacing the grosso. By law Venetians were required to settle all small payments in silver, while the gold coins were used for large scale and foreign transactions. Consequently, silver circulated faster in Venice in the multitude of small transactions. The grosso itself was reissued in 1379, lighter in weight and less pure, by Doge Andrea Contarini during the crisis of the War of Chioggia when the existence of Venice itself was threatened by the Genoese occupation of the town of Chioggia at the southern extremity of the Venetian lagoons. The same doge also debased the soldino.[32] Subsequent doges continued to debase the silver coinage until 1472 when the silver grosso was permanently withdrawn from circulation, leaving other silver coins of small denomination.[33] Concurrently, the gold ducat maintained the same weight and fineness.

The standards of value in fourteenth century Venice were based on the gold ducat, and other coins and moneys of account were expressed in terms of the ducat.[34] As the silver coins were more debased, and the grosso replaced by the soldino, the Venetian lira di grossi came to be payable in ever greater and greater numbers of the silver coins. The continuing debasement of Venetian silver money in relation to the gold ducat is well illustrated by Table 3, drawn up by Domenico Brusasete on March 22, 1703.

★ ★ ★

The previous section has defined the Venetian coins and the Venetian moneys of account from 1150 to 1400. At this point we shall discuss where the Venetians got the gold and silver.

---

31. Lane, "Infidelities," p. 58 and n. 52; Papadopoli, *Le Monete*, Vol. 1, pp. 157-164; Cessi, *Prob. Mon. Ven.*, pp. lxii-lxiii. For the tornesello, see below.

32. Luzzatto, "L'oro e l'argento," p. 270; and Papadopoli, *Le Monete*, Vol. 1, pp. 206-210.

33. Luzzatto cited 1472 as the Venetian turning point from a bimetallic to a gold standard. Lane argued more forcibly for 1328. See above, n. 29.

34. The lira di grossi, later called the lira di grossi a oro, was worth 10 gold ducats. Compare Lane, "Infidelities," pp. 60-61, where he also describes the lira di grossi in monete, based on the soldino. The ratio kept changing, in response to the changing bimetallic ratio.

Table 3. Relationship between Ducat and Venetian silver[35]

Date	Gold coin	Lire di piccoli
1285	1 ducat = 40 soldi de grossi (sic!)	
1399	1 ducat = 93 soldi or	4:13
1417	1 ducat = 100 soldi or	5
1429	1 ducat = 104 soldi or	5:4
1443	1 ducat =	5:14
1472	1 ducat =	6:4
1517	1 ducat =	6:10
1520	1 ducat =	6:16
1529	1 ducat =	7:10
1572	1 ducat =	8
1573	1 ducat =	8:12
1597	1 ducat =	10
1608	1 ducat =	10:16
1638	1 ducat =	15
1643	1 ducat =	15
1683	1 ducat =	17

Evidence for the northern origin of the silver used by the Venetian mints is derived from a number of documents which date between 1180 and 1230. At this time, international payments in Europe were made in terms of marks of silver, a measure of weight. For example, in January 1181, officials of the Knights Hospitalers of Jerusalem, coming from Bohemia and Lombardy, delivered 200 marks of silver of Cologne weight and 72 marks of gold in Venice to the chief Venetian Hospitaler.[36] He, in turn, acting at the mandate of Roger of Les Moulins, master of the Hospital in Jerusalem, and of Count Rudolf, conveyed this bullion to the procurator of San Marco, a high-ranking Venetian official who regularly transferred funds between Venice and the

---

35. *A.S.V. Procuratorie de Citra*, B. 280, c. 88-89v. This seventeenth century Venetian in the offices of the Procuratorie di S. Marco di Citra evidently did not understand the complexities of the medieval Venetian money of account. He quoted the 1285 ratio correctly in numerical terms, but used the lira di grossi, instead of the lira a grossi, in which this ratio was stated in the original decree published by Cessi, *Prob. Mon. Ven.*, doc. 40. This also may explain his omission of all subsequent quotations of ratios for the next 114 years. After 1399 Brusasete correctly quotes the ratios in terms of the libra di grossi. Brusasete, also, in his introductory paragraph, explained the coinage of the ducat as it was done in 1703, not in 1284. See another comment on Brusasete by Lane, "Vecchie Monete," p. 78.

36. Ercimbald, prior of San Egidio in Venice. R. Morozzo della Rocca and A. Lombardo, eds., *Documenti del Commercio Veneziano nel Secolo XI-XIII*, Vol. 2 (Documenti e Studi per la Storia del Commercio e del Diritto Comerciale Italiano, Vol. 20) (Turin, 1940), doc. 324. For a discussion of international payments in marks of silver see Balint Homan, "La circolazione delle monete d'oro in Ungheria dal x al xiv secolo e la crisi europea dell'oro nel secolo xiv" *Rivista italiana di numismatica*, Vol. 35, Ser. 2, Vol. 5, (1922), pp. 113-114.

Holy Land. In 1201, the French Crusaders promised to pay the Venetians 85,000 marks of pure silver, Cologne weight, for their transportation on the Fourth Crusade.[37] In 1203, Wolfger, who was traveling to Rome to receive the pallium as newly elected bishop of Passau, brought silver in bullion to the Veneto and exchanged it for local currencies.[38] Passau on the Danube was one of the principal markets for Bohemian silver. In October, 1202, in Venice, the leaders of the Fourth Crusade—Baldwin Count of Flanders, Count Louis of Blois, and the Marshal of Champagne—borrowed Venetian money from a consortium of Venetians, promising to repay them in silver at the next fair of Lagny in Champagne.[39] In June 1225, the Duke of Austria and Styria and Count Menegard, on their way to Rome, borrowed Venetian pennies from the Venetian Doge and promised to repay him about September 1 in 50 marks of good silver. Collateral was given by the count, and the agreement confirmed when they again passed through Venice in August, on their return from Rome.[40]

Evidence has recently come to light of a German merchant, Bernardus Teotonicus, who became unusually wealthy by bringing silver and copper into Venice.[41] His will, dated 1213, reveals him to have been one of Venice's richest inhabitants, worth 27,500 Venetian pounds. He had lent 15,000 Venetian pounds to Doge Pietro Ziani sometime between 1205 and 1213, in time of extraordinary need of the state. His house in Campo S. Bartolomeo, near the later Fondaco dei Tedeschi and the later bankers' tables sold for the unusually large sum of 3,500 Venetian pounds. He had personal contact with the Hohenstaufen, entertaining both Barbarossa and Henry VI in his home in Venice. Probably he helped finance Barbarossa's expedition on the Third Crusade and, probably also, helped transfer the funds that ransomed Richard the Lionheart from his Austrian prison. Bernardus had business contacts with rulers and merchants in the silver producing regions of Friesach in Carinthia, of the Rodna mines in Hungary, as well as continuous commercial relations with Munich and with Aquileia in Friuli.

---

37. G. F. Tafel and G. M. Thomas eds. *Urkunden zur älteren Handels- und Staatsgeschichte der Republik Venedig*, (Fontes Rerum Austriacarum, Historische Commission der Kaiserlichen Akademie der Wissenschaften in Wien, Abt. II, Diplomataria et Acta, Bd. XII,) Vol. I (Vienna, 1856), pp. 366-371; and Geoffrey of Villehardouin, *La Conquête de Constantinople*, E. Faral ed. and tr., (Paris, 1938), Vol. 1, par. 21, 22.

38. *Reiserechnungen Wolfger's von Ellenbrechtskirchen*, ed. I. von Zingerle (Heilbronn, 1877), pp. 58-59.

39. Morozzo della Rocca and Lombardo, *Documenti*, Vol. 1, doc. 462.

40. "Liber Comunis quo vulgo nuncupator Plegiorum," ed. Roberto Cessi in *Deliberazioni del Maggior Consiglio di Venezia* (Accademia dei Lincei, Commissione per gli Atti delle Assemblee Costituzionali Italiane) (Bologna, 1950), p. 83, par. 137.

41. Wolfgang von Stromer, "Bernardus Teotonicus e i rapporti commerciali fra la Germania Meridionale e Venezia prima della istituzione del Fondaco dei Tedeschi," *Centro Tedesco di Studi Veneziani*, Quaderni 8 (Venezia, 1978). Prof. von Stromer has studied the business of Bernardus Teotonicus with great thoroughness after finding his will, its five later copies, and 24 protocols in the archives of the Procuratori di S. Marco in Venice.

In addition to obtaining silver from Carinthia and Hungary, German merchants in the thirteenth century brought silver from the mines of Bohemia. Janàcek suggested that the Bohemian mines of Jihlava and Německý Brod produced an important part of the silver used in Europe during the first half of the thirteenth century. Also important for providing silver for the Italian mints in this century were the mines of Silesia and Saxony (Meissen).[42]

During the Fourth Crusade and immediately after, some silver bullion also arrived in Venice from Constantinople. The booty from the sack of Constantinople in the Fourth Crusade was distributed, as agreed upon in April 1204, 3/8 to the Venetians, 3/8 to the other Crusaders, and 1/4 to the new Latin emperor of Constantinople. From the total booty collected, the Venetian share amounted to 400,000 marks of silver.[43] In April 1207, also in Constantinople, two brothers from Lombardy asked that the procurator of San Marco in Venice should receive their two investments totalling forty-nine marks of pure silver, which Venetian merchants were to transport from Constantinople to Venice. They requested payment in Venice in good old Milanese pounds.[44] This was done and quittance given in October, 1207, in Venice. These examples demonstrate that up to 1225 Frenchman, Germans, and Lombards brought silver into Venice or promised to do so.

After 1225, foreign merchants as well as Venetians brought large quantities of gold and silver bullion to Venice. The Venetian state encouraged and closely regulated the import of bullion. Venice, as early as 1217, had agreed with King Andrew of Hungary that gold, precious stones, silks and drugs should be exempt from all customs dues in both states. This permitted Hungary to pay for her Venetian luxury imports with gold from Hungarian mines. Venetians also traveled north to Hungary with their luxury goods from the Levant, which they exchanged there for Hungarian gold or silver. For example: gold, silver and jewels valued at 800 marks silver were reported stolen from two Venetian merchants in Hungary by King Bela, son of King Andrew of Hungary, according to a report of March, 1224. On the same occasion, a third Venetian accused the Hungarian king of stealing fourteen marks of silver and four marks of gold. The Hungarian silver mines flourished greatly during the last quarter of the thirteenth century.[45] Some silver came to Venice in the thirteenth century from the Balkans, also, as the mines of Serbia began producing silver about 1250, and the Venetians dominated Ragusa, the chief Dalmatian seaport which received this silver ore.[46] Howev-

---

42. Janàcek, "L'argent tchèque," pp. 246-7; Homan, "Circolazione," pp. 126, 132-135.

43. Bertelè, "Moneta veneziana," p. 74.

44. Libri bonorum imperialium veterum. This was the money of account based on the silver penny of Milan. Morozzo della Rocco and Lombardo, *Documenti*, Vol. 2, docs. 487, 493.

45. Cessi, *Prob. Mon. Ven.*, docs. 14, 33, 54, 62, 63, 70, 91, 94, 170; Homan, "Circolazione," pp. 134-5 and n. 14, 130-1; *Liber Plegiorum*, ed. Cessi, p. 56, par. 41 and p. 55.

46. Desanka Kovàcevic, "Dans la Serbie et la Bosnie médiévales: Les Mines d'or et d'argent," *Annales: Economies, Sociétés, Civilisations*, Vol. 40, No. 2 (1960), p. 249.

er, Serbian silver output was probably small before 1300, and the Venetian source of silver bullion probably continued to be the northern mines, especially those in Hungary and Bohemia.

When the ducat was authorized in 1284, the Venetians must have found a supply of gold. The mines of Hungary did produce some gold, as did the Bohemian mines in the thirteenth century. A Venetian decree of 1284 expressly allowed German merchants to bring gold freely to the Venetian mint.[47] Nothing suggests that the issues of gold ducats were very large during the half century after 1284. Gold must have not been very plentiful, as all over Europe the price of gold rose relative to the price of silver in the last decade of the thirteenth and the first decades of the fourteenth century. In response to this rise in the price of gold the Byzantine emperor Andronicus II, about 1295, devalued the gold hyperperon and supported it with new issues of silver, of the same weight and fineness and in direct competition with the Venetian grosso.[48]

Silver suddenly became more abundant between 1295 and 1300 when the Kutna Hora mines in Bohemia were discovered. It has been suggested that the output of this mine increased the total European silver production by 45%. This would explain the cheapening of silver as evidenced by the gold and silver ratios. German merchants from Ratisbon were active in bringing the Bohemian silver to Venice in the first quarter of the fourteenth century.[49] About 1300 the character of the silver changed. Instead of exporting silver bars, the Bohemian rulers, for their own profit and prestige, began to insist that their silver be struck into coins before export.

The role of the Hungarian silver and gold mines between 1300 and 1340 is not clear. During the dynastic struggles following the extinction of the native Hungarian royal house of Arpad, 1301 to 1308, the Hungarian mining industry suffered and long distance commerce was disrupted. The strong Hungarian king, Charles Robert of Anjou 1308-1342, rather than insisting on his regalian rights, encouraged local Hungarian landlords to develop the mines and pay a tax of one-third of the output to him. He also decreed that only Hungarian coins and not bullion could be exported. Gold bullion became so scarce in Hungary that the papal collectors could not collect any gold at all there between 1332 and 1337. On the other hand, some silver must have left Hungary because a Venetian merchant's manual, the *Zibaldone da Canal*, which has been dated between 1310 and 1320, specified that silver from Germany and *from Hungary* was brought to money changers in Venice for assaying. German and possibly Hungarian merchants would bring this gold and silver to their shelter in Venice, the Fondaco dei Tedeschi. In 1314

---

47. Grierson, "Moneta Veneziana," pp. 78-9 has suggested the gold came from the Balkans, from German merchants and from the Kingdom of Naples. The decree of the Great Council of Venice in Cessi, *Prob. Mon. Ven.*, doc. 39.
48. Bertelè, "Il *Libro dei Conti* di G. Badoer," pp. 250-1.
49. Janáček, "L'argent tchèque," pp. 247-249.

the import of gold and silver into Venice through the Fondaco dei Tedeschi was closely regulated by the Venetian state.[50]

In the 1340s gold became more abundant in Venice. This parallels Venice's problems with its silver coinage and the Venetian change to the gold standard, as was discussed earlier. By 1350 the Venetian mint output of ducats had probably increased enormously. Cessi believed that gold was brought to the Italian mints because they would pay more for it than the mints of the Levant. In other words, in Venice and Florence silver was underpriced in terms of gold.[51]

Where did this gold originate? From the western Sudan, from the Black Sea, and from Hungary. The gold from the western Sudan had funnelled north to the Moroccan and Algerian coast for some centuries, where the western Italian cities received it. Luzzatto indicated that in the decade, 1325 to 1335, it came increasingly to Venice, in addition to, or instead of, its earlier travel to Genoa and Florence. Certainly Venetian merchants in the early fourteenth century frequently traded with Tunis. The Venetian merchant's manual, noted above, recorded more market information about Tunis and neighboring north African cities, than any other area.[52] Some evidence suggests the Black Sea as another point of origin for Venetian gold in the mid-fourteenth century. Venetian mints, like the Florentine, received great quantities of Black Sea gold beginning in 1342. The Venetian mint in 1342 and 1343 had to increase its personnel to handle the extra gold brought to Venice "on the galleys."[53] The Venetians, in their annual muda to Romania and the Black Sea, imported gold from Trebizond on the southeast coast of the Black Sea, and perhaps also from Tana, at the mouth of the Don, directly east of the Crimean peninsula.[54]

Hungary again after 1337 began to export gold through German, Hungarian, and possibly Venetian merchants. Certain Venetians took up residence in fourteenth century Hungary and became judges, mint masters, and financial

---

50. Homan, "Circolazione" pp. 139-141; the merchants' manual is *Zibaldone da Canal: Manoscritto Mercantile del sec. XIV*, ed. Alfredo Stussi, with studies by F. C. Lane, Th. E. Marsten, and O. Ore (Fonti per la Storia di Venezia, Sez. V. Fondi Vari) (Venice, 1967), p. 7. For regulation, Cessi, *Prob. Mon. Ven.*, doc. 70.

51. Cessi, *Prob. Mon. Ven.*, pp. lxvii-lxxxvii.

52. Luzzatto, *Storia econ. di ven.*, p. 98. For the Tunis trade see *Zibaldone da Canal*, ed. Stussi, pp. 42-52 and Lane, *Venice, Mar. Rep.*, p. 130.

53. Cessi, *Prob. Mon. Ven.*, pp. lxx-lxxi and docs. 96, 97, 99. "cum aurum ducatur Venetias per mare per mercatores in maxima quantitate. . . ."

54. F. Thiriet, *Régestes des Délibérations du Senat de Venise concernant la Romanie*, Vol. 1 (Paris, 1958), nos. 17 and 178. Gold, silver, and precious stones obtained at Trebizond in 1344 and 1345. See also Marie Nystozopoulou, "Venise et la Mer Noire du XI au XV Siècle" in A. Pertusi, ed., *Venezia e il Levante fino al Sec. XV*, Vol. 1, p. 2, (Florence, 1973), pp. 559, 561-563, 568-569. See also Luzzatto, *Storia econ. di Ven.*, p. 64. This author has found no evidence of the Venetian Flanders galleys bringing gold or silver into Venice from the north. They filled their holds with cloth and raw wool. Lane, *Venice, Mar. Rep.*, p. 126 and Luzzatto, *Storia econ. di ven.*, p. 149.

advisors. The increasing output of the Hungarian gold mines in the 1330's was stimulated by royal edicts which set the gold and silver ratio at 1:15.6 in 1335 and 1336, and at 1:14.7 in 1342. The kings of Hungary and Bohemia from 1336 also encouraged merchants to avoid the staple market of Vienna and export gold and silver freely to the west by a northern route. Gold production increased in Hungary in these years and by 1344 the gold and silver ratio in Hungary had returned to its "normal medieval figure" of 1:11 or 1:10. The increase in gold supplies in Italy in 1344 can also be explained by the Italian visit of Queen Elizabeth of Hungary in that year. She brought with her a vast amount of gold and silver, equal to the royal income from the Hungarian mines for six years, and she spent it in Naples and Rome trying to purchase the kingdom of Naples for her son, Andrew. Cessi believed that this increasing quantity of imported gold, from all sources, was coined into Venetian ducats and then much of it was immediately exported from Venice.[55]

It is not clear whether or not *silver* supplies in Venice sufficed to fill the demand for silver at the mint throughout the fourteenth century. In the first half of the fourteenth century, it is possible that supplies of silver were decreasing and that in Europe the price of silver increased and gold decreased: i.e. the gold and silver ratio which in Venice had been 1:13 or 1:14 in the years 1300 to 1308 went to 1:10.65 in Byzantium in 1340 and to 1:9.4 in Venice in 1350. After 1350, the output of nearly all European mines probably declined. Certainly, the silver output from the Bohemian mines did decrease after 1350.[56] Only the Serbian silver mines increased production and by 1375 the Serbian and Bosnian mines were beginning their most productive decades. Possibly these Balkan mines tried to supply much of the European demand for silver, but could not equal the total European silver production of earlier years. The Venetians maintained a near monopoly of the output of the Balkan mines, because most Serbian silver, brought from the inland mines by Ragusan merchants to their coastal city, was shipped directly to Venice. Ragusa kept only a small portion of the silver for her own coinage. Ragusan merchants had all the rights to develop the mines and market their ores. In

---

55. For Hungarian gold and the ratios, Homan, "Circolazione," pp. 131, 150-152 and Gy Székely, "Wallons et Italiens en Europe centrale aux XI-XVI Siècles" *Annales*. (Universitatis Scientiarum Budapestinensis de Rolando Eötvös Nominatae. Sectio Historica.) Vol. 6 (1964), pp. 8, 10, 11, 18, 29. Homan quoted the contemporary chronicler, John of Küküllö, who stated that the Hungarian queen brought 27,000 marks of pure silver and 17,000 marks of pure gold and one-half bushel of Hungarian gold florins. Later she wrote her son Louis the Great of Hungary requesting an additional 4,000 marks of gold. For Venetian gold coinage and its export see Cessi, *Prob. Mon. Ven.*, p. c and doc. 133.

56. Janáček, "L'argent tchèque," p. 255 and John Nef, "Mining and Metallurgy in Medieval Civilisation" in M. Postan and E. E. Rich, eds., *The Cambridge Economic History of Europe*, Vol. 2 (Cambridge, 1952), pp. 456-458. For the ratios see Bertelè, "Il *Libro dei Conti* di G. Badoer," p. 253 and Watson, "Back to Gold," p. 24. Frederic Lane has communicated with this author that in his forthcoming article in the Festschrift for Hermann Kellenbenz, he will present evidence that the Venetians felt oversupplied with silver in the 1340's.

addition to the great quantities of silver, the mines of Serbia and Bosnia also produced some gold, lead, copper, iron and cinnabar.[57]

In the closing decade of the fourteenth century, Venice became the main market to which German merchants brought the silver and gold mined in Bohemia. Recent studies have shown that the Rutingers of Ratisbon and the Kress of Nuremberg brought to Venice gold ingots, gold chains, and also silver bars, silver objects, and the Prague silver groschen, which was sold in Venice by weight and fineness.[58]

Throughout the centuries, 1150 to 1400, the Venetian mints had been receiving most of their silver from European sources. Gold was brought to the Venetian mints by German merchants coming from the Hungarian mines, or by Venetian merchants coming from northwest Africa or the Black Sea. The following discussion of the shipments of gold and silver out of Venice will reveal that Venice exported great quantities of gold ducats and some silver coins in the fourteenth century. For this, the supply of gold was sufficient in the last half of the fourteenth century, but the evidence is not sufficient to affirm when the Venetian mints received adequate supplies of silver.

★ ★ ★

By the year 1400 Venetian mints produced most of the coins in circulation throughout the eastern Mediterranean in Muslim as well as in Christian principalities, in international trade as well as on domestic markets. The following paragraphs will consider the evidence of Venice as a major supplier of coins for the Near East. The outflow of Venetian coins was probably not very great until the later thirteenth century.

Venetian merchants before 1250 carried with them away from Italy, in addition to merchandise, Venetian and other Italian coins, silver bullion, and rarely gold. One example from the twelfth century demonstrates this.[59] The same pattern of coins, bullion, and merchandise in Venetian businessmen's baggage is evident from the surviving government books from 1224 to 1228. For example, in 1224 a Venetian, Petrus Abramo of San Marziale, gave the chamberlain of the city of Ancona one mark of silver and three samite cloths worked in gold in order to redeem the oars and sail of a ship which the chamberlain had confiscated. Eventually Petrus sent his agent off with the ship and its cargo of two French horses, iron, and textiles on a trading voyage to the Morea, while Petrus himself stayed in Ancona to bargain with the

---

57. Kovàcevic, "Dans la Serbie," pp. 250-252, 254-258.
58. Janàček, "L'argent tchèque," pp. 253-254.
59. See my "L'inventario di Gratiano Gradenigo" forthcoming in *Studi Veneziani* which describes a Venetian merchant dying at sea, whose goods were brought to Pisa for auction in 1176. He carried textiles and tailors' supplies, personal effects, and some North African gold coins (Massamutini) and a few Byzantine coins (purse with 44 solidis in follaris and 5 stamini.) *Biblioteca Nazionale Marciana*, Lat. Ms. Cl. XIV. Cod. 71.

chamberlain.[60] The next year, 1225, the Venetian Little Council recorded a long list of claims by Venetian merchants against thefts by the Marquis of Ferrara, his men, and other pirates in the waters immediately south of Venice. These claims cover twelve sides of the manuscript of the *Liber Plegiorum*.[61] Most of the stolen goods enumerated were trading merchandise and Italian silver coins, but one claim listed: "15 shillings in grossi coins and sterlings and manulatos."[62] In 1238 the men of Ferrara robbed two other Venetians as they sailed down the Adriatic to Brindisi. From the one they took fifteen shillings in sterlings, fourteen shillings and one penny in grossi, £9 in Veronese and £7 in Venetian pennies, and other silver coins worth £25, plus 4½ marks of silver-foil. From the other Venetian, they took fourteen shillings in grossi, fifteen shillings di grossi in sterlings, a good knife worth £3 and three fortonos of silver.[63] In another final example, when the Venetian nobleman, Marco Minoto of San Cassian, was shipwrecked on the island of Lefca in the Ionian Sea in June, 1228, the Duke of Corfu came and stole from him 450 arms-lengths of linen cloth and arms with draperies, all valued at £75 Venetian. When the Duke of Corfu realized that he and his companions were merchants, he also demanded money, and took from Marco "1500 sterlings, one hyperperon of twenty-five carat gold (sic!), and one silver belt valued at ten hyperpera."[64] These examples demonstrate that before 1250 Venetian merchants in foreign parts carried silver coins from Venice, Verona, and from Champagne and England; and also silver bullion and some gold, in addition to other merchandise.

It has long been assumed by historians that the grosso became the principal silver coin circulating throughout the Aegean and the Levant in the thirteenth century.[65] However, until the surviving documents and numismatic evidence are carefully studied anew, we may wonder whether Venetian merchants traveled east with money bags bulging with grossi, or whether they and the other Levantine businessmen did not often adopt as their standard the Venetian money of account whose values derived from the unvarying silver content of the grosso. Gold coins and a gold standard of

---

60. *Liber Plegiorum*, ed. Cessi, p. 55.
61. A.S.V. *Liber Plegiorum*, cc. 56r-61r; or Cessi edition, pp. 113-127.
62. A.S.V. *Liber Plegiorum*, c. 59v; Cessi edition, p. 120. Manulatos were electrum coins issued by the Emperor Manual I Comnenus, 1143-1180, Bertelè, "Moneta veneziana", p. 79, n. 2. Sterlings would be English pennies, at this time weighing 1.46 grams at 925/1000 pure, Spufford, "Coinage and Currency," p. 599.
63. *Liber Plegiorum*, ed. Cessi, p. 194, par 87. The fortone was a silver coin circulating at the fairs of Champagne and in Flanders. In Flanders, 4 fortone were equal to 1 mark, Pegolotti, *La pratica*, pp. 235, 239.
64. *Liber Plegiorum*, ed. Cessi, pp. 196-7, par. 94.
65. Papadopoli, *Le Monete*, Vol. 1, pp. 89, 251; Grierson, "Moneta Veneziana," pp. 80-81; Bertelè, "Il *Libro dei Conti* di G. Badoer," p. 251; and Josef Müller, "Venezianer Münzen im XIII Jahrhunderte und ihr Einfluss auf das mitteleuropäischen Geldwesen," *Numismatische Zeitschrift*, Vol. 51 (1883), passim, esp. pp. 233-238.

value did continue in use in international trade throughout the thirteenth century; these were not altogether displaced by silver.

One narrative statement to support the theory of the widespread circulation of Venetian grossi is the oft-quoted sentence from the Venetian chronicler, Martino da Canale. Writing between 1267 and 1275, the Venetian chronicler who had much first-hand information about Syria and Romania speaks of the grosso as, "the noble money of silver called ducato, which circulates in all the world because of its excellence."[66]

Merchants' manuals of the early fourteenth century demonstrate that merchants traveling from Venice used Venetian gold coins only rarely in their business. Rather, the Venetians used some grossi coins and the Venetian money of account based on the grosso, but most often they traded with gold or silver measured by weight and exchanged for local currencies; or they bartered. Both the Venetian manual, *Zibaldone da Canal*, and the Florentine manual by Pegolotti, *La Pratica della Mercatura*, record which gold, silver, coins, and merchandise were traded in each of many cities and also what the exchange value of these commodities would be in terms of the local weights or coins or local money of account. Many cities in the Near East had money systems based on gold coins; European cities had money systems based on silver coins. The Venetian *Zibaldone da Canal* presents the earlier evidence, since its contents can be dated about 1310. From the thirty-two Mediterranean cities and areas to which Venetian merchants using this handbook might travel, only eight are noted as trading in gold: Tunis in West Africa, Apulia, Thebes in Greece, Lajazzo in Armenia, Tripoli in Syria, Constantinople, Alexandria in Egypt, and perhaps Acre. Silver was acceptable in Apulia, Thebes in Greece, Lajazzo in Armenia, Acre, Constantinople, and Soldaia in the Crimea.[67] The Venetian regulations in this merchants' manual for buying gold and silver on the Rialto correspond to the year 1310.[68] Grossi coins and the Venetian money of account based on the grosso were commonly cited in this manual for exchange rates with foreign coins, demonstrating that the Venetian merchants often traveled with grossi coins. The gold ducat was only mentioned in exchanges with two places: Tunis and Cyprus.[69] This suggests that the great Venetian export of gold ducats had not yet begun by 1310. This merchant's manual demonstrates that the grosso coin could be used abroad by the Venetians, but not everywhere. Venetian merchants in this decade used many foreign coins and moneys of account in foreign markets, and thus

---

66. Martin da Canal, *Les Estoires de Venise*, ed. A. Limentani (Florence, 1972), cap. xxxvii, pp. 46-47. See also his Introduction pp. xxi-xxxii and G. Fasoli, "La *Cronique de Veniciens* di Martino da Canale" *Studi Medievali*, ser. 3, Vol. 2 (1961), pp. 1, 44-47, 67-74.

67. *Zibaldone da Canal*, ed Stussi, pp. 42-72, 108-9, for the instructions of market activity in foreign cities. For the dating see pp. xii-xiv, xli-xliii, liii-lviii; and Lane, "Vecchie Monete," pp. 50-52. For gold and silver trading, Ibid., pp. 22, 42, 58, 109, 64, 69, and 70.

68. *Zibaldone da Canal*, ed. Stussi, p. 71; Lane, "Infidelities," p. 53.

69. Grosso in Zara, Milan, Genoa, Chiarenza in the Morea, Negroponte, and Alexandria in Egypt, *Zibaldone da Canal*, ed. Stussi, pp. 54, 57, 65, 61, 72. Ducat, Ibid., pp. 51, 56.

their minds had to be agile in figuring many exchange rates and in recognizing many different types of money.

The Florentine manual by Pegolotti, dated 1340 but probably compiled at least ten years earlier, presents a more organized and sophisticated overview of market conditions in areas touched by the Florentine merchants one generation later. As late as this the Venetian ducat was mentioned in exchanges with only three cities: Famagosta on Cyprus, Constantinople, and London.[70] The Venetian grossi were only named in circulation in the following areas where Venice had political prerogatives: the kingdom of Rascia in Slavonia; Modon, Coron, Chiarenza, Thebes, and Negroponte in Greece; Cyprus; Constantinople; and Tana east of the Crimea.[71] The Venetian lira a grossi was an accepted standard of value in a more widespread area: in Apulia, in Famagosta on Cyprus, Negroponte in Greece, in Tabriz, in Constantinople, in Alexandria and in Bruges in Flanders. The lira di grossi was mentioned for London.[72] In these cities and elsewhere the Florentine merchants of the Bardi often brought their own gold florins or they carried quantities of gold and silver to be exchanged by weight for local currencies. Pegolotti distinguished in each city between the value in local measures of gold bullion, gold in bars, gold plate, gold thread, and gold coins; as well as between silver plate, silver bars, silver with the seal of Venice (or of other cities), and silver coins.[73]

When a single coin achieves popularity and circulates in a wide area beyond the limits of the principality issuing it, that coin becomes envied and ultimately copied by other states. Would not a catalogue of such imitations also describe the geographical extent of the circulation of such a coin? Would not the laws against imitations also catalogue the awareness of such competitor coins on the part of the originator of that coin?

The Venetian silver grosso was frequently imitated in the thirteenth century, for example: by Italian rulers in Vicenza, Verona, Mantua, and Montferrat, and by the transalpine rulers of Hungary, the Tyrol, and France. Although the Venetians were aware that some eastern mints tried to copy the grosso in the first quarter of the century and passed laws against counterfeiting in the second quarter of the century, it is only about 1275 that Venetian legislation became shrill against such practices. The mints of Rascia and Ragusa on the Dalmatian coast, of Chiarenza in Achaia, and elsewhere in the Levant, and of Brescia in Italy were named as offenders.[74] Dante in his

---

70. Pegolotti, *La Pratica* p. 50, 97-98, and 151-2.
71. Ibid., pp. 40, 87, 91, 117, 119, 150, 153.
72. Ibid., pp. 31, 50, 75, 97-98, 149, 151-2, 172-3, 248-9.
73. For example, for Venice, he lists, Ibid., p. 138, "oro in verghe, oro in piatti, oro in buglione d'ogni ragione . . . argento in piatti, argento in verghe, argento in buglione d'ogni ragione." Under the list of fineness of silver, Ibid., p. 291, he also lists "argento della bolla di Vinegia."
74. For copies Robbert, "Reorganization," pp. 58-59; Müller, "Venezianer Münzen," p. 228. For Venetian legislation, Papadopoli, *Le Monete*, Vol. 1, pp. 89-91, 96-7, 101, 115, 142-143.

Paradiso singled out the rulers of Balkan Rascia who counterfeited the coins of Venice. Between 1295 and 1304 the Palaeologan rulers of Constantinople began to strike a new silver coin identical to the grosso in weight and fineness.[75] Between 1314 and 1328 Venice passed another series of laws against counterfeit grossi, in which the making, passing, or owning of counterfeit grossi were declared criminal acts. False soldini were included in the ban in 1339.[76]

Throughout the fourteenth century Venetians continued to export silver bullion and silver coins. The Venetian state encouraged the export of silver in July, 1328, by giving special privileges to those merchants who wished to load silver on the galleys. Those who had brought silver to the mint were permitted to get the newly minted coins in time to load them on the galleys.[77] Soon after their appearance in 1332, the soldini and mezzanini were exported along with the grosso. Few if any grossi must have been exported during those years when the mint did not issue grossi, 1354 to 1379.

New silver coins were also struck for export only. The Venetian mint initiated the silver tornesello in 1354 for use in the Frankish east. It did not circulate in Venice but competed directly with the large issues of base silver denier tournois from Chiarenza and other mints in the principality of Achaia. This Chiarenza denier tournois was much in demand throughout the Aegean as a useful coin for small transactions. The Palaeologan emperors of Constantinople also issued a base silver and copper tornese which became the smallest fractional coin in their silver system.[78] The Venetian tornesello gave the Venetian merchants engaged in the overseas trade an addition to the recently debased soldini and the discontinued mezzanini and for the hard-to-find grosso. The Venetian tornesello, weighing 0.724 grams at only 13/100 fine silver, (after 1368 at 111/1000 fine) had a similar intrinsic value to the Chiarenza denier tournois, and it seems to have successfully revived the political and mercantile image of Venetian strength in the Greek east. It is reported on January 25, 1386, that the Venetian mints struck as many as 12,000 marks of silver into torneselli, valued at 14,000 gold ducats, annually, during the reign of Doge Antonio Venier (1382-1400). The coin continued to be struck in every reign at least until the late fifteenth century.[79] After the beginning of the fifteenth century, the Venetian mints struck other small silver coins for the Dalmatian coast, which, like the Venetian tornesello, did not circulate in Venice.[80]

In Egypt and the areas under Mamluk control, Venetian silver coins circulated widely according to Egyptian sources of the late fourteenth and

75. *Paradiso*, Canto XIX, lines 140-141. See also Papadopoli, *Le Monete*, Vol. 1, p. 91. For Constantinople, Bertelè, "*Ill Libro dei conti* di G. Badoer," p. 251.
76. Papadopoli, *Le Monete*, Vol. 1, pp. 150-2, 160-1.
77. Lane, "Infidelities," p. 56.
78. Bertelè, "*Ill Libri dei Conti* di G. Badoer," pp. 255-6.
79. Papadopoli, *Le Monete*, Vol. 1, pp. 177-179, 227; 404, 408; and Cessi, *Prob. Mon. Ven.* docs. 117, 119, and pp. xc-xci.
80. Papadopoli, *Le Monete*, Vol. 1, pp. 293-299.

fifteenth century. In addition to three types of gold coin, al Maqrizy named the "boundougy" or Venetian grosso as a fourth type of coin imported from the Latin west which circulated in Egypt. Niccolò Frescobaldi, the Florentine, reported in 1384 from Egypt that the Venetian grosso equalled the Egyptian dirhem and was the only foreign silver money acceptable there.[81] This identifies the demand for Venetian silver coin in the Levant; and, to satisfy this demand, Venetian merchants had argued for the continued issues of grosso coins, and probably also for the Venetian tornesello. To summarize, Venetian silver coins continued to be exported to the Near East throughout the fourteenth century.

The export of silver coins from Venice continued into the fifteenth century. The famous deathbed oration of Doge Tommaso Mocenigo in 1423 speaks of the export into Egypt and Syria of the silver mezzanini and grossi. He also reported that Venice exported one hundred thousand ducats' worth of silver coins, mezzanini and soldini, annually, to her mainland possessions.[82]

In addition to this export of silver, the Venetians began to export large quantities of gold ducats in the mid-fourteenth century.[83] A witness in 1369 complained that the good new ducats were exported, while the old ones remained at home. An official report of 1359 complained that Venice was experiencing a great inconvenience because of the lack of ready cash, not only at home but also abroad, and that the gold money was becoming especially scarce because it was being exported.[84] The Venetian merchants were bringing ducats to Mediterranean ports and were keeping their accounts in ducats, but other peoples used the ducat, also. The ducat circulated in western Europe, alongside the florin. Beginning in the first quarter of the fifteenth century, in France and surrounding countries, the term ducat began to be universally used when referring to full-weight gold coins, instead of the term florin. Doge Tommaso Mocenigo's "Deathbed Oration" and interpretations of it, inform us that in 1423 the Venetian mints issued annually 1,200,000 ducats of gold and 800,000 ducats' worth of silver coins. From these, Venice annually exported 300,000 ducats' worth to Egypt, to Syria, and to her own Aegean possessions, and to England, while the rest remained in Venice.[85] It

81. quoted in A. Rauge van Gennep, "Le Ducat Venitien en Egypte," *Revue Numismatique*, Ser. 4, Vol. 1 (1897), pp. 377-378, 380-381.

82. Tommaso Mocenigo, "Deathbed Oration" ed. H. Kretschmayr. in his *Geschichte von Venedig*, Vol. 2 (Gotha, 1920, reprinted Stuttgart, 1964), p. 618.

83. The following discussion of the export of Venetian gold ducats is based on published materials, and can be considered tentative until amended by further research in the manuscripts and among archaeological remains.

84. Cessi, *Prob. Mon. Ven.*, p. c and doc. 133.

85. A. Dieudonné, "Des espèces de circulation internationale en Europe, depuis saint Louis," *Revue suisse de numismatique* Vol. 22 (1920), esp. pp. 13-17; Mocenigo, "Deathbed oration" ed. Kretschmayr, pp. 617-619. Gino Luzzatto confirmed and interpreted the Mocenigo statistics in "Sull'Attendibilità di alcune Statistiche Economiche Medievali," in his *Studi di Storia Economia Veneziana* (Padua, 1954), pp. 271-284.

has been estimated that in the fifteenth century every Venetian great galley on the muda to Alexandria in Egypt carried gold and silver coins and silver bullion in its cargo valued at 100,000 ducats.[86]

In fourteenth and fifteenth century Mamluk Egypt gold ducats were the principal coin of international trade. Gold ducats also were perhaps the only gold coin circulating alongside silver in internal Egyptian transactions, at least from 1399 to 1427. In addition, Egyptian sources show that around the year 1400 Venetian ducats (known as "dinar ifranty") were in general circulation within Syria and along its coast, in the Hejaz, in Yemen, "as well as in the important Greek and Latin Countries." Their popularity in the Muslim world was apparently due to their constant weight and fineness. "Thus, unlike Moslem coins, they were circulated by number, not by weight." Ducats continued to circulate widely in Egypt for the rest of the Mamluk period, although the Mamluk rulers tried to issue their own gold coins, with Muslim inscriptions, to compete with the ducats. Italian coins were the only source for Egyptian gold in the fifteenth century.[87]

Not only the Venetians, but also the Ragusans used the gold ducat in their Levantine trade after 1360. Ragusa, like other Dalmatian cities, had been dominated by Venice up to 1358. In that year Venice made peace with the King of Hungary and was forced to cede him a major portion of the Dalmation coast, including Ragusa. Thereafter the virtually independent merchants of Ragusa increased their trade with the Levant, and, from mid-century, this trade was carried on by means of the Venetian ducat. Surviving documents of Ragusan trade with the Levant demonstrate that they had used the Venetian silver grosso throughout the thirteenth century, and from 1284 to 1350 they had used their own money of account (the Ragusan hyperperon) and often thereafter for statements of value. The ducat, beginning with a single reference in 1284 was stipulated for actual payments only rarely until 1345. Thereafter ducats were stipulated much more frequently and finally exclusively by Ragusan merchants in their payments in the Levantine trade.[88] Ragusa's modern historian spoke of that city's unfavorable balance of trade with Alexandria and Syria, and concluded that Ragusans paid for the balance in precious metal or silver coins.[89] This widespread use of Venetian ducats (and possibly also of Venetian silver coins) by Ragusa, a competitor of Venice, demonstrates the widespread use of Venetian coin in the Levant.

The existence of counterfeit ducats is further evidence to support the

---

86. Lane, *Venice, Mar. Rep.* p. 338.

87. A. Udovitch, "England to Egypt: 1350-1500" joint article with R. S. Lopez and H. Miskimin in M. A. Cook ed., *Studies in the Economic History of the Middle East* (London, 1970), pp. 125-126; Jere L. Bacharach, "The Dinar Versus the Ducat," *International Journal of Middle East Studies*, Vol. 4 (1973), pp. 77-96; and Van Gennep, "Le Ducat Vénitien," pp. 373-381, 494-495, 504-508.

88. Barissa Krekić, *Dubrovnik (Raguse) et le Levant au Moyen Âge* (Paris, 1961), Regestes, 167-409.

89. Ibid., p. 122 and doc. 1246. But few references are found to such payments.

thesis that Venetian ducats were exported in large quantities and became the principal money of international trade in the Levant. Herbert E. Ives in his fundamental study of the imitations of the Venetian ducat, which was prepared for publication by the eminent numismatist, Philip Grierson, pointed out that the imitation ducats were produced "south and east of Venice, in the eastern Mediterranean, the Levant, and out as far as India." In addition, Grierson noted that Portugal, Spain, and Egypt copied the weight and fineness, though not the design, of the ducat.[90] Ives described and catalogued imitations of the Venetian ducat by the Roman Senate from 1350 to 1439, by Amadeus of Savoy from 1416 to 1439, by the Grand Masters of the Knights of St. John of Jerusalem on the island of Rhodes in the years between 1346 and 1522, and by their successors on the island of Malta from the sixteenth to the eighteenth centuries. The ducat was also imitated by the Genoese and Milanese princes of the Aegean island of Chios from 1415 to 1461, and by the Genoese on Mytilene from 1376 to 1462. Venice reproached one ruler of Mytilene in 1357 for counterfeiting the ducat.[91] Ives also recorded imitation ducats made by the Genoese rulers of Phocaea and Pera outside Constantinople and by the Angevin duke of Achaia in Chiarenza from 1346 to 1364. Other imitations of the ducat which were grouped together as "Anonymous Levantine Imitations" copied various Venetian ducats from the fourteenth century to the nineteenth. The ducat was also imitated in India. In addition, Papadopoli reproduced a letter dated 1370 from Venetian officials in Crete who forced the Muslim ruler of Ephesus in Asia Minor to agree not to strike any more ducats.[92]

During the years 1350 to 1360 many laws were passed in Venice against false or foreign imitations of Venetian coins. Punishments became more severe: clipping was now punishable by loss of the right hand, or of the eyes, or by life imprisonment. The doge's coronation oath, which since 1249 had included a clause whereby the doge would promise to act against counterfeiters, now in 1355 was strengthened by a clause whereby the doge promised to punish counterfeiters with death at the stake.[93] The Venetians did not want competition from foreign mints either at home or abroad.

In conclusion, Venetian mint output grew from modest amounts in the twelfth century until they produced all the coin for Venetian domestic use and most of the coins circulating in the eastern Mediterranean. Silver came to

---

90. Herbert E. Ives, *The Venetian Gold Ducat and Its Imitations*, Philip Grierson, ed., in *Museum Notes and Monographs* The American Numismatic Society (New York, 1954). For Egyptian attempts to replace the ducat with a Muslim gold coin see Van Gennep, "Le Ducat Vénitien," pp. 373-381, 494, 508; Bacharach, "Dinar versus Ducat," and Ives, *Venetian Gold Ducat*, pp. 3-4.

91. The rebuke in Papadopoli, *Le Monete*, Vol. 1, p. 274. The counterfeits in Ives, *Venetian Gold Ducat*, pp. 9-12, 13-14, 18-21, 22-23, and Plates IX-X.

92. Ives, *Venetian Gold Ducat*, pp. 24-26, 29ff; Papadopoli, *Le Monete*, Vol. 1, p. 214.

93. Papadopoli, *Le Monete*, Vol. 1, pp. 101, 167, 172, 186, 190, 195.

Venice from Germany, Bohemia, Hungary, the Balkans and Constantinople; gold arrived from Tunis, Hungary, and the Black Sea. In the later middle ages, the eastern Mediterranean became the Venetian monetary zone, where all merchants used the Venetian money of account as their standard, as well as making payments in Venetian silver and gold coins. The Venetian silver grosso was readily accepted and copied in the thirteenth century. From 1350 the gold ducat dominated the Levant and was accepted in western Europe and copied as far as India. By 1400 Venetian money circulated from London to Yemen.

3

Money and Money Movements in France and England at the end of the Middle Ages

HARRY A. MISKIMIN

Only by means of comparative history which elicits cooperation between many scholars of differing but clearly related specialties can we begin to perceive world bullion movements on a global scale and thus eradicate those errors that have become embedded in monetary history as a consequence of our normal, narrow and regional approach. We may, for example, through comparative study become aware of and perhaps even reconcile what now appear to be manifest contradictions in the literature. To cite but one example of such a problem may be helpful. In 1970, R. S. Lopez writing on the balance of payments of Italy in the second half of the 14th century observed that "the shortage of silver coins aroused alarm in Milan, Bologna, Florence and Lucca simultaneously."[1] He continued by noting that "it is certain that trade with the Levant in the fourteenth and fifteenth centuries drained from Italy an ever growing amount of gold." Both gold and silver were in short supply in Italy during the late 14th century according to Lopez, but let us consider the view of Ashtor. As set forth in his version, "the most striking feature of economic life in Egypt and Syria at the turn of the fourteenth century was the disappearance of silver coin."[2] This phenomenon is explained, he says, by the increase in the value of silver in Italy that caused Italian merchants to import silver from the Near East to Europe. Thus with two eminent scholars and two differing sets of regional evidence carefully exploited, we are left with two evidently contradictory analyses. The absence of precious metal in each region during the late 14th century is explained by bullion transfers to the other.

The statistical material generated by the comparative method should go far toward providing the evidence that will allow us to eliminate such disparities in our studies and permit us all to write with greater assurance. Until that evidence is forthcoming, however, I believe caution is a virtue and that it must be supplemented by an honest recognition that monetary history is in

---

1. Lopez, R., H. A. Miskimin and A. Udovitch, "England to Egypt, 1350-1500: Long-term Trends and Long-distance Trade" in M. A. Cook, ed., *Studies in the Economic History of the Middle East* (London, Oxford University Press, 1970), pp. 93-128.

2. Ashtor, E., *A Social and Economic History of the Near East in the Middle Ages* (Berkeley, University of California Press, 1976), p. 305.

its formative stages. With this caveat in mind, we may, though tentatively, draw some conclusions from the tables here presented and perhaps even shed some light on the nature and causes of the Lopez-Ashtor disagreement.

What is immediately evident from our tables is that the quantity of metal struck by both French and English mints during the early 14th century, roughly until 1360, was substantially larger than the amount struck in the following years. In the English case, recovery is not complete until the late 15th century. While our French figures terminate in 1395, there is other, more circumstantial evidence—contemporary comment, bullionist legislation, the manifest concern for the balance of payments expressed in the proceedings against Jacques Coeur and in the Pragmatic Sanction—that suggests that French monetary difficulties paralleled those of the English during much of the 15th century. The decline in mint output in both countries is most pronounced in the silver coinage, but it is also clearly present in that of gold. In England, annual silver coinage output from 1363 to 1464 is only 3.81% of the average from 1273 to 1322; in the case of gold, the average during the same hundred year period is only 28.44% of that attained during the years 1344-1362.[3]

Based on these figures, it seems certain that less metal was being coined in both France and England during the later middle ages than was the case earlier. Further elaboration and analysis, however, becomes slightly more speculative. Elsewhere I have suggested that the reduction in mint output resulted from an adverse balance of payments in northwestern Europe, caused in large part by plague-induced demographic decline.[4] As population fell, more land was available and average agricultural productivity probably increased; agricultural prices declined—relatively in nominal money, absolutely in silver value—while the prices of manufactured goods rose, thus providing townspeople with more free funds. These funds were supplemented through the concentration of wealth as a result of the inheritance effects of the plagues. Specie flowed toward the towns, and stimulated by sudden affluence and fear of death, people spent freely either on luxury or salvation. In one case, bullion was shipped to Italy and the Levant for the purchase of exotic goods while in the other, it was transmitted to Rome or Avignon in papal remittances.

Though stripped bare of both detail and the appropriate caveats in the interest of brevity, this hypothesis regarding the nature and causes of the bullion deficiency during the later middle ages seems to accord with the facts and to account for the broadest spectrum of observable phenomena. There are, however, in the literature alternative hypotheses although they are, in my opinion, less satisfactory. One option is simply to deny the existence of a

---

3. Miskimin, H. A., "Monetary Movements and Market Structure—Forces for Contraction in Fourteenth and Fifteenth Century England", *The Journal of Economic History*, XXIV (1964), 470-90.

4. Miskimin, H. A., *The Economy of Early Renaissance Europe, 1300-1460* (Cambridge, Cambridge University Press), 1975.

bullion shortage; this appears to have been Postan's position in his 1959 debate with Robinson in which he maintained that England's balance of payments situation was "probably more favorable" during the 14th and 15th centuries than it had been during the 13th.[5] I believe that this position is no longer tenable in the face of the mint output declines shown in our tables and the directly contradictory testimony of so many contemporary observers.[6] Denial of the bullion shortage is rendered especially difficult if one discounts hoarding, as does Postan (and I believe rightly). It is, in fact, theoretically possible to explain the dearth of coinage in terms of hoarding, but problems arise. What, for example, would simultaneously motivate vast numbers of Englishmen to bury large sums of money and then to petition the monarchy to alleviate the resulting scarcity of coin? How did the French catch the identical disease? The argument for hoarding is not compelling on its face and as Postan points out, the financial exigencies of the monarchy and the rising expenditure patterns of the magnates—I would add of the townspeople as well—do not support the thesis that bullion was simply being squirreled away.[7]

Some alternative explanations of late medieval monetary phenomena begin by postulating a variant mechanism for motivating monetary movements. A. M. Watson argues that differentials in regional gold-silver ratios, especially between western Europe and the Middle East, were such as to cause massive shipments of metal from one region to the other, and thus that when one region was rich in gold, the other was apt to be innundated with silver.[8] A corollary proposition, subordinate to this view, also denies the reality of the bullion famine in the later middle ages since, although silver is acknowledged to have become scarce, gold is presumed to have flowed into western Europe in return for the silver that departed.[9] A variety of difficulties spring from the too mechanical application of modern economic formulations to the historical past; one of the more serious is to give the theory primacy over the evidence. Thus, for example, Watson informs us that "France was afflicted with an acute silver crisis, such that for years on end (e.g., from 1308 to 1314) no good silver coins were struck; although gold was becoming increasingly plentiful, there was an acute shortage of silver in circulation."[10] In fact, as our tables of French coinage indicate, something over 72,000 kilograms of fine silver were struck by French mints between 1308 and 1312 while the level of debasement was at its 14th century nadir.[11] Recorded

5. Postan, M. M., "Note", *The Economic History Review*, 2nd ser., XII (1959), 77-82.
6. See Miskimin, "Monetary Movements" for a discussion of the value of contemporary comment.
7. Postan, "Note", p. 78.
8. Watson, A. M., "Back to Gold—and Silver", *The Economic History Review*, 2nd ser., XX (1967), 1-34.
9. Ibid., pp. 31-32.
10. Ibid., p. 15.
11. Miskimin, H. A., *Money, Prices and Foreign Exchange in Fourteenth Century France* (New Haven, Yale University Press, 1963), p. 37.

annual silver output during the period indeed ranked among the peak years of the entire 14th century. Gold output from 1311 to 1318 was also remarkably high and compares favorably with other years of peak production.

This enormously complex field under the most favorable circumstances is filled with snares and traps for the unwary. Even with better comparative evidence, however, considerable subtlety and caution will be required in its interpretation. It is true, for example, as Watson points out, that many regions of Europe adopted gold coinages during the last two and a half centuries of the middle ages. It is equally true that in England where figures are firm, the value and even the weight of the gold coinage exceeded that of the silver coinage in the century or so following 1357, but what does that really mean?[12] Some of the literature interprets the rising proportion of gold in the total coinage as evidence of the operation of international arbitrage predicated on variants in regional gold-silver ratios. Thus it is argued that when the silver coin was exported, gold flowed into the exporting country and that in consequence, the proportion of gold in the total bullion output of the mints rose. This interpretation is, of course, seductive when the perspective is limited to the short term—to a year or to a small cluster of years, but what if we consider a somewhat more extended period? In France, the maximum annual output of gold during the 14th century occurs in the year 1339 and seems more likely to have been the result of dishoarding induced by the early stages of the Hundred Years War than of any abstract calculation based on gold-silver ratios. Two other clusters of active gold minting occur in 1311-18 and again in 1356-68. Output during the remainder of the 14th century falls sharply below these levels and the decrease is cotemporal with that in the silver coinage. In England, the gold coinage begins only in 1344, but during the next thirty-one years more gold is struck than was minted during the entire 16th century. After the early 1370's, English gold output falls sharply below these levels and only briefly recovers during the period from 1412 to 1425; thereafter it again declines. In the cases of both English peaks, political events would appear to have been critical in determining the level of the gold coinage. The first surge in output covers the period from the battle of Sluis to the ransom of Jean le Bon, whereas the subsequent crest seems related to the successes of Henry V and victory at the battle of Agincourt.

Now it is certainly true that gold becomes increasingly important in the coinage of western Europe during the later middle ages, but the pattern does not provide support for the hypothesis that it did so because of the direct exchange of metal in response to stimuli deriving from regional variations in gold-silver ratios. Both gold and silver outputs exhibit a pattern of secular decline after the middle years of the 14th century, and that decline is only occasionally relieved through the benignity of specific political events. While,

---

12. Watson, "Back to Gold", p. 32 and note 2. The same pattern occurs in France in the years from 1362 to 1390.

for example, the gold coinage expands sharply in the years after 1412, so too does that of silver. The recovery of the gold coinage is greater, but the joint expansion does not suggest the export of silver to obtain gold. Instead, it appears that both gold and silver were being persistently exported and that gold became more significant in the overall coinage, not because there was more of it in western Europe, but rather because there was less silver. Watson himself observes this phenomenon, noting that "by the end of the fourteenth century gold may have been used to supplement the silver that was being sent eastward."[13] He further comments that since silver was more highly valued in the east, that metal was exported preferentially in commerce and that gold was retained when there was a free choice. Here we are in complete agreement. A merchant, choosing which metal to employ in purchasing goods from the east, would clearly have selected silver during the later middle ages; when there was no alternative, he would export gold. In this sense, variations in the relative valuations of the two metals in the separate regions would indeed encourage the initial export of silver in settlement of the adverse balance of accounts arising from commerce. This is not, however, equivalent to the direct exchange of one metal for another.

Watson, judiciously and with commendable thoroughness, describes the difficulties inherent in the calculation of gold-silver ratios and even in determining which among the many possible ratios it is appropriate to use.[14] Does one want the market rate for unminted metals, the market ratio for coins, the official ratio for coins, mint bullion purchase ratios, official exchange ratios for domestic and foreign coins, or ratios established for official accounting purposes such as taxation? Very different results may be obtained depending on the ratio selected and this fact might well affect our perception of the role of ratios in motivating monetary movements. It is not, I believe, sufficient to point to *"general movements"* in the ratio, i.e., to some non-mathematical, impressionistic average of the mass of possibilities and to attempt to apply the resulting abstraction to monetary history. We must know in depth the nature of the ratio that we are employing and we must explore in detail the institutional and economic structure that gives or denies validity to the gold-silver ratio as a factor in economic calculations.

Elsewhere, I have attempted to outline some of the significant factors that affect the credibility of assigning to gold-silver ratios the role of determining monetary movements during the later middle ages.[15] Even at the risk of repetition, it is, perhaps, proper to reiterate here since the matter is of great importance to our collective enterprise. Gresham's law states that "bad" money drives out "good," i.e., that when the intrinsic value of different coins

---

13. Ibid., p. 18.
14. Ibid., pp. 22-23.
15. Miskimin, H. A., "The Enforcement of Gresham's Law". To appear in the *Proceedings* of the Quarta Settimana di Studio, Instituto Internazionale di Storia Economica "Francesco Datini", Prato, April, 1972.

does not accord with the value assigned in money of account, the coins of low intrinsic value will circulate while the higher value coins will disappear. This assumes that the government possesses enough political force to insist upon the legal tender value of the coinage and to decree circulation at par. There is, however, substantial evidence that neither the French nor the English monarchies gained this power until the end of the middle ages.[16] If money did not circulate at par, one of the most easily calculated gold-silver ratios—the official rate between minted coins—becomes meaningless. Coins are weighed and circulate as bullion; the market rate for bullion then dominates over all official rates. Since weighing the coin and defying the king's will were illegal practices, it is especially unlikely that adequate records would be kept of those transactions that set the market ratios, so it is wise to exercise considerable skepticism with regard to such fragmentary glimpses of market ratios that have chanced to survive. Furthermore, since these ratios were determined by supply and demand in a very thin market, we would expect them to be highly sensitive to small changes in the quantity of bullion sought or offered. Unlike the official rate, made rigid by an inept and cumbersome bureaucracy, these rates would adjust swiftly to small changes in market conditions and thus rather quickly limit the regional variations in the gold-silver ratio and its power to inspire international metal movements.

Several additional conditions must also be met before we may assume that gold-silver ratios guide monetary flows. Among the most important is the existence of free and open exchange between the two metals in each of the regions considered. I am not persuaded that such conditions prevailed in most of western Europe. If mint ratios are at issue, we must note that there was no legal obligation on the part of governments to exchange one metal for another and further, that it was common practice to purchase bullion in coins of the same metal after a delay and after the subtraction of seigniorage charges.[17] Where then could an enterprising arbitrager exchange metals? If he used the official royal changers, he was at once subject to a fee and to the risk that his foreign coin might be forced into the mints. In the latter case, his return would be reduced by seigniorage charges which in 14th century France sometimes amounted to fifty per cent. If he went to the free market, such as it was, he was exposed to legal risk, but a probable greater deterrent was the necessity of locating persons who wished to exchange one metal for the other and who were sufficiently liquid to be able to produce large enough sums of specie to make the costs and dangers of international bullion shipments worthwhile.

16. Ibid.
17. As for example: Lettres patentes. 15 June 1359. "... vous faciez, faire et ouvrer deniers d'or fin au mouton, de 52. de poiz, ou royauls d'or fin de 66. de pois aud. marc, lesquelx yceulx changeurs auroient plus agréables, et leur donnant pour marc d'or fin, 63 rouaulx ou 50 moutons..." L. F. J. C. DeSaulcy, *Recueil de documents relatifs à l'historie des monnaies frappées par les rois de France depuis Philippe II jusqu'à François Ier* (Paris, Imprimerie Nationale, 1879), I, p. 161.

Beyond these legal and institutional impediments to international arbitrage founded upon comparative gold-silver ratios, there exist some significant technical factors that could hinder the free working of such exchange. If the gold-silver ratio in question is the ratio that arises out of the government's specification of the circulation value of coin, the quality both of the average coin at the time of its manufacture and of the circulating coinage as a whole become critically relevant. In the case of the 14th century French coinage, we are fortunate to possess evidence regarding the quality control standards of the mints, both from the surviving coins and from documentary statements of the legal tolerance and of the sampling techniques applied in the mints. Elsewhere I have attempted to consider this question in detail; a brief summary will suffice here.[18] The weights of coins leaving the mints may safely be presumed to have been normally distributed around the legal mean. We can translate the legal tolerance specified in mint ordinances into statistical terminology and refer to the standard deviation—a measure of the spread of weights around the mean. By arbitrarily assigning a standard deviation of five per cent of the legally established coin weight to a hypothetical normal distribution, it can be shown that a mint master, meeting these standards, would fail the actual royal assay more than two-thirds of the time. Since punishment was draconian, it is unlikely that a mint master would lightly have indulged himself in so generous a level of tolerance. In short, our hypothetical distribution permits far more variation than the law allowed. Numerous coins—some 25.6% of all issues—preserved in the Bibliotheque Nationale are overweight, some by more than 30%. Working only with the upper tail of our hypothetical distribution, we find that 2% of the surviving coins are more than four standard deviations overweight. The probability of a coin appearing in that range is 0.00003 so we must conclude that our hypothetical distribution is in fact narrower than the actual mint standard. It follows from the above that roughly the heaviest sixth of the coins struck in any given issue would weigh at least 10% more than the lightest sixth. Thus if the legally established gold-silver ratio were 10 to 1, the lower sixth of the coins issued would entail a ratio of 9.5 to 1, while the upper sixth would be at a ratio of 10.5 to 1. There is thus not one gold-silver ratio based on legal standards, but a wide continuum of ratios that depends on the specific coin exchanged. It is, of course, true that, given a substantial number of coins, their average weight at the time of release from the mint would approximate the legally established weight, but what would happen after a period of public circulation? Since simply by culling out the heavy coins and trading light for heavy, one could make a ten per cent profit at virtually no risk, we may presume that if people were sensitive to international ratios, they would not have foregone this opportunity. We may also presume that the average quality of the coinage would quickly deteriorate as heavy coins disappeared, thus rendering the official gold-silver ratio a meaningless abstraction.

18. Miskimin, "Gresham's Law".

Given the many difficulties impeding the free operation of arbitrage based on gold-silver ratios, it seems obvious that great caution should be exercised before we assume that such arbitrage explains the monetary movements of the late 14th and 15th centuries. Arguments based on arbitrage do not appear more compelling than the hypothesis predicated on demographic decline, changing consumption patterns, and consequent shifts in the balance of payments. Before opting for the balance of payments–population hypothesis, however, it is necessary to consider one further challenge. In 1974, N. J. Mayhew, after reviewing the literature, concludes that "the supply of money current in England was sharply reduced in the thirty years before the Black Death."[19] If true, and if the shortage were of the same nature as that which followed the first plague, the case for making plague-related depopulation part of the engine behind monetary movements would be weakened. Let us consider the evidence. Mayhew, citing a 1964 article of mine, writes that "strangely, when Miskimin examined the mint records of Edward I, II and III, he concentrated on the period of reduced activity after 1360, ignoring the twenty barren years 1322-1342."[20] He is, of course, right that there are barren years, but I believe that if he had considered my earlier writings on this question, my oversight would have appeared less "strange."[21] Prior to 1360, the coinage outputs of the mints of France, Flanders and England exhibit a pattern of alternation.[22] Years of peak coinage in one country are very often years of dearth in one or both of the others. It is as if the bullion, particularly silver bullion, were being driven from one country's mints to another's and never finding a permanent resting place. But, and the point is of crucial significance, the total amount of bullion distributed among the three countries does not appear to weaken until after 1360. At that point, there is a simultaneous decrease in the outputs of all three countries, and it is at that point that I maintain that demographic decline, through a complex set of events, led to an increasingly adverse balance of payments for western Europe as a whole. The barren years that Mayhew notices may be explained within the boundaries of western Europe. The hundred barren years after 1360 are not explicable in this way, and since both the silver and, although slightly later, gold coinages are afflicted, I believe the balance of payments argument remains the most persuasive explanation. Gold-silver ratios in this period offer theoretical evidence at one remove from hard facts, but they are sometimes useful—particularly, as Watson notes, in determining which metal a merchant might choose for settling his trade accounts. Yet when given the choice, as in the Lopez-Ashtor debate, the actual presence or absence of bullion itself must take precedence over hypothetical propositions regarding bullion flows.

19. Mayhew, N. J., "Numismatic Evidence and Falling Prices in the Fourteenth Century", *The Economic History Review*, 2nd ser., XXVII (1974), 1-15.
20. Ibid., p. 10.
21. See Miskimin, *Money, Prices*, Chapter 7.
22. Ibid., Figure 15, p. 106.

Appendix I

Output of the French mints in kilograms of fine metal: 1308-1395

The information pertinent to the French coinage has been calculated from the documents printed in DeSaulcy's *Recueil de documents relatifs à l'histoire des monnaies frappées par les rois de France*, I and II. The Gold output of the Paris mint for the period August, 1392 to 16 September, 1394 is taken from M. Rey, "Les Emissions d'écus à la Couronne à l'Hôtel des Monnaies de Paris vers la fin du XIVe siècle et dans la premières années du XVe, 1385-1413", *Melanges d'histoire du Moyen Age dédiés à la mémoire de Louis Halphen* (Paris, Presses Universitaires de France, 1951), pp. 595-603.

Unlike those for the London mint, these figures can make no claim to completeness. On the one hand, French records have suffered more mishaps than the English, and on the other, there are some additional documents that DeSaulcy overlooked. In my opinion, however, the series is full enough to provide an index of the overall level of French coinage and to identify crests and troughs. Where documents give an output total for a period longer than one year, I have allocated the total in proportion to time. Mint profits were very high in some years in which there is no record of coinage; this information must be used to supplement the figures presented here. High mint profits occurred in 1298-99, 1301, 1324-25, 1327 and 1349. It is not possible to calculate output from profits nor to determine the proportion of gold to silver in the total struck. See Miskimin, *Money, Prices*, p. 43.

In converting from marks to kilograms, I have used the conventional factor, 1 mark = 244.75 grams, despite my impression that this understates the amount struck. The large number of surviving overweight coins suggests either that this weight is too low or that there was more than one mark in use in France during the fourteenth century. See H. Miskimin, "Two Reforms of Charlemagne? Weights and Measures in the Middle Ages", *The Economic History Review*, 2nd ser., XX (1967), 35-52 for a history of medieval weights. I have weighed the gold coins preserved in Brussels and found that, unlike the French case, there are virtually no overweight coins.

Table 1. Output of the French mints in kilograms of fine metal—1308-1395

Year	Silver 0.958 fine	Gold 24k
1308	1591.6	
1309	15146.1	
1310	10167.9	
1311	32682.0	611.6
1312	14982.9	1613.6
1313		2238.7
1314		1647.4
1315		1453.1
1316		1125.6
1317		2049.8
1318		1056.1

Table 1. (continued)

Year	Silver 0.958 fine	Gold 24k
1319	200.9	385.7
1320	200.9	257.5
1321	200.9	257.5
1322	200.9	257.5
1323–1330[a]		
1331		2.0
1332		23.0
1333		23.0
1334–1337[b]		
1338	3740.0	1153.0
1339	25185.8	6326.1
1340		127.5
1341		195.1
1342		234.0
1343		260.7
1344		265.1
1345		99.6
1346		782.2
1347–1349[c]		
1350	1041.4	
1351	1617.6	
1352	207.8	
1353[c]		
1354	10839.2	44.3
1355	28922.8	1499.1
1356	18723.4	1462.6
1357	19274.6	1486.1
1358	18779.4	977.3
1359	15898.0	1562.5
1360	15172.8	1716.2
1361	5828.2	2458.8
1362	512.8	1787.9
1363	221.5	1269.8
1364	739.9	794.0
1365	1093.8	2233.6
1366	1241.1	2342.0
1367	754.3	1202.2
1368	618.2	1212.2
1369	548.7	903.4
1370	553.6	1234.8
1371	843.2	910.2
1372	1372.3	598.6
1373	2065.2	615.5
1374	833.1	643.0
1375	362.7	833.1

Table 1. (continued)

Year	Silver 0.958 fine	Gold 24k
1376	458.9	651.3
1377	352.0	715.4
1378	167.7	599.1
1379	231.5	502.2
1380	167.7	275.6
1381	134.9	851.0
1382	456.9	598.4
1383	61.9	337.0
1384	133.6	467.5
1385	149.3	1399.7
1386	87.1	704.4
1387	4.7	41.1
1388	18.8	–
1389	616.0	253.3
1390	6695.4	351.2
1391	2909.3	254.3
1392	453.3	173.8
1393	2226.0	200.5
1394	663.0	142.7
1395	587.6	123.4

[a] No record of coinage; mints closed 10/10/29 to 9/20/30.
[b] No record of coinage; mints closed 3/9/25 to 2/13/37.
[c] No record of coinage.

Appendix II

Output of the London Mint in Kilograms of Fine Metal: 1273-1600

The information concerning the output of the London mint has been calculated from the tables printed in Sir John Craig's *The Mint* ([Cambridge, Cambridge University Press, 1953], pp. 410-15). Output from the Calais, Canterbury, and Durham mints is not included. Where Craig gives an output total for a period longer than one year, I have allocated the total in proportion to time.

Table 2. Output of the London mint in kilograms of fine metal—1273–1600

Year	Silver 0.925 fine[a]	Gold 24K (994.8)	Year	Silver 0.925 fine[a]	Gold 24K (994.8)
1273	2288.7		1299	4647.0	
1274	3576.7		1300	37421.5	
1275	2760.4		1301	14556.2	
1276	7011.3		1302	2458.0	
1277	7011.3		1303	2007.0	
1278	7011.3		1304	5954.9	
1279	32410.2		1305	24595.9	
1280	60809.8		1306	22434.9	
1281	31954.3		1307	31135.5	
1282	13639.8		1308	24721.5	
1283	13639.8		1309	34117.0	
1284	6570.8		1310–11[b]		
1285	6570.8		1312	4904.2	
1286	25473.1		1313	2567.6	
1287	24273.6		1314	10733.2	
1288	12930.6		1315	4208.9	
1289	3579.1		1316	396.1	
1290	3579.1		1317	2736.6	
1291	839.1		1318	4675.7	
1292	1500.4		1319	3136.2	
1293	888.4		1320	3021.4	
1294	2203.0		1321	3304.1	
1295	2318.4		1322	—	
1296	1612.0		1323	358.3	
1297	2342.9		1324	606.7	
1298	702.9		1325	40.2	

Table 2. *(continued)*

Year	Silver 0.925 fine[a]	Gold 24K (994.8)	Year	Silver 0.925 fine[a]	Gold 24K (994.8)
1326	49.0		1352	24603.9	1684.4
1327	82.2		1353	31438.9	1249.5
1328	52.5		1354	13082.4	2888.4
1329	252.3		1355	13518.4	1963.6
1330	172.2		1356	7930.1	192.1
1331	176.0		1357	5077.0	1791.8
1332	143.8		1358	3447.2	2602.2
1333	229.2		1359	2917.5	2271.2
1334	133.7		1360	1719.8	1549.4
1335	237.6		1361	1685.1	4822.3
1336	1082.6		1362	3964.7	3042.4
1337	434.2		1363	881.0	875.8
1338	490.6		1364	795.0	475.2
1339	619.3		1365	415.7	363.2
1340	638.9		1366	–	383.1
1341	373.0		1367	–	257.9
1342	1772.9		1368	614.1	584.7
1343	4877.3		1369	429.7	1686.9
1344	12757.0	936.0	1370	544.4	515.1
1345	7808.0	263.5	1371	224.3	358.3
1346	2406.6	214.1	1372	48.6	506.3
1347	1432.8	910.4	1373	126.7	338.7
1348	2558.8	1077.3	1374	130.2	223.6
1349	1321.9	273.6	1375	1166.6	241.4
1350	2923.4	903.1	1376	816.0	130.9
1351	7224.4	2219.8	1377	63.0	95.2

Table 2. (continued)

Year	Silver 0.925 fine[a]	Gold 24K (994.8)	Year	Silver 0.925 fine[a]	Gold 24K (994.8)
1378	353.7	113.4	1404	126.7	109.2
1379	353.7	113.4	1405	24.5	77.0
1380	353.7	113.4	1406	28.3	125.3
1381	353.7	113.4	1407	22.4	69.3
1382	353.7	113.4	1408	2.1	50.4
1383	353.7	113.4	1409–11[b]		
1384	353.7	113.4	1412	678.5	3125.3
1385	305.1	265.6	1413	1272.9	2895.1
1386	305.1	265.6	1414	1635.1	1634.4
1387	305.8	265.9	1415	1635.1	1634.4
1388	—	—	1416	1635.1	1634.4
1389	99.0	634.7	1417	1635.4	1634.4
1390	627.7	566.1	1418	367.7	497.9
1391	765.9	534.3	1419	367.7	497.9
1392	114.8	589.6	1420	517.5	903.1
1393	62.3	302.7	1421	775.0	2053.6
1394	51.4	314.6	1422	806.9	2287.6
1395	51.4	314.6	1423	806.9	2287.6
1396	59.5	186.8	1424	806.9	2287.6
1397	205.4	398.5	1425	563.3	1204.4
1398	205.4	398.5	1426	944.7	1070.0
1399	401.3	385.9	1427	209.2	264.9
1400	80.1	170.1	1428	406.6	587.8
1401	80.1	170.1	1429	809.7	425.1
1402	80.1	170.1	1430	202.2	106.4
1403	45.1	103.9	1431	986.4	452.1

Table 2. (continued)

Year	Silver 0.925 fine[a]	Gold 24K (994.8)	Year	Silver 0.925 fine[a]	Gold 24K (994.8)
1432	512.6	227.1	1458	1279.6	29.4
1433	300.9	170.4	1459	1084.7	6.6
1434	199.1	220.4	1460	2461.5	39.2
1435	155.0	140.0	1461–1462[b]		
1436	155.0	140.0	1463	2077.4	51.1
1437	191.0	117.9	1464	2077.4	51.1
1438	516.5	109.5	1465	9674.7	2155.4
1439	1500.7	189.6	1466	9674.7	2155.4
1440	286.6	103.9	1467–1468[b]		
1441	286.6	103.9	1469	2848.9	711.3
1442	286.6	103.9	1470	2848.9	796.7
1443	286.6	103.9	1471	2537.2	526.6
1444	54.2	84.7	1472	2537.2	526.6
1445	72.1	56.3	1473	2537.2	526.6
1446	—	—	1474	2537.2	526.6
1447	900.6	95.2	1475	2537.2	526.6
1448	30.8	30.4	1476	1100.3	391.9
1449	245.3	71.0	1477	1100.3	392.2
1450	1620.4	124.2	1478	731.6	389.4
1451	—	—	1479	649.6	364.2
1452	3771.6	144.2	1480	682.0	479.7
1453	1430.7	91.0	1481	341.2	276.4
1454	1260.3	43.0	1482	598.3	267.3
1455	955.9	25.9	1483	1085.4	162.0
1456	955.9	25.9	1484	2485.3	255.8
1457	2328.9	44.4	1485	865.7	135.1

Table 2. (*continued*)

Year	Silver 0.925 fine[a]	Gold 24K (994.8)	Year	Silver 0.925 fine[a]	Gold 24K (994.8)
1486	1434.6	164.8	1518	186.5	743.9
1487	665.9	120.0	1519	2655.0	848.5
1488	927.2	138.6	1520	11.9	560.9
1489	952.4	82.2	1521	346.8	384.5
1490–1494[b]			1522	2655.4	227.1
1495	1699.8	284.8	1523	3274.7	141.4
1496	1699.8	284.8	1524–1526[b]		
1497	1699.8	284.8	1527	10815.4	599.7
1498	1700.2	284.8	1528	10815.4	599.7
1499	4188.0	311.1	1529	10815.4	599.7
1500	4188.0	311.1	1530	10815.8	599.7
1501	3666.6	429.7	1531–1532[b]		
1502	3666.6	429.7	1533–34	7653.7	371.6
1503	2718.0	442.6	1535–1536[b]		
1504	4829.7	560.9	1537	7066.9	357.2
1505	8420.0	734.8	1538	4594.9	192.4
1506	7182.7	1477.6	1539	10225.5	611.3
1507	5651.2	1324.0	1540	4295.0	260.0
1508	4473.5	1897.5	1541	3089.3	158.5
1509	1723.3	1844.0	1542	2290.1	64.7
1510	615.8	1069.3	1543	2290.1	64.7
1511	232.7	780.6	1544	2290.1	64.7
1512	1936.3	416.4	1545	17446.7	1979.7
1513	2529.4	1131.2	1546[b]		
1514	832.4	493.7	1547[d]	21415.2	4413.6
1515	191.0	649.4	1548	37786.8	Missing
1516	33.6	834.2	1549–1551[c]		

Table 2. *(continued)*

Year	Silver 0.925 fine[a]	Gold 24K (994.8)	Year	Silver 0.925 fine[a]	Gold 24K (994.8)
1552	1868.5	6.3	1578	10602.0	126.0
1553	6992.8	202.9	1579	10602.0	126.0
1554	6992.8	202.9	1580	10602.0	126.0
1555	6992.8	202.9	1581	10602.0	126.0
1556–1559[b]			1582	9803.8	367.0
1560	32521.5	222.2	1583	12318.9	221.1
1561	32521.5	222.2	1584	12318.9	221.1
1562	32521.5	222.2	1585	12318.9	221.1
1563	10783.2	222.2	1586	12318.9	221.1
1564	10783.2	222.2	1587	12318.9	221.1
1565	10783.2	222.2	1588	12318.9	221.1
1566	10783.2	222.2	1589	12318.9	221.1
1567	10783.2	222.2	1590	12318.9	221.1
1568	10783.2	222.2	1591	12318.9	221.1
1569	10783.2	222.2	1592	12318.9	221.1
1570	10783.2	222.2	1593	12318.9	221.1
1571	10783.2	222.2	1594	31438.2	160.6
1572	10783.2	222.2	1595	35375.9	282.7
1573	10602.0	126.0	1596	13531.3	187.2
1574	10602.0	126.0	1597	16011.4	87.5
1575	10602.0	126.0	1598	3363.6	158.9
1576	10602.0	126.0	1599	2510.5	166.2
1577	10602.0	126.0	1600	2955.3	177.7

[a]The reader will note that this series differs by 8.1% from the figures presented by Prof. Munro. The disparity results from his conversion to pure silver whereas my series preserves the 0.925 fineness of sterling.
[b]Missing.
[c]Fineness is unclear during this period.
[d]J. D. Gould, *The Great Debasement*, presents somewhat different figures for this year.

95

Note to Appendix II

The table of English coinage output printed above differs from that presented in Professor Munro's chapter in several ways. My figures record kilograms of sterling silver, .925 fine; as a result, they are consistently 8.1% greater than Professor Munro's. The figures for gold are comparable and not affected by this difference.

The backward extension by Blunt and Brand of the series to the period prior to 1273 is a valuable addition to our knowledge, unpublished at the time of my writing. They support my general thesis; they do not alter the material here presented. Similarly, it is useful to have new material for the provincial mints, since they reinforce my contention that coinage levels in the thirteenth and early fourteenth centuries were far greater than those in the following one-hundred-fifty years. However, I thought it wise to exclude production from the Calais mint, since my purpose was to estimate changes in the money supply within England. Inclusion of these figures would confuse, rather than aid, the investigation, since money struck in Calais may be presumed to have circulated in France particularly after 1420.

Finally, in my series, I have purposely avoided the use of decennial averages. Output levels varied greatly from year to year, and in consequence, the output of a single year may distort the average for a decade as in the cases of the years 1279 and 1280. When considering international bullion movements and questions of gold-silver arbitrage, the evidence is at once more visible and more capable of supporting refined argument when presented on an annual basis. The remaining differences between the two sets of figures are slight.

4

Bullion flows and monetary contraction in late-medieval England and the Low Countries

JOHN MUNRO

Bullion flows and monetary contraction

For almost a century now, many economic historians have contended that Europe's aggregate stock of precious metals was declining in the late Middle Ages.[1] But even amongst those who support this proposition in general, there is still no agreement on four vital questions: (1) What were the principal causes of this reduction in Europe's precious metal stocks—in particular of

---

1. The major studies in chronological order are: W. A. Shaw, *The History of Currency, 1252-1894* (London, 1896), pp. 1-60; Sir Albert Feavearyear, *The Pound Sterling: A History of English Money* (Oxford, 1931); 2nd rev. edn. by E. Victor Morgan (Oxford, 1963), pp. 10-15; Marc Bloch, "Le problème de l'or au moyen âge," *Annales d'histoire économique et sociale*, V (1933), 1-34 [reprinted as "The Problem of Gold in the Middle Ages," in *Land and Work in Medieval Europe: Selected Papers by Marc Bloch*, ed. J. E. Anderson (London, 1967), pp. 186-229]; Marc Bloch, *Esquisse d'une histoire monétaire de l'Europe* (Paris, 1954), pp. 35-77 [but written before 1940]; Henri Laurent, "Crise monétaire et difficultés économiques en Flandre aux XIVe et XVe siècles," *Annales d'histoire économique et sociale*, V (1933), 156-61; Henri Laurent, *La loi de Gresham au moyen âge* (Brussels, 1933), pp. 3-36; A. Girard, "Un phénomène économique: la guerre monétaire, XIVe-XVe siècles," *Annales: E. S. C.*, II (1940); F. Graus, "La crise monétaire du XIVe siècle," *Revue belge de philologie et d'histoire*, XXIX (1951), 445-54; Johan Schreiner, "Wages and Prices in the Later Middle Ages," *Scandinavian Economic History Review*, II (1954), 61-73; W. C. Robinson, "Money, Population, and Economic Change in Late-Medieval Europe," *Economic History Review*, 2nd ser. XII (1959), 63-76; Harry A. Miskimin, "Monetary Movements and Market Structure: Forces for Contraction in Fourteenth and Fifteenth Century England," *Journal of Economic History*, XXIV (1964), 470-90; R. Cazelles, "Quelques reflexions à propos des mutations de la monnaie royale française (1295-1360)," *Le moyen âge*, LXXII (1966), 83-105, 251-78; Harry A. Miskimin, *The Economy of Early Renaissance Europe, 1300-1460* (Englewood Cliffs, N.J., 1969; reissued Cambridge, 1975), especially pp. 25-72, 132-57; E. Fournial, *Histoire monétaire de l'occident médiéval* (Paris, 1970); R. S. Lopez, H. A. Miskimin, and A. L. Udovitch, "England to Egypt, 1350-1500: Long-Term Trends and Long-Distance Trade," *Studies in the Economic History of the Middle East*, ed. M. A. Cook (London, 1970), pp. 93-128; C. C. Patterson, "Silver Stocks and Losses in Ancient and Medieval Times," *Economic History Review*, 2nd ser. XXV (1972), 205-35; Nicholas Mayhew, "Numismatic Evidence and Falling Prices in the Fourteenth Century," *Ibid.*, 2nd ser. XXVII (1974), 1-15; Nicholas Mayhew, "The Monetary Background to the Yorkist Recoinage of 1464-1471," *British Numismatic Journal*, XLIV (1974), 62-73; John Day, "The Decline of a Money Economy: Sardinia in the Late Middle Ages," in *Studi in memoria di Federigo Melis*, 3

monetized metals? (2) When did this monetary contraction commence: in the late-thirteenth, early-, mid-, or late fourteenth century? And when did it terminate? (3) Was this contraction ever serious enough to constitute a "bullion famine:" i.e., was there a serious scarcity of coined money in relation to the transactions demand for money? Did the frequent coinage debasements and the spread of credit instruments counteract any deflationary effects of reductions in bullion supplies? (4) What was the relationship between monetary factors and general economic movements — *conjoncture* — in late-medieval Europe? In particular, were monetary factors a primary or even a secondary cause of the so-called "Great Depression" of late medieval Europe? Or was a contraction in monetary flows merely a passive response to, an adaptation to, general demographic and economic contraction? Did monetary factors behave, as Marc Bloch suggested, like a "seismograph that not only registers earth tremors but sometimes brings them about"?[2]

This last question, the most complex of all, is quite beyond the scope of this modest study, and indeed transcends the theme of this volume on international monetary flows. But one can hardly pose the first three questions without being aware of the fourth; and any subsequent answers to the fourth question will ultimately depend on answers given to the first three. Answers to these questions in turn will depend upon how we interpret the admittedly meagre quantitative data at our disposal, chiefly mint statistics. In doing so, it is imperative to distinguish clearly between those forces that instigated bullion flows and those that resulted in monetary contraction; between transient and purely regional monetary scarcities produced by trade and bullion flows and a long-term, secular decline in Europe's total specie stocks.

The causes of monetary contraction and specie outflows

For many historians, the earliest and principal cause of a general European monetary contraction per se was a severe slump in European silver mining,

---

vols. (Florence, 1978), III, 155-76; John Day, "The Great Bullion Famine of the Fifteenth Century," *Past and Present*, no. 79 (1978), 3-54. The most impressively documented study so far on this issue, Day's article appeared the year following the presentation of my paper to the 1977 Wisconsin Conference, and after I had prepared a paper on "Monetary Contraction and Industrial Change in the Late-Medieval Low Countries" for the Third Oxford Symposium on Coinage and Monetary History, held in September 1978. My arguments, essentially unchanged from those two conferences of 1977-78, are similar to Day's on monetary contraction itself, though we differ on minor points.

2. Bloch, "Problem of Gold," p. 186. See also Carlo Cipolla, "Currency Depreciation in Medieval Europe," *Economic History Review*, 2nd ser. XV (1963), 413-22 [reprinted in *Change in Medieval Society: Europe North of the Alps, 1050-1500*, ed. Sylvia Thrupp (New York, 1964), pp. 227-36]; R. S. Lopez and H. A. Miskimin, "The Economic Depression of the Renaissance," *Economic History Review*, 2nd ser. XIV (1962), 408-26; Carlo Cipolla, R. S. Lopez, and H. A. Miskimin, "Economic Depression of the Renaissance? Rejoinder and Reply," *Ibid.*, XVI (1964), 519-29; and the writings of Miskimin in n. 1 above.

from the late thirteenth or early fourteenth centuries, variously in Saxony, Bohemia, the Tyrol, Alsace, England, and Sweden. Prof. John Nef has argued that because there had been no real progress in mining technology since Classical times, indeed because medieval mining and drainage technology were generally inferior to that of the Romans, diminishing returns would soon afflict any mining ventures.[3] Whether or not mines became physically depleted, rising marginal costs—or quite simply, rising waters and flooding —were bound to force many mines to halt production in this era. To be sure, as some other historians have countered, new mines, especially gold mines, were opened in Hungary, Bosnia, Serbia, Bohemia (Kutna Hora for silver), and Sardinia.[4] But Nef does not agree that their combined outputs offset the sharp decline in production from older centres in Saxony, Bohemia, and elsewhere in Europe. More recently Prof. John Day has contended that production from new mines compensated for the decline of the old, if at all, only until the 1360s, when some of the new mines, especially at Kutna Hora and in Sardinia, also began to decline.[5] Europe did not in fact enjoy a genuine silver mining boom, comparable to that of the twelfth and thirteenth centuries, again until the 1460s, when South German mining engineers achieved two crucial technological breakthroughs: mechanical drainage pumps to eliminate flooding, and a lead-chemical process to separate the metals in those argentiferous-cupric ores that were so very plentiful in Central Europe.[6] Naturally one is tempted to believe that a "bullion famine" and high metal prices provided the necessary economic incentives to produce these innovations.

In the post-war years, however, this "mining slump" theory did not fare so well (until recently) in the literature on late-medieval economic changes, thanks in particular to the persuasive writings of Prof. Michael Postan.[7]

---

3. John U. Nef, "Mining and Metallurgy in Medieval Civilisation," *Cambridge Economic History of Europe*, II (ed. M. M. Postan, Cambridge, 1952), 456-69; and his "Silver Production in Central Europe, 1450-1618," *Journal of Political Economy*, XLIX (1941), 575-91. See also Schreiner, "Wages" (n. 1 above), pp. 67-71.

4. D. Kovačević, "Les mines d'or et d'argent en Serbie et en Bosnie médiévales," *Annales: E. S. C.*, XV (1960), 248-58; Graus, "Crise monétaire," pp. 450-2; Day, "Bullion Famine," pp. 35-7; Day, "Sardinia," pp. 156-65; Andrew Watson, "Back to Gold—and Silver," *Economic History Review*, 2nd ser. XX (1967), 30-2; Marian Malowist, "Problems of the Growth of the National Economy of Central-Eastern Europe in the Late Middle Ages," *Journal of European Economic History*, III (1974), 331-57.

5. Day, "Bullion Famine," p. 35.

6. Nef, "Mining," pp. 458-73 and "Silver Production," pp. 575-91.

7. In particular, Michael Postan, "Rapport," *IXe congrès international des sciences historiques* (2 vols. Paris, 1950), I, 225-41 [revised and republished as "The Economic Foundations of Medieval Society," in his *Essays on Medieval Agriculture and General Problems of the Medieval Economy* (Cambridge, 1973), pp. 3-27, especially 7-13]; "The Trade of Medieval Europe: The North," *Cambridge Economic History*, II (1952), 191-222, especially 211-16; and his "Note" to W. C. Robinson's "Money, Population, and Economic Change in Late-Medieval Europe," *Economic History Review*, 2nd ser. XII (1959), 77-82. Another of his arguments, a contentious one too complex to discuss here, is that only demographic and not monetary factors would explain divergent trends in grain prices and industrial prices and wages. See also n. 22 below.

While in no way denying that medieval silver mining did suffer a slump, he firmly rejected the contention that any decline in mined output could in itself have produced a bullion famine. In the first place, he argued, no mining slump could have seriously diminished the European money supply when Europe's total stock of precious metals, amassed over two or three centuries of economic expansion, was "two hundred to five hundred times its annual accretions." In the second place, most countries did not possess gold or silver mines and were consequently dependent upon foreign trade and their international balance of payments for their monetary stocks. Indeed for much of the medieval and early-modern era Europe generally received much more gold from its African trade than from its own mines. In any event, Postan observed that in late-medieval Europe "changes in trade balances did not synchronize with the ups and downs of silver mining" and that England in particular continued to enjoy a large influx of precious metals long after the mining slump had commenced.

Then came his final point, constituting the most devastating refutation of the "mining slump" thesis: the very severe declines in European population, of 30 to 40 percent during the fourteenth and fifteenth centuries, which must have increased the per capita supplies of precious metals. Or in the *aggregate* terms of the Fisher Identity, M.V. = P.T: the fall in the volume of transactions (T) resulting from depopulation and various economic dislocations of the era would have been much greater than any decline in money stocks (M) attributable to the mining slump. Consequently, unless coinage velocity of circulation (V) also declined sharply, the general price level (P) should have risen. In fact, during the quarter-century following the Black Death, much of Europe did experience very considerable inflations. As David Herlihy has so aptly observed, "men were dying, but coins were not."[8]

But Postan's fundamental contention that late-medieval Europe possessed vast accumulated stocks of immortal moneys, more than adequate for a contracted, depopulated economy, has recently been undermined by several studies contending that medieval coins were indeed highly perishable. The principal authors, C. C. Patterson, Marion Archibald, and Nicholas Mayhew, thus suggest a second reason for a late-medieval monetary contraction, a more powerful one when combined with the mining-slump thesis: precious metal losses from normal wear and tear in coinage circulation, from ship-

---

8. David Herlihy, *Medieval and Renaissance Pistoia: The Social History of an Italian Town, 1200-1430* (New Haven-London, 1967), pp. 125, 122-47; Karl Helleiner, "Population Movement and Agrarian Depression in the Later Middle Ages," *Canadian Journal of Economics and Political Science*, XV (1949); E. J. Hamilton, *Money, Prices, and Wages in Valencia, Aragon, and Navarre, 1351-1500* (Cambridge, Mass. 1936), appendices; Léopold Genicot, "Crisis: From Middle Ages to Modern Times," *Cambridge Economic History of Europe*, I (rev. edn. 1966), 677-94 (tables 5-17); Harry Miskimin, *Money, Prices, and Foreign Exchange in Fourteenth Century France* (New Haven, 1963), pp. 53-71; Schreiner, "Wages," pp. 69-73; E. H. Phelps Brown and Sheila Hopkins, "Seven Centuries of the Prices of Consumables, Compared with Builders' Wage-Rates," *Economica*, new ser. XXIII (1956) [reprinted in *Essays in Economic History*, ed. E. M. Carus-Wilson, II (London, 1962), 193-6.]

wrecks, from unretrieved hoards, indeed from "losses in the fire" during recoinages.[9] Patterson believes that the average annual rate of precious metal loss from such causes was as much as 1.0 percent per annum. Subsequently Mayhew, citing earlier analyses of Sir John Craig, adopted a much more conservative estimate of 0.2 percent weight loss per annum; nevertheless he calculates that in early fourteenth century England such a loss rate would have meant that "seven tons of silver vanished into thin air through wear alone every decade."[10] Consequently these authors contend that unless a country's money supply was frequently and abundantly replenished with fresh mintings it would inevitably, and over a century drastically, diminish.

With insufficient mining production, was late-medieval Europe collectively able to fuel its mints and fully replenish its coinage supplies from its balance of payments with the rest of the world? The simple answer, according to many historians, is no. Indeed in the most recent analysis of this issue, John Day contends that the principal if not sole cause of the late-medieval "Great Bullion Famine" was a severe balance of payments deficit with the Islamic Near East.[11] I believe Day's thesis—in fact a quite traditional thesis—to be a thoroughly sound one, particularly because it can be buttressed by some additional statistical evidence from the meticulously researched publications of the eminent Orientalist Prof. Eliyahu Ashtor.[12] By his calculations for the later fifteenth century, Western Europe's annual deficit amounted to about 400,000 gold ducats: the difference between total oriental imports from Syria and Egypt valued at 660,000 ducats per annum on the average (= 2,345 kg. pure gold) and total European exports to the Islamic East valued at an annual average of 260,000 ducats (=925 kg. gold). For the second half of the fifteenth century, that average annual specie drain to the East, the equivalent of 1,420 kg. pure gold or about 16,500 kg. pure silver, exceeded by some 30 percent the *combined* annual average mint outputs of both England and the Burgundian Low Countries: the equivalent of 1,093 kg. gold or 12,703 kg. silver (Tables 7, 9).

9. Patterson, "Silver Stocks," pp. 205-34; Mayhew, "Numismatic Evidence," pp. 1-15; Marion Archibald, "Wastage from Currency: Long-Cross and the Recoinage of 1279," in *Edwardian Monetary Affairs (1279-1344)*, ed. Nicholas Mayhew (*British Archeological Reports* 36, Oxford, 1977), pp. 167-86; Sir John Craig, *The Mint: A History of the London Mint from A. D. 278 to 1948* (Cambridge, 1953), pp. xvi, 60.

10. Mayhew, "Numismatic Evidence," p. 3. It should be noted, however, that an annual average weight loss of 0.2 percent would mean a loss of 1.83 percent in the first decade—not 2.0 percent; and in the second decade, a loss of 3.32 percent, not 4.0 percent.

11. Day, "Bullion Famine," pp. 5-12, 35-40.

12. For the following see Eliyahu Ashtor, *Les métaux précieux et la balance des payements du Proche-Orient à la basse époque* (Paris, 1971), pp. 65-96; Ashtor, "The Venetian Supremacy in Levantine Trade: Monopoly or Pre-Colonialism?," *Journal of European Economic History*, III (1974), 5-53; Ashtor, "The Volume of Levantine Trade in the Later Middle Ages (1370-1498)," *Ibid.*, IV (1975), 573-612; Ashtor, "Observations on Venetian Trade in the Levant in the XIVth Century," *Ibid.*, V (1976), 533-86; Ashtor, *A Social and Economic History of the Near East in the Middle Ages* (London, 1976), pp. 319-31. See also Day, "Bullion Famine," pp. 7, 10-11. The Venetian ducat = 3.56 g. pure gold.

Nevertheless, before agreeing that such a specie outflow produced or contributed to a bullion famine in late-medieval Europe, we must answer a number of rather crucial questions. First, since Day himself dates the onset of this "bullion famine" only from the 1390s—much later than does any other monetarist historian in this debate—must we assume therefore that trade deficits with the Near East became a serious problem only from that era? That question, with its overtones of negative doubt, naturally arises from a long tradition of historical scholarship maintaining that Europe in almost every era from the ancient Classical period to modern times has suffered a chronic payments deficit with "The East"—with the possible exception of the mid-thirteenth century.[13] Ashtor, however, does cautiously suggest that the European deficits resulted in serious specie drainages only from the late fourteenth or early fifteenth centuries. His investigations indicate that Venetian imports from Syria and Egypt grew by at least a third during the fifteenth century.[14] Even though some of that expansion was at the expense of other ports in the Levantine trade, principally Genoa, Marseilles, and Barcelona, western specie exports may well have expanded considerably, to reach those very high levels in the late fifteenth century. Secondly, if western imports of oriental goods evidently increased during the fifteenth century, did European exports to the Near East also decline? Both Ashtor and Prof. Abraham Udovitch have concluded that the Levantine market in this era seriously diminished as the result of depopulations, deflation, and depression that afflicted Mamluk Egypt in particular. Ashtor notes that Italian exports to the Near East changed in their composition from higher priced goods to become "mainly olive oil, cheap and dried fruits, copper (for small coins), and cheap cloth." Udovitch believes that the Levant itself also suffered an overall balance of payments deficit with a growing outflow of specie to India and other Asian states, principally in payment for oriental luxury goods and spices—as much for domestic consumption as for re-export.[15] Prof. Jere Bacharach (in this volume) would also agree that Mamluk Egypt suffered from economic depressions and crises, a growing deficit in its balance of payments, severe monetary difficulties, a "silver famine" in particular, from ca. 1400. Bacharach also ascribes part of Mamluk Egypt's

13. See Robert Lopez, "Back to Gold, 1252," *Economic History Review*, 2nd ser. IX (1956), 219-41; and also in this volume Thomas Walker, "The Italian Gold Revolution of 1252: Shifting Currents in the Pan-Mediterranean Flow of Gold" and Louise Robbert, "Monetary Flows: Venice, 1150 to 1400 A.D."

14. See n. 12 above, especially Ashtor's "Volume of Levantine Trade," pp. 605-12; "Venetian Supremacy," pp. 5-17, 23-40, 46-53.

15. A. L. Udovitch, "England to Egypt, 1350-1500," *Studies in the Economic History of the Middle East*, ed. M. A. Cook (London, 1970), pp. 115-28: Ashtor, "Venetian Supremacy," and his *Economic History*, pp. 280-331. In his *Métaux précieux* (1971), pp. 13, 41-53, 97-110, Ashtor had however attributed the increasingly severe monetary scarcity of fifteenth-century Egypt to Mamluk military expenditures abroad and especially to hoarding, not to the balance of trade, which he believed remained favourable overall for the Levant. But subsequent views expressed in "Volume of Levantine Trade" (1975) are more in accordance with Udovitch's (see p. 612).

balance of payments deficit to imports of European manufactured goods, in apparent contradiction of Ashtor's observations.[16] But the Levant was much more than just Egypt; and if fifteenth-century Egypt was as economically depressed and crisis-ridden as all these authorites state, its market could hardly have consumed enough European goods to offset Europe's evident deficits.

Next, one must inquire whether Western Europe had earlier been able to finance its (previously smaller) trade deficits with the Levant from trade surpluses with other parts of the world; and then found itself unable to earn such surpluses, from the later fourteenth century. In particular did Italian revenues from their trade with Tunisia, the Maghreb, and Morocco—once Europe's chief source of gold—diminish in the later Middle Ages?[17] Did other indigenous African factors sharply diminish the flow of 'Sudanese' gold from the Niger basin: in particular, as Day contends, the disintegration of the Mali Empire and nomadic disruptions of Saharan trade routes, from the later fourteenth century?[18] Furthermore, did Western Europe also develop a trade deficit with the eastern Baltic and Russia, especially during the fifteenth century, in purchasing grain, lumber, naval stores, and luxury furs, while failing to expand its textile exports?[19] Day himself believes that, although Europe's growing balance of payments deficit with the Near East remained the principal cause of the late-medieval monetary contraction, that scarcity became an acute "bullion famine" only from the 1390s when the combination of a severe diminution in Sudanese gold supplies and the worsening European mining slump further reduced precious metal stocks.

Another vigorous proponent of the thesis of a late-medieval bullion famine who assigns an even greater responsibility to Europe's balance of payments deficit with the East, indeed the rest of the world, is Prof. Harry Miskimin, also a contributor to this volume.[20] Of all the participants in the current debate he is the most original. For he offers a tightly organized economic model that explains *why* that balance of payments deficit became so

---

16. Jere Bacharach, "Circassion Monetary History: Silver," *Numismatic Chronicle*, 7th ser. XI (1972), 267-81; Bacharach, "The Dinar versus the Ducat," *International Journal of Middle East Studies*, IV (1973), 77-96; Bacharach, "Monetary Movements in Egypt, 1171-1517" in this volume; Boaz Shoshan, "Exchange-Rate Policies in Fifteenth-Century Egypt," forthcoming (draft version kindly communicated to me, 1981).

17. See Lopez, "Back to Gold," pp. 232-40; Bloch, "Problem of Gold," pp. 212-18; Ashtor, *Métaux précieux* (n. 12), pp. 65-96; Marian Malowist, "Quelques observations sur le commerce de l'or dans le Soudan occidentale au moyen âge," *Annales: E. S. C.*, XXV (1970), 1630-6; E. W. Bovill, *The Golden Trade of the Moors*, 2nd ed. (London, 1968), pp. 13-44, 79-131.

18. Day, "Bullion Famine," pp. 36-40.

19. As suggested by Miskimin in "England to Egypt," pp. 102-04 and in his *Economy* (n. 1), pp. 138-42. See also Artur Attman, *The Russian and Polish Markets in International Trade, 1500-1650* (Goteborg, 1973), pp. 103-93 (especially 103-18 on influxes of specie into Russia from the 11th century).

20. See the works of Miskimin cited in n. 1 above; and also in this volume, his "Money and Money Movements in France and England at the End of the Middle Ages."

severe, from perhaps the mid-fourteenth century, in the context of a thesis that also seeks to explain the nature of the late-medieval "Great Depression." He has adroitly avoided the pitfalls of other, earlier monetarists such as W. C. Robinson, who rashly argued (1959) that an earlier monetary contraction, ca. 1290-ca. 1330, was directly responsible for that Depression.[21] His arguments were devastatingly demolished by Postan, who has long maintained, just as strongly, that demographic decline, various fourteenth-century dislocations and other "real" factors were instead the chief culprits—and that the Depression commenced well before the Black Death of 1348.[22] In almost full accordance with Postan's thesis, Miskimin also assigns prime responsibility to demographic decline, though much more to the Black Death, demonstrating with economic models how falling population first produced a European-wide agrarian depression, whose chief economic consequence was to alter the barter terms of trade sharply in favour of the urban industrial sector, thus "making it a magnet for the nation's bullion." Without contradicting the logic of the Postan thesis, he further argued that another, more immediate consequence of the mid-century depopulations was to increase per capita wealth, in particular permitting some survivors to become very prosperous from inherited cash balances. Miskimin contends that the general reaction of such survivors to the various plagues and other catastrophes of this era was to spend those cash balances rapidly in a spirit of unbridled hedonism. The previously discussed inflations following the Black Death are certainly evidence supporting his thesis. Miskimin's model in particular seeks to demonstrate that such a hedonistic consumption pattern, as it became firmly implanted in late-medieval European society, led to an increased importation of costly luxury goods: from Italy into northern Europe, and from the Near East into Europe as a whole, chiefly by Italian trade. The result was "a growing outflow of bullion that in turn generated the monetary shortage so much commented upon after the Black Death." Miskimin's ultimate conclusion is that a protracted, worsening monetary scarcity exacerbated and prolonged the medieval "Great Depression" into the late fifteenth century.[23] This second part of his thesis has not, of course, won favour with the "real school" followers of Postan, who firmly deny that any monetary factors contributed

---

21. Robinson, "Money, Population, and Economic Change," pp. 63-76, followed by Postan's devastating "Note," pp. 77-82.

22. See above n. 7, and also M. M. Postan, "Some Economic Evidence of Declining Population in the Later Middle Ages," *Economic History Review*, 2nd ser. II (1950), 221-46 [reprinted in his *Essays* (n. 7), pp. 186-213]; Postan, "Medieval Agrarian Society in Its Prime: England," *Cambridge Economic History*, I (2nd edn. 1966), 560-70; Postan, *The Medieval Economy and Society: An Economic History of Britain, 1100-1500* (London, 1972), pp. 27-40, 224-46. See also n. 24 below.

23. Quotations from Miskimin, "Monetary Movements," p. 490. Some additional support for Miskimin's theses of the "inheritance effect" and "hedonistic consumption pattern," along with a more highly skewed income distribution after the depopulations, can be found in Herlihy, *Pistoia* (n. 8), pp. 86-105, 134-44; and in R. S. Lopez, "Hard Times and Investment in Culture," *The Renaissance: A Symposium* (New York, 1953), pp. 19-32.

to that Depression.[24] But such a debate, beyond the scope of this study, cannot be resolved here—or indeed elsewhere, I suspect.

Miskimin's thesis, so directly based upon English evidence, in particular contradicts Postan's assertion that late-medieval England consistently enjoyed favourable trade balances, so that there was "every reason why silver should have continued to be imported."[25] More recently Terence Lloyd has supported Postan's contentions, at least for most of the fourteenth century. I have found Lloyd's analysis of the English customs accounts to be both brilliant and provocative; but having myself worked with the enrolled customs accounts in the London Public Record Office from the 1270s to the 1520s I cannot accept his conclusions for reasons that I have elaborated elsewhere.[26] In any event, Lloyd himself admits that England's "favourable" trade balances were probably diminishing from the 1360s; and he thus implicitly also admits the possibility that England did suffer trade deficits in the fifteenth century. For in that century England's chief exports, raw wool and woollen cloths, indisputably suffered a severe decline, despite temporary upsurges in the 1420s and early 1440s. Such exports reached their dismal nadir in the 1460s, with an aggregate volume that was only 47 percent of the exports achieved in that Indian Summer boom of the early 1390s—and just 38 percent of the late 1350s' export volume.[27]

While it remains impossible to make definitive statistical assertions about English balance of payments in the late Middle Ages, Miskimin did support his contentions with impressive and voluminous citations of sumptuary, anti-luxury, anti-Italian, and bullionist legislation. Indeed from both official ordinances and popular writings one can amplify his list quite considerably. In my view, the period of most intense bullionist legislation was from the 1360s, when Parliament first banned the export of both precious metals without royal licences, to about the 1490s.[28] As early as 1381, a parliamentary commission concluded that a combination of foreign coinage debasements

24. See in particular John Hatcher, *Plague, Population, and the English Economy, 1348-1530* (London, 1977), pp. 47-54; and M. M. Postan and John Hatcher, "Population and Class Relations in Feudal Society," *Past and Present*, No. 78 (1978), 24-37.
25. Postan, "Trade of Medieval Europe," pp. 212-13.
26. T. H. Lloyd, "Overseas Trade and the English Money Supply in the Fourteenth Century," *Edwardian Monetary Affairs (1279-1344)*, ed. Nicholas Mayhew (Oxford, 1977), pp. 96-124; John H. Munro, "Monetary Contraction and Industrial Change in the Late-Medieval Low Countries, 1335-1500," in *Coinage in the Low Countries (880-1500): Third Oxford Symposium on Coinage and Monetary History*, ed. Nicholas Mayhew (*BAR* International 54, Oxford, 1979), pp. 97-8, 126 nn. 13-14.
27. Calculated from E. M. Carus-Wilson and Olive Coleman, eds. *England's Export Trade, 1275-1547* (Oxford, 1963). Such a decline could not have been offset by higher values for wool and cloth, nor by wool export duties, which were already at virtually their highest levels by the late 1330s, and which were in fact reduced in the 1420s. See n. 60.
28. John Munro, "Bullionism and the Bill of Exchange in England, 1272-1663: A Study in Monetary Management and Popular Prejudice," in *The Dawn of Modern Banking* (Center for Medieval and Renaissance Studies, University of California; New Haven-London, 1978), pp. 169-243.

(see below) and a deficit in the balance of payments—excessive imports, excessive papal taxes, and other specie transfers by Italian merchants—were primarily responsible for the current and truly serious dearth of coinage (Tables 1-3).[29] Parliament subsequently responded by enacting two sets of remedies to stem this (supposed) external haemorrhaging of specie. The most sophisticated and impressive were a series of Employment Acts (1390-1487) that required all aliens to export as much in value in English goods as they imported or remitted abroad by bills of exchange. More deleterious and counter-productive were several bullionist impositions on the wool export trade, which variously forced English merchants (Staplers) to exact full payment in ready English money and bullion to be coined at the king's mints.[30] There is no evidence that such measures were at all successful in protecting and augmenting England's money supplies; and by often hindering trade seriously they may actually have curbed English bullion inflows. But such legislation remains very significant in demonstrating late-medieval England's very genuine concerns about the money supply—and in blaming Italian importers for that dearth.

Further evidence that both England and the cross-Channel Low Countries experienced chronic balance of payments deficits with Italy in the later fourteenth and fifteenth centuries can be found in the late Raymond De Roover's analyses of Italian commercial and bills of exchange transactions. But he also discovered that Flanders often enjoyed a payments surplus in its Spanish trade, thus permitting some merchants to settle Italian debts by tri-lateral bills that drew on accumulated credits in Spain.[31] Yet it hardly seems likely that Spanish trade could compare in value with the Italian trade of this era, to permit north-western Europe as a whole to balance its deficits with the rest of the Mediterranean—above all, the Near East![32]

As sympathetic as I am, however, to the Miskimin thesis on specie outflows, I would not ignore the importance, albeit much lesser importance, of other factors in explaining a late-medieval monetary contraction. Apart from the two previously discussed, the European mining slump and the constant "leakage" from physical losses of metal, there are two more factors

---

29. *Rotuli Parliamentorum*, III, 126-27: nos. 1-2. See Munro, "Bullionism," pp. 169-90 for its long-term significance.

30. Munro, "Bullionism," pp. 169-243; J. H. Munro, *Wool, Cloth, and Gold: The Struggle for Bullion in Anglo-Burgundian Trade, 1340-1478* (Brussels-Toronto, 1973), chapters 1, 2, 4; Eileen Power, "The Wool Trade in the Fifteenth Century," in *Studies in English Trade in the Fifteenth Century*, ed. E. Power and M. Postan (London, 1933), 79-90.

31. Raymond De Roover, "La balance commerciale entre les Pays Bas et l'Italie au XVe siècle," *Revue belge de philologie et d'histoire*, XXXVII (1959), 374-86; De Roover, *L'evolution de la lettre de change, XIVe-XVIIIe siècles* (Paris, 1953), pp. 64-5; De Roover, *The Bruges Money Market Around 1400* (Brussels, 1968).

32. See also Wendy Childs, *Anglo-Castilian Trade in the Later Middle Ages* (Manchester, 1978), chapters 3-4; G. A. Holmes, "Florentine Merchants in England, 1346-1436," *Economic History Review*, 2nd ser. XIII (1960), 193-208; E. B. Fryde, "Anglo-Italian Commerce in the Fifteenth Century," *Revue belge de philologie et d'histoire*, L (1972), 345-55.

to consider, both variations on one theme: hoarding and other circumstances that reduced coinage velocity; and an increased industrial use of precious metals. According to J. D. A. Thompson and others, proportionately more European coin hoards can be attributed to the late Middle Ages, the later fourteenth century in particular, than to the adjacent eras.[33] Yet for the specific shorter period of 1440-1465, which Day and others (myself included) believe to be the most severe phase of the bullion famine, very few coin hoards have been documented.[34] Certainly in the later fourteenth and early fifteenth centuries hoarding can be regarded as perfectly natural reactions to the various plagues and epidemics, famines, wars, peasant jacqueries and urban rebellions, economic dislocations and general insecurity that afflicted so much of Europe. Indeed one would expect to find a significant decline in coinage velocity in general in late-medieval Europe with such depopulations, such a large drop in the number of spenders, with market contractions and periodically severe depressions. Hence the previous question in revised form: did coinage velocity fall just passively in response to this economic contraction, in direct proportion to the decline in the volume of transactions (T); or did a fall in V exacerbate that economic contraction? Thus increased hoarding may have been again a rational response to the deflations (however caused) that beset many parts of Europe from the 1370s and 1380s. But extensive hoarding would then have accelerated that deflation, which in turn would have squeezed profit margins, bred a pessimistic business psychology, and curtailed investments and production. In particular, monetary scarcity and deflation would have constricted investment by increasing the real burden of debt, both the cost of previously borrowed capital in the repayments and of future borrowing in possibly higher interest rates.

Furthermore, various government reactions to depressed trade, monetary scarcity, and fiscal stringency in protectionist and bullionist legislation, especially the almost universally imposed bans on precious metal exports, undoubtedly interfered with the international flow of specie and depressed coinage velocities in the late Middle Ages. On the other hand such bullionist restrictions undoubtedly also encouraged a greater use of bills of exchange, letters

33. J. D. A. Thompson, *Inventory of British Coin Hoards, 600-1500 A.D.* (Oxford, 1956), p. xxvi and passim. See also Jean Lafaurie, *Les monnaies des rois de France: Hughes Capet à Louis XII* (Paris, 1951); Peter Spufford, *Monetary Problems and Policies in the Burgundian Netherlands, 1433-1496* (Leiden, 1970), pp. 55-73 (but see also Appendix D: Coin Hoards, 1425-40, pp. 203-13); Day, "Bullion Famine," pp. 44-5. See also Erik Aerts and Herman Van der Wee, *The Leuven Coin Find of 1851 and the Currency of the Burgundian Netherlands in the Middle of the Fifteenth Century: A Case Study* (Katholieke Universiteit te Leuven, Centrum voor Economische Studiën, 1980, in mimeo), pp. 1-37.

34. For late medieval price trends and variations, see the sources cited in n. 8 above; and also, for the Low Countries, Herman Van der Wee, *The Growth of the Antwerp Market and the European Economy, Fourteenth to Sixteenth Centuries*, 3 vols. (The Hague, 1963), I (Tables, from the 1360s) and III (Graphs); and his "Prices and Wages as Development Variables: A Comparison between England and the Southern Netherlands, 1400-1700," *Acta Historiae Neerlandicae*, X (1978), 58-78.

obligatory, and other credit instruments to obviate the physical transport of specie; and such credit instruments would admittedly have increased coinage velocity by economizing on the use of specie within countries. But surely the use of such international credit instruments in the fourteenth and fifteenth centuries was largely restricted to a very few international centres dominated by Italian commerce. It has yet to be proven that the use of credit instruments, most of them still tied to specie payments, expanded sufficiently and universally to counteract fully the various forces of monetary contraction.[35]

Such arguments concerning hoarding and coinage velocity are not, however, necessarily a contradiction of the Miskimin thesis, provided that one specifies differences between periods and social classes. Thus the widespread, hedonistic spending sprees in the inflationary post-Black Death era of ca. 1350-1375 may have been followed by a dolorous parsimony and hoarding amongst some of the population in that subsequent, deflationary era from the late 1370s. Some segments of society—nobles, upper echelons of the Church, gentry, wealthy bourgeoisie, even "kulak" peasants may have continued to engage in conspicuous consumption of imported luxuries throughout both eras. Poorer peasants and artisans, on the other hand, likely responded to crises and chronically bleak times by increased hoarding: the more so that plagues, epidemics, and wars became more fixed in the rhythm of later medieval life.

Furthermore, the increased industrial consumption of gold and silver in church and other building construction and in resplendent luxury display, including brocaded textiles, which so many art historians of late-medieval Europe have described, is fully in accord with the Miskimin thesis.[36] Such display still constitutes hoarding of metals, albeit in a more aesthetic and useful form. Of course hoarding, even in this respect, differs sharply from all other forms of metal losses discussed earlier in that hoarded metals can easily be restored to monetary circulation as minted coins (with some obvious exceptions). As the mint accounts discussed below will demonstrate, and as Postan rightly reminds us, there were occasional, perhaps frequent circumstances at times, in some regions, that stimulated or forced dishoarding:

---

35. Even for the 16th-17th centuries, Frank Spooner contends that "the arrival and distribution of bullion . . . was enough to set in motion a series of credit transactions under the virtual monopoly of the great fairs and commercial centers. And yet the opposite conditions could also occur: an unexpected delay in the arrivals of bullion could precipitate a sudden spate of bankruptcies." *The International Economy and Monetary Movements in France, 1493-1725* (Cambridge, Mass. 1972), p. 3. See the writings of De Roover in n. 31 above; Munro, "Bullionism," pp. 169-243; Paul Einzig, *History of Foreign Exchange*, 2nd edn. (London, 1970), pp. 76-112; Michael Postan, "Credit in Medieval Trade' and 'Private Financial Instruments in Medieval England," both reprinted in his *Medieval Trade and Finance* (Cambridge, 1973), pp. 1-27, 28-64.

36. See Miskimin, *Economy*, pp. 92-104, 134-44; the works of Lopez and Herlihy in n. 23 above; Johan Huizinga, *The Waning of the Middle Ages* (London, 1926), pp. 140-52; Philip Ziegler, *The Black Death* (London, 1969), pp. 240-88; Françoise Piponnier, *Costume et vie sociale: la cour d'Anjou, XIVe-XVe siècle* (Paris, 1970), chapters 7-10; Agnes Geijer, *A History of Textile Art* (London, 1979), pp. 141-55.

government taxes and forced loans, purveyances, etc. in time of war, ransoms, booty and plunder, and even those late-medieval cyclical movements that temporarily restored some prosperity, as in the 1420s.

Short-term and long-term forces in international bullion flows

Insofar as we all must rely heavily on mint and coinage statistics for evidence of late-medieval Europe's monetary contraction and bullion flows, we must distinguish carefully between those forces that merely circulated metals amongst regional mints, or caused only temporary regional coinage scarcities, and those that produced a permanent outflow of bullion from Europe. And insofar as we assert that a drainage of specie to the East was the major if not the sole cause of that contraction, we must understand how the mechanics of minting influenced the form or nature of those specie outflows.

First we return to the issue of "leakage," of physical deterioration of coins through wear and tear, and especially also from the nefarious but common practice of clipping and "sweating" coins to extract bullion. Such metallic losses could further contract the domestic money supply by curtailing, even halting, domestic mint production and by instigating bullion outflows. Thus since silver coins in particular normally circulated by "tale," at decreed face values, and not by weight, merchants would collectively respond to a significant presence of underweight or fraudulent coins in circulation by discounting the entire coinage, bad and good alike: that is, by bidding up all prices. Once the market price for bullion had risen above the mint price, the premium that coins normally commanded over bullion, by virtue of their portability and convenience, would disappear. Without the premium, necessary to cover at least the minting costs (brassage and seignorage), merchants would instead export their bullion to foreign mints offering better prices, and would cull out good, full weight coins for foreign trade.[37] The coinage standard and the money supply would thus continue to diminish until the prince either: (1) imposed a total recoinage on his subjects at their expense, as a *renforcement* (strengthening) that would also further contract the money supply—with better but necessarily fewer coins; or more likely, (2) he debased the coinage in weight, perhaps also in fineness, to match the inferior standard of the current circulating coinage. Such a debasement, in striking more but inferior coins from the same amount of metal, necessarily raised the mint's money-of-account price for bullion (and the value of metal in the coin) above the market price, thereby restoring the coinage premium. If the new mint price was

---

37. See Feavearyear, *Pound Sterling* (n. 1), pp. 10-15; John Munro, "An Aspect of Medieval Public Finance: The Profits of Counterfeiting in the Fifteenth-Century Low Countries," *Revue belge de numismatique*, CXVIII (1972), 127-48; N.J. Mayhew and D. R. Walker, "Crockards and Pollards: Imitation and the Problem of Fineness in a Silver Coinage," *Edwardian Monetary Affairs*, ed. N.J. Mayhew (Oxford, 1977), pp. 125-46; and Mavis Mate, 'Monetary Problems in England, 1272-1307', *British Numismatic Journal*, XLI (1972), 37-45 (with analyses of such discounting and consequent price increases).

raised high enough, it would succeed in attracting foreign bullion and specie as well. Foreign princes, faced with a consequent outflow of bullion and a closure of their own mints, could normally respond by similar debasements. But debasements were frequently undertaken also for aggressive as well as for these defensive reasons: to produce seignorage revenues, as the prince's own tax on coinage.[38] Unfortunately, debasements, whether defensive or aggressive in origin, provoked retaliatory *guerres monétaires*. Indeed Western Europe, after having enjoyed a century of monetary stability until Philip the Fair's aggressive debasements of 1295-1305, was thereupon plagued by two centuries of such monetary wars and mint manipulations.[39]

One aspect of medieval mint manipulation that is crucially important for understanding the nature of international bullion flows was the deliberate alteration of the so-called mint-ratio in order to attract either gold or silver by "favouring" the one metal over the other. The mint could do so either by physically debasing the one coinage *more* than the other, in weight and/or fineness, by physically strengthening that coinage in question *less* than the other (in a *renforcement*), or simply by raising its money-of-account and exchange value. But any influx of the one metal so "favoured" would normally be mirrored, though not necessarily in proportion, but an efflux of the other metal in effecting international payments. For that other metal would now be relatively "favoured" by those foreign mints (and money-changers) that had maintained the former mint-ratio, or had raised it in favour of that other metal. Such flows would seem to be a reflection of Gresham's Law in which the cheap, i.e., debased and "favoured," metal drives out the dear metal.

Thus in that seminal article of 1967 discussed in the editor's introduction, "Back to Gold—and Silver," Andrew Watson contended that late-medieval Europe, by adopting strongly pro-gold mint ratios, minted large quantities of this metal, while suffering large outflows of silver to the Islamic Near East, whose mints generally "favoured" that metal just as strongly. Watson also

---

38. See A. Blanchet and A. Dieudonné, *Manuel de numismatique française*, II (Paris, 1916), chapters 3-7; A. Grunzweig, "Les incidences internationales des mutations monétaires de Philippe le Bel," *Le moyen âge*, LIX (1953), 117-72; Cazelles, "Mutations," pp. 83-105; Girard, "Guerre monétaire;" Hans Van Werveke, "Currency Manipulation in the Middle Ages: The Case of Louis de Male, Count of Flanders," *Transactions of the Royal Historical Society*, 4th ser. XXXI (1949), 115-27 [reprinted in his *Miscellanea Mediaevalia* (Ghent, 1968), pp. 255-67]; Munro, *Wool, Cloth, and Gold* (n. 30), chapter 1. In that work, on pp. 14-23, I had argued against the theses of Carlo Cipolla (n. 3) and Henri Laurent (n. 1) that later medieval coinage debasements were undertaken chiefly to remedy scarcities of specie. Citing Postan et al to deny that later-medieval Europe ever experienced any such scarcity, I instead contended that the princes' principal motivation for aggressive, autonomous debasement was to acquire seignorage profits. Obviously my subsequent researches, providing me with a greater spatial and temporal perspective, have forced me to alter my views on late-medieval money supplies, by 180 degrees; nevertheless, I still maintain my former views on the principal causes of aggressive and defensive debasements.

39. Watson, "Back to Gold," pp. 1-34 (esp. 31-4).

Bullion flows and monetary contraction

contended that, while late-medieval Europe might have experienced a "silver famine" of its own making, there was no general monetary scarcity, as silver outflows were generally marched by gold inflows. On both issues, Watson's views have been attacked by several historians—not surprisingly by Miskimin in particular.[40] I myself have argued that any medieval international arbitrage trade in precious metals had to surmount a number of quite formidable barriers.[41] In the first place, no mint would act as a market to exchange metals freely in either bullion or coined form. The operative mint ratio was not that generally given in the literature—the ratio of official coined values (the *traite* ratio)—nor of the two mint prices for bullion: instead it was the ratio of the mint's price for bullion in the one metal to the coined or *traite* value of the other metal, to take account of the mintage fees. Furthermore any merchant seeking to trade in metals would have had to pay money changers' fees and transport costs; and he would have had to face the prospect of receiving worn or clipped coins in return; or of having his bullion and coin seized by customs inspectors and frontier guards, or by pirates and brigands— would have had to pay insurance costs, in effect. Consequently the difference between two countries' effective mint ratios had to be very large indeed before merchants could have hoped to cover all these costs and to profit from a purely arbitrage trade in precious metals. Therefore I believe that very, very rarely indeed were such mint-ratio alterations and arbitrage trades in themselves able to instigate international bullion flows in the late-medieval world.[42]

Nevertheless it would be equally wrong, in my view, to dismiss the significance of mint-ratios in medieval bullion flows. For even small differences in such ratios, or even *traite* ratios, indisputably had a powerful influence in determining which of the two metals would predominate in a country's foreign trade receipts and expenditures, and thus in its annual mint outputs. For merchants buying goods in several countries would seek to have

---

40. Miskimin, *Money*, pp. 117-19 (on this aspect of Gresham's Law) and "Money and Money Movements" in this volume; Ashtor, *Métaux précieux* (n. 12), pp. 31-6 (though a qualified agreement with Watson on p. 51); Ashtor, "Recent Research on Levantine Trade," *Journal of European Economic History*, II (1973), 193; see also Day, "Bullion Famine," pp. 5-6. But a position similar to Watson's, though for a more limited time and area, can be found in R. H. Bautier, "L'or et l'argent en Occident de la fin du XIIIe siècle au début du XIVe siècle," *Académie des inscriptions et belles lettres: comptes rendus* (1951), 169-74.

41. See John Munro, "Mint Policies, Ratios, and Outputs in the Low Countries and England, 1335-1420: Some Reflections on New Data," *Numismatic Chronicle*, 8th ser. I (1981), 71-116 (especially Appendix A, 95-9); *Wool, Cloth, and Gold* (n. 30), pp. 29-32, and "Bullionism," pp. 185-7, in which I am much less resolute in denying that mint-ratio alterations could instigate bullion flows.

42. Watson (nn. 4, 39 above) does cite evidence, particularly from Jacques Coeur's trial of 1453 (pp. 20-2), to demonstrate that silver was being shipped to the Levant precisely to profit from such an arbitrage trade in metals; but the evidence does not prove that such bullion shipments were divorced from concurrent trading ventures in goods. See n. 41 above.

their gold and silver bullion coined in those mints offering the higher purchasing power for each metal. Thus their arbitrage profits would be realized not directly through an exchange of metals but indirectly through normal channels of international trade: exchanging metal for goods and ultimately goods for metal. In a recent article, I have sought to demonstrate this indirect arbitrage in an economic model employing actual trade, price, and coinage statistics for England and Flanders in 1396.[43] Eventually, one would suppose, the market would adjust the prices of both the metals and the goods, would impose equilibrium prices, to eliminate such disparities and thus the arbitrage profits. But with the primitive transport and communication facilities of the late Middle Ages, with so many impediments to free trade, these very imperfect markets made equilibrium adjustments very, very slowly indeed.

Finally, to complete this theoretical section, we must realize that the effect of any given debasement (or *renforcement*) upon mint outputs, the country's "net" money supply, and international bullion flows are all quite unpredictable. The consequences of a debasement could depend upon a complex combination of the following factors: (1) how much of the current coinage was actually surrendered and reminted into the more plentiful but inferior new coinage; (2) how much domestic bullion from dishoarding, and how much foreign bullion and specie would be attracted by higher mint prices (as in the previous analysis); (3) how much of a balance of trade surplus might then be earned, if any, from any increase in exports and any reduction in imports resulting from an accompanying fall in the exchange rate; (4) how much inflation then ensued from the debasement and increase in the money supply to curb exports and encourage imports; (5) how much of the newly debased coinage might be more profitably spent abroad, in places as yet unaware of the debasement, especially if such debased coins were counterfeits of much more respectable earlier coins or of good, popular foreign coins:[44] (6) how much of the other, now "disfavoured" metal was sent abroad; (7) how much inflation and monetary instability from excessive debasements encouraged hoarding or a flight of capital. All of these caveats must be kept clearly in mind in considering the following, final section.

Monetary contraction and bullion outflows: Trends in the mint outputs of late-medieval England and the Low Countries

Faute de mieux, we are all necessarily dependent upon mint production statistics for quantitative evidence of late-medieval monetary contractions and bullion flows. In the Appendix, I discuss a number of technical problems in utilizing these mint data; and explain why I believe that my mint statistics

---

43. Munro, "Mint Policies," pp. 95-9.
44. On this point see Munro, *Wool, Cloth, and Gold* (n. 30), chapter 2; and Munro, "Profits of Counterfeiting."

for England (1235-1500) and the Low Countries (1335-1500) are more complete and accurate than those that were employed in the articles of Harry Miskimin (1964) and John Day (1978). My English mint statistics, furthermore, cover a far longer time-span than did Miskimin's (1273-1470) and Day's (1340-1500).

A comparison of the mint outputs for these two regions in the late Middle Ages is particularly illuminating because of the similarities and differences in their economic and monetary experiences. On the one hand, their economies were closely intertwined, more so than with any of their other neighbours.[45] The Low Countries were by far the most important market for English wools, and by the mid-fifteenth century also became the most important market for English woolens as well (except in Flanders, where they had long been banned). Through the fairs of Flanders and Brabant, Bruges and Antwerp, the Low Countries were also one of England's chief sources of imported manufactured goods. On the other hand, their "states" of economic development were certainly different. The Low Countries were the most highly industrialized, commercialized, and urbanized region in northern Europe; England, despite its important cloth industry and trade, was still essentially agrarian. Furthermore, the monetary experiences of the two regions were also radically different. England's was indeed unique. No other late-medieval European country (of importance) maintained stable coinages for such long periods—fifty to sixty years at a time, so rarely engaged in mint-manipulation, and so consistently favoured a gold coinage, from its adoption in the 1340s. The Low Countries had a much more typically medieval European monetary history, with cycles of often frenetic debasements, sometimes terminating in brutal *renforcements*, and with periodic alterations of the mint-ratio to favour one or the other metal. Consequently, we wish to determine if these economically related regions demonstrated similar long term trends in their mint outputs, despite different stages of economic development and despite their radical monetary differences.

In testing a hypothesis of monetary contraction and bullion outflows, we must use the mint statistics as only very general indicators of trends, since they can hardly serve as year-to-year statistical proxies of the money supply, for the reasons already explained. Above all, we must clearly distinguish between those mint outputs that resulted from balance of payments surpluses and from increased levels of economic activity on the one hand, and those, on the other hand, from mint-manipulations in debasements, *renforcements* and mint-ratio alterations. Sometimes the differences are obvious; sometimes the distinctions cannot be clearly made.

For England, with its remarkable monetary stability, the relatively few mint alterations are simply summarized: (1) recoinages to eliminate underweight and fraudulent coins in 1247-49, 1279-82, 1300-02, and 1421-22

---

45. See J. H. Munro, "The Costs of Anglo-Burgundian Interdependence," *Revue belge de philologie et d'historie*, XLI (1968), 1228-38; *Wool, Cloth, and Gold* (n. 30), pp. 1-10, 181-7.

(partially); and (2) coinage debasements in 1335, 1345-46, 1351, 1411, and 1464-65; only those of 1335 and 1411 were purely defensive. For Flanders and then the Burgundian Low Countries, as Table 10 summarizes, the forty-year period 1349-1389 was punctuated by 22 debasements of the gold coinage and 18 debasements of the silver, most of them aggressive, with one abortive *renforcement* in 1384. A highly successful if brutal *renforcement* was imposed in 1389-90 (gold by 41.7 percent and silver by 31.8 percent). Thereafter, Flanders maintained a perfectly stable, strong money policy (indeed with another small *renforcement* in 1409) until 1416.[46] From then until 1433, the Burgundian Low Countries suffered from another round of often drastic debasements (12 for gold, 7 for silver), especially in the years 1426-33. Then a *renforcement* and general unification of the Burgundian coinage in 1433-34 was followed by 33 years of strong, stable money, interrupted only in 1454 by a minor debasement of gold. During the last third of the century, from 1466 to 1497, the Low Countries were again plagued by aggressive debasements, 22 of them, especially during the misrule of Archduke Maximilian and the civil war years of ca. 1483-92. Two *renforcements* intervened, in 1484 and 1489; a final modest debasement of 1496 (adjusted 1497-1500) successfully ushered in a quarter-century of monetary stability (until February 1521).[47]

As a cursory examination of the tables and graphs reveals, all of these mint alterations were associated with some increases in coinage outputs, some great and some small. Indeed the truly massive English silver outputs following the recoinages of 1247-49, 1279-82, and 1300-02 are, so to speak, especially striking.[48] Certainly they produced far and away the largest silver outputs recorded in our entire series. But in this period England had no gold to compete with silver; and since English sterlings then served as an international medium of exchange in northern Europe, these large outputs may reflect for this reason an international demand for English silver coinage.[49] Nevertheless the generally high levels of coinage output, in terms of both kilograms of metal and sterling values, in the thirteenth century may be taken as evidence of that era's plentiful money supply.

46. Munro, "Mint Policies," pp. 78-95.
47. Spufford, *Monetary Problems*, pp. 141-6; Munro, *Wool, Cloth, and Gold*, pp. 65-180; Van der Wee, *Antwerp Market*, I, 127-8.
48. The thirteenth-century silver coinages and especially the outputs resulting from the recoinages of 1247-49 and 1279-82 are even greater than those indicated in Table 1, because we lack the accounts for a number of provincial mints. Marion Archibald has estimated that such missing provincial mints accounted for perhaps 50 percent of total English coinage in 1247-50; for about 5 percent, from 1250-1278; and for 20-25 percent from 1278 to 1290: "Wastage from Currency," pp. 167-86.
49. Jules Chautard, *Imitations des monnaies au type esterlin frappées en Europe pendant le XIIIe et XIVe siècle* (Nancy, 1871); S. E. Rigold, "The Trail of the Easterlings," *British Numismatic Journal*, XXVI (1949-51), 31-55; Mayhew and Walker, "Crockards and Pollards," pp. 125-46; N.J. Mayhew, "The Circulation and Imitation of Sterlings in the Low Countries," *Coinage in the Low Countries (880-1500)*, ed. N. J. Mayhew (Oxford, 1979), pp. 54-68.

Bullion flows and monetary contraction 115

As the graph shows, there are three other major, if less impressive, coinage upsurges common to both England and the Low Countries; and these upsurges are only partly explained by mint policies: those of ca. 1350-70, ca. 1415-35, and ca. 1465-80. In each of them, coinage alterations initially played a significant role. For Flanders in particular, during the first of these upsurges, the mints' debasements were very drastic: gold by 36.5 percent and silver by 43.3 percent, from 1350 to 1369.[50] But the same cannot be said for England, which terminated its brief experimentation with aggressive debasement at the very outset of this period, in July 1351: thereafter England maintained a perfectly stable coinage for 60 years exactly, until 1411.[51] Nor can the Flemish debasements in themselves be called upon to explain the equally impressive English mint outputs, since all foreign coins imported into England had to be reminted anyway, whether they were good or bad.[52] Thus since the mint outputs for both countries in this period ca. 1350-70 are so very large and so remarkably similar, we must look for additional explanations. These should include the following: (1) the Miskimin thesis concerning the "inheritance effect" of the Black Death and subsequent depopulations: the aforementioned hedonistic spending sprees of inherited cash balances, associated with Death, Disaster, and Display; (2) a remarkably strong boom in both English wool and cloth exports and also in Flemish cloth production for most of this period: in part stimulated by devaluations, by English royal finances, especially royal intervention in the wool export trade to generate large revenues, and also by a general boom in the European economy;[53] (3) bullion flows from royal taxation and military expenditures in this very active phase of the Hundred Years War; (4) the profits of war and associated piracy, in ransom and plunder; (5) dishoardings periodically encouraged by these factors. Finally, since England had only just recently adopted a gold coinage (1344), fully establishing it in 1351, the veritable explosion in English gold

50. See Munro, "Mint Policies," pp. 76-8, 110-11 (Table 8); Van Werveke, "Currency Manipulation," pp. 255-67; and Tables 4-6, 10, below.
51. In January 1352, Parliament responded to Edward III's recent debasements (1344-51) by forbidding the crown to impair the coinage in any way (25 Ed. III stat. 5 c. 13, in *Statutes of the Realm*, I, 322; *Rot. Parl.* II, 240: n. 32). Subsequently, after a long and severe deterioration of the coinage, an influx of foreign counterfeits (Galley Halfpence), and "la graunde Escarcité de Money" finally forced the crown to adopt a defensive debasement, Henry IV first had to obtain Parliament's assent, in November 1411. (*Rot. Parl.* III, 658-9: no. 28; *S. R.* II, 163, 168). See Feavearyear, *Pound Sterling* (n. 1), pp. 15-39; Munro, "Bullionism," pp. 186-91.
52. See Munro, "Bullionism," pp. 187-96. One of the purposes of establishing the Calais mint in 1363 was to facilitate such recoinages of specie from Anglo-Flemish trade.
53. See Munro, "Low Countries," pp. 110-12, 138 (Table 1); Hans Van Werveke, "De economische en sociale gevolgen van de munt politiek der graven van Vlaanderen (1337-1433)" and his "Currency Manipulation" both in *Miscellanea Mediaevalia* (n. 38), pp. 243-54, 264-6; E. M. Carus-Wilson, "Trends in the Export of English Woollens in the Fourteenth Century," *Economic History Review*, 2nd ser. III (1950) [reprinted in her *Medieval Merchant Venturers* (London, 1954), pp. 247-52 (though I do not agree with all her references to Flanders).] Miskimin, furthermore, has shown that for France the years 1355-75 were also ones of very large coinage outputs: *Money*, pp. 87-8, 90-4, 100-15.

mint outputs might be regarded as that generational "one shot" coinage of England's long accumulated stocks of gold plate, jewelry, bullion, and foreign coins, some of which had already served a monetary function in the economy. But such a hypothesis could hardly apply to Flanders, where a gold coinage had been implanted much longer, since 1335.

For the second of these major coinage upsurges, somewhat similar explanations can be offered as for the first, except the "inheritance effect" thesis. For England, the initial major factor was undoubtedly the debasement (gold by 10 percent, silver by 16.7 percent) and general recoinage from 1411-12 which possibly took several years to complete. That was followed in 1421-22 by an official (though unverified) recoinage of all underweight or fraudulent specie.[54] For the Burgundian Low Countries, certainly the debasements of 1426-33 and the *renforcement* recoinages of 1433-34 again played a very major role in their large coinage outputs. But again, with such a similarity in the upsurges, we must look to additional factors, for both countries:[55] (1) a significant expansion in Anglo-Burgundian trade in general, following upon the English conquest of Normandy (1415-17), Burgundian seizure of Paris (1417), and their Treaty of Troyes alliance, in 1420; (2) a boom in the English cloth trade, while both the English wool trade and the Low Countries' textile industries temporarily halted a steep decline that had begun in the 1370s; (3) taxation and military expenditures in this very active phase of the Hundred Years' War; (4) coinage of booty and ransoms from both English and Burgundian campaigns in northern France; (5) for England, the re-opening of the Calais mint from July 1422 (closed since 1404). One of its chief tasks, following upon the partial recoinage of that year, was to prevent the importation of fraudulent coin by requiring that all foreign specie received from wool sales at the Calais Staple be reminted there into good English coin.[56] In 1430 the English crown, responding to Burgundian debasements and counterfeiting of English gold nobles, imposed its most onerous bullionist laws yet on the Calais Staple: forbidding the use of credit, requiring payment in good English coin for two thirds of the wool price, and in bullion for the remainder.[57] As Table 1 indicates, the Calais mint had been spectacularly successful from the very beginning, vastly exceeding the London Tower Mint in coinage outputs in the 1420s and early 1430s. But the 1430 bullion laws had only a very limited, short term success in maintaining coinage outputs. Insofar as they severely damaged the English wool trade, which plunged by over 50 percent, provoked a Burgundian ban on English cloth,

---

54. 9 Hen. V stat. 1 c. 11 in *S.R.* II, 208. The recoinage, principally of underweight gold nobles, was mintage-free.

55. See Munro, *Wool, Cloth, and Gold*, chapter 3; Munro, "Low Countries," pp. 112-14 and 151 (Table 15).

56. *Calendar of Patent Rolls, 1422-29*, p. 337; Munro, *Wool, Cloth, and Gold*, pp. 75-92.

57. 8 Hen. VI c. 18 in *S.R.* II, 254-5; *Rot. Parl.* IV, 359: no. 60. See Power, "Wool Trade," pp. 82-3; Munro, *Wool, Cloth, and Gold*, pp. 74-126; T. H. Lloyd, *The English Wool Trade in the Middle Ages* (Cambridge, 1977), pp. 257-72.

and then became an issue in the Anglo-Burgundian war of 1436-40, they were clearly self-defeating. Calais mint outputs dropped dramatically from 1433, ceasing entirely by 1440.[58] Other reasons for that latter decline will be considered subsequently.

For the third and final upsurge in mint outputs under consideration, ca. 1465-1480, coinage debasements and monetary ordinances possibly played a relatively greater role, particularly because warfare was so much less a factor than in the two previous periods—apart from the duke of Burgundy's ultimately fatal campaigns against Louis XI of France and the Swiss. For England certainly the single most important cause of its coinage upsurge in the 1460s was Edward IV's debasements of 1464-65: silver by 20 percent and gold by 26 percent (in total). His reimposition of the Calais bullion laws, in modified form, in 1463 may have contributed to those high mint outputs.[59] According to many historians, these debasements were also important in stimulating, through foreign exchange devaluation, a revival and then strong boom in English cloth exports, which continued almost unabated into the sixteenth century.[60] That cloth export boom may have contributed to, but did not long sustain any high level of English mint outputs after the 1470s.

For the Low Countries, obviously one must cite the impact of the previously mentioned, very drastic coinage debasements and *renforcements* from 1466 to 1497. But in Prof. Herman Van der Wee's view, the initial silver debasements of the 1460s and 1470s were especially important for another reason: in offering such a high price and exchange-ratio for silver, they were very instrumental in attracting South German merchants with large quantities of silver from their recently developed Central European mines to the Brabant Fairs.[61] According to Prof. Nef, South German mining technology and enterprise succeeded in increasing that silver production five-fold from the 1460s and the 1530s.[62] Table 4 indicates a dramatic upsurge in Brabantine silver coinage outputs (at Antwerp) from the 1470s, surpassing the Flemish coinage outputs for the last two decades of the fifteenth century. These events

---

58. Munro, *Wool, Cloth, and Gold*, pp. 93-126; Lloyd, *Wool Trade*, pp. 261-70. See also Table 1 below.

59. *C. P. R. 1461-67*, pp. 370-1; Craig, *The Mint*, pp. 72-5; Munro, *Wool, Cloth, and Gold*, pp. 155-79; Lloyd, *Wool Trade*, pp. 278-81.

60. Van der Wee, *Antwerp Market*, II, 82-3. The English cloth trade to the Low Countries, however, was interrupted by a Burgundian ban from 1464 to 1467. By this era, lost markets or lost access routes to other European regions had forced the English to concentrate more and more of their cloth exports on the Antwerp market. See Munro, "Industrial Protectionism in Medieval Flanders: Urban or National?" in *The Medieval City*, ed. H. A. Miskimin, et al (New Haven-London, 1977), pp. 246-53 and Tables 4-5; Munro, "Low Countries," pp. 116-22; Munro, *Wool, Cloth, and Gold*, pp. 155-86; M. M. Postan, "Economic and Political Relations of England and the Hanse, 1400-1475," in *Studies in English Trade*, pp. 151-3.

61. Van der Wee, *Antwerp Market*, II, 80-3, 104-05.

62. See n. 3 above.

also benefited the English insofar as their cloth trade boom was directly dependent upon expansion of the Brabant Fairs and South German trade. Undoubtedly the Low Countries' mint outputs could have been even higher had it not been for the very destructive effects of the revolt against Maximilian (1483-1492). Indeed Maximilian's debasements and *renforcements* were both so drastic that they may well have encouraged a flight of bullion and capital.[63] At the same time, in view of the previous discussion of contemporary Mediterranean economic and monetary history, and of the Miskimin and Day theses, one wonders whether the influx of this new South German silver into the Low Countries was still not exceeded by an outflow of bullion through trade to Italy and the Islamic East.

What conclusions may be drawn about these three coinage upsurges? The strong emphasis given to the role of coinage debasements may well lead one to believe that they so distorted late-medieval mint statistics that no trends can be safely gleaned from them.[64] As the tables and graphs suggest, so great were the fluctuations evidently engendered by debasements, even when combined with other factors, that the standard deviations of mint outputs for various periods were generally greater than their corresponding means, especially (and surprisingly) in the English mint data. Nevertheless the most striking feature of the three upsurges as viewed together in the graphs is the very dramatic diminution in the intensity of those upsurges. Indeed, if one believes that debasements hopelessly bias the mint data, it is most instructive to compare the relative power of such debasements to draw metals into the English and Flemish mints at various periods. To make such a comparison comprehensible, with two metals and fluctuations in their money-of-account values—with the problem of combining monetary "apples and oranges"—I have converted all the mint output data into a "constant" pound sterling, using the official English gold and silver values from 1351 to 1411.[65] I have chosen this period because it neatly bisects the whole period under discussion (from 1235) and because it represents the longest period of stability (sixty years) in medieval English monetary history. The consequent gold to silver ratio of 11.16:1 is certainly not unrealistic for late-medieval Europe. All of the means are comparable: decennial outputs of aggregate gold and silver coinages; and the decades chosen are those of the most intensive monetary changes in debasements or recoinages. The decade 1420-9 has been included for England because of all the special monetary circumstances discussed above:

63. Spufford, *Monetary Problems*, pp. 141-6; Van der Wee, *Antwerp Market*, I, 127-8; and II, 105-11.

64. Such is the criticism that R. D. Ware made of Miskimin's "Monetary Movements," in the appended "Discussion," *Journal of Economic History*, XXIV (1964), 491-5.

65. From 24 June 1351 to 29 November 1411, 1 Tower Pound (= 349.9144 g.) of gold at 23 7/8 carats (99.479166% pure) was worth £15.0s.0d. ster., so that 1 kilogram of pure gold = 1000 × £15.00/349.9144 × 0.99479166 = £43.092065. Similarly, 1 Tower Pound of silver at 11 oz. 2 dwt = 92.50% pure was worth 25s. (£1.25), so that 1 kilogram of pure silver = £3.861948. See n. 51 above and Table 10 below.

Bullion flows and monetary contraction

A comparison of mint outputs during monetary changes

Decade	ENGLAND Mean Annual Value of Total Mint Outputs in Constant £ Sterling	Percentage of 1350-9 Mean
1300–09	113,054	95.9
1350–59	117,867	100.0
1410–19	63,579	53.9
1420–29	85,004	72.1
1465–74	65,392	55.5

Decade	LOW COUNTRIES Mean Annual Value of Total Mint Outputs in Constant £ Sterling	Percentage of 1355-64 Mean
1355–64	141,026	100.0
1390–99	41,040	29.1
1425–35	83,035	58.9
1470–79	45,112	32.0
1480–89	33,141	23.5

Those who would deny that there was any monetary contraction in the late Middle Ages, and also that there was any relation between mint outputs and money supplies, would have to explain why so much less metal passed through the mints of both countries during recoinages or debasements in the fifteenth than in the fourteenth century—and in the case of the Low Countries so much less than through the *Flemish* mints alone in the mid-fourteenth century. Surely Archduke Maximilian and Edward IV were as avaricious as were Count Louis de Mäle of Flanders and Edward III of England a century earlier?

If we are to contend that there was such a late-medieval monetary contraction (a contention I myself had once opposed),[66] when did it commence? By the late thirteenth century, as Robinson suggested? Or by the early fourteenth century, well prior to the Black Death, as Nicholas Mayhew contends?[67] Tables 1, 2, 3, and 9 do offer some support for those contentions:

66. See Munro, *Wool, Cloth, and Gold*, pp. 14-23; and n. 38 above.
67. Robinson, "Money," pp. 63-76; Mayhew, "Numismatic Evidence," pp. 1-15.

that the mean English coinage output (constant pounds sterling) in the half century 1250-99 was 39.2 percent higher than the corresponding mean in 1300-49; that English mint outputs were very low indeed in the 1290s and especially in the 1320s and 1330s. When measured in terms of the above-defined constant pound sterling, the mean English mint ouput in the 1330s was lower than in any decade of the fifteenth century. But the first of these disastrous minting slumps, in the 1290s, a very brief one, can be attributed to: (1) an influx of counterfeit sterlings from the Low Countries, the famous Pollards and Crockards; and (2) Edward I's diplomatic and military expenditures abroad, estimated by Prestwich at £350,000 in 1294-98.[68] The subsequent general recoinage from 1300, to eliminate those counterfeit sterlings, produced the largest silver and the second largest aggregate mint output in English medieval history. The severe minting slump in the second period, the 1320s and 1330s, may also be explained in part by another influx of counterfeit sterlings and other debased coins from the continent.[69] A more complete explanation would include Edward III's very lavish expenditures abroad, first on diplomacy and then on warfare, at the outset of the Hundred Years' War. One historian has estimated that the bullion outflow in the later 1330s from such financing was about one million pounds; another termed this outflow a veritable "sterling crisis."[70] But there is no hard evidence that such a dearth of coinage was at all general in Europe. Flemish silver coinage outputs in 1335-39 (the earliest accounts available) averaged 91 percent of those in 1350-54; incomplete French mint accounts do show a very large silver coinage at the end of the 1330s.[71] The fact remains in any event that

---

68. Michael Prestwich, "Edward I's Monetary Policies and Their Consequences," *Economic History Review*, 2nd ser. XXI (1969), 409-12; Mavis Mate, "Monetary Problems," pp. 57-8, 63-6. On counterfeit sterlings, see n. 49 above. For the subsequent, massive coinage outputs, from 1300, see also Mavis Mate, "High Prices in Early Fourteenth-Century England: Causes and Consequences," *Economic History Review*, 2nd ser. XXVIII (1975), 1-16.

69. Mate, "High Prices," pp. 12-15; Cazelles, "Mutations," pp. 88-94, 256-62 (French debasements of 1323-9, 1337-43); Miskimin, *Money* (n. 8), pp. 143-44 (Appendix C); Joseph Ghyssens, "Le monnayage d'argent en Flandre, Hainaut, et Brabant au début de la guerre de Cent Ans," *Revue belge de numismatique*, CXX (1974), 109-91; Hans Van Werveke, "Currency Manipulation," pp. 255-67. For English complaints about foreign counterfeit coins, see *Foedera, conventiones, litterae et acta publica, 1066-1383*, ed. Thomas Rymer et al (4 vols. Record Commission, London, 1816-69), II.i, 544 (1324), II.ii, 814 (1331); *S.R.* I, 273-4 (1335: 9 Ed. III stat. 2, "Statute of York"); *Calendar of Close Rolls 1341-43*, pp. 685-6 (1342); *S.R.* I, 299 (1343: 17 Ed. III c. 1); *Foedera* III.i, 16-17 (1344); *Rot. Parl.* II, 160: nos. 15-16 (1346).

70. See Michael Prestwich, "Currency and the Economy of Early Fourteenth Century England," in *Edwardian Monetary Affairs (1279-1344)*, ed. N. J. Mayhew (*BAR* 36, Oxford, 1977), pp. 45-58 (who also attributes the silver coinage slump to silver exports in exchange for gold, not yet legally monetized); Mate, "High Prices," pp. 12-15; Mayhew, "Numismatic Evidence," pp. 11-15; E. B. Fryde, "Financial Resources of Edward III in the Netherlands, 1337-40," *Revue belge de philologie et d'histoire*, XL (1962), 1168-87; XLIV (1967), 1142-1216; E. B. Fryde, 'Loans to the English Crown, 1328-1331', *English Historical Review*, LXX (1955); Feavearyear, *Pound Sterling*, pp. 15-20; and especially Edward Ames, "The Sterling Crisis of 1337-39," *Journal of Economic History*, XXV (1965), 496-522.

71. Miskimin, *Money*, pp. 90 (fig. 13), 93 (fig. 14), 158 (App. D); Cazelles, "Mutations," pp. 83-105. See Tables 4-6 below.

this deep trough in English minting was followed by a sharp upswing in the later 1340s and then by truly enormous outputs of coinage in the period 1350-69, as demonstrated earlier, with the largest gold and one of the largest silver outputs recorded in medieval England. Certainly the available supply of precious metals had remained most abundant.

Consequently, I would date the onset of a late-medieval monetary contraction, in terms of a definite secular trend, from the late 1360s or early 1370s, certainly for both England and the Low Countries. Although Day would prefer to commence a definite bullion famine from the 1390s, my dating of the onset of general contraction seems to be quite consistent with his published graphs; only the Florentine downswing on his graph commences later, from the 1390s.[72] Similarly, despite Miskimin's strong emphasis on the role of the Black Death, and a suggested commencement of contraction in the 1350s, my dating does not conflict with the logic of his thesis. I certainly agree that a general demographic and economic contraction, a secular depression, may well have had its true beginnings long before this monetary contraction. Nevertheless, my other research in late-medieval economic history also indicates to me that whatever the true origins of such a "depression," the most severe and prolonged phase of economic contraction also commenced about the 1370s and lasted to perhaps the late 1460s in northern Europe, though in some regions certainly (e.g., Flanders) until the early sixteenth century.[73]

All of my statistical investigations suggest that in such secular terms the century 1370-1470 also marked the most severe monetary contraction for both England and the Low Countries, when their aggregate gold and silver coinages are evaluated in "constant" pounds sterling (1351-1411 values) and in current moneys-of-account. The graphs and Tables 1-6 lend strong support to the views of Day and Peter Spufford that the most acute bullion famine, the true nadir of this monetary contraction was in the 1440s and 1450s (ignoring the Burgundian gold debasement of 1454).[74] From the late 1440s to the mid 1460s also appears to be the period of the most severe, acute depression in northern Europe during the fifteenth century.[75] But monetary contraction itself continued to ca. 1500.

72. Day, "Bullion Famine," Graph II, pp. 50-1.
73. See Munro, "Low Countries," pp. 95-161; Munro, "Industrial Protectionism," pp. 247-53; Carus-Wilson, "Trends," pp. 252-61; Van der Wee, *Antwerp Market*, II, 7-18, 28-30, 289-95, 309-14; and G. A. Holmes, *The Estates of the Higher Nobility in Fourteenth-Century England* (Cambridge, 1957), pp. 115-20; and especially for the 1370s as the watershed, A. R. Bridbury, "The Black Death," *Economic History Review*, 2nd ser. XXVI (1973), 577-92 (esp. 584-5, 592).
74. Day, "Bullion Famine," pp. 40-6; Spufford, *Monetary Problems*, pp. 115-21. See also n. 83 below.
75. Munro, "Low Countries," pp. 109-10, 115-18, 151-2 (Tables 12-3); Munro, *Wool, Cloth, and Gold*, pp. 131-54, 181-6; Van der Wee, *Antwerp Market*, II, 61-73; H. L. Gray, "English Foreign Trade from 1446 to 1482," in *Studies in English Trade*, pp. 1-38; Postan, "Economic and Political Relations of the Hanse," *Ibid.*, pp. 149-53; E. M. Carus-Wilson, "The Iceland Trade" and "The Overseas Trade of Bristol," *Ibid.*, pp. 155-246; Postan, "Trade," pp. 246-56.

My statistical comparisons of mint outputs in England and the Low Countries are summarized in Table 9, which presents their long-term trends in terms of both fifty- and hundred-year means. Note that for England, even in terms of debasement-inflated current pounds sterling, the annual mean for 1370-1469 of the total value of coinages struck is only 71.0 percent of the corresponding mean for the preceding century 1270-1369; in terms of the constant pound sterling, to eliminate the effects of debasement, the English mean of total coinage struck in 1370-1469 is only 55.8 percent of 1270-1369 mean. If one were to agree that the *greater* part of the demographic decline had already occurred by the 1370s, from the major plagues, then it becomes very difficult to defend the notion that per capita money supplies were maintained during the later fourteenth and fifteenth centuries. But admittedly there are eminent historians who do contend that English population continued to plunge drastically, not just to decline slowly, until the later fifteenth century.[76]

If we now turn, however, to more useful fifty-year means for both England and the Low Countries, the pattern of decline for both from the mid- to later-fourteenth century through both halves of the fifteenth century is much clearer, much starker. (Since Flemish mint statistics begin only in 1335, we obviously cannot make the same century-long comparisons as for England). Indeed when we combine gold and silver coinages together in constant pounds sterling values (1351-1411), we find that the extent of decline is on the same drastic order of magnitude for both countries. Thus in these terms the mean annual such value of total English coinage output for 1450-99 is only 39.5 percent of the comparable mean for 1350-99, while in the Low Countries the same mean for 1450-99 is just 37.0 percent of the corresponding 1350-99 mean. When we remember that in fact we are comparing combined outputs for all the Burgundian mints in the fifteenth century with chiefly Flemish mint outputs alone in the fourteenth century, the extent of that decline appears to be all the greater. Would anyone contend that the population of the two countries declined in the same proportion as evidently did the aggregate coinage outputs—by some 60-65 percent from 1350-99 to 1450-99? In terms of current moneys-of-account values, of course, the minting slump appears to be less severe. In England, the mean value of total coinage struck in these terms for 1450-99 is a perhaps more respectable 57.4

---

76. Hatcher, *Plague*, pp. 55-73. Even so, his graph on p. 71 does indicate that by his estimates by far the greatest depopulation had occurred by the late 1370s. For the Low Countries, see especially W. P. Blockmans, "The Social and Economic Effects of the Plague in the Low Countries, 1349-1500," *Revue belge de philologie et d'histoire*, LVIII (1980), 833-63, in particular rejecting the widely-cited contention that the Black Death spared much of this region, as expounded in Hans Van Werveke, *De Zwarte Dood in de zuidelijke Nederlanden, 1349-1351* (Brussels, 1950). In the fifteenth-century Low Countries, the rate of decline in their mint outputs exceeds the most severe estimates of population decline there. See also Joseph Cuvelier, *Les dénombrements de foyers en Brabant, XIVe-XVe siècles* (Brussels, 1912); Maurice Arnould, *Les dénombrements de foyers dans le comté de Hainaut, XIVe-XVIe siècles* (Brussels, 1956); Van der Wee, *Antwerp Market* (n. 34), I, 545-48 (Appendix 49), II, 7-9, 31-7, 61-7.

percent (!) of the corresponding 1350-99 mean. In the Low Countries, thanks to the severe debasements of Charles the Rash and Maximilian, the mean values of total coinage outputs in current pounds *groot* Flemish are about the same for these two periods. The significant decline in current money-of-account values of total coinage struck is instead in the first half of the fifteenth century. Thus, as suggested, the fifteenth-century remains overall a century of significant monetary contraction. It is a moot point whether the decline was more severe in the hundred years 1370-1469 or 1400-1499, in terms of the weight of both metals struck, and their values in constant pounds sterling or current moneys-of-account. But clearly the influx of Central European silver into north-western Europe in the second half of the fifteenth century did little to offset the overall trend of monetary contraction—and presumably of the bullion outflows.

If we now consider the two precious metals separately, and look at the first, far-left column of Table 9, we might think that the most drastic monetary contraction of all, over the 250-year period 1250-1499, is that of English silver: for the mean output in 1450-99 is just a miserable 14.2 percent of the mean silver output in 1250-99. But some considerable proportion of that silver decline is illusory and is accounted for by England's adoption of a gold coinage in January 1344-July 1351. Once the existing stocks of gold bullion, plate, and foreign coin had been minted, further gold coinages necessarily had to be struck at the direct expense of silver. Gold was either purchased directly with silver, or more commonly, gold was received instead of silver in foreign trade transactions.

That question necessarily returns us to the vexing problem of bimetallic ratios. When most countries in north-west Europe adopted gold coinages in the fourteenth century, they did so with a vengeance: Flanders in 1335, England in 1344, the Rhine Electors in 1356 (France had already done so in 1266).[77] For England and the Low Countries in particular the impact of imposing a strongly pro-gold mint ratio upon the relative mintings of the two precious metals can be readily discerned in Tables 3 (England) and 6 (Low Countries). Thus in England, in the first full decade of gold coinage, the 1360s, gold accounted for almost two-thirds (64 percent) of the total value of coinage struck. In the next decade, gold's share of total coinage values leaped fantastically to 95.7 percent (mean value), reaching a peak of 97.2 percent of total coinage values in the decade 1400-09. No wonder historians have commented on a "silver famine" in the later fourteenth century! England consistently maintained this pro-gold mint policy for the rest of the fifteenth century, though not always at the same drastic expense of the silver coinage.

At first glance, from Table 6, a similar if less drastic pattern emerges in Flanders' and Brabants' fourteenth-century mint outputs. Count Louis de Mäle also imposed a pro-gold mint policy on Flanders, so that gold coinages

---

77. See Watson, "Back to Gold," pp. 9-21; Shaw, *History of Currency*, pp. 1-60; Lopez, "Back to Gold," pp. 219-40.

accounted for an average of almost 75 percent of total mint output values in the 1350s, 78 percent in the 1360s, and 71 percent in the 1370s. But almost immediately after Count Louis' death in 1384, his successor Duke Philip the Bold of Burgundy made an abrupt volte-face in this policy by adopting a much more pro-silver mint-ratio. He did so as part of a sweeping monetary reform: a *renforcement* and the striking of new common coinages for both Flanders and Brabant, in fact adopting Brabant's current mint-ratio. Although that reform failed, and was temporarily followed by more pro-gold debasements, the pro-silver *renforcement* of 1389-90 fully succeeded, at the very time that France also shifted to a more pro-silver mint ratio.[78] That pro-silver mint policy was rigidly maintained in the Low Countries for the next 35 years; and the results show clearly in Table 5. Thus silver's share of total mint output values in the Low Countries rose from 32 percent in the 1380s to 56 percent in the 1390s, when the policy had been fully implemented, to a peak of 87 percent of total coinage values in the decade 1410-19.

Then in 1425-26, Duke Philip the Good of Burgundy (1419-67) made another monetary volte-face, abruptly switching back to a strongly pro-gold mint ratio, for reasons discussed elsewhere, chiefly military.[79] Despite his withdrawal from the French war and his *renforcement*-monetary reform in the 1430s, Duke Philip maintained that pro-gold mint policy until the very eve of his death, to 1466. Again the table demonstrates that between 1425 and 1465 the pattern of Burgundian minting dramatically altered in accordance with this change in ratios: so that gold's share of total mint output values rose from a mean of 37 percent in the 1420s to just over 98 percent in the 1450s. The Burgundian mint ratio in this forty-year period was in fact even more strongly pro-gold than England's, so that the English mints, against their will, were consequently favouring silver. Thus in the 1430s and again in the 1450s, silver accounted for an average of 71-72 percent of total English mint output values. Indeed England's mint at Calais, almost entirely at the mercy of Burgundian commerce, coined little gold after 1425, none from 1431, but vast quantities of silver, until the outbreak of the Anglo-Burgundian war in 1436.[80]

Then, in May 1466, the Burgundian mints effected the third, and for our period final, change in the bimetallic ratio, again to favour silver.[81] As suggested earlier, that mint-ratio alteration, with such a high price for silver, was a significant factor in luring South German silver and commerce to Antwerp. Similarly Tables 3 and 6 demonstrate once more that after this mint-ratio alteration the Low Countries were coining predominantly silver,

78. See Munro, "Mint Policies," pp. 65-88.
79. Munro, *Wool, Cloth, and Gold*, chapters 3-4, especially Graph III, on p. 71 (See p. 51 n. 28 for an explanation of the ratio lines.)
80. *Ibid.*, pp. 188 (Table A: gold), 191 (Table B: silver); and in the text, pp. 81-126.
81. *Ibid.*, pp. 155-73, especially Graph VI on p. 156. See also the Chi-Square Contingency tables, to demonstrate the importance of mint-ratios (1425-99) in Munro, "Low Countries," pp. 147-8, Table 9 (with commentary on pp. 101-02).

accounting for an average of 77 percent of total mint output values in 1470-99; and England was coining predominantly gold, though by no means in the overwhelming proportions of the previous century.

Several important conclusions can be drawn from these analyses and from another consideration of Table 9 on long-term means of the mint productions. First, even though mint-ratio alterations in themselves rarely if ever instigated international bullion flows, clearly they played a decisive role through international payments in determining which metal would predominate in a country's bullion flows and mint outputs. Secondly, the much discussed "silver famine" of late medieval Europe was as much a product of these strongly pro-gold mint ratios as it was a feature of that era's general monetary contraction. In England indeed that "silver famine" preceded the general monetary contraction by almost a generation.

Insofar as late-medieval countries did favour gold so strongly, they undoubtedly did their domestic economies a disservice. While gold, by its high and generally stable values, was the preferred metal for international trade and finance, for diplomacy and warfare, such high values severely limited its usefulness in domestic transactions. Consider that in England the smallest gold coin, the rarely struck ferlin (quarter-noble), at 20d was worth three times the daily wage of a master mason.[82] Since most countries' money-of-account pricing systems were based directly on the silver penny, not on gold, the effect of such mint policies in driving away silver was clearly deflationary. Witness again the general deflation in England during the later fourteenth and fifteenth centuries—and also in the Burgundian Low Countries during much of the pro-gold period, ca. 1435-1465.[83]

But thirdly, the so-called "silver famine" was much more a transitory and then a regional problem than is generally stated. As stressed earlier, England was the only major northern European country to pursue so consistently for so long a pro-gold policy. As we have seen, Flanders, Brabant, and France all switched to a more pro-silver mint ratio in the late 1380s. Thereafter, the Low Countries maintained that ratio for 70 of the next 110 years (i.e., except 1425-1465). Insofar as they successfully attracted silver, that policy may have contributed to supposed "silver famines" elsewhere—perhaps in the Near East (early fifteenth century), as well as in England?[84]

Fourthly and finally, as Table 9 demonstrates, while the mint outputs of both metals declined sharply in both England and the Low Countries from the later fourteenth century, in both the decline was by the greater in gold outputs! Thus in the Low Countries, the mean silver output for 1450-99 is

---

82. E. H. Phelps Brown and Sheila Hopkins, "Seven Centuries of Building Wages," *Economica*, new ser. XXII (1955) [reprinted in *Essays in Economic History*, ed. Carus-Wilson (London, 1962), 177-8: ca. 1360-1412, at 5d. per day; 1412-1532, at 6d. per day (plus food and drink on occasion). See also Bloch, "Problem of Gold," pp. 186-229.

83. Van der Wee, "Prices and Wages," pp. 58-78; Munro, "Low Countries," pp. 109-10 and Table 18 (p. 157: regressions of prices on coinage outputs). See also n. 34 above.

84. See above pp. 14-15 and n. 16.

virtually identical (99.1 percent) to that for the first half of the century, and 85.7 percent of the silver mean for 1350-99. The Low Countries' mean gold output for the period 1450-99, on the other hand, is just 62.4 percent of the mean output for 1400-49—and just a miserable 19.3 percent of the mean for 1350-99. Even in pro-gold England, the mean gold coinage output for 1450-99 was just half that for 1400-49, 49.2 percent, and only 31.8 percent of the mean gold output in 1350-99. (The English mean silver output for 1450-99 was, however, a much higher 75.0 percent of the 1350-99 mean silver output). Such a situation should have been deduced from the fact that, despite frequently cited complaints about silver scarcities in the fifteenth century, the most consistent monetary concern of that era in both countries was rising gold prices.[85]

Where was all the northern gold going? Into hoards, jewelry, plate church ornamentation? Into thin air, from wear? To Italy? To the East? Was indeed Asia's appetite for European gold just as voracious as for its silver? Presumably if Europe had a chronic deficit in its balance of payments with Asia, both metals would flow there, and not just silver. But not all questions can be fully answered by the papers published in this volume.

85. See Spufford, *Monetary Problems*, pp. 115-21; Munro, *Wool, Cloth, and Gold*, pp. 134-6; and Munro, "Low Countries," pp. 144-5, Table 7: values of fine gold on the Antwerp Market and official gold values, 1420-99. See also the perspicacious comments of John Day concerning the rising bimetallic ratios as indicators of increasing scarcity of gold relative to the silver supplies, in "Bullion Famine," pp. 40-3. I do not believe, however, that this situation can be fully explained by the opening of new European silver mines; for the decline in gold minting comes much too early and is much too drastic. More rapid declines in European gold mining and in imports of African gold are reasonable conjectures, but only conjectures.

Appendix: The mint production statistics for England, 1235-1500 and the Low Countries, 1335-1500

Miskimin's seminal article of 1964 on late-medieval monetary contraction was based fundamentally on English mint statistics for 1273-1470, specifically:[1] (a) Sir John Craig's occasionally careless rendition of the London Tower Mint accounts (in *A History of the London Mint*, 1953) published by C. G. Crump and C. Johnson (1272-1377) and then by G. C. Brooke and E. Stokes (1377-1550) in the 1913 and 1929 volumes respectively of the *Numismatic Chronicle*; (b) Crump and Johnson's publication of the Calais mint accounts from 1363 to 1384 in the same journal. Since then a number of important additional mint accounts have appeared. First and foremost, in the *British Numismatic Journal* of 1980, C. E. Blunt and J. D. Brand published the London and Canterbury mint accounts of Henry III (from the Pipe Rolls), 1220 to 1272; they are virtually complete for both mints from 1234. (The complete Canterbury mint accounts from 1272 to the mint's final closure in December 1346 were also published by Crump and Johnson). Then in my 1973 monograph *Wool, Cloth, and Gold*, I published the complete set of Calais mint production figures from July 1422 to September 1440, extracted from the K.R. and L.T.R. Exchequer accounts in the London Public Record Office. From the same sources, T. H. Lloyd supplied the hitherto unknown Calais mint accounts from January 1387 to March 1403, in his 1977 monograph *The English Wool Trade in the Middle Ages*. Subsequently, in the *Numismatic Chronicle* for 1981, I amended some of his Calais mint figures, filled a nonexistent gap in the accounts from February 1388 to January 1390, and extended the Calais mint figures to March 1404. Finally, in his 1978 monograph *The Tudor Coinage*, C. E. Challis has presented a new set of London Tower Mint accounts from November 1485 to July 1603, correcting and expanding on those accounts previously published by Brooke and Stokes in 1929.

Therefore, on the basis of these newer studies, I decided that Miskimin's hypotheses should be retested with a new, more complete, and much longer run of English mint statistics, from 1235 to 1500 (265 years). The tables presented here provide production statistics for London's Tower Mint for this entire period and from all the extant provincial mint accounts as follows: Canterbury, 1235-1346; Exeter, Newcastle, Bristol, Chester, and York, 1300-02; York, again from 1353 to 1355, 1423-24, and 1470-75; Bristol again, from 1470-72; and Calais, from 1363-1384, 1387-1404, and 1422-1440. The only important provincial mint accounts that appear to be missing are from lesser mints partaking in the recoinages of 1274-49 and 1280-82. The London mint accounts themselves are remarkably complete from 1235, with lacunae for only 15 years from then to 1500: for 1310-11, 1322, 1409-11, 1461-62, 1467-68 and 1490-94. I have deduced that only a trickle of coinage, if any, flowed from the Mint in 1409-11 and 1461-62, in each case just prior to a major debasement.

For the 1977 Wisconsin conference I also presented a set of mint statistics for the

---

1. For complete citations of all the following publications, see the sources for the ten monetary tables.

Low Countries from 1335 to 1500. As it happened, the very next year an apparently similar set of mint statistics for the Low Countries, though in graphic form, was published in John Day's article, "The Great Bullion Famine of the Fifteenth Century." A singularly brilliant tour de force, in my opinion, Day's article documented the most convincing case yet for a late-medieval monetary contraction, with coinage production statistics for France (or various regions), Navarre, Barcelona, Valencia, Sardinia, Florence, Genoa, and Lübeck, as well as for England and the Low Countries—representing a very broad geographic sweep. But I am compelled to note that his mint production figures for both England and the Low Countries are incomplete. His English statistics, which commence only in 1340, a full century after mine, are based on Sir John Craig's London figures; they differ from those of Miskimin only in that Day has used my Calais mint figures for 1422-1440. No other provincial English mint accounts seem to have been used.

Secondly, while those for the Burgundian Low Countries seem to be quite complete, based on my published figures (to 1480) and those of Peter Spufford (to 1499), they are regrettably defective in many places for the preceding century.[2] For the Flemish outputs from 1334 to 1384, Day utilized the forty-seven mint accounts of Counts Louis de Crécy and Louis de Mäle, preserved in the Algemeen Rijksarchief at Brussels, and published in 1856 by Victor Gaillard in his *Recherches sur les monnaies des comtes de Flandre*. But Gaillard was unaware that another nine mint accounts for Louis de Mäle (for various years, 1353-1365) were to be found in France's Archives départementales du Nord at Lille; and Day furthermore was evidently also unaware that Prof. Hans Van Werveke had published synopses of these nine accounts in his *De muntslag in Vlaanderen onder Lodewijk Van Male*, in 1949. Much more recently, in 1979, Wim Blockmans has published synopses of three more important Flemish mint accounts, found much earlier by his late father Frans: for Ghent, from 28 June 1335 to 16 June 1336 (?); and for Mechelen, from 7 August 1380-28 April 1381, and from then to 3 May 1382, in his essay "Devaluation, Coinage, and Seignorage Under Louis de Nevers and Louis de Male, Counts of Flanders, 1330-84," in Nicholas Mayhew, ed. *Coinage in the Low Countries, 880-1500*. This essay also corrects a number of errors by Gaillard and Van Werveke, and provides much additional information on Flemish coinage in this era.

Elsewhere, in the 1949 *Transactions of the Royal Historical Society*, Hans Van Werveke rather gloomily reported that "more than one fifth of the original [Flemish mint] records are still lacking."[3] Perhaps that comment had inspired Day to extrapolate Flemish mint data for all those years not covered by the Gaillard collection. Apart from the fact that we can now utilize another 12 accounts without such extrapolation, I do not believe that the remaining lacunae up to 1384 are that serious:

(1) 7 November 1338-16 April 1343: 4.44 years
(2) 17 October 1343-20 January 1346: 2.26 years
(3) 11 August 1347-24 December 1347: 0.37 year
(4) 21 February 1349-6 May 1349: 0.18 year
(5) 24 November-20 February 1356: 0.23 year

2. John Day, "The Great Bullion Famine of the Fifteenth Century," *Past and Present*, no. 79 (1978), 52.
3. Hans Van Werveke, "Currency Manipulation in the Middle Ages: The Case of Louis de Male, Count of Flanders," *Transactions of the Royal Historical Society*, 4th ser. XXXI (1949), 116.

(6) 11 May 1364-10 February 1365: 0.70 year
(7) 18 June 1368-21 April 1369: 0.84 year
*(8) 5 August 1370-18 June 1373: 2.87 years
(9) 26 June 1376-6 November 1376: 0.36 year
*(10) 27 June 1377-30 January 1380: 2.59 years

I fully accept Blockman's contention that gaps nos. 1-2 almost certainly reflect actual mint closures during the Artevelde revolt and Count Louis' exile.[4] Since the next five gaps (nos. 3-7) and no. 9 are all so short, for less than a year, one may also reasonably conclude that the Flemish mints were closed in those periods as well. For all these periods I have assigned a nil value in computing quinquennial or decennial means of mint outputs. The only significant gaps remaining are nos. 8 and 10, marked by an asterisk, totalling just 5.46 years. From a comparison of Flemish and Brabantine monetary history in this period, which reveals no references at all to Flemish coinages during these gaps, I strongly suspect that they also reflect mint closures.[5] In the tables on Flemish mint production, nos. 4-7, I have computed two sets of means, quinquennial and decennial, for the period of these gaps: (a) by again assigning a value of zero to these periods, dividing total output by 5 and 10 respectively, and (b) by computing a mean based only upon the actual number of recorded years of mint activity: i.e. dividing total output in 1370-74 by 2.13 and in 1375-79 by 2.74 years. These means, necessarily larger than the former, are given in square brackets. For the long-term means in Table 9, I have assigned zero values to all the gaps. In many places, therefore, my Flemish mint statistics will differ considerably from Day's. (He also used semi-log or ratio-scale graphs, which are good for demonstrating comparative rates of change, relative change; but not so good for showing absolute changes — the actual extent of decline, my primary concern here).

For fourteenth-century Brabant, the 1380s and 1390s, Day unfortunately had to rely on texts and précis in Henri Laurent's often inaccurate *La loi de Gresham au moyen âge* (1933), which provided figures only for September 1384-May 1386 and May 1394-January 1395.[6] Laurent was surprisingly unaware of some other Brabant mint accounts, and neglected to supply data from some already published accounts. I currently have in press synopses of all the available Brabantine mint accounts, from the earliest extant archival records up to the Burgundian coinage unification of 1433-35, as follows: July 1375-March 1383 (Leuven), September 1384-May 1386 (Leuven), June 1392-January 1395 (Vilvoorde and Leuven), September-October 1405 (Antwerp), April 1409-June 1412 (Vilvoorde and Leuven), June 1417-December 1417 (Vilvoorde), December 1418-October 1421 (Maastricht and Brussels), September 1429-March 1432 (Leuven). Figures for 1420-21 and 1429-32 have been given in Day's graphs.[7]

---

4. Willem Blockmans, "Devaluation, Coinage, and Seignorage Under Louis de Nevers and Louis de Male, Counts of Flanders, 1330-84," *Coinage in the Low Countries (880-1500)*, ed. N. J. Mayhew (*BAR* International 54, Oxford, 1979), pp. 72-4.

5. See John H. Munro, "Le monnayage, monnaies de comptes, et mutations monétaires au Brabant à la fin du moyen âge," in *Etudes d'histoire monétaire médiévale*, ed. John Day (Cahiers Jussieu, Université Paris 7, Presses Universitaires de Lille, forthcoming, 1982).

6. Henri Laurent, *La loi de Gresham au moyen âge: essai sur la circulation monétaire entre la Flandre et le Brabant à la fin du XIVe siècle* (Brussels, 1933), pp. 80-1 (précis only), 149-50, doc. no. 10; Day, p. 52.

7. See n. 5 above.

In my tables, all data on mint production in the Low Countries, except for Flanders 1334-1384 (as above), are based upon my own research in the Belgian and Dutch archives. All data have been computed to run for Michaelmas years ending 29 September, in accordance with the English mint accounts, using monthly averages where necessary to apportion the mint data into Michaelmas years.

The mint statistics for the Low Countries, commencing in 1334-34 with the earliest Flemish account, are necessarily restricted to Flanders until 1375, when the first surviving Brabantine account becomes available. From the 1380s, we may legitimately combine the Flemish and Brabantine mint outputs in our tables. For in 1384, with accession of the ducal house of Burgundy in Flanders (Philip the Bold), the two principalities agreed to strike a common, unified coinage; and in 1388, to have the Flemish mints strike all the coinage for both, sharing mint profits equally.[8] Even when the duchess of Brabant temporarily broke that agreement in the early 1390s (restoring it in 1399), the Brabantine coinages remained within the dominant Flemish monetary orbit, as did subsequent Brabantine coinages, after the duchy passed into Burgundian hands in 1406.[9] From 1421 to 1430, we must also add the neighbouring mints in Namur, Hainaut, and Holland-Zeland, as the third duke of Burgundy (Philip the Good) personally acquired these provinces (and also Brabant in his own name). Indeed having promised the Flemish Estates in 1418 not to alter the coinage again for fifteen years, Philip proceeded to utilize these new mints in order to strike debased imitations of Flemish coins.[10] Finally, from October 1433, Duke Philip unified the coinages of all the Burgundian Low Countries on the basis of the newly reformed Flemish coinage. Thereafter, as Tables 4 and 5 demonstrate, minting leadership in the Low Countries alternated between Flanders and Brabant.

Although I have published elsewhere parts of these statistical series on mint ouputs in England and the Low Countries, this is the first time that they are presented completely, correctly, and in their entirety for the Late Middle Ages: in kilograms of *pure* metal, in current moneys-of-account values (pounds sterling and pounds groot Flemish), and in constant pounds sterling of 1351-1411 values.[11]

---

8. John Munro, "Mint Policies, Ratios, and Outputs in the Low Countries and England, 1335-1420: Some Reflections on New Data," *Numismatic Chronicle*, 8th ser. I (1981), 78-90.

9. See n. 5 above, and n. 10 below.

10. John Munro, *Wool, Cloth, and Gold: The Struggle for Bullion in Anglo-Burgundian Trade, 1340-1478* (Brussels-Toronto, 1973), pp. 65-103.

11. See nn. 5, 8, 10, and also "Monetary Contraction and Industrial Change in the Late-Medieval Low Countries, 1335-1500," in *Coinage in the Low Countries*, pp. 139-44. The mint statistics published in this volume differ in some places from those previously published because of: (a) discovery of new mint accounts, especially for Calais, Mechelen, and the 14th-century Brabantine towns; (b) different methods of handling lacunae in mint-data; except for London in 1409-11 and 1461-62 (when output was evidently almost nil), I have not this time extrapolated for missing years; instead I have taken averages of extant data only, as explained above; (c) different methods of apportioning uneven accounts into consistent Michaelmas years (ending on 29 September).

Table 1. Silver coinage outputs of the English mints: decennial means in kilograms of pure metal, current and constant pounds sterling, 1240–9 to 1495–99

Decade	London kg. of silver	Provincial kg. of silver	Total kg. of silver	Value in current £ sterling	Value in constant £ sterling*
1240–9	10,906.25	8,366.89	19,273.14	60,041.73	74,431.90
1250–9	10,655.67	10,500.85	21,156.52	65,908.98	81,705.36
1260–9	6,923.69	5,192.88	12,116.57	37,746.79	46,793.55
1270–9	6,219.53	312.98	6,532.51	20,379.62	25,228.20
1280–9	19,223.84	6,958.80	26,182.64	81,889.30	101,116.01
1290–9	1,925.87	162.24	2,088.11	6,531.98	8,064.18
1300–9	18,425.41	10,848.36	29,273.77	91,573.58	113,053.79
1310–9[a]	3,858.70	5,050.61	8,909.31	27,869.91	34,407.30
1320–9[b]	798.81	813.87	1,612.68	5,044.73	6,228.06
1330–9	320.72	20.39	341.11	1,180.17	1,317.34
1340–9	3,304.83	65.51	3,370.34	11,809.47	13,016.05
1350–9	10,323.96	668.50	10,992.46	42,333.94	42,452.29
1360–9	971.73	99.29	1,071.02	4,136.23	4,136.23
1370–9	354.69	–	354.69	1,369.79	1,369.79
1380–9	256.99	–	256.99	992.47	992.47
1390–9	233.94	–	233.94	903.48	903.48
1400–9[c]	46.96	–	46.96	181.35	181.35
1410–9[c]	869.40	–	869.40	4,028.78	3,357.57
1420–9	588.49	3,547.53	4,136.02	19,167.69	15,973.08
1430–9	424.62	4,936.23	5,360.85	24,843.99	20,703.32
1440–9	250.64	–	250.64	1,161.53	967.94
1450–9	1,360.46	–	1,360.46	6,304.81	5,254.01
1460–9[d]	3,834.76	–	3,834.76	20,747.32	14,809.65

Table 1. (continued)

Decade	London kg. of silver	Pronvincial kg. of silver	Total kg. of silver	Value in current £ sterling	Value in constant £ sterling*
1470–9	1,792.17	113.27	1,905.44	11,038.11	7,358.74
1480–9	926.38	–	926.38	5,366.46	3,577.64
1495–9[e]	2,123.97	–	2,123.97	12,303.99	8,202.66

*Values of constant pound sterling (1351–1411 values)
1 kg. of silver = £3.861948
[a]Mean of 8 years: 1312–19 (1310–11 missing)
[b]Mean of 9 years: 1320–21, 1323–29 (1322 missing)
[c]Accounts for 1409–11 missing: amounts have been extrapolated
[d]Mean of 8 years: 1460–66 and 1469 (1467–68 missing); 1461–2 extrapolated
[e]Mean of 5 years: accounts missing for 1490–94

Table 2. Gold coinage outputs of the English mints: decennial means in kilograms of pure metal and in current and constant pounds sterling, 1340–9 to 1495–9

Decade	London mint kg. gold	Provincial mints kg. gold	Total mints kg. gold	Value in current £ sterling	Value in constant £ sterling*
1340–9 [a]	395.07	–	395.07	15,963.27	17,024.20
1350–9	[658.44]	–	[658.44]	[26,605.44]	[28,373.67]
1360–9	1,750.11	–	1,750.11	75,224.50	75,415.57
1370–9	1,395.53	715.34	2,110.87	90,961.88	90,961.88
1380–9	263.14	339.70	602.84	25,977.63	25,977.63
1390–9	200.45	88.57	289.02	12,454.62	12,454.62
1400–9[b]	397.68	146.71	544.39	23,458.78	23,458.78
1410–9[b]	111.09	34.91	146.00	6,291.41	6,291.41
1420–9	1,397.52	–	1,397.52	66,865.30	60,221.85
1430–9	1,302.11	299.83	1,601.94	76,701.38	69,031.25
1440–9	191.40	8.22	199.62	9,557.86	8,602.08
1450–9	78.72	–	78.72	3,768.99	3,392.08
1460–9[c]	53.77	–	53.77	2,574.57	2,317.10
1470–9	657.73	–	657.73	41,294.58	28,342.97
1480–9	499.03	14.00	513.03	33,160.84	22,107.24
1495–9[d]	206.77	–	206.77	13,365.40	8,910.28
	296.41	–	296.41	19,159.40	12,772.92

*Values of constant pound sterling (1351–1411 values)
1 kg. of gold = £43.092065
[a] Gold first struck in January 1344: mean of 1344–49
[b] Accounts missing for 1409–11: amounts extrapolated
[c] 8-year mean for 1460–66 and 1469 (1467–68 missing); amounts extrapolated for 1461–62 (also missing)
[d] 5-year: mean only: accounts missing for 1490–94

Table 3. Silver and gold coinage outputs of the English mints: decennial means in current and constant pounds sterling, 1240–49 to 1490–99

Decade	Silver current £ sterling	% of total	Gold current £ sterling	% of total	Total coinage current £ sterling	Outputs in: constant £ sterling*
1240–9	60,041.73	100.0	–	–	60,041.73	74,431.90
1250–9	65,908.98	100.0	–	–	65,908.98	81,705.36
1260–9	37,746.79	100.0	–	–	37,746.79	46,793.55
1270–9	20,379.62	100.0	–	–	20,379.62	25,228.20
1280–9	81,889.30	100.0	–	–	81,889.30	101,116.01
1290–9	6,531.98	100.0	–	–	6,531.98	8,064.18
1300–9	91,573.58	100.0	–	–	91,573.58	113,053.79
1310–9[a]	27,869.91	100.0	–	–	27,869.91	34,407.30
1320–9[b]	5,044.73	100.0	–	–	5,044.73	6,228.06
1330–9	1,180.17	100.0	–	–	1,180.17	1,317.34
1340–9	11,809.47	42.5	15,963.27	57.5	27,772.74	30,040.25
1350–9	42,333.94	36.0	75,224.50	64.0	117,558.44	117,867.86
1360–9	4,136.23	4.3	90,961.88	95.7	95,098.11	95,098.11
1370–9	1,369.79	5.0	25,977.63	95.0	27,347.42	27,347.42
1380–9	992.47	7.4	12,454.62	92.6	13,447.09	13,447.09
1390–9	903.48	3.7	23,458.78	96.3	24,362.26	24,362.26
1400–9[c]	181.35	2.8	6,291.41	97.2	6,472.76	6,472.76
1410–9[c]	4,028.78	5.7	66,865.30	94.3	70,894.08	63,579.42
1420–9	19,167.69	20.0	76,701.38	80.0	95,869.07	85,004.33
1430–9	24,843.99	72.2	9,557.86	27.8	34,401.85	29,305.40
1440–9	1,161.53	23.6	3,768.99	76.4	4,930.52	4,360.02
1450–9	6,304.81	71.0	2,574.57	29.0	8,879.38	7,571.11
1460–9[d]	20,747.32	33.4	41,294.58	66.6	62,041.90	43,152.62

Table 3. (*continued*)

Decade	London mint kg. gold	Provincial mints kg. gold	Total mints kg. gold	Value in current £ sterling	Value in constant £ sterling*
1470–9	11,038.11	25.0	33,160.84	44,198.95	29,465.98
1480–9	5,366.46	28.6	13,365.40	18,731.86	12,487.92
1495–9[e]	12,303.99	39.1	19,159.40	31,463.39	20,975.58

*Constant pounds sterling values, of 1351–1441:
1 kg. of gold = £43.092065   1 kg. of silver = £3.861948
[a]Mean of 8 years: accounts missing for 1310–11
[b]Mean of 9 years: accounts missing for 1322
[c]Accounts missing for 1409–11; but outputs have been extrapolated
[d]Mean of 8 years: accounts missing for 1467–68; accounts also missing for 1461–62, but outputs have been extrapolated.
[e]Mean of 5 years: accounts missing for 1490–94.

Table 4. Silver coinage outputs of the Low Countries' mints: decennial means in kilograms of pure metal and in current pounds groot Flemish, 1335–9 to 1490–9

	Kilograms					Values in pounds groot
Decade	Flanders	Brabant	Holland	Namur & Hainaut	Total silver	
1335–9[a]	5,244.71					6,814.58
1340–9[b]	1,982.18					3,943.57
1350–9	6,800.46					15,417.83
1360–9	6,669.80					19,335.41
1370–9[c]	2,708.55					9,882.69
1380–9	[5,559.80][d] 2,539.11[e] [2,654.86][f]	[558.45][g] 580.44[h]			3,119.55	13,121.03
1390–9	5,004.76	577.05			5,581.81	22,696.94
1400–9	731.13	156.14			887.27	3,617.37
1410–9	1,845.48	926.64 [855.63][i]			2,772.12	9,794.69
1420–9	9,000.26	438.16		449.70	9,888.12	49,086.20
1430–9	4,792.21	1,088.91	844.50	554.99	7,280.61	38,167.37
1440–9	119.78	–	42.67	–	162.46	849.34
1450–9	52.23	29.93	4.67	27.51	114.34	644.31
1460–9	1,036.73	501.11	36.83	51.83	1,626.50	9,767.56
1470–9	4,284.99	3,637.40	–	–	7,922.39	52,339.77
1480–9	1,962.03	4,032.04	1,540.26	–	7,534.34	73,534.33
1490–9	1,292.68	1,959.39	191.05	168.15	3,611.27	28,616.41

[a]5-year mean only; Flemish accounts Sept. 1334 to Sept. 1339.
[b]10-year mean; but no record of minting in 1340–42; 1344–45.

Table 4. (*continued*)

---

^cNo record of Flemish minting from 5 Aug. 1370 to 18 June 1373 (34.44 months) and from 27 June 1377 to 30 Jan. 1380 (31.07 months). Zero values have been assigned to these gaps, and the mean is based upon the full ten years.

^dMean is calculated on the basis of the 4.872 years of recorded mint activity.

^eFor Mich. 1379 to 30 Jan. 1380, see note c above. There are no mint accounts available for the period 2 May 1386 to 30 Sept. 1388 (28.94 months), when the mints were evidently still active. The mean is therefore calculated on the basis of the remaining 91.06 months in the decade = 7.58833 years.

^fMean is calculated on the basis of 7.26 years of recorded mint activity, as explained in note c and e above: i.e. from 30 Jan. 1380 to 2 May 1386, and from 1 Oct. 1388 to 30 Sept. 1389.

^gLeuven mint account commences only on 1 July 1375; the mean is therefore calculated on the remaining 4.25 years to 30 Sept. 1379. The value of this output is not included in the totals in the two final columns.

^hMean computed on the same basis as the one for Flanders, as in e above. Approximately the same number of months are also missing for the Leuven mint accounts.

ⁱTo be consistent with the Flemish mean, Brabant mean computed on basis of full ten years, even though accounts are missing from 21 March 1411 to 25 June 1412. On basis of 3.737 years of recorded output, mean would be 855.63 kg.

Table 5. Gold coinage outputs of the Low Countries' mints: decennial means in kilograms of pure metal and in current pounds groot Flemish, 1335–9 to 1490–9

| Decade | Kilograms ||||| Values in pounds groot |
	Flanders	Brabant	Holland	Namur & Hainaut	Total gold	
1335–9[a]	295.06					4,389.93
1340–9[b]	48.02					1,002.08
1350–9	1,906.79					45,790.63
1360–9	2,330.41					68,997.42
1370–9[c]	640.75					24,679.11
1380–9[e]	[1,315.26][d] 538.29	[734.23][g] 69.99[h]			608.28	28,289.20
[f]	[562.83]					
1390–9	426.58	25.57			452.15	18,057.78
1400–9	32.09	5.25			37.34	1,464.57
1410–9	13.59	30.56[i] [28.08]			44.15[i]	1,505.25
1420–9	201.19	2.77	121.69	135.77	461.43	28,586.05
1430–9	275.55	389.57	154.88	427.06	1,247.06	78,742.35
1440–9	62.88	—	3.35	—	66.24	3,808.70
1450–9	239.34	160.65	29.34	107.31	536.64	32,888.71
1460–9	82.18	24.32	1.71	3.50	111.71	7,194.08
1470–9	177.48	159.39	—	—	336.87	25,017.80
1480–9	16.37	59.01	18.46	—	93.84	12,930.40
1490–9	39.06	39.90	0.76	—	79.71	8,127.79

[a]5-year mean only; Flemish accounts Sept. 1334 to Sept. 1339.
[b]10-year mean; but no record of minting in 1340–42; 1344–45.

Table 5. (*continued*)

[c] No record of Flemish minting from 5 Aug. 1370 to 18 June 1373 (34.44 months) and from 27 June 1377 to 30 Jan. 1380 (31.07 months). Zero values have been assigned to these gaps, and the mean is based upon the full ten years.

[d] Mean is calculated on the basis of the 4.872 years of recorded mint activity.

[e] For Mich. 1379 to 30 Jan. 1380, see note c above. There are no mint accounts available for the period 2 May 1386 to 30 Sept. 1388 (28.94 months), when the mints were evidently still active. The mean is therefore calculated on the basis of the remaining 91.06 months in the decade = 7.58833 years.

[f] Mean is calculated on the basis of 7.26 years of recorded mint activity, as explained in note c and e above: i.e. from 30 Jan. 1380 to 2 May 1386, and from 1 Oct. 1388 to 30 Sept. 1389.

[g] Leuven mint account commences only on 1 July 1375; the mean is therefore calculated on the remaining 4.25 years to 30 Sept. 1379. The value of this output is not included in the totals in the two final columns.

[h] Mean computed on the same basis as the one for Flanders, as in e above. Approximately the same number of months are also missing for the Leuven mint accounts.

[i] To be consistent with the Flemish mean, Brabant mean computed on basis of full ten years, even though accounts are missing from 21 March 1411 to 25 June 1412.

Table 6. Silver and gold coinage outputs of the Low Countries' mints: decennial means in current pounds groot Flemish and in constant pounds sterling,* 1335–9 to 1490–9

Decade	Silver current £ groot	% of total	Gold current £ groot	% of total	Total coinage current £ groot	Outputs in: constant £ sterling*
A. *Flanders, 1335 to 1399*						
1335–9[a]	6,814.58	60.8	4,389.93	39.2	11,204.51	32,969.68
1340–9	3,943.57	79.7	1,002.08	20.3	4,945.65	9,724.53
1350–9	15,417.83	25.2	45,790.63	74.8	61,208.46	108,430.50
1360–9	19,335.41	21.9	68,997.42	78.1	88,332.83	126,180.72
1370–9[b]	9,882.69	28.6	24,679.11	71.4	34,561.80	38,071.54
[c]	[20,286.06]		[50,658.45]		[70,944.51]	[78,148.86]
1380–9[d]	10,682.33	29.9	25,073.94	70.1	35,756.27	33,002.05
[e]	[11,169.28]		26,216.93		[37,386.21]	[34,506.44]
1390–9	20,346.65	54.4	17,028.22	45.6	37,374.87	37,710.21
B. *Low Countries, 1380–1499*						
1380–9[d]	13,121.03	31.7	28,289.20	68.3	41,410.23	38,259.73
1390–9	22,696.94	55.7	18,057.78	44.3	40,754.72	41,040.47
1400–9	3,617.37	71.2	1,464.57	28.8	5,081.94	5,035.60
1410–9	9,794.69	86.7	1,505.25	13.3	11,299.93	12,608.63
1420–9	49,086.20	63.2	28,586.05	36.8	77,672.25	58,071.36
1430–9	38,167.37	32.6	78,742.35	67.4	116,909.72	81,855.47
1440–9	849.34	18.2	3,808.70	81.8	4,658.04	3,481.76
1450–9	644.31	1.9	32,888.71	98.1	33,533.02	23,566.54
1460–9	9,767.56	57.6	7,194.08	42.4	16,961.64	11,095.60
1470–9	52,339.77	67.7	25,017.80	32.3	77,357.57	45,112.03
1480–9	73,534.33	85.0	12,930.40	15.0	86,464.73	33,141.03
1490–9	28,616.41	77.9	8,127.79	22.1	36,744.20	17,381.41

Table 6. (continued)

*Values in constant pounds sterling (1351-1441 Values):
1 kg. of gold = £43.092065   1 kg. of silver = £3.861948
[a] 5-year mean only, for 1335–9 (first account commences in Sept. 1334).
[b] There are no records of Flemish minting from 5 Aug. 1370 to 18 June 1373 (34.44 months) and from 27 June 1377 to 30 Jan. 1380 (31.07 months). Zero values have been assigned to these gaps, and the mean is based upon the full ten years.
[c] Mean calculated on the basis of 4.872 years of recorded mint activity.
[d] For Mich. 1379 to 30 Jan. 1380, see note b above. There are no mint accounts available for the period 2 May 1386 to 30 Sept. 1388 (28.94 months), when the mints were evidently active. The mean is therefore calculated on the basis of the remaining 91.06 months in the decade = 7.58833 years.
[e] Mean calculated on the basis of 7.26 years of recorded activity, as explained in notes b and d above: i.e. from 30 Jan. 1380 to 2 May 1386, and from 1 Oct. 1388 to 30 Sept. 1389.

Table 7. Output of silver and gold coinages in England and the Low Countries: quinquennial means in kilograms of pure metal, 1335–9 to 1495–9

	Silver kilograms			Gold kilograms		
Years	England	Flanders	Low Countries	England	Flanders	Low Countries
1335–9	483.08	5,244.71		–	295.06	
1340–4	3,629.67	[176.76][b]		187.30	[1.32][b]	
1345–9	3,111.00	3,787.61		602.84	94.73	
1350–4	15,583.84	5,761.30		1,734.56	914.93	
1355–9	6,401.07	7,839.63		1,765.65	2,898.65	
1360–4	1,858.76	4,107.57		2,403.54	2,575.95	
1365–9	283.28	9,232.03		1,818.20	2,084.87	
1370–4	198.82	4,213.83[c]		953.77	815.35[c]	
		[9,891.61][d]			[1,913.96][d]	
1375–9	510.55	1,203.27[e]	1,761.72	251.91	466.15[e]	1,200.38[j]
		[2,194.41][f]	[2,752.86]		[850.13][f]	[1,584.36]
1380–4	326.15	3,029.10[g]	3,737.39	155.43	546.74[g]	625.91
		[3,243.73][h]	[3,952.02]		[585.48][h]	[664.65]
1385–9	187.82	1,592.58[i]	1,926.05	422.62	521.97[i]	574.23
1390–4	300.14	4,072.16	5,055.16	646.86	484.31	528.03
1395–9	167.75	5,937.36	6,108.47	441.92	368.84	376.25
1400–4	79.22	1,462.25	1,462.25	217.58	64.18	64.18
1405–9	14.70	–	312.28	74.42	–	10.50
1410–4	665.04	3,508.66	4,926.55[k]	1,553.28	24.55	72.28[k]
1415–9	1,073.75	182.30	617.70	1,241.75	2.64	16.04
1420–4	1,953.69	12,910.53	13,882.67	2,375.36	11.27	14.39
1425–9	6,318.34	5,089.98	5,893.58	828.54	319.11	908.46
1430–4	9,006.56	4,812.88	7,654.38	266.30	189.52	1,731.18
1435–9	1,715.14	4,771.55	6,906.84	132.94	361.58	762.93

142

Table 7. (continued)

Years	Silver kilograms England	Silver kilograms Flanders	Silver kilograms Low Countries	Gold kilograms England	Gold kilograms Flanders	Gold kilograms Low Countries
1440–4	271.56	226.65	311.99	107.00	102.73	109.44
1445–9	229.71	12.92	12.92	50.43	23.04	23.04
1450–4	1,578.57	15.06	42.45	83.32	161.99	346.22
1455–9	1,142.35	89.41	186.22	24.22	316.69	727.06
1460–4	2,169.42	–	–	81.97	11.12	13.75
1465–9	6,610.33[a]	2,073.46	3,253.00	1,617.34[a]	153.24	209.69
1470–4	2,567.15	5,117.94	7,407.07	595.17	74.31	96.04
1475–9	1,243.75	3,452.06	8,437.72	430.88	280.64	577.69
1480–4	955.91	2,853.73	6,793.23	284.92	30.68	65.32
1485–9	896.86	1,070.34	8,275.45	128.62	2.07	122.37
1490–4	n.a.	950.51	2,894.03	n.a.	8.90	46.45
1495–9	2,123.97	1,634.85	4,328.51	296.41	69.21	112.97

[a]3-year mean: 1465–66, 1469.
[b]Flanders: gold and silver coinages for 16 Apr. to 17 Oct. 1343 only.
[c]There are no records of Flemish minting from 5 Aug. 1370 to 18 June 1373 (34.44 months); but the means computed here are based on 60 months (5 years).
[d]The Flemish means for 1370–4 are computed on the basis of the 25.56 months of recorded mint activity, as explained in note c.
[e]There are no records of Flemish minting from 27 June 1377 to 30 Jan. 1380 (27.10 months for this quinquennium; 3.97 for the next); but the means computed here are based on 60 months for the full quinquennium 1375–9.
[f]The Flemish means are computed on the basis of 32.9 months of recorded mint activity from Mich. 1374 to 27 June 1377.
[g]See note e. Mean based on the full 60 months of the quinquennium.
[h]Flemish means are computed on the basis of the 56.03 months of recorded mint activity, from 30 Jan. 1380 to Mich. 1384. See note e.
[i]The Flemish mint accounts are missing from 2 May 1386 to 30 Sept. 1388, 28.94 months. The Flemish means are therefore computed on the basis of the remaining 31.06 months, including 4.93 months when the mints were definitely closed, from 4 Dec. 1385 to 2 May 1386.
[j]Mean of 4.25 years. Account available only from 1 July 1375 (9 months missing).
[k]For Brabant (Leuven), mean based on 3.737 years of recorded mint activity: accounts missing from 21 Mar. 1411 to 25 June 1412.

Table 8. Values of the coinage outputs of England and the Low Countries: quinquennial means in current moneys-of-account and in constant pounds sterling (1351–1411 values)

Years	Total coinage values in current £			Total coinage values in constant £		
	England £ sterl.	Flanders £ groot	Low Countries £ groot	England £ sterl.*	Flanders £ sterl.*	Low Countries £ sterl.*
1335–9	1,737.43	11,204.51		1,865.64	32,969.68	
1340–4	20,628.44	[337.46][b]		22,088.53	[739.61][b]	
1345–9	34,917.03	9,553.85		37,992.00	18,709.44	
1350–4	134,310.62	31,918.46		134,929.78	61,676.03	
1355–9	100,806.24	90,498.46		100,806.23	155,184.97	
1360–4	110,752.03	83,249.37		110,752.05	126,866.38	
1365–9	79,444.17	93,416.29		79,444.17	125,495.06	
1370–4	41,867.94	45,985.67[c]		41,867.94	51,408.61[c]	
		[107,947.58][d]			[120,677.47][d]	
1375–9	12,826.91	23,137.93[e]	54,537.83[j]	12,826.91	24,734.45	58,530.68[j]
		[42,196.83][f]			[45,108.42][f]	
1380–4	7,957.28	36,481.82[g]	42,881.01	7,957.28	35,258.48[g]	41,405.53
		[39,066.73][h]			[37,756.71][h]	
1385–9	18,936.90	34,354.67[i]	38,569.06	18,936.90	28,643.20[i]	32,182.88
1390–4	29,033.43	36,184.84	41,959.32	29,033.43	36,596.42	42,276.69
1395–9	19,691.10	38,564.87	39,550.11	19,691.10	38,824.00	39,804.25
1400–4	9,681.97	8,478.82	8,478.82	9,681.97	8,412.66	8,412.66
1405–9	3,263.54	–	1,685.06	3,263.54	–	1,658.53
1410–4	77,356.99	13,062.75	19,606.13[k]	69,502.54	14,608.00	22,140.72[k]
1415–9	64,431.16	806.11	2,993.73	57,656.31	817.79	3,076.54
1420–4	122,786.43	63,093.00	67,946.44	109,904.20	50,345.45	54,234.51
1425–9	68,951.70	50,153.92	87,398.05	60,104.45	36,510.98	61,908.22
1430–4	54,489.94	35,938.47	155,836.00	46,258.34	26,753.90	104,160.96
1435–9	14,313.75	44,611.30	77,983.43	12,352.47	34,008.71	59,549.98

144

Table 8. (continued)

	Total coinage values in current £			Total coinage values in constant £		
Years	England £ sterl.	Flanders £ groot	Low Countries £ groot	England £ sterl.*	Flanders £ sterl.*	Low Countries £ sterl.*
1440–4	6,381.66	7,077.30	7,901.58	5,659.60	5,302.16	5,920.67
1445–9	3,479.38	1,414.50	1,414.50	3,060.44	1,042.84	1,042.84
1450–4	11,305.14	10,009.21	21,442.17	9,686.90	7,038.64	15,083.42
1455–9	6,453.63	19,947.79	45,623.89	5,455.33	13,992.12	32,049.65
1460–4	14,557.49	681.41	842.41	11,910.37	479.18	592.30
1465–9	141,182.56[a]	22,403.18	33,080.89	95,223.05[a]	14,611.03	21,598.91
1470–4	53,341.60	35,818.13	51,002.14	35,561.09	22,967.39	32,744.41
1475–9	35,056.29	45,299.40	[101,867.52] 103,712.98[f]	23,370.88	25,425.04	57,479.64
1480–4	23,954.35	25,535.80	[57,264.52] 59,458.24[f]	15,969.58	12,343.02	29,049.63
1485–9	13,509.38	10,205.67	113,471.24	9,006.26	4,222.80	37,232.43
1490–4	n.a.	7,507.11	26,466.58	n.a.	4,054.34	13,178.36
1495–9	31,463.39	20,241.50	47,021.80	20,975.58	9,296.11	21,584.46

*Values of constant pound sterling (1351–1411 values)
1 kg. of gold = £43.092065 1 kg. of silver = £3.861948
[a]3-year mean: 1465–66, 1469.
[b]Flanders: gold and silver coinages for 16 Apr. to 17 Oct. 1343 only.
[c]There are no records of Flemish minting from 5 Aug. 1370 to 18 June 1373 (34.44 months); but the means computed here are based on 60 months (5 years).
[d]The Flemish means for 1370–4 are computed on the basis of the 25.56 months of recorded mint activity, as explained in note c.
[e]There are no records of Flemish minting from 27 June 1377 to 30 Jan. 1380 (27.10 months for this quinquennium; 3.97 for the next); but the means computed here are based on 60 months for the full quinquennium 1375–9.
[f]The Flemish means are computed on the basis of 32.9 months of recorded mint activity from Mich. 1374 to 27 June 1377.
[g]See note e. Mean based on the full 60 months of the quinquennium.

Table 8. (*continued*)

[h]Flemish means are computed on the basis of the 56.03 months of recorded mint activity, from 30 Jan. 1380 to Mich. 1384. See note e.
[i]The Flemish mint accounts are missing from 2 May 1386 to 30 Sept. 1388, 28.94 months. The Flemish means are therefore computed on the basis of the remaining 31.06 months, including 4.93 months when the mints were definitely closed, from 4 Dec. 1385 to 2 May 1386.
[j]Mean of 4.25 years. Account available only from 1 July 1375 (9 months missing).
[k]For Brabant (Leuven), mean based on 3.737 years of recorded mint activity: accounts missing from 21 Mar. 1411 to 25 June 1412.
[l]Coinage of Brabant as 'weak coinage' (by 12.5%).

Table 9. A comparison of long-term averages of mint outputs in England and the Low Countries: in kilograms of pure metal, constant pounds sterling (1351–1411 values), and current moneys, 1250–1499

England (1250–1499)

Period of the mean	Silver kg.	Gold kg.	Total coinage values Constant £ sterl.	Total coinage values Current £ sterl.
1250–99	13,615.3	—	52,581.5	42,491.3
1300–49	8,843.4[a]	84.1[a]	37,775.0[a]	31,353.8[a]
1350–99	2,581.8	1,059.5	55,624.6	55,562.6
1400–99	2,132.8	684.8	37,744.4	42,513.7
1450–99	1,935.4[b]	336.7[b]	21,984.9[b]	31,901.3[b]
1300–99	5,615.8[a]	586.8[a]	46,975.8[a]	43,832.6
1400–99	2,041.5[b]	523.9[b]	30,457.7[b]	37,606.9[b]
1270–1369	9,116.6[c]	438.8[c]	54,115.1[c]	48,332.0[c]
1370–1469	1,626.3[d]	555.1[d]	30,201.2[d]	34,310.0[d]

Low Countries (1350–1499)

Period of the mean	Silver kg.	Gold kg.	Total coinage values Constant £ sterl.	Total coinage values Current £ groot Flemish
1350–99	4,856.3[e]	1,200.5[e]	70,487.0[e]	52,242.0[e]
1400–99	4,198.1	371.2	32,210.6	43,124.4
1450–99	4,161.8	231.8	26,059.3	50,212.2
1400–99	4,179.9	301.5	29,134.9	46,668.3
1370–1469	3,450.0[f]	453.6[f]	32,868.5[f]	39,815.9[f]

*Gold values (1351–1411): 1 kg. = £43.092065; Silver values (1351–1411): 1 kg. = £3.861948.

[a]Mean of 47 (97) years: data missing for 1310–11 and 1322.
[b]Mean of 43 (53) years: data missing for 1467–68 and 1490–94.
[c]Mean of 97 years (see a).
[d]Mean of 98 years (see b).
[e]Mean of 47.5883 years: data missing for 1386–88, when mints were active.
[f]Mean of 97.5883 years: data missing, see e.

Table 10. Official values of gold and silver in the Low Countries and England, 1335–1504: in pounds groot Flemish and pounds sterling English, respectively, per kilogram of pure metal, with the official gold to silver ratios (ratios of traites)

Date	Low Countries (Flanders)* Gold £ groot	Silver £ groot	Ratio	England Gold £ ster.	Silver £ ster.	Ratio
May 1335					3.601	
Jun 1335	14.641	1.196	12.24			
Jun 1337	16.394	1.433	11.44			
Apr 1343		1.759				
Jan 1344				43.092	3.476	12.40
Aug 1344				37.825	3.424	11.05
Jun 1345				37.825	3.450	10.96
Jan 1346		1.835				10.96
Aug 1346				40.219	3.476	11.57
Nov 1346		2.016				11.57
May 1349	20.877	2.016	10.36			11.57
May 1351	20.877	2.164	9.65			11.57
Jul 1351			9.65	43.092	3.862	11.16
Jan 1352	22.300	2.164	10.30			11.16
Sep 1352	22.775	2.164	10.52			11.16
Sep 1353	22.775	2.221	10.25			11.16
Dec 1354	23.945	2.385	10.04			11.16
Jul 1356	24.787	2.385	10.39			11.16
Sep 1357	25.682	2.385	10.77			11.16
Oct 1357	24.787	2.385	10.39			11.16
Jul 1358	25.558	2.385	10.72			11.16
Oct 1359	25.833	2.487	10.39			11.16
Nov 1359	26.966	2.487	10.84			11.16
Feb 1360	27.251	2.487	10.96			11.16
Sep 1361	28.201	2.558	11.02			11.16

Table 10. (continued)

	Low Countries (Flanders)*			England		
Date	Gold £ groot	Silver £ groot	Ratio	Gold £ ster.	Silver £ ster.	Ratio
Dec 1361	29.111	2.558	11.38			11.16
Dec 1363	31.387	2.771	11.33			11.16
Mar 1364	29.111	2.771	10.51			11.16
Apr 1365	30.813	3.038	10.14			11.16
Jan 1368	32.405	3.240	10.00			11.16
Apr 1369	32.405	3.391	9.56			11.16
Sep 1369	32.865	3.553**	9.25**			11.16
Apr 1370	33.124	3.553**	9.32**			11.16
Jun 1373	39.530	3.739	10.57			11.16
	39.896	3.739	10.67			
Jan 1380	41.692	4.121	10.12			11.16
	42.598	4.121	10.34			
Sep 1383	45.570	4.300	10.60			11.16
Sep 1384	35.120	3.553	9.89			11.16
Apr 1386	41.727	4.050	10.30			11.16
Oct 1386	41.727	4.050	10.30			11.16
	41.727	4.107	10.16			
Apr 1387	49.811	4.910	10.14			11.16
Oct 1388	55.566	5.337	10.41			11.16
Jan 1390	39.223	4.050	9.68			11.16
Jan 1391	39.223	4.086	9.60			11.16
Jun 1393	39.223	4.050	9.68			11.16
Jul 1407	39.636	4.050	9.79			11.16
Aug 1409	33.443	3.482	9.60			11.16
Nov 1411			9.60	47.880	4.634	10.33
Dec 1416	37.554	4.263	8.81			10.33

149

Table 10. (continued)

	Low Countries (Flanders)*			England		
Date	Gold £ groot	Silver £ groot	Ratio	Gold £ ster.	Silver £ ster.	Ratio
Jun 1418	47.290^a	4.832^a	9.79			10.33
Oct 1421		4.832^c				10.33
Aug 1423		4.263^c				10.33
Jun 1425	56.784^a	4.832^a	11.75			10.33
Jul 1425	56.784^c	4.519^c	12.57			10.33
Jul 1426	64.412^c	4.558^c	14.13			10.33
Nov 1426	64.412^a	4.832^a	13.33			10.33
Sep 1427	50.672^a	4.832^a	10.49			10.33
Jun 1428	62.054^c	5.116^c	12.13			10.33
Nov 1428	58.835^a	5.474^a	10.75			10.33
Feb 1429	64.052^c	5.525^c	11.59			10.33
Jun 1429	65.834^c	5.756^c	11.44			10.33
Sep 1429	64.248^c	5.453^b	11.78			10.33
Jan 1430	65.834^c	5.756^b	11.44			10.33
May 1430	64.248^b	5.453^b	11.78			10.33
Dec 1430	65.985^c	5.756^c	11.46			10.33
Mar 1431	64.722^b	5.453^b	11.87			10.33
Mar 1431	65.985^c	5.756^c	11.46			10.33
Apr 1431	66.504^c	5.815^c	11.44			10.33
Sep 1431	67.031^c					10.33
Dec 1431	70.381^c	6.139^c	11.46			10.33
Oct 1432	73.110^c	6.139^c	11.91			10.33
May 1433	77.176^c	6.634^c	11.63			10.33
Aug 1433	57.883^c	4.975^{a,b}	11.63			10.33
Oct 1433	55.591^{a,b}	5.116	10.87			10.33
Mar 1443	57.907	5.116	11.32			10.33

Table 10. (continued)

Date	Low Countries (Flanders)* Gold £ groot	Silver £ groot	Ratio	England Gold £ ster.	Silver £ ster.	Ratio
Jan 1454	61.286	5.116	11.98			10.33
Aug 1464			11.98	59.850	5.793	10.33
Mar 1465			11.98	64.638	5.793	11.16
Jun 1466	63.480	5.862	10.83			11.16
Oct 1467	65.029	6.004	10.83			11.16
Dec 1474	74.316	6.821	10.90			11.16
Sep 1477	83.607[a]	6.821[a]	12.26			11.16
Sep 1477	83.607[b]	7.674[b]	10.89			11.16
Jul 1482	92.898[a]	8.527[a]	10.89			11.16
Mar 1484	74.316	6.821	10.90			11.16
Apr 1485	92.898[a]	10.232[a]	9.08			11.16
	102.189[a]	10.232[a]	9.99			
Dec 1485	92.898[b]	8.527[b]	10.89			11.16
Mar 1486	92.898[a]	8.527[a]	10.89			11.16
Aug 1486	102.189[b]	9.379[b]	10.89			11.16
Jul 1487	121.347[b]	10.386[b]	11.68			11.16
	121.347[b]	11.750[b]	10.33			
Jun 1488		14.244[d]				11.16
Nov 1488	168.832[d]	17.480[d]	9.66			11.16
	204.643[d]	17.480[d]	11.71			
Jun 1489		13.643[a]				11.16
		15.000[a]				
Jul 1489		15.348				11.16
		15.804				
Jan 1490	61.932	5.683	10.90			11.16
Mar 1492	74.316	7.248	10.25			11.16

151

Table 10. (*continued*)

	Low Countries (Flanders)*			England		
Date	Gold £ groot	Silver £ groot	Ratio	Gold £ ster.	Silver £ ster.	Ratio
Dec 1492	76.196[a]	7.248[a]	10.51			11.16
Sep 1493	74.316[a]	7.248[a]	10.25			11.16
Sep 1493	74.316[b]	8.156[b]	9.11			11.16
Aug 1494	[74.316]	7.248	10.25			11.16
Jan 1495	74.316	7.248	10.25			11.16
	86.704	8.156	10.63			
May 1496	90.704[a]	8.420[a]	10.77			11.16
May 1496	90.704[b]	8.527[b]	10.64			11.16
	98.262[b]	9.593[b]	10.24			
Sep 1497	90.704[b]	8.420[b]	10.77			11.16
May 1499	94.483	8.527	11.08			11.16
Feb 1500	94.977	8.527	11.14			11.16
May 1503	94.977	8.527	11.14			11.16
	94.977	8.898	10.67			
Jan 1504	94.977	8.527	11.14			11.16

*The monetary values for the Low Countries for the period 1335 to 1418 are for Flanders only; thereafter, the principality concerned is indicated as follows:

[a]Flanders.
[b]Brabant.
[c]Namur.
[d]Mechelen (seigneurie of).

Where there is no such indication, the data refer to all of the Burgundian Low Countries, whose coinages were unified after 1433; but as indicated, regional mints occasionally differed from the 1470s, particularly during the era of revolt against Maximilian.
**silver values estimated (interpolated) for 1369 and 1370.

Table 10. (*continued*)

Note: all of these monetary data have been taken directly from the mint accounts or mint indentures, not from official proclamations (except for the Low Countries in March 1443, when the value of the gold *rijder* or *philippus* was raised from 48d. to 50d. groot Flemish). When one or the other metal was not struck, no value was given for that metal.

All of these values are in terms of 100% pure metal, not in terms of official standards of fineness. In England sterling silver was 92.5% pure; in the Low Countries silver *argent-le-roy* was 23/24th or 95.833% pure. The English gold noble at 23 7/8 carats was thus 99.47916% pure gold. In both England and the Low Countries this was often taken as the standard for 'fine' gold.

When two sets of values are given for one date, that means that an alteration in the coinage(s) took place some time after the mint account commenced.

Sources for the Monetary Tables

A. *ENGLAND*
  1. *London and Canterbury: 1235-1272*
     C. E. Blunt and J. D. Brand, "Mint Output of Henry III," *British Numismatic Journal*, XXXIX (1970), 61-6.
  2. *London, Canterbury, and Other Provincial Mints: 1272-1377*
     C. G. Crump and C. Johnson, "Tables of Bullion Coined Under Edward I, II, and III," *The Numismatic Chronicle*, 4th ser. XIII (1913), 200-45.
  3. *London, 1377-1485 (and Bristol, 1470-72; York, 1470-75)*
     G. C. Brooke and E. Stokes, "Tables of Bullion Coined from 1377 to 1550," *The Numismatic Chronicle*, 5th ser. IX (1929), 27-69.
     The London statistics for 1432 and 1433 have been corrected by information in:
     Sir John Craig, *The Mint: A History of the London Mint from A. D. 287 to 1948* (Cambridge, 1953), pp. 408, 412.
  4. *London, 1485-1499*
     C. E. Challis, *The Tudor Coinage* (Manchester, 1978), pp. 305, 307.
  5. *Calais*
     a) 1363-1384: Crump and Johnson, "Tables of Bullion Coined," *N.C.*, XIII (1913), 235, 243-45.
     b) 1387-1404: T. H. Lloyd, *The English Wool Trade in the Middle Ages* (Cambridge, 1977), p. 241: Table 15, as verified, corrected, and extended, from: Public Record Office, Lord Treasurer's Remembrancer Exchequer, E. 364/22, 25, 28, 30, 34, 36, 37; King's Remembrancer Exchequer, E. 101/184/10
     c) 1422-1440: Public Record Office, L. T. R. Exchequer, E. 364/59, 61-3, 65-6, 69, 72; K. R. Exchequer, E. 101/192-3; and *Calendars of Patent Rolls, 1422-29*, pp. 337-8, 502; *Calendar of Patent Rolls, 1429-1436*, pp. 256-7, and 259.
     F. A. Walters, "Supplementary Notes on the Coinage of Henry VI," *Numismatic Chronicle*, 4th ser. XI (1911), 171-2 (for Calais in 1436 and 1439-40; for York in 1423-24).

B. *THE LOW COUNTRIES*
  1. *Flanders, 1334 to 1384:* Ghent, Bruges, and Mechelen
     Victor Gaillard, ed. *Recherches sur les monnaies des comtes de Flandre, II: Sous les règnes de Louis de Crécy et de Louis de Mâle* (1856); Hans Van Werveke, *De muntslag in Vlaanderen onder Lodewijk van Male* (Mededeligen van de koninklijke Vlaamse Academie voor Wetenschappen, Letteren, en Schone Kunsten van België, Klasse der Letteren, no. XI: 5, Brussels, 1949).
     Frans and Willem P. Blockmans, "Devaluation, Coinage, and Seignorage Under Louis de Nevers and Louis de Male, Counts of Flanders, 1330-84," in *Coinage in the Low Countries (880-1500)*, ed. Nicholas Mayhew (*BAR* International series 54, Oxford, 1979), pp. 69-94 (mint accounts of Ghent, 1335-6, and Mechelen, 1380-82).
  2. *Flanders, 1384-1499*
     a) *Ghent:*
        Archives générales du royaume (Belgique), Comptes en rouleau nos. 824-26 (1388-90), nos. 827-31 (1410-19); Acquits de Lille, nos. 936: 1-8 (1410-22) and nos. 937: 1-26 (1422-47); Chambre de Comptes, reg. no. 580 (1419-29) and reg. nos. 18,195-200 (1459-89); Archives départmentales du Nord (Lille), Chambre de Comptes, Série B, no. 1606 (1441-42)
     b) *Bruges:*
        A. G. R., Comptes en rouleau nos. 776-87 (1392-1402); Acquits de Lille, no. 1512: 1-4 (1454-56, 1466-67); Chambre de Comptes, reg. nos. 18,103-124 (1455-1499); A. D. N.

(Lille), Chambre de Comptes, Série B., no. 644/15,939 (1454-55); no. 19,960/19,312 (1454-55)
   c) *Mechelen*
   A. G. R., Comptes en rouleau nos. 2142-43 (1384-85); Chambre de Comptes, reg. nos. 48,976-77 (1390-92); Comptes en rouleau nos. 2145-46 (copies of 1390-92); Acquits de Lille nos 1512: 2-3 (1454-56) and 1512: 5 (1459-60); Chambre de Comptes, reg. no. 18,242: 1-3 (1485, 1488-89)
   d) *Fauquemont*
   A. G. R., Comptes en rouleau nos. 2586-87 (1396-98)
3. *Brabant, 1375-1499*
   a) *Leuven (Louvain)*
   A. G. R., Comptes en rouleau no. 2591 (1375-1383); nos. 2144 and 2589 (1384-86); no. 2590 (1393-94), no. 2588 (1394-95); Chambre de Comptes, no. 18,064 (1410-11), nos. 18,065-68 (1429-32); nos. 18,069-72 (1466-74);
   Frederic Verachter, ed. *Documents pour servir à l'histoire monétaire des Pays Bas* (Antwerp, 1841), pp. 108-10; and Alphonse de Witte, *Histoire monétaire des comtes de Louvain, ducs de Brabant*, 2 vols. (Antwerp, 1894-96), I, 185-86:Jan.-July 1410 (both from private MSS).
   b) *Vilvoorde*
   A. G. R., Comptes en rouleau nos. 2590, 2592 (1392-93); Chambre de Comptes, no. 18,094 (1417); Verachter, *Documents monétaires*, pp. 107-8 (1409: from A. G. R., Chambre de Comptes no. 48,967, missing since June 1940)
   c) *Maastricht*
   A. G. R., Chambre de Comptes, nos. 18,071 and 18,073 (1419)
   d) *Brussels*
   A. G. R., Chambre de Comptes no. 17,985 (1420-21); nos. 17,986-89 (1434-37)
   e) *Antwerp*
   Alphonse de Witte, *Histoire monétaire*, I, 181 (1405)
   A. G. R., Chambre de Comptes, reg. nos. 17,880-82 (1474-99)
4. *Holland, 1425-1493:* Dordrecht, Zevenbergen, The Hague
   Algemeen Rijksarchief der Nederlanden, Rekeningen van de grafelijkheids-rekenkamer, nos. 4937: I-XVIII (1425-27, 1434-40, 1454-55, 1466-67, 1481-88, 1489-93);
   Accounts in précis in P.O. Van der Chijs, *De munten der voormalige graafschappen Holland en Zeeland* (Haarlem, 1858), corrected by Peter Spufford, *Monetary Problems and Policies in the Burgundian Netherlands, 1433-1496* (Leiden, 1970), appendices (which I used for the Dordrecht account for 1489-93, unavailable to me).
5. *Namur, 1421-1499*
   A. G. R., Chambre de Comptes reg. nos. 18,203: 1-21 (1421-33); reg. nos. 18,204 (1497-1500)
6. *Valenciennes (Hainaut)*
   A. D. N. (Lille), Chambre de Comptes, Série B., no. 17,651 (1434); no. 31 and 19,975/19,402 (1454-55); A. G. R., Acquits de Lille, no. 1512: 4 (1466-67); Spufford, *Monetary Problems*. p. 184 (for 1455-56: from Acquits de Lille, no. 937bis, now missing from the A. G. R.) The Valenciennes mint accounts for Feb. 1437-Mar. 1439 and July 1441-Oct. 1444 are missing (Spufford, p. 177)

Graph I: England's Coinage Outputs
Silver and gold coinage outputs of the English mints, 1300-1499: Decennial means in kilograms of pure metal

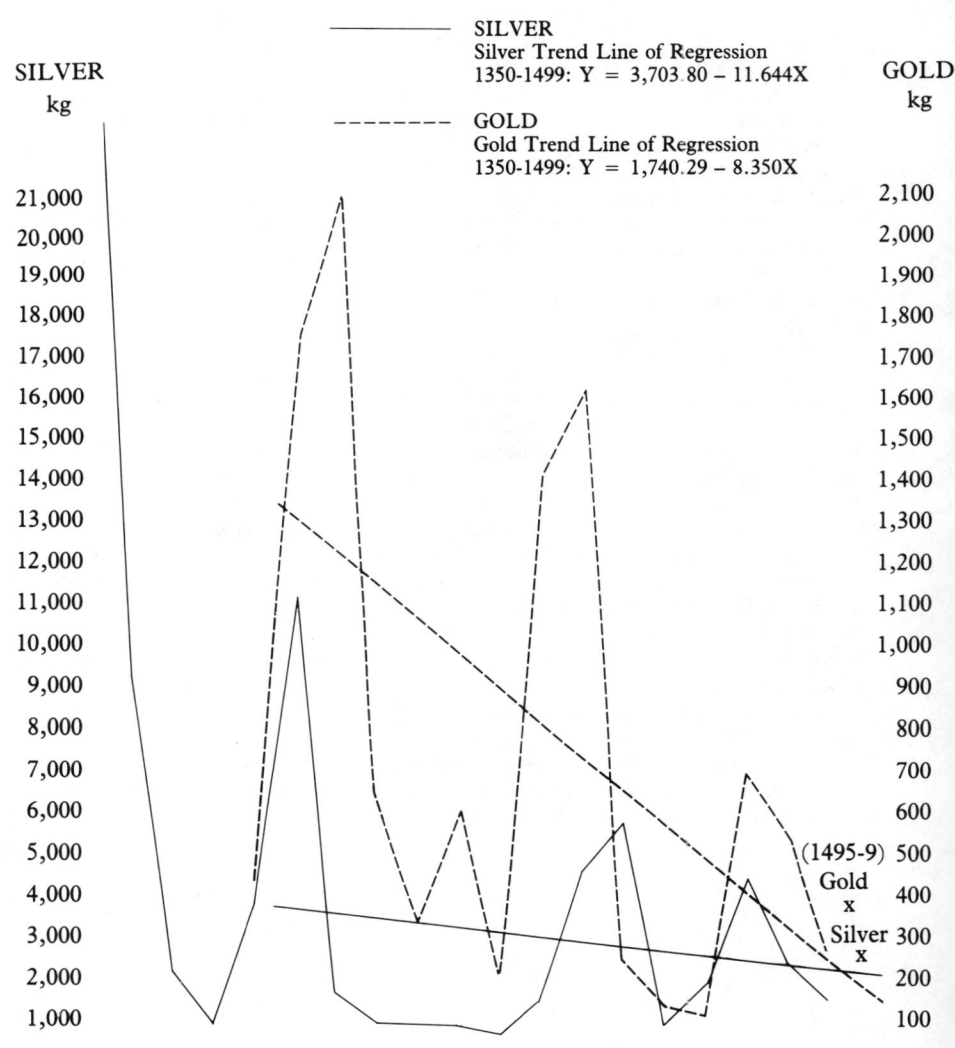

SILVER
Silver Trend Line of Regression
1350-1499: Y = 3,703.80 − 11.644X

GOLD
Gold Trend Line of Regression
1350-1499: Y = 1,740.29 − 8.350X

Bullion flows and monetary contraction 157

Graph II: The Low Countries' Coinage Outputs

Silver and gold coinage outputs of the Low Countries' mints, 1350-1499: Decennial means in kilograms of pure metal

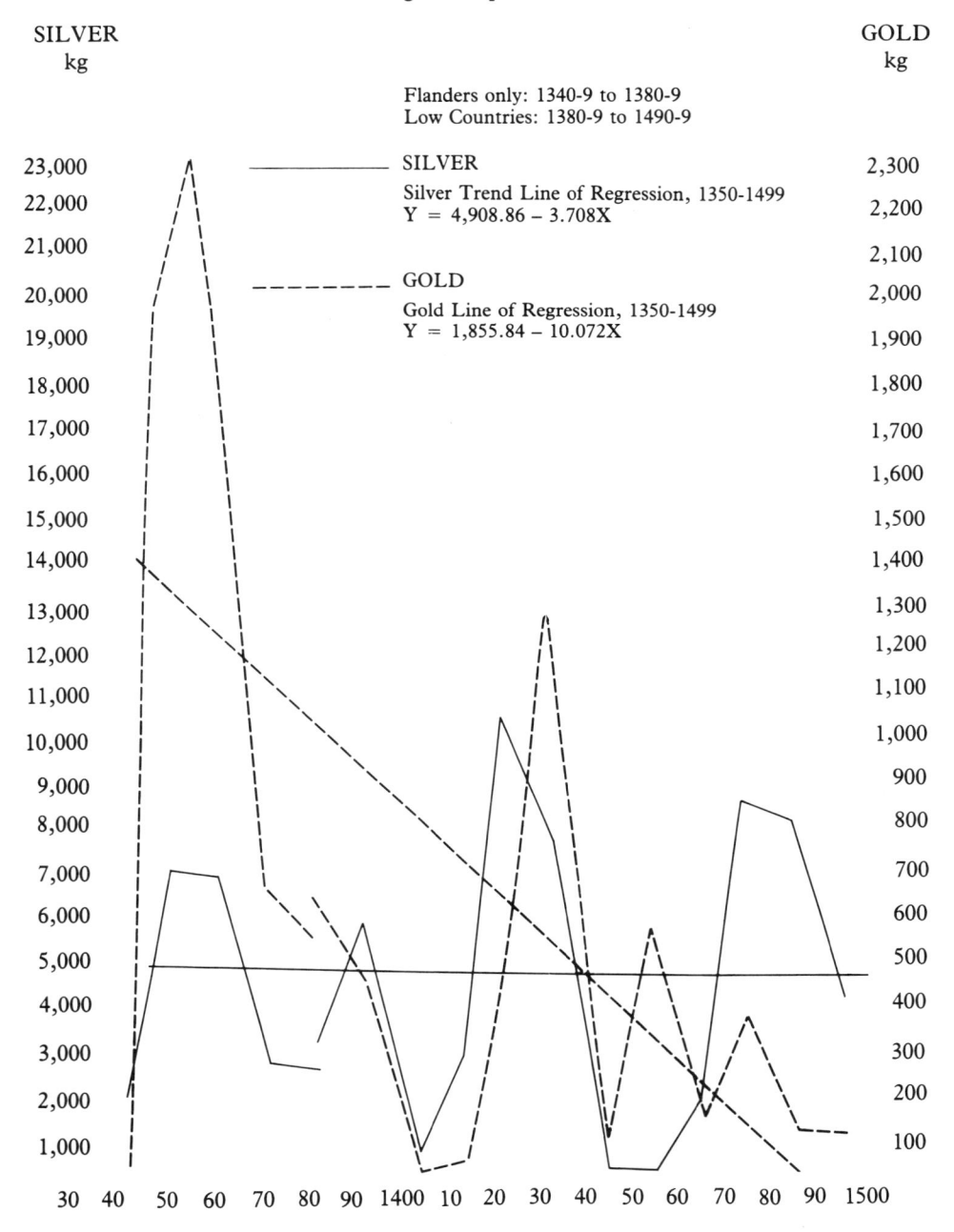

Flanders only: 1340-9 to 1380-9
Low Countries: 1380-9 to 1490-9

——————— SILVER
Silver Trend Line of Regression, 1350-1499
$Y = 4,908.86 - 3.708X$

- - - - - - - GOLD
Gold Line of Regression, 1350-1499
$Y = 1,855.84 - 10.072X$

Graph III: Coinage Outputs of England and the Low Countries

The aggregate silver and gold coinage outputs of England (1290-1499) and the Low Countries (1340-1499): Decennial means in constant pounds sterling of 1351-1411 values

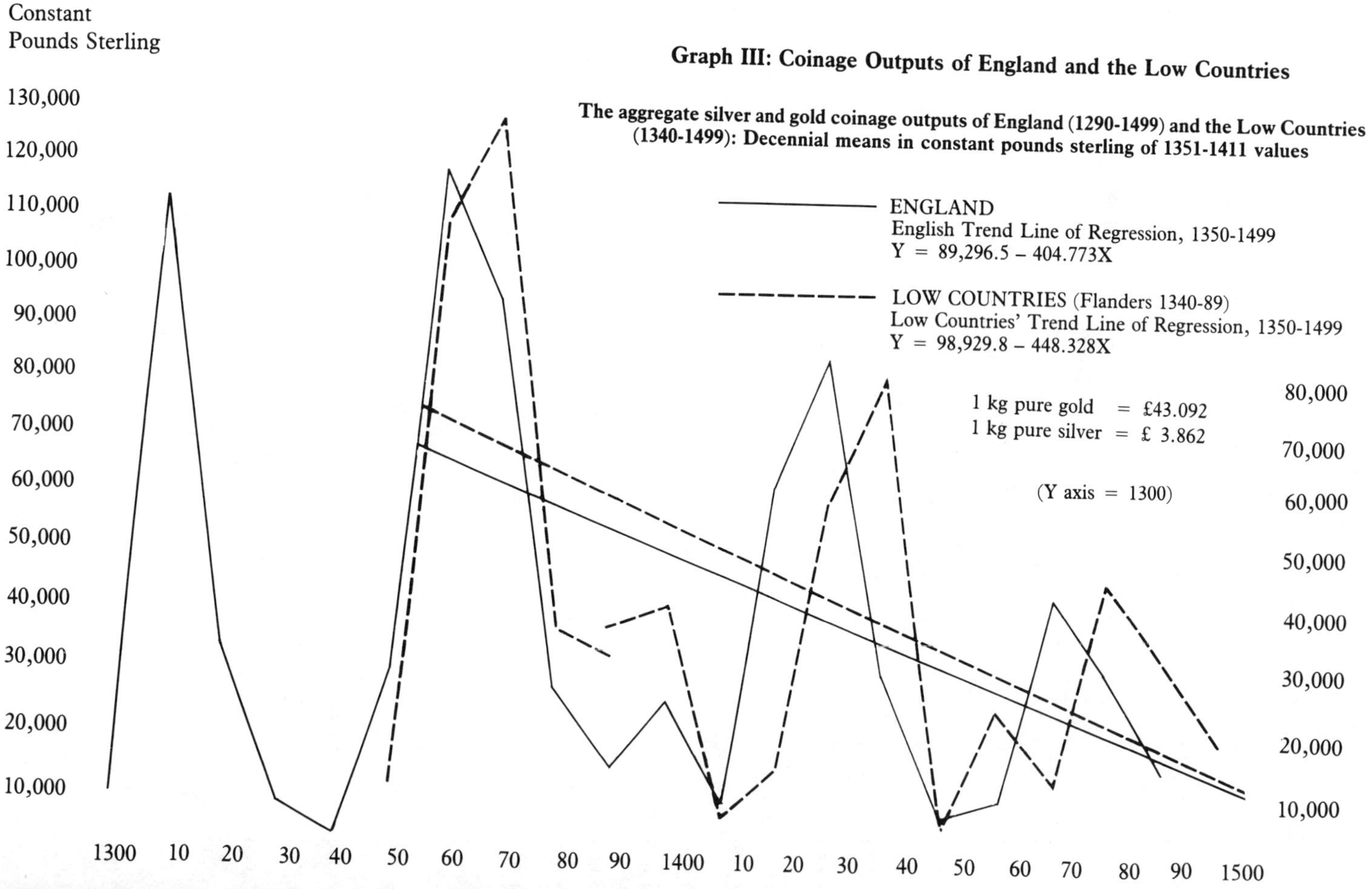

ENGLAND
English Trend Line of Regression, 1350-1499
Y = 89,296.5 − 404.773X

LOW COUNTRIES (Flanders 1340-89)
Low Countries' Trend Line of Regression, 1350-1499
Y = 98,929.8 − 448.328X

1 kg pure gold = £43.092
1 kg pure silver = £ 3.862

(Y axis = 1300)

5

Monetary movements in medieval Egypt, 1171-1517

JERE L. BACHARACH

Geography is the *sine qua non* for understanding monetary movements in medieval Egypt.[1] First, the Nile Valley was a crossroads for trade routes from West Africa, South Asia, Europe and Greater Syria-Iran-Central Asia. Egyptian coins were an important item carried in this international trade. Sometimes metals left Egypt in the form of raw or worked gold, silver or copper. Trade also brought gold, silver and copper into Egypt, and this fact illustrates the second geographic factor. During the years under study, 1171-1517, Egypt, which lacked indigenous raw materials to produce *dīnārs* (gold coins), *dirhams* (silver coins) and *fals* (copper coins), was dependent upon earlier stocks or imports for these metals.

The primary purpose of this essay (within the obvious limitations created by the data found in medieval sources) is to trace the flow of coins and metals into and out of Egypt. Quantitative data are lacking. A second goal is to identify significant turning points within Egypt in terms of its medieval monetary history.

The medieval scholar al-Maqrīzī wrote the first systematic examination of the monetary and numismatic changes in Egypt.[2] His treatise, composed in 1415, has been the basis for almost every modern study.[3] In the mid-1950s Professor Andrew S. Ehrenkreutz broke new ground by combining textual data with the numismatic evidence.[4] By 1970 new information and interpre-

---

1. The term medieval is derived from European usage and implies a time period which ended in the late 15th and early 16th century. It does not imply any particular characteristics to Muslim society.

2. For a brief sketch of al-Maqrīzī's contributions to monetary history, see Jere L. Bacharach's "Circassian Mamluk Historians and Their Quantitative Economic Data," *JARCE*, XII (1975): 77-78.

3. The best edition of the *Shudhūr al-ʿUqūd fī Dhikr al-Nuqūd* is Daniel Eustache, "Études de numismatique et de metrologie musulmanes: II," *Hesperis-Tamudd*, X (1969): 95-190. See p. 189 for a list of earlier editions of this work. One of the first major studies to draw extensively from al-Maqrīzī was M. de Bouard, "Sur l'evolution monetarie de l'Égypte medievale," *EC*, XXX (1939): 427-459.

4. Andrew S. Ehrenkreutz, "The Standard of Fineness of Gold Coins Circulating in Egypt at the Time of the Crusades," *JAOS*, LXXIV (1954) 162-166, and "The Crisis of dinar in the Egypt of Saladin," *JAOS*, LXXV (1956) 178-84.

tations had been published by Ashtor, Balog and Watson, among other scholars.[5] Within the last half decade the number of publications directly and indirectly related to Egyptian monetary developments for the years under study has increased dramatically.[6]

Scholars have divided the 350 years under survey into three periods which coincide with major political divisions:

- the Ayyubid Period (1171-1250),
- the Bahri Mamluk Period (1250-1382), and
- the Circassian Mamluk Period (1382-1517).

The first begins with the Sunni Muslim Kurd, Saladin (1171-1193), who deposed the Fatimid Shi'ite rulers of Egypt (969-1171). Until 1174 Saladin owed allegiance to an overlord in Mosul. After this date he spent most of his career bringing Syria and Northern Iraq under his control, and then fighting the Crusaders. Although the senior member of the Ayyubid family ruled from Cairo after Saladin's death, the Ayyubids of Egypt rarely controlled Syria.

The second or Bahri Mamluk Period begins in 1250, but it was only with the rule of the Sultan Baybars (1260-1277) that a strong government was established over Egypt and Syria. In fact, from 1260 until the Ottoman conquest in 1517, Syria can be considered a province of Egypt. Almost all of

---

5. For the most thorough bibliography for the pre-1963 years, see Subhi Labib, *Handelsgeschichte Ägyptens im Spatmittelalter (1171-1517)* (Wiesbaden, 1965). Prof. Eliyahu Ashtor's contributions include two books: *Histoire des prix et des salaires dans l'orient médiéval* (Paris, 1969); and *Les métaux précieux et la balance des payements du Proche-Orient a la basse epoque* (Paris, 1971). Besides numerous articles, Dr. Paul Balog published *The Coinage of the Mamluk Sultans of Egypt and Syria* (New York, 1964). This is the single most important work on Mamluk numismatics. Dr. Balog has generously shared with the author the manuscript of his forthcoming corpus of "The Coinage of the Ayyubids." I am most grateful for his continuing help and support.

6. Professor Eliyahu Ashtor has published many articles in addition to the two books cited in note 5, works which are directly related to monetary developments. Some of them are "Banking Instruments Between the Muslim East and Christian West." *JEEH*, I (1972): 553-573; "Etudes sur le systeme monetaire des Mamluk Circassiens," *Israel Oriental Society*, VI (1976): 264-287; "Observations on Venetian Trade in the Levant in the XIVth century," *JEEH*, V (1976): 533-586; "Profits from trade with the Levant in the fifteenth century," *BSOAS*, XXXVIII (1975): 250-275; "Quelques problemes que soulève l'histoire des prix dans l'orient médiéval," *Studies in Memory of Gaston Wiet*, ed. by Myriam Rosen-Ayalon, Jerusalem, 1977, pp. 203-234; and "The Volume of Levantine Trade in the Later Middle Ages (1370-1498)" *JEEH*, IV (1975), 573-612. The most detailed study of the Ayyubid and Bahri periods is Hassanein Rabie, *The Principal System of Egypt A.H. 564-741/A.D. 1169-1341* (London, 1972). A number of articles have been written by Gilles Hennequin, including " 'Bonne' ou 'mauvaise' monnaie? Mutations monétaires et loi de Gresham avant l'epoque moderne," *L'information historique*, XXXIX, No. 5, (1977): 203-212; "Mamlouks et métaux précieux," *AI* XII (1974): 37-44; "Nouvequx aperçus sur l'histoire monétaire de l'Égypte a la fin du moyen age," *AI* (1977): 179-215; "Points de vue sur l'histoire monetaire de l'Egypte musulmane au moyen âge, " *AI* (1974): 1-36; and "Problèmes théoriques et pratiques de la monnaie antique et médiévale," *AI* (1972): 1-51. There are three articles directly related to the Circassian period by this author. Jere L. Bacharach, "The Dinar Versus the Ducat," *IJMES*, IV (1973): 77-96, "Circassian Monetary History: Silver," *Numismatic Chronicle*, 7th Ser., XI (1972): 267-281, and "Circassian Monetary Policy: Copper," *JESHO*, XIX (1976): 32-47.

the Bahri sultans were descendants of Sultan Qalā'ūn (1280-1290), although more and more power rested with the Mamluks serving the court. (A Mamluk was a male of slave origin, usually a speaker of a Turkish language who had been converted to Islam, trained as a cavalryman and manumitted. Mamluks formed a foreign-born elite in Egypt and Syria.) In 1382 the "dynastic" rule ended when Barqūq (1382-1399) seized power. The third or Circassian Period, 1382-1517, was marked by periods of rule by a strong sultan, followed by an interregnum, and then another powerful sultan.

This study will follow the standard political periodization, but conclude that for Egyptian monetary history different chronological divisions are more significant. The reign of Saladin (1171-1193), which introduced into Egypt a new dynasty, new military and taxing systems, and for monetary historians, a new era—the age of silver—is the logical place to begin. However, a number of points concerning monetary developments under the preceding dynasty, the Shi'ite Fatimids (969-1171), must first be made.

Egyptian Fatimid coinage was dominated by the use of gold coins.[7] Among these *dīnārs* of exceptionally fine quality many met the canonical *mithqāl* weight standard (4.25 grams). Until the 1070s West Africa was the primary supplier of bullion to Egypt.[8] After that date the source for gold bullion needs, is uncertain. Supplies from local mines, from Ethiopia, possibly some trade with West Africa (undocumented), and existing stocks appear to have been adequate. The origin of the relatively small quantities of silver bullion used for coins, vessels and artifacts is very difficult to determine. It is possible some came from Spain[9] and some through Greater Syria from Central Asia or Anatolia.

Although extant Egyptian Fatimid *dirhams* (silver coins) are very rare, a number of exchange rates between Fatimid *dīnārs* and *dirhams* can be found in the sources.[10] Individual market rates could always vary, but the majority of rates can be converted into a gold: silver ratio of 1:93. In view of the many pitfalls in calculating gold: silver ratios, we should explain that this ratio derived from the following calculation: one "pure" *dīnār* of a *mithqāl*

---

7. P. Balog, "History of the Dirham in Egypt from the Fatimid Conquest Until the Collapse of the Mamluk Empire," *Revue Numismatique*, 6ᵉ Ser. III, (1961): 123.
8. R. Messier, "The Almoravids: West African Gold and Gold Currency of the Mediterannean Basin," *JESHO* 17 (1974): 31-47. Material on later West African-Egyptian trade ties can be found in Ashtor, *Les métaux précieux*, pp. 18-22; Nehemia Levtzion, *Ancient Ghana and Mali*, (New York, 1973), pp. 124-135; and Jean Devisse, "Routes de commerce et échanges en afrique occidentale en relation avec la Méditerranée: Un essai sur le commerce africain médiéval de XIᵉ au XVIᵉ siècle," *Revue d'histoire economique et sociale*, L (1972): 42-73, 357-397.
9. The most important work for the Egyptian/Indian trade is S.D. Goitein's "Letters and Documents on the Indian Trade in Medieval Times," (reprinted) in *Studies in Islamic History and Institutions* (Leiden, 1966), pp. 329-350. Additional data can also be found in S. D. Goitein *Letters of Medieval Jewish Traders*, (Princeton, 1973); especially pp. 175-230. For Spanish silver see Goitein, *Letters*, p. 26.
10. S. D. Goltein, "The Exchange Rate of Gold and Silver Money in Fatimid and Ayyubid Times," *JESHO*, VIII (1965): 30-35.

weight was worth 40 *dirhams* of 33% silver, and the *dīnār* and *dirhams* have a 7:10 ratio in terms of their respective weights. When the gold: silver ratio of 1:9.3 was first established in Egypt is not clear, but during the Fatimid era this ratio was the most comon, and as this essay argues, was the standard to which all later minor or temporary changes returned until the 15th century. In the first decades of that century the 1:9.3 gold: silver ratio, which had been the norm for at least four hundred years, was replaced by new ratios that reflect a significant shift in the relative value of gold and silver. (This argument does not affect the corrolary statement that short-term variations in the ratio did have a very real impact on merchants, government business, etc.)

During the last years of Fatimid rule the Egyptian treasury was faced with a serious bullion shortage. The local gold mines had run dry, the tombs of the Pharoahs had been sacked, and the treasury was empty. The rulers had created a financial drain by maintaining a lavish court, and, more importantly, paying large sums of gold to placate or fight the Crusaders. There may even have been a disruption of the gold trade from the Western Sudan because of the activities of the Almohads in North Africa. One result of the crisis was the fact that Fatimid *dīnārs* no longer weighed the canonical *mithqāl* weight standard (4.25 grams). However, historians know from specific gravity tests that the *dīnār*'s fineness was not lowered.[11]

The monetary policies of Saladin have been analyzed by more scholars than the policies of any other Egyptian ruler. Numismatists have focused upon the Sunni inscriptions and styles of his coinage, the diminution of the weight standard for the *dīnār*, and its debasement. Students of monetary history have stressed the last two aspects, as well as Saladin's use of the *dirham* as the basic unit of the monetary system. There has also been a debate over the quality of Saladin's Egyptian *dirhams*. The impact of all these developments has always been studied from the short-run point of view; that is, the reign of this ruler. This essay re-examines Saladin's policies in the context of a longer time span.

The medieval scholar Al-Maqrīzī, described the monetary crisis associated with Saladin's reign as follows: "Gold and silver left the country . . . [and that] to say the name of a pure dinar was like mentioning the name of a wife to a jealous husband, while to hold such a coin in one's hand was like crossing the doors to paradise."[12] It is true that Saladin's need for money was at least as great as that of the last Fatimids. He had to maintain a large army, and until 1174, he was expected to send money to his overlord in Mosul. He met his needs by using less gold in each *dīnār*.

Professor Ehrenkreutz was the first to prove that Saladin debased the *dīnār*.[13] With very few exeptions, Egyptian *dīnārs* had always been be-

---

11. Ehrenkreutz, *JAOS*, 1954.
12. *Shudhur*, p. 127.
13. Ehrenkreutz, *JAOS*, 1956.

tween 98% and 100% gold. All *dīnār/dirham* exchange rates had been calculated as if the *dīnār* weighed a *mithqāl* and was pure gold. Even when *dīnārs* did not weigh a *mithqāl* (as under the last Fatimids), it was easy to weigh a number of them and calculate how many *mithqāls*' worth of stamped pure gold were present. Thus, debasing the *dīnārs* created a far more serious problem for all who used the coin than just dropping a fixed weight standard. It was not easy to determine the exact percentage of gold in the coin.[14]

There probably was an immediate inflationary effect created by Saladin's debased *dīnārs*, but the very limited records on prices for his reign do not indicate that Egypt suffered a major crisis.[15] It is even possible that after his initial windfall profits, Saladin mixed good and bad *dīnārs*, and exchange rates reflected the value of the good ones.[16] In any case, immediately after the reign of Saladin, the Ayyubid rulers of Egypt minted almost pure gold coins, and for the next two hundred years the Egyptian *dīnārs*, with a very few exceptions, were of a very high degree of fineness (see Table 1 below).

Besides debasing the Egyptian *dīnārs*, Saladin changed the basis of record-keeping from *dīnārs* to *dirhams*:

In the Judaeo-Arabic documents dating from the Fatimid period which have been found in the Cairo geniza, prices and values are indicated in dinars, whereas in those of the Ayyubid period they are fixed in dirhams.[17]

In addition, according to al-Maqrīzī, Saladin issued new *dirhams* in 1187 composed of 50% silver to replace the Fatimid and early Ayyubid black *dirhams* (*dirham waraq*), which contained only 33% silver.[18] Modern scholars, such as Rabie and Ashtor, believe that this event marked the introduction of a silver standard in Egypt, and that gold "considered from then on only as a commodity with a daily fluctuation market price to be calculated in silver dirhams, lost its position as the standard of currency."[19]

The history of Saladin's silver issues is difficult to reconstruct because of the lack of data. The number of known Syrian Saladin pieces is almost four times the number of extant Egyptian ones.[20] Among the latter, only 10% of

---

14. A. S. Ehrenkreutz, "Extracts From the Technical Manual on the Ayyubid Mint in Cairo," *BSOAS*, XV (1953): 502-514. J. L. Bacharach, "Foreign Coins, Forgers and Forgeries in Fifteenth Century Egypt," *Actes du 8ᵉᵐᵉ congres international de numismatique* (Paris-Bale, 1976), pp. 504-505.
15. Ashtor, *Historie des prix*, pp. 124-133.
16. The limited data for *dīnār/dirham* exchange rates is to demonstrate major changes in the monetary conditions of Saladin's Egypt. Before he began debasing the *dīnārs*, the *dīnār/dirham* exchange rate was 1:40 as a gold/silver ratio of 1:9.3. The one datum from the middle of his reign is for a *dīnār/dirham* rate of 1:37.5 (*Shudhūr*, p. 131). If it is assumed that the *dīnār* is pure gold, then the gold/silver ratio is 1:8.8, but if the exchange rate adjusted to a debased dīnār of 94% gold, then the gold/silver ratio remained constant at 1:9.3. By the end of Saladin's reign the ratio was definitely 1:9.3.
17. Ashtor, *A Social and Economic History*, p. 239.
18. *Shudhur*, p. 129.
19. Ashtor, *ibid.*, p. 239; and Rabie, p. 174.
20. Balog, *Catalogue of Ayyubid Coins*, unpublished.

them have the shape of the new 1187 issue. None of these pieces have been tested for their degree of fineness, and there is no contemporary market record indicating that they were purer than the "black *dirhams*" of one-third silver. Thus, we may assume that there were few silver coins in Saladin's Egypt at the beginning of the "age of silver" and they were of relatively poor quality.

The switch to a monetary system, based on silver rather than gold, does not necessitate major economic or monetary changes. Assuming a relatively free market—and it is only under the Circassian Mamluks that this state cannot be assumed—then exchange rates and pieces can adjust easily to a new bookkeeping system. Saladin's policy of using *dirhams* for recordkeeping did not have to be disruptive. He probably initiated the policy for the following reasons: By calculating wages, prices, etc., in *dirhams* and paying in debased *dīnārs*, the state acquired a windfall profit; and as Saladin spent his early career in Northern Iraq where silver was relatively abundant, he may have come to think of it as the standard currency and simply applied his earlier experience to Egypt.

If, as argued, Saladin's internal monetary policy was not significant in its long-term effect, did his policies have a major effect on monetary and bullion flows and/or Egypt's balance of payments? The type of data used by modern economists to answer either question is lacking, but possible answers can be suggested.

Saladin's active military policies against Muslims and Crusaders in Greater Syria necessitated increasing demands on the Egyptian treasury. The debasing of his *dīnārs* is one piece of evidence. These and other available *dīnārs* (and gold) were shipped to Syria to pay his troops. This Syrian drain probably ended with his death, since most of his successors in Egypt spent their revenues locally. Using the number of existing Egyptian Ayyubid *dirhams* as evidence, there was a relative increase in the flow of silver to Egypt. The silver, as booty and trade, probably came from Syria, now under Ayyubid control, where it was more abundant. The role of the Indian and European trade in the balance of payments is impossible to calculate. Professor Ashtor states that trade in both areas picked up.[21] This means that more spices were traveling East to West, while the Indian market demanded finished goods, chemicals and metals, especially copper and gold.[22] The position of this author is that Egypt had a temporary shortage of gold, but not a balance-of-payments deficit. In short, from the details cited, we may argue that Saladin's monetary policies were neither radical nor revolutionary, nor, in the long term, significantly different from those of the later Fatimids.

According to al-Maqrīzī and modern scholars, the next major development in the monetary history of Egypt took place during the reign of al-Kāmil (1218-1238). The medieval historian informs us that al-Kāmil

21. Ashtor, *ibid.*, p. 241.
22. Goitein, *Letters*, p. 175.

issued, in 1225, new silver coins which were to be composed of two-thirds silver.[23] In the words of Dr. Balog: "The reform was a huge fraud. Instead of being better, the average silver content of the round dirham was now even a little less than that of the dirham waraq."[24] One possible explanation for al-Kāmil's claim that he issued a finer *dirham* was that he could pay some debts with them at a favorable rate before the market realized that these *dirhams* were in fact the same quality as the older ones. This windfall profit policy would have hurt some individuals. However, this policy and that concerning copper coins had no significant effect on monetary flows or the balance of payments.

In the same year of 1225 al-Kāmil issued copper coins (*fals;* plural: *fulūs*) as an official currency in order to facilitate small-scale exchanges. The quantity of copper available was very limited, however, and within a short time the circulating coins were being debased by even cheaper materials. Egyptian copper was probably being shipped to India.[25] Therefore, by 1233 the *fulūs* were officially withdrawn from circulation.[26] In fact, copper as a local Egyptian currency came to play a major role only in the latter half of the 14th century.

A more fundamental change in the Egyptian monetary situation transpired during the period encompassing the reigns from that of the Ayyubid al-Ṣāliḥ (1240-1249) through that of the Bahri Mamluk Baybars (1260-1277). Gold *dīnārs* continued to be minted in varying weights, but of high quality. The new element was a significant change in the quantity of silver in circulation in Egypt and in the *dirhams*'s degree of fineness. The new Egyptian silver coins, composed of approximately two-thirds silver, were closer to the theoretical weight of a silver *dirham* (2.975 grams) than any earlier Ayyubid Egyptian coins. While many scholars have noted the improvement in quality and changes in inscription and design, there has been no discussion of why silver became relatively abundant.

During the reign of al-Ṣāliḥ Ayyūb, Syria came under his control. As ruler of Damascus, he was able to gain control of the locally circulating Syrian, Crusader and Central Asian silver coins. This money was the bullion source for his Egyptian silver coins. In addition, the weight and degree of fineness of his new Egyptian pieces were similar to those that were circulating in Syria.

Between the death of al-Ṣāliḥ Ayyūb and the reign of Baybārs, Syria was, once more, independent of Egyptian control. Based on numismatic studies, the quality—and probably the quantity—of Egyptian silver issues declined.[27] Egypt lacked Syria's silver resources. Thus it is probable, based

23. *Shudhur*, p. 131.
24. Balog, *RN*, p. 130.
25. Goitein, "India Trade," p. 340.
26. Bouard, p. 451; and Rabie, p. 183.
27. The data were gathered from Dr. Balog's catalogs on Ayyubid and Mamluk coinages. Some silver came from Louis IX's troops who fought in Egypt. Ashtor, *ibid.*, p. 292.

on one indirect datum, that silver was still more highly valued in Syria than it was in Egypt.[28] The result was that, although Egypt minted *dirhams* using bullion which it had derived from Syria, the coinage tended to flow back to Syria.

Baybārs reconquered Syria after the defeat of the Mongols in Palestine in 1260. He confiscated Syrian and Crusader silver coins, carried them back to Egypt, and had them reminted as his own new-style coinage.[29] The relative abundance of Egyptian *dirhams* for his reign may have been the cause for a shift in the gold: silver ratio for Egypt, whereby gold increased in its relative value.[30] If, as has been argued, silver was more valued in Syria than in Egypt when the latter's gold: silver ratio was 1:9.3, then the new situation only increased the profitability of shipping silver back to Syria. Baybārs' own policies of sending large sums in *dirhams* to Syria as gifts, or to pay for garrisons, only intensified the flow of silver from Egypt.[31] The net result was that the older Egyptian gold: silver ratio was re-established.

Unfortunately, data are lacking for a more precise analysis of monetary changes, bullion flows and balance-of-payments conditions. We can assume that enough gold flowed into Egypt and remained to meet its needs, or at least to avoid debasement of the *dīnār*. Copper was not a major metal for coinage. Egypt probably retained a balance-of-payments surplus. A relatively new development pattern under Baybārs however was the influx of Syrian silver and its apparent return to the Levant. Thus, it is not surprising that the reigns of succeeding Bahri sultans are noted for their increasing lack of *dirhams*.

The next major attempt to remedy Egypt's silver shortage took place during the third reign of al-Nāsir Muḥammad (1310-1341). Following a series of military campaigns into Anatolia, the Kingdom of Little Armenia began, in 1323, sending an annual tribute of 100,000 *dirhams* to Egypt. The need for *dirhams* within the Mamluk Empire was so great that large numbers of the Armenian coins were merely overstruck with Muslim inscriptions rather than being reminted.[32] At the end of 1339 Sultan al-Nāsir Muḥammad emptied his treasury of 2,000,000 *dirhams*, but according to an Arab chronicler, this still did not meet the market needs.[33] The number of extant Egyptian silver coins continued to decline.

The disappearance of Egyptian *dirhams*, after the Sultans al-Ṣāliḥ Ayyūb, Baybārs and al-Nāsir Muḥammad were able to import large quantities of

---

28. Balog, *RN*, p. 129; and Ashtor, *ibid*.
29. The most detailed study of his coinage is Michael L. Bates, "The Coinage of the Mamluk Sultan Baybars I: Additions and Corrections," *ANSMN*, 23 (1977): 151-174.
30. Ashtor, *Les métaux précieux*, pp. 37-38.
31. *Ibid.*, p. 38.
32. Balog, *Mamluk Coinage*, pp. 146-147; and Ashtor, *Israel Oriental Society*, p. 274.
33. Rabie, p. 194. Al-Maqrīzī, *Sulūk*, 11: 488, records an exchange rate of one *dīnār* for *dirhams*. If these *dirhams* were only 50% silver, which is possible based on the tests of Dr. Balog, then the gold/silver ratio would be about 1:9.

coins and bullion, raises the question of the causes for this loss. Perhaps the simplest explanation is that Egypt valued gold more than silver relative to its immediate trading partners. Thus Ayyūbid Egypt found it profitable to export silver to Syria and possibly to Europe for gold, local products and future Mamluks who were carried by Italian slave merchants from the Black Sea. In the opposite direction, some of the silver flowed to Yemen and was used in the Indian trade.[34]

The reign of al-Nāṣir Muḥammad is also associated with a series of developments affecting the *dīnār*, but, as in the case of silver, the basic long-term relationships remained unchanged. The first change in the status of the Egyptian gold coins took place during the second reign of al-Nāṣir Muḥammad (1291-1309). In order to buy the loyalty of his troops and to meet the expenses of raising, in 1299, a large army to fight the Mongols of Iran, al-Nāṣir issued large numbers of gold coins. The Arabic sources record an exchange rate of one *dīnār* equalling 17 *dirhams*.[35]

The standard interpretation for this new rate is that the market, flooded with *dīnārs*, saw their value declined relative to silver. This may be true, but we may suggest an alternate interpretation. Specific gravity tests of al-Nāṣir's *dīnārs* for this period prove that they were more debased than almost any Egyptian *dīnār* since Saladin's reign. Therefore, the drop in exchange rates could be interpreted as market adjustment on the basis of a gold: silver ratio of 1:9.3 to debased *dīnārs*. One other datum can be used to support this argument. According to Professor H. Rabie, the Venetian ducat issued in 1284 was first accepted in Egypt in 1302.[36] The temporary interest in the Venetian gold coin may have been due to al-Nāṣir debasing the *dīnār*. The ducat may have been the source of bullion for al-Nāṣir's 1299 issue. In any case, the Venetian ducat did not play a major role within the Mamluk Empire of Egypt, Syria and the Hijaz until after 1399.

In one other period—this time during al-Nāṣir's third reign—the value of gold relative to silver may have dropped due to the flooding of the market with large quantities of gold. This event is associated with the 1324 pilgrimage of the West Sudanic (Takrūr) leader Mansa Musa. A number of West African pilgrimages are recorded in the Arab chronicles after Mansa Musa, but none was on as grand a scale. All of these African caravans carried ample supplies of gold.

How much gold was imported from Africa during the years surveyed, and whether there were major changes in the trade pattern, cannot be documented. It is possible that an increasing quantity of African gold dust was

---

34. Fourteenth century Yemini Rasulid *dirhams* were found in the Broach, India hoard of 1880. W. F. Prideaux "Coins of the Benee Rasool Dynasty of South Arabia," *JBBRAS* XVI (1883-85): 8-16.

35. Al-Maqrīzī, *Sulūk*, 1:899.

36. Rabie, p. 189. The coin hoard found in Broach, India in 1880 also included Mamluk and Venetian pieces. The earliest ducat was from the reign of Doge Gradenigo (1339-1342). O. Codrington, "On a Hoard of Coins Found at Broach," *JBBRAS* XV (1881-1882): 360.

diverted north in the 14th century to meet Europe's need for bullion for their new gold currencies, but then, European traders also carried more gold coins to Alexandria where they bought spices. Therefore, a shift in the Sudanic trade did not necessarily mean a significant decline in the quantity of gold metal available for the Mamluk sultans to remint into their own coinage, or to export.

The monetary history of Egypt for the last half of the 14th century represents a transitional period. Many of the patterns which date from before Saladin's reign continued through the end of the Bahri "dynasty."

In 1382 a major political change occurred in Egypt. The Circassian era, beginning with the reign of Barqūq (1382-1399), was marked with alternating periods of stability and instability until the Ottoman conquest of 1517. But the general monetary picture did not change. Mamluk *dīnārs* continued to be almost 100% gold and of varying weight. *Dirhams* were still one-third silver and in relatively short supply. The gold: silver ratio remained, on the average, 1:9.3. The new trends from mid-century were the increasing appearance of European coins, especially ducats. Data from the Arabic sources indicate that these coins traded at a rate relative to their intrinsic value,[37] while hoard evidence proves that some of them were carried to India.[38] Copper was becoming a more important European export item.[39] Copper was used for Mamluk *fals*,[40] but some of it was probably exported to India. It is possible that, due to a number of factors including the recurring impact of the Black Death, the increasing rapacity of the ruling Mamluks, significant levels of hoarding, ect., Egypt suffered a balance-of-payments deficit.

A critical turning point in Egyptian monetary history took place immediately following the death of Barqūq. His son Faraj (1399-1412) became sultan, but was challenged in that position by numerous rivals. There was a sudden lack of confidence in the government. Faraj needed vast quantities of gold as quickly as possible to buy the support of his troops. He probably used money stored in the treasury, including European gold coins which had not been reminted. The net result of his action was that the European coinage—and in particular the Venetian ducat—began to circulate widely within the Mamluk Empire. In fact, the ducat became the most important gold coin within Egypt until the mid-1420s.

As this writer has argued earlier, the triumph of the ducat was not due to

37. Bacharach, "Dinar," 79-80.
38. The most recent Venetian ducats in the Broach hoard are dated from the reign of Doge Contarini (1367-1283).
39. Professor Ashtor discusses European copper exports to Egypt in a number of his works, e.g., *Les métaux précieux*, p. 85.
40. The trend toward the use of copper and the dropping of silver intensified in 1391 when an official named Jamāl al-Dīn Maḥmūd al-Ustadār (d. 1397), with the unofficial support of the sultan, sold Europeans Egyptian silver and flooded the market with copper. The ultimate result was that the quality of the *dirham* repreatedly declined, but at the same time the *dīnār/dirham* exchange rate increased, with a net result that the 1:9.3 gold/silver ratio was maintained. Bacharach, "Copper," p. 35 and "Silver," p. 270.

its having a higher degree of purity than that of the Muslim coin, nor even a uniformity of weight which was lacking in the Muslim *dīnārs*, but because, in terms of Gresham's Law, it was "bad money."[41] Specifically, the Venetian ducat was overvalued in relation to the existing Muslim *dīnārs*. At the very most, the ducat contained a little over 80% of the gold in a theoretical *mithqāl* Muslim coin, but was valued at 90% to 95% of the theoretical Muslim gold coin.[42] Because of the political crisis of 1399, large numbers of the overvalued ducats were circulating in the interior of Egypt and Gresham's Law went into effect.

As indicated before, silver coins were being constantly debased because of a shortage of the metal. This situation reached a point of no return in 1403 when Egypt suffered one of its worst famines. Prices skyrocketed. This pressure, together with the continued debasement of the *dirham*, broke the traditional relationship between gold and silver, which had existed at least from the Fatimid era. According to a contemporary Arab source, all transactions were being undertaken in copper, not silver.[43] The problem was that there had never been a traditional relationship between copper and gold or silver. The opportunity to manipulate the exchange rates, as well as pressures from price rises, created a situation in which even copper exchanged as a commodity. All accounts came to be calculated in a new money of account called "*dirhams* of copper" or "trade *dirhams*"—from the Arabic *darāhīm min al-fulūs* (*dirhams* of *fulūs*). Unlike many other systems in which a money of account is used, the "trade *dirhams*" had no fixed relationship to any coinage, and therefore all coins were calculated in as many "trade *dirhams*" as market conditions warranted.[44]

Silver coins were in short supply[45] and even copper began to disappear from the Egyptian market. In this case it is more difficult to determine the effective causes for this deficiency. Into the 1430s we find references to an increased shortage of copper and its continual debasement by cheaper metals.[46] A decline may have occurred in copper imports from Europe, but far more significant was the patterns of internal use and exportation. The Arab historians do speak of the domestic use of copper for vessels and other

41. Bacharach, "Dinar," 80.
42. Ashtor, *Israel Oriental Society*, pp. 267-271. The percentage of gold in Venetian ducats is based on an analysis of a collection acquired in Egypt and reflects only one sample of pieces minted for the Egyptian trade.
43. al-ʿAynī, *ʿIqd al-Jumān fī Taʾrīkh Ahl al-Zamān* (MS. Cairo: Dār al-Kutub al-Miṣriyya, Taʾrīkh, No. 1104), IV: 187.
44. Bacharach, "Silver," p. 270.
45. Professor Boaz Shashon of Beersheva University, Israel, has shared with me a number of his forthcoming studies on Circassian monetary and economic history. I am convinced by his arguments for the impact of copper and the differences between official and market rates. He also argues that the gold: silver ratio for 1403-1410 should be 1:7 and not 1:14 (Table III). I am not yet convinced by this argument but greatly appreciate the opportunity to have read his material in manuscript. I look forward to its publication.
46. Bacharach, "Copper," pp. 41-42.

utensils. Of far greater importance is the shipment of this metal eastward. As part of the trade to India, copper was shipped through the Hijaz and Yemen. Calls to stop the exportation appear to have been unsuccessful, as merchants recognized that copper had a greater value further East than it did as domestic currency.[47] By the end of the first decade of the 15th century, Egypt appeared to be losing its gold, silver and copper coinage, due to an unfavorable balance-of-payments and poor domestic monetary policies.

Muslim gold, silver and copper coins were to return to the Egyptian scene, but at different times and in different circumstances. Silver, reappearing first, did not re-establish the monetary preeminence it had during the Ayyubid and Bahri Periods. A new Muslim gold coin became the major gold piece for the Mamluk Empire and greatly influenced its Muslim neighbors. During the last quarter of the 15th century copper coins became available in abundant number. A detailed analysis of all three developments has been given by this author in other works and the following should only be considered as a summary of more elaborately detailed and fully documented arguments.

The silver shortage in Egypt ended during the reign of al-Mu'ayyad Shaykh (1412-1421). In early 1412 a rival, Nawrūz (d. 1414), controlled Syria. Nawrūz began to issue a new silver coin of roughly 1.45 grams of almost 100% silver. The degree of fineness and weight were based on the Venetian *grosso* which was circulated in Syria. Nevertheless, the weight of the new coin at a half *dirham* (or *niṣf*) fit into the traditional Muslim *dirham* weight system. However the exchange rates established for the *niṣf* and *grosso* made the *niṣf* overvalued in relation to the Venetian coin.

Mu'ayyad Shaykh attempted to issue a similar coin in Egypt but lacked the bullion resources to do so. In December 1414 he defeated Nawrūz and seized Syria as well as Nawrūz' treasury. The Egyptian market suffered from a brief inflation when it was flooded with the new Syrian *niṣf* coinage, but Shaykh now had enough bullion to issue his own silver in large quantity. Despite the Sultan's attempt to make silver the basis of all calculations, inadequate silvers stocks meant that the "trade *dirham*" system continued to be used in Egypt.

For the next fifty years good quality silver coins were issued in Egypt. However, in order to make a limited supply of bullion go further, the weights of the coins were systematically lowered by each succeeding ruler. In fact, one Sultan even inscribed 3/8ths on his coin, which had been modeled after the *niṣf* (4/8ths) coin of his predecessor. Other problems, at least according to one medieval author, were that such silver was being hoarded in the form of domestic vessels or was being exported to the Hijaz,[48] though in what quantity is unknown.

---

47. *Ibid.*, p. 43.
48. al-Maqrizi, *Sulūk*, IV: 977. Ibn Hajar al-ᶜAsqalāni, *Inbā' al-Ghumr bi-Anbā' al-ᶜUmr* (MS. Cairo: Dār al-Kutub al-Miṣriyya, Ta'rīkh, No. 2476), II: 326, v.

The return of a Muslim gold coin is associated with Sultan al-Ashraf Barsbāy (1422-1437) and his coin, the *ashrafī*. Earlier attempts to supplant the ducat failed because the Muslim coins were always undervalued in relation to the ducat. In the few cases when they were not, there was not enough Muslim gold minted to put Gresham's Law into effect to reverse the dominance of the ducat. In 1425 Barsbay began another attempt to supplant the ducat. His new gold issue, called the *ashrafī* was based on the weight of the ducat, or rather it weighed slightly less than the ducat, and was to be traded at par. His initial attempts to have his coin accepted in the market appear to have failed.

In 1429 after he reminted 50,000 ducats which had been paid as ransom by the Cypriot King James I, Barsbāy was able to place large numbers of his coins on the market. By enforcing exchange rates that made his coin overvalued in relation to the Venetian money, Gresham's Law went into effect. For all intents and purposes, the ducat disappeared from the interior of the Mamluk Empire. Ducats continued to circulate in Egypt after 1429. Sometimes they were overvalued or undervalued in relation to the *ashrafī*, but the basic relationship that Barsbāy had established, and which his successors maintained, was that, in market terms, the Muslim coin acted as the bad money and the Venetian money as the good.

The export pattern of the ducat and the *ashrafī* appear to stabilize for the rest of the fifteenth century. Gold coinage moved from Egypt (and probably Syria) into the Hijaz to Aden, and then on to India.[49] The quantities can never be calculated, but it appears that gold and copper played a significant role in paying for the spices which were in ever-increasing demand in the West.

The re-appearance of copper coins and the end of the Mamluk gold/silver monetary system can be associated with the long reign of Sultan Qayitbāy (1468-1496). We find no evidence of a major disruption in the supply of gold from Europe and the Western Sudan during his reign. It is even possible that the quantity of coins imported by the Venetians grew as the value of the trade in spices increased.[50] Qayitbāy also imported large quantities of copper which were re-issued as *fulūs*, probably to meet local needs for small change.[51] Presumably the eventual state monetary crisis was not due primarily to supply patterns.

Qayitbāy perceived himself as a pious Muslim ruler. In particular, he translated this belief into an extremely active building policy, endowing a large number of mosques and schools and assigning them local revenues. In other words, a growing percentage of Egypt's internal economy was geared to economically non-productive activities. State revenues from internal sources

---

49. Abu Muḥammad al-Tayyib, *Qilādah al-Nahr fī Wafyāh aʿyān al-Dahr* (MS. Cairo: Dār al-Kutub al-Miṣriyya, Ta'rīkh, No. 167), II: 1163; and Bacharach, "Dinar," pp. 88-89.
50. Eliyahu Ashtor, "The Volume of Levantine Trade," pp. 609-612.
51. Ashtor, *Les métaux précieux*, pp. 82-86.

declined. At the same time, and often for reasons of religious prestige Qayitbāy was involved in a series of wars against the Ottomans in Anatolia. Millions of *dīnārs* were spent on his numerous campaigns. Troops had to be paid; weapons and horses and slave manpower (future Mamluks) had to be imported. As most of the campaigns occurred outside his empire, even the money spent by his troops did not stay to be recirculated. The result was that the *dīnār* was reduced in weight and debased, a policy which his successors intensified during the first decade-and-a-half of the next century. Whatever importance the Portugueses rounding of the Cape may have had on spice prices, and earlier European activities in Africa on the possible flow of Sudanic gold to Egypt, it was Qayitbāy's own policies which caused the final, significant disruption of the Mamluk gold system.

Dirhams, which had been in short supply anyway, but which had been since 1415 an almost pure silver coin, were debased. This trend continued until the Ottoman conquest of 1517. Finally, in order to meet local domestic needs and possibly the demands of the Indian trade, copper became a major import item and large numbers of *fulūs* circulated. From the numismatic evidence, even these coins appear to be of poor quality after his reign. On the eve of the Ottoman conquest, all three currencies—gold, silver and copper —were of poor quality and in limited supply. It was the Ottomans, often pictured in modern literature as desecrators of Egypt, who brought political stability, economic development, and an improved monetary situation to Egypt.

The monetary history of Egypt from 1171 to 1517 can be divided into two periods which do not conform precisely with the traditional political divisions. The first period began before Saladin and continued to 1399. One feature was the relative stability of the gold: silver ratio. Despite a few short-term disruptions—e.g., under Saladin, al-Ṣāliḥ Ayyūb, Baybārs and al-Nāṣir Muḥammad, and daily variations—the vast majority of exchange rates reflected a 1:9.3 ratio. A second feature was the relative stability of the coins themselves in terms of their quality and relative quantity. *Dīnārs* were almost always of the highest purity, but had no fixed weight standard for individual pieces. They could not be traded by number. The weight of individual *dirhams* was closer to a fixed standard, but the quality was significantly poorer. With minor exceptions, the supply of gold relative to silver remained constant, permitting the gold: silver ratio to remain constant. Finally copper did not play a major role as a local currency, although it, along with the gold and silver, was an important metal carried in the international trade.

When the period under study began, Egypt received its gold from local sources, Ethiopia and possibly West Africa. By the second half of the 14th century West Africa was the primary source for gold, with Europe playing an increasingly important role. More silver reached Egypt from Syria than from any other area. By the end of this period direct shipments of silver from Europe were steadily increasing. Europe had also become extremely important for Egypt's copper supply.

When gold, silver and copper reached Egypt, they served a number of purposes. Besides being a source for minted coin, they were used for manufactured goods and decoration. The largest quantities were probably used to pay for goods and services imported into Egypt. As a general rule, gold and copper flowed through the Hijaz and/or Yemen to India. The silver flow is more difficult to trace, but most of it was shipped to Syria, and to a lesser degree to Yemen. Finally, Egypt probably held a balance-of-payments surplus for most of this period.

If the first chronological division is noted for stability and relative economic prosperity, then the second period, 1399-1517, was the reverse. While the new trends date to the 14th century, the 15th century was one of political, economic and monetary crisis and chaos. European gold, which dominated the domestic market, was only supplemented by a Muslim coin based on a Venetian standard. The new silver pieces were also derived from a Venetian model, although they fit into a Muslim system of weights. Copper became a major local currency. Rates of exchange changed rapidly and significant fluctuations in the gold: silver ratio took place.

The external drain of coin and metal became so severe that the Circassian sultans attempted to outlaw it. The direction in which the metals flowed did not change, only their relative quantity. Egypt suffered a growing balance-of-payments deficit. The cost of Qayitbāy's wars and the Portuguese disruption of the spice trade only exasperated Egypt's poor monetary and trade position. The Ottomans conquered an impoverished nation, a condition very different from the Egypt Saladin seized.

Table 1. Purity of gold coins minted in Egypt (1171–1517)*

Ruler	Mint	Muslim Date	Christian Date	Degree of fineness
Ayyubids				
Saladin				Under 90(13), 90, 91(3), 92,
(569–89/1171–93)				94(2), 95(6), 96(6), 97(5),
				98(2), 99(5)
al-ʿAzīz				93, 98, 99, 100(6)
(589–95/1193–98)				
al-ʿĀdil I				98, 99, 100(6)
(596–615/1198–1200)				
al-Kāmil				93, 94, 95(4), 96(4), 97(2),
(615–35/1200–18)				98(4), 99(6), 100(5)
al-ʿĀdil II	Cairo	635	1238	97
(635–37/1238–40)	"	635	1238	97
al-Ṣāliḥ	Cairo	637	1240	99
(637–47/1240–49)	"	638	1241	98
	"	639	1242	98

Table 1. (continued)

Ruler	Mint	Muslim Date	Christian Date	Degree of fineness
	"	640	1242	98, 99
	"	641	1243	97
	"	642	1244	98
	"	643	1245	98
	"	644	1246	97
	"	646	1248	97
Bahri Mamluks				
Shajar al-Durr (648/1250)	Cairo	648	1250	93, 98
Aybak (648–55/1250–57)	Cairo	654	1256	93
	Alex.	654	1256	94, 99
ᶜAlī (655–57/1257–59)	Cairo	657	1259	97
	Alex.	657	1259	96, 97, 98, 99, 99
Quṭuz (657–58/1259–60)	Alex.	658	1260	97, 97, 98, 98, 99, 99, 100
Baybars (658–76/1260–77)	Cairo	659	1261	99
	"	660	1262	97, 99
	"	661	1263	96
	"	666	1268	99
	Alex.	659	1261	98
	"	660	1262	96, 97
	"	667	1269	97, 98
	"	N.D.	N.D.	97, 98, 99, 99, 99
Baraka Qan (676–78/1277–79)	Alex.	676	1277	95, 97
Qalā'ūn (678–93/1279–90)	Cairo	689	1290	98
	"	N.D.	N.D.	98
	Alex.	N.D.	N.D.	98, 98, 99
Khalīl (689–93/1290–93)	Cairo	690	1291	91
	"	691	1292	87, 92
	"	692	1293	99
	Alex.	692	1293	98
Kitbughā (694–96/1294–96)	Cairo	695	1295	97, 97
Lājīn (696–98/1296–99)	Cairo	697	1297	96, 98
	"	N.D.	N.D.	96, 96
al-Nāṣir Muḥ. (698–708/ 1299–1309) (709–41/1310–41)	Cairo	705	1305	97
	"	707	1306	92
	"	N.D.	N.D.	92, 92, 94
	Cairo	711	1311	99
	"	713	1313	98
	"	722	1322	99
	"	733	1332	98

Table 1. (continued)

Ruler	Mint	Muslim Date	Christian Date	Degree of fineness
	"	739	1338	98, 99
	"	740	1339	96
	"	741	1340	99
	"	N.D.	N.D.	99
Ismāʿīl	Cairo	743	1342	98
(743–46/1342–45)	"	746	1345	98, 99
Shaʿbān I	Cairo	747	1345	99
(746–47/1345–46)				
Hājji I	Cairo	747	1346	97
(747–48/1346–47)				
Hasan (1st reign)	Cairo	750	1349	97, 99
(748–52/1347–51)	"	751	1350	97, 99
	"	752	1351	99
Ṣāliḥ	Cairo	752	1351	99
(752–55/1351–54)	"	753	1352	96, 99
	"	754	1353	97, 98
Ḥasan (2nd reign)	Cairo	756	1355	99
(755–62/1354–61)	"	757	1356	99
	"	758	1357	97
	"	760	1359	99
	Alex.	756	1355	97
	"	758	1357	95
	"	759	1358	98
	"	N.D.	N.D.	98, 99
Muḥammad	Cairo	763	1361	99
(762–64/1361–63)	"	763	1362	99
	"	764	1363	94
	Alex.	762	1361	96
	"	764	1363	97
Shaʿbān II	Cairo	768	1366	99
(764–78/1363–77)	"	769	1367	94, 99
	"	771	1369	97, 99, 99
	"	772	1370	98, 99
	"	773	1371	99
	"	774	1372	99, 99
	"	777	1375	99
	"	778	1376	98
	Alex.	769	1367	99, 99
	"	773	1371	98, 99
	"	775	1373	99
ʿAlī	Cairo	778	1377	99
(778–83/1377–81)	"	779	1378	97, 99
	"	781	1380	99, 99
Ḥājjī II	Cairo	783	1381	96
(783–84/1381–82)				

Table 1. (continued)

Ruler	Mint	Muslim Date	Christian Date	Degree of fineness
Circassian Mamluks				
Barqūq	Cairo	784	1382	96
(784–91/1382–89)	"	785	1383	96
and 792–801/	"	786	1384	98
1390–99)	"	788	1386	96
	"	789	1387	93
	"	792	1390	95
	"	794	1392	95
	"	795	1393	95
	"	797	1394	98
	"	798	1395	96, 99
	Alex.	788	1386	99
	"	791	1389	97
Faraj	Cairo	801	1399	99, 98
(801–09/1399–1405	"	805	1402	95, 98, 99, 99
and 809–15/	"	806	1403	96, 96, 98, 99
1406–12)	"	807	1404	97, 97
	"	808	1405	98
	"	809	1406	99
	"	810	1407	95, 95, 96, 97, 98, 98, 99
	"	812	1409	98
	"	814	1411	98
	Alex.	N.D.	N.D.	95, 95, 98
Shaykh	Cairo	815	1412	95, 95
(815–24/1412–21)	"	816	1413	92, 95, 96
	"	818	1415	95
	"	823	1420	99
	Alex.	N.D.	N.D.	93
Barsbāy	Cairo	829	1425	93, 94, 99, 99, 99
(825–41/1422–38)	"	834	1430	97, 98, 99
	"	835	1431	97
	"	836	1432	94, 96
	"	837	1433	95, 98, 99
	"	838	1435	97, 97, 98
	"	840	1437	99, 99
	"	841	1438	99
Yūsuf	"	N.D.	1438	97, 98, 99
(841–42/1438)				
Jaqmaq	"	N.D.	N.D.	95, 96, 97(3), 98(9), 99(7)
(842–57/1438–53)				
Uthmān	"	857	1453	98
(857/1453)				
Aynāl	"	N.D.	N.D.	97, 98(3), 99(3)
(857–65/1453–61)				

Table 1. (continued)

Ruler	Mint	Muslim Date	Christian Date	Degree of fineness
Aḥmad (865/1461)	"	865	1461	94(2), 96(2)
Khushqadam (865–72/1461–67)	"	N.D.	N.D.	93, 95(2), 96(4), 97(3), 98(4)
Temirbughā (872–73/1461–68)	"	N.D.	N.D.	99
Qaytbāy (873–901/1468–96)	"	N.D.	N.D.	Under 70, 88(2), 90(2), 91, 92(4), 93(8), 94(5), 95(9), 96(12), 97(5), 98(5), 99
al-Nāṣir M (901–04/1496–98)	"	N.D.	N.D.	91, 93(2), 97(2), 98
al-Ẓāhir Qānsūḥ (904–05/1498–1500)	"	N.D.	N.D.	93, 94(2)
Ṭumānbāy (906/1501)	"	906	1501	84
al-Qānsūḥ al-Ghawrī (906–22/1501–16)	"	N.D.	N.D.	Under 70(3), 70(4), 77(4), 79(6), 80(6), 81(5), 87, 89(5), 90(1), 92(3), 93, 95
Ṭumānbāy (922/1517)	"	922	1517	Under 70, 70(2), 75

*The data for the coins from al-ʿĀdil II were obtained by the author from specific gravity tests on specimens in New York, London, Paris and the Horowitz collection. The data on the earlier coins were done by Andrew S. Ehrenkrentz, "The Crisis of 'Dinar' in the Egypt of Saladin" *JAOS* LXXVI (1956): 181.

Table 2. **Purity of silver coins minted in Egypt, 1171–1517 (and *Syria)**

Ruler	Percentage of purity
Ayyūbids	
Saladin (566–89/1171–1193)	
(dirhams waraq)	27, 28(2), 29(2), 30
al-ʿAzīz (589–95/1193–98)	
(dirhams waraq)	26, 28(3)
al-Manṣūr (595–46/1198–1200)	
(dirhams waraq)	27, 28
al-ʿĀdil I (596–615/1200–18)	
(dirhams waraq)	28
al-Kāmil (615–35/1200–18)	
(dirhams waraq)	28(2), 29

Table 2. (*continued*)

Ruler	Percentage of purity
al-ʿĀdil II (635–37/1238–40)	
(globular dirhams)	26, 27, 28
al-Ṣāliḥ Ayyub (637–47/1240–49)	
(globular dirhams)	26, 27, 28
(normal flan)	70, 75
Baḥrī Mamlūks	
Aybak (648–55/1250–57)	66(2), 70(3), 71(3), 72(2), 74, 75
ʿAlī (655–57/1257–59)	59, 61(2), 62, 64, 65(4), 66, 68, 69(3), 71
Quṭuz (657–58/1259–60)	53, 54, 56, 57, 58, 59, 60(2), 61
Baybars (658–76/1260–77)	57, 60, 62(4), 63(2), 65(2), 66(2), 67, 73, 77
Baraka Qan (676–78/1277–79)	59, 67, 77
Salāmish (678/1279)	68
Qalāʾūn (675–89/1279–90)	65(2), 68
Khalīl (689–93/1290–94)	59, 65, 66, 67, 68, 69
Kitbughā (694–96/1294–96)	61, 63
Lājīn (696–98/1296–99)	56(2), 61
al-Nāṣir Muḥammad (709–41/1310–41)	
(dirhams waraq)	46, 50, 52, 64, 65, 66(3), 68, 73, 78
(normal flan)	55, 68, 72, 74(2)
(Armenian overstruck pieces)	74(2), 77
Ḥasan (755–62/1354–61)	68
Shaʿbān II (764–78/1363–77)	65, 66
ʿAlī (778–63/1377–81)	59, 70
Ḥajji II (778–84/1381–82)	84
Circassian Sultans	
Barqūq (784–801/1382–99)	27(2), 46, 48, 55, 58, 60, 61
al-Mustaʿīn (815/1412)	90
Shaykh (815–24/1412–21)	90(2), 92, 93, 95
al-Ṣālih M. (824–25/1421–22)	84, 88
Barsbāy (825–41/1422–37)	66, 73, 74, 75, 83(3), 84(4), 85(3), 86, 87(2), 89(4), 90(4), 91(2), 92(2), 93(5), 94, 95, 98, 99
Yūsuf (841–42/1437–38)	94
Jaqmaq (842–57/1438–53)	83, 85, 88, 92, 93, 95
Aynāl (857–65/1453–61)	84, 86, 88, 90, 91(3), 92, 93(2), 96(8), 97
Khushqadam (865–72/1461–67)	83, 92, 93
Qayitbāy (873–901/1468–96)	81, 83, 89, 90, 92, 93, 94, 95
al-Nāṣir M. (901–04/1496–98)	86, 87, 91
al-Qānṣuh al-Ghawrī (906–22/1501–16)	62, 82, 80, 85

*The data are derived from Jere L. Bacharach and Adon A. Gordus "Studies on the Fineness of Silver coins" JESHO XI (1968): 314–17 and Paul Balog "History of the Dirham in Egypt" *RN*, III (1961): 109–46.

Table 3. Gold: silver ratios*

Christian year	Gold: silver ratio
Ayyūbid period*	
Fatimid period average	1:9.3
1179 or 83	1:8.8 (or 1:93)
1194 or 99	1:9.3
1217	1:9.3
1223	1:9.3
1232	1:9.3
1233	1:9.0
1239	1:8.6
Baḥrī Mamlūks	
Baybars (1260–77)	1:13.4 (or 1:11.9)
Qalā'ūn (1280–90)	1:9.9
1289	1:9.3
1296	1:9.3
End of 13th cent.	1:11.9
Beg. of 14th cent.	1:8
1313	1:9.3
1314	1:9.3
1316	1:9.3
1317	1:9.3
1318	1:9.3
1323	1:9.3
1324	1:11.7
1324–36	1:10.3
1324	1:8.1
1332	1:9.3
1333	1:9.3
1338	1:9.3
1339	1:9.3
1339	1:11.7
1340	1:11.7
1341	1:9.3
1344	1:9.3
1348	1:9.3
1348	1:7.1
1350–52	1:9.3
1356–59	1:9.3
1359	1:9.3
1360	1:9.3
1382 to 1403	1:9.3**
Circassian Mamlūks	
1403	1:14
1404	1:14

Table 3. (continued)

Christian year	Gold: silver ratio
1405	1:14
1406	1:12.6
1407	1:14
1410	1:14
1411	1:10.8–11.6
1412	1:10.7
1413	1:11.6
1414	1:7
1415	1:8.9
1416	1:9.6
1417	1:10.1
1418	1:11.9
1419	1:10.6
1420	1:10.6
1421	1:10.6
1421	1:8.8
1423	1:8.8
1424	1:9.2
1430	1:12.2
1430	1:9.8
1432	1:11.1
1433	1:11.2
1439	1:10.2
1443	1:10.1
1456	1:10.1
1463	1:10.1
1466	1:10.3
1470	1:10.3
1472	1:10.3
1479	1:10.3
1483	1:11.1
1498	1:12.5
1507	1:8.5

*The gold: silver ratios are calculated from data in Eliyahu Ashtor, *Les Métaux Precieux et la balance des pavements du proche-orient a la basse époque* (Paris, 1971), Jere L. Bacharach "A Study of the Correlation between Textual Sources and Numismatic Evidence for Mamluk Egypt and Syria A. H. 784–872/A.D. 1382–1468" Unpublished Ph.D. Thesis, University of Michigan, 1967 and S. D. Goitein "The Exchange Rate of Gold and Silver in Famimid and Ayyubid Times," *JESHO* VIII (1965): 1–46.

**This ratio assumes that the increasing market rate of dirhams for the dinar paralleled a decrease in the purity of the dirhams.

List of Abbreviations

JARCE—Journal of the American Research Center in Egypt
EC—L'Égypte contemporaire
JAOS—Journal of the American Oriental Society
JEEH—Journal of European Economic History
BSOAS—Bulletin of the School of Oriental and Africa Studies, London University
AI—Annales Islamoloque
JESHO—Journal of the Economic and Social History of the Orient
IJMES—International Journal of Middle East Studies
RN—Revue numismatique
ANSMN—American Numismatic Society Museum Notes
JBBRAS—Journal of the Bombay Branch of the Royal Asiatic Society

6

Outflows of precious metals from early Islamic India

J. F. RICHARDS

For millenia India has been regarded in the Mediterranean and European worlds as a distant, exotic land populated by a highly sophisticated civilization— but a civilization remote in distance and in custom to that of Christianity and Islam. Equally wonderous was the manifest wealth and productivity of the densely-settled population of the Indian subcontinent. India's coastal ports made available drugs, spices and textiles in great quantities which her enterprising merchants sold in East Africa, Southeast Asia and even in Central Asia. Marvelous also was the sustained inflow into India of precious metals, required to pay for these goods, and the voracious appetite of the people of the land for specie. In their economic self-sufficiency, Indians would accept little else in payment. As a result, the great question which has puzzled economic thinkers for centuries has been the operation of the Indian exchange mechanism and its role in the complex economy of the subcontinent.

One theory evolved early in the medieval Mediterranean world ascribed the continuing Indian exchange surplus to an inordinate tendency on the part of Indians to hoard precious metals—a cultural quirk. Writing in the fourteenth century, the Damascene scholar 'Umari, that shrewd interrogator of returned travellers and expatriates, commented in his famous *Masalik*:

I must add that the inhabitants of India have the character of liking to make money and hoard it. If one of them is asked how much property he has, he replies, "I don't know, but I am the second or third of my family who has laboured to increase the treasure which an ancestor deposited in a certain cavern, or in certain holes, and I do not know how much it amounts to." The Indians are accustomed to dig pits for the reception of their hoards. Some form an excavation in their houses like a cistern, which they close with care, leaving only the opening necessary for introducing the gold pieces. Thus they accumulate their riches. They will not take worked gold, either broken or in ingots, but in their fear of fraud refuse all but coined money.[1]

---

1. H. M. Elliot and John Dowson, *The History of India as told By Its Own Historians*, 8 vols., reprint of original edition, II, p. 584. I have used the Elliot version of this passage in preference to the rendition of a more recent translation of the Indian section of Al-'Umari's work. Cf. the version by Iqtidar Husain Siddiqi and Qazi Muhammad Ahmad, *A Fourteenth Century Arab Account of India Under Sultān Muhammad bin Tughlaq*, Aligarh, 1971, p. 62: "It is

'Umari's description of excessive individual (ultimately) non-productive obsessive saving or hoarding has persisted to the present day as an explanation for what seems to baffled economic historians, an insatiable demand for precious metals. The image of the world's gold and silver flowing gradually toward the subcontinent, itself essentially a non-producing area (with the exception of alluvial gold washings in peripheral areas) can thus be easily accounted for.

That most Indians of the pre-modern period had what to the non-Indian appeared to be an excessive liking for gold and silver (as well as for jewelry) would be hard to deny. The ubiquitous display of female ornaments in families of even the most limited means, and presentation at weddings of such jewelry has long been noticed. That this usage might well be 'rational' in that it was a source of economic security for the Hindu widow who retained her personal belongings, but little else, after her husband's death is on the other hand also easy to see. Lavish use of gold and silver in decoration and construction of temples and idols in an economy where treasure was abundant was not non-productive hoarding, but rather served an aesthetic purpose. The practice of private hoarding in India (essentially unmeasureable) as in other insecure pre-modern polities, threatened by sudden war and plunder, was also a form of security far better than deposit banking. Nonetheless, both argument and counter argument suffer from the problem of timelessness: if used exclusively they ultimately undercut any need for or incentive to look at changing monetary patterns in the history of the subcontinent.

The general argument or thesis put forward in this essay is that the early Muslim period in India (*circa* 1200 to 1400 A.D.) brought with it substantial changes in the economy and monetary systems. The earliest Indo-Muslim Conquest regimes created new demand mechanisms and new ties with the greater Muslim world beyond the subcontinent. These linkages fostered substantial, but not equivalent, return flows or outflows of precious metals from India. The provenance of North India may be defined somewhat crudely as the mountain valleys of Kashmir and Peshawar; the Indo-Gangetic plain, including Panjab, Hindustan and Bihar; the riverine expanse of eastern North India, Bengal, and Orissa; and on the west coast Rajasthan, Gujarat. In essence this is the territory subject to some degree of Muslim political control by the mid-14th century (apart from the Maratha areas of the western Deccan first fallen under the domination of the Sultans of Delhi and later (by 1347 A.D.) under the breakaway Bahmani regime. The Dravidian south with its divergent gold based monetary system and later subordination to Islamic Conquest demands a separate treatment.

---

known about the Indians that they are fond of hoarding money. If any one is asked how much wealth he has, he replies: I can not tell exactly but I am of the second or third grandson to hoard the wealth of my grandfather in this hole and I do not know how much it amounts to.

People of India dig well (*sic*) to hoard wealth. Some of them make cisterns with holes inside the tanks to hoard gold coins therein. For fear of fraud they accept neither worked gold nor broken pieces nor gold bars but *dinars* (gold coins)."

The best available evidence for most of North India, so defined (with the possible exception of mercantile Gujarat), suggests that in the three to four centuries prior to the Muslim occupation of Delhi and the key points of the Indo-Gangetic plain (1193-1210 A.D.), an economic and monetary malaise prevailed, coterminous with political decentralization under Rajput dynasties. Surviving highly localized bardic genealogical and epigraphic traditions mesh with traces of distinctive local weights, measures, and other signs of political and economic fragmentation. The extant coins are sparse, badly minted and of low grade billon—especially by comparison with previous centuries. The general impression conveyed by research carried out by historians of the period is that of a ruralized society in which urban places were relatively unimportant. Instead the rural peasantry was tied to estates run by Hindu and Buddhist temples and monasteries, by royal officials and by local aristocrats settled on the land.[2] Similar evidence of declining commercial activity from the ports of the Persian Gulf, usually linked with the western Indian trading cycle, indirectly corroborates this view of the pre-Muslim North Indian economy. Thus "An important though negative testimony to the decline in prosperity is the virtual cessation of coinage on the Persian Gulf over a period of some two centuries, from soon after the Seljuk conquest (early 9th century) until the establishment of a Mongol dynasty (early 12th century) in Iran."[3] Even more dramatically, in Nepal's mountain valley, no coins are extant from the 8th century A.D., although by the 12th century manuscript colophons and inscriptions confirm minting and use of indigenous coins. The first surviving Nepali coins date from the 1600's. In other words the derivative trade economy of the kingdoms of Nepal seems to have declined as did that of the larger North Indian economy.[4]

Plunder, tribute and dishoarding on the military frontier

One characteristic societal response to the localized Rajput North Indian political system, does seem to have been diversion of gold and silver from monetary and exchange uses to immobilized savings by *rajas*, their vassals (*sāmantas*), temples and monasteries, merchants and more well-to-do free peasants or even tenantry on estate lands. Payment of the land or harvest tax with a share of the crop rather than in cash presumably added to this tendency. Trade slackened, bringing a lessened flow of precious metals into the subcontinent, although some exchange did continue. Moreover, long-

---

2. See R. S. Sharma, *Indian Feudalism c. 300-1200 A. D.* (Calcutta, 1965).
3. Nicholas M. Lowick, "Trade Patterns on the Persian Gulf In the Light of Recent Coin Evidence," *Near Eastern Numismatics, Iconography and History. Studies in Honor of George C. Miles* (Beirut, 1974), p. 321.
4. See Luciano Petech, *Mediaeval History of Nepal (c. 750-1480)* Roma, 1958, pp. 177-78. The above speculation is mine, not that of Petech. Literary evidence for Nepali coins and the circulation of Delhi Sultanate "tangas" begins in the 12th century A.D. See N. G. Rhodes and C. Valdettaro, "Coins in Medieval Nepal," *Numismatic Chronicle*, vol. CXXVI (1976), 158-166.

standing sources of alluvial gold workings in Tibet, in the streams of the Himalaya *terai* and along the Brahmaputra River in Assam probably supplied considerable uncoined gold dust for commercial transactions.[5] Holdings of treasure constituted a sacred trust built up over the generations of a dynasty or family or the centuries of a temple. Indian hoards probably continued to expand in size and number in the pre-Muslim period.

Within the Muslim world, which had early expanded to the eastern borders of Iran and conquered Sind, the hoarded wealth of Ind or Hind became part of the proverbial Muslim knowledge of the subcontinent. Far more offensive than merely being infidels, Hindus worshipping idols ministered by Brahmin priests of the region were engaged in polytheism, that abomination attacked by Muhammad in the sanctuary at Mecca. North Indian society was also known to be defended by ferociously war-like Rajputs, mounted warriors who would prove a worthy opponent to Muslim warriors. Thus, seventh century Arab Muslim military leaders and rulers initiated an aggressive raiding process of constant military action against stubborn Rajput defenses—a process which was to last almost without let-up for well over a millenium and end in the conquest of virtually the entire subcontinent by the 1690's A.D. The idiom of this long crusade was that of the external holy war, the *jihād* fought by men who considered themselves *ghazis* or fighters for the faith against heterodoxy and polytheism.[6] On occasion, the appearance of an especially strong Muslim leader, coincident with a temporary weakening of Hindu defenses allowed the conquest of another extended portion of territory—often after a set-piece battle.[7]

The pattern of Muslim assault from the west in Iran and the northwest of Hindu Kush and Transoxania thus became set. The stated objective of course was expansion of the faith, and the Muslim Community *(Umma)* by conquest and conversion. The signs and tangible symbols of victory included plunder: long lines of enslaved captives, who had not accepted Islam; files of horses, elephants, camels; arms, but above all else jewels and precious metals. Preferably the latter were ripped from the desecrated temples of the enemy.

5. For example the early 18th century chronicler Khafi Khan observed that in Assam the main medium of trade and state revenues was gold dust washed from the Brahmaputra river. A Mughal count revealed that the group of hereditary gold washers required by the Assamese ruler to pay one *tola* (12.05 grams [Hinz, p. 34]) annually to the state was 20,000, for a total return of 241 kg. of gold annually. Khafi Khan further noted that he had seen Chinese merchants buy up this dust and offer it for sale in the port of Surat in Gujarat in the last years of the 17th century. Obviously, much later evidence such as this can only be suggestive but the same impression of trans-Himalayan supplies from Tibet also persists. Khafi Khan, *Muntakhab-ul Lubab*, ed. by K. D. Ahmad and Woseley Haig, 3 vols. (Calcultta, 1860-74), II, p. 131.

6. Cf. the entry *djihad* and the entry for *ghazī* in the *Encyclopaedia of Islam*, second revised edition, for a succinct discussion of these important concepts.

7. Recent discussions of the South Asian frontier of Islam, comparable to that in the west, include A. L. Strivastava, "A Survey of India's Resistance to Medieval Invaders from the North-West: Causes of Eventual Hindu Defeat," *Journal of Indian History*, XLIII (1965), 349-68; and J. F. Richards, "The Islamic Frontier in the East: Expansion into South Asia," *South Asia* (1974), pp. 91-109.

Captured treasure heaped in the courtyard of the raiding Sultan was used for largess, for strengthening his currency, and for augmenting his own vaults. Wealth taken in this manner was a tangible sign of God's favor toward the faithful.

Treasure forcibly seized and carried back from India to the cities of central Asia and Iran and infused into those monetary systems must be accounted a recurrent mechanism for dishoarding of Indian treasure and for outward flows of specie. While probably never so regular, nor so large in the aggregate as specie moving toward India within trade channels, certainly the total amount over the centuries was far from nominal. The great legendary Muslim raiders and conquerors who swept from the west returned with vast amounts of treasure; but more numerous smaller victories, often unrecorded, also sent a steady stream of specie to the older Muslim lands.[8]

Before considering episodes within the larger historical process we should distinguish between the various mechanism contributing to the larger process of expropriation of portable wealth and its transfer out of Muslim-held India in a reverse flow toward Central Asia and the Middle East. First, the return of plunder including human and animal captives as well as other forms of wealth from raids into Hind (i.e., the lands defined by the Muslims as Dar-al Harb or the "Land of War" beyond the reach of the Sharia) to whichever Muslim city was the base for these incursions was unquestionably of great magnitude.

The profits from such expeditions were both irregular and variable. Thus, profit and loss accounting for a series of such incursions (if this were only possible) could well reveal proceeds varying from a substantial loss, in the case of stiff resistance, to a staggering profit, in the case of the successful (but limited and hurried) sack of a city or temple. However, the total extant body of direct and indirect evidence supports the conclusion that in the aggregate this aggressive military policy was immensely profitable in the course of years and decades of consistent assault. That Muslim commanders could rely upon the essentially unpaid services of irregular, but zealous, companies of *ghazis*, as well as professional troops, added to their possibility of a surplus of booty over expenditure. Moreover, when a decisive eastward shift in the military frontier took place (as in the Ghaznavid occupation of Kabul and Peshawar in the 11th century) seizure and resulting thorough, unhurried pillage of private, kingly, and temple vaults pushed extraordinary amounts of long-accumulated treasure west to the Muslim redoubt from which the invasion army departed.

A concomitant source of profit was exaction of ransom in return for the release of kings, princes or other high-status captives. Of course, this latter source of revenue bore the same characteristics as that from plunder. The sums demanded for ransom were quite large, but the opportunity for such negotiated transactions was infrequent. Those monies paid by beset rulers as 'tribute' to ward off Muslim assaults by invading armies also fall into the

---

8. Cf. Richards, *op. cit.* for a partial listing of recorded encounters, pp. 94-98.

category of ransom payments—for a state rather than a person. Especially damaging campaigns into the land of polytheists tended to produce demands for tribute and formal submission from those (primarily Rajput) kings and their vassals on the losing end of this struggle, (which, for individual Hindu rulers or entire dynasties was far from inevitable or necessary). Within the evolving Indo-Muslim political system, regular payment of tribute and various statements of submission did constitute a diplomatic settlement which, at least for a time, ensured a reprieve in military pressure and occasionally support for such tributary rulers against internal or external threats.

Tribute is used here in the sense of a term denoting a formal demand for payment imposed on the weaker party after a negotiated settlement. Form, amount and frequency of payment was based upon considerations of relative power and negotiating skill: e.g., the location of the tributary kingdom; its capacity to offer troublesome resistance; the king's willingness (or lack thereof) to submit to personal humiliation, etc. Like taxation, tributary payments did bear a crude relationship to the total size, wealth and productivity of the kingdom, but in the Indo-Muslim system the tributary ruler, keeping most of his internal freedom of action, distributed the burden of payment in accordance with custom, the existing tax/tribute structure and his own policies. Prior to formal conquest of new tracts of territory and after, as well, tributary payments clearly augmented the movement of treasure west to successive capitals on the Hind frontier.

After formal conquest, (e.g., after the seizure of Lahore in the 11th century or Delhi in the late 12th century, or Lakhnavauti in the early 13th century) we find that the mechanisms of plunder, ransom, and tribute continued even within what ostensibly became the Dar-ul Islam. In other words, the initially violent process of political definition, of pacification, of delineation of nascent administrative units such as the *sadi* (later the *shiq*) of a 'hundred' of villages (reflecting usually a pre-existent Rajput or Jat or other similar lineage territory) varied little in style, and mode of operation from its analogue in the trans-frontier struggle. Rural pacification contributed to the enrichment and massive concentration of wealth found within the growing Indo-Muslim colonial metropolii and their satellite towns. The number of polities (or petty lineage states), of temples, of towns subject to military pressure and dishoarding increased, while the level, status, size and resources of each unit now under pressure diminished. That is, there occurred a great increase in the number of transfers of treasure to the center as the new Indo-Muslim rulers slowly moved toward taxation and systematic administration.

Concentration of wealth and power in the great Indo-Islamic centers created markets and a corresponding demand for commodities and human talent. Payment for these, as well as for maintenance of linkages with the major and minor cultic centers and for continuing recognition as an Islamic land also generated steady westward transfers of treasure. Another more dramatic device came into play when the Indo-Muslim conquerors themselves hit by Mongol raids and invasions as a result of the unpredictable nature of steppe military power fell back on the defensive. Thus, in the second phase of extended conflict the mass of accumulated wealth in Ghazna, Lahore, and Delhi, was stripped off and brought back to the Muslim west from India.

To illustrate this point and to provide some idea, however incomplete, of the quantitative scale of plunder (and tribute) taken we may turn briefly to the exploits of four of the greatest heroes of Indian Islam—Sultan Mahmud of Ghazni (early 11th century), Sultan Muizzuddin of Ghur (or Muhammad of Ghur) (late 12th and first decade of 13th century), Sultan Alauddin Khilji of Delhi (early 14th century) and Amir Timur (late 14th century).

Mahmud of Ghazni, son of a Persianized Turkish military-slave ruler offered to successive generations of orthodox Islamic kings and generals, as well as ordinary Muslims, the model of a consummate *ghazi* king. He built a magnificent capital at the smallish city of Ghazni (in present-day Afghanistan) and a sprawling saddle empire stretching from Lahore in the Panjab in the east to Nishapur and beyond in Khurasan in the west. Following the example of his father, Amir Subuktigin, Mahmud put much of his extraordinary energy and ability into a series of deep wracking raids into North India during his reign (997 A.D. to 1027 A.D.).[9]

Mahmud's primary opponent in the east was Jayapal, member of a powerful Brahmin dynasty known as the Hindu Shahis, which had been pushed east back to Peshawar and the Panjab. The end result of Mahmud's campaigns was to extend the military and political frontier of Islam into the North Indian plain past Lahore, the capital of the five rivers or Panjab area. In the course of this bitterly fought conflict, the Sultan defeated a Hindu Shahi prince and pursued him to the heavily fortified fortress of Bhimnagar in the Panjab. After the assault, capture and sack of the fortress the usual stock-taking took place. Al-'Utbi, a Ghaznavid chronicler commented:

> The kings, nobles and pious Sultans of Hindustan (i.e. Rajas) used to collect treasures for the big idol installed therein. From time to time they sent treasures of every sort there. . . . They considered it an act of piety, which would bring them religious merit and earn them a place near God. The Sultan acquired here so much of the best that had been gathered here for ages, that there were not enough camels to carry it nor means of transport either. The clerks could not prepare an inventory and the imagination of the accountants failed to grasp it in their number.[10]

When some solution to these difficulties had been overcome, Mahmud "him-

---

9. For a full description of the Ghaznavid empire see C. E. Bosworth, *The Ghaznavids* (Edinburgh, 1963); and for the Indian campaigns see Muhammad Nazim, *Mahmud of Ghazna* (Cambridge, 1931).

10. S. R. Sharma, "A Contemporary Account of Sultan Mahmud's Indian Expeditions," *Journal of the Aligarh Research Institute*, I (1941), 127-165 and 141. The *Tarikh-i Yamini* was written by Abu Nasr Muhammad bin 'Abd al Jabbar al-'Utbī, a Ghaznavid official, some years after these events which he witnessed. Rather than rely upon the usual Persian translation of the

self took of it (the treasure) as much as his camels could carry. The remainder was appropriated by his chiefs. The silver coins numbered 7 crore royal rupees and the gold objects weighed 700,400 maunds."[11] If, as was probably the case, the "royal rupees" rendered by Sharma were the silver "Bull and Horseman" coins of the penultimate phase of Hindu Shahi coinage (averaging 2.9 to 3.3 grams at 61 and 70% purity) the total amount of pure silver taken was somewhere between 123,200 kg. and 154,700 kg. (with certain obvious problems in such a calculation).[12] The gold seized in this episode amounted to 1,190,680 kg. or 1190.7 metric tons.[13] On at least one occasion after his triumphal return from an Indian raid, Mahmud flamboyantly ordered his captured treasure to be publicly displayed at Ghazni.[14] Well might Mahmud celebrate, for as Bosworth has pointed out, the inflow of Indian treasure was one of the major economic resources of the early Ghaznavid regime (the other being the trade and agricultural prosperity of Khurasan).

Shortly after the death of Mahmud, his son Masud (d. 1041 A.D.), disastrously defeated at the desert battle of Dandanqan in 1040 A.D., lost the Persian provinces of the empire to the Seljuk king Chaghri Beg Daud and his Turkoman followers.[15] Thereafter although the later Ghaznavids continued to hold Ghazna, Lahore became even more important as an urban center of power of the diminished empire, now virtually a regional kingdom in the northwest Indian subcontinent.[16] Throughout the 11th and 12th centuries the Muslim-Rajput military political frontier line remained virtually stationary—coterminous with the Ghaznavid boundary stretching along the

---

original Arabic, Sharma provides a translation direct from the original. Cf. Charles Rieu, *Catalogue of the Persian Manuscripts in The British Museum*, 1966 reprint of the 1897 edition, I, 157-158.

11. Sharma, p. 142. The plunderers also found a house made entirely of silver 30 by 15 yards in size.

12. David W. Macdowell, "The Shahis of Kabul and Gandhara," *The Numismatic Chronicle*, VIII, Seventh Series (1968), 189-224. Macdowell has completed a persuasive numismatic analysis of the surviving Hindu Shahi coinage. He concludes that the coinage of Jayapala is the "conch-shell issue of Samanta," the last issue in silver before the appearance of a billon coinage after the failure of the celebrated silver mines at al Panjhir, pp. 210-211.

13. The calculation for the weight of a *man* in this period is based on Walter Hinz, *Islamische Masse Und Gewichte*, Leiden, 1955. A similar assault on the holy city of Mutra produced "98,300 *misqals* of gold" or 431.7 kg. and "200 times this weight in silver." According to 'Utbi the silver had to be broken for weighment," which suggests that many of these figures were based on more than a guess. The value of the *misqal* for Iran is given in Hinz, p. 5 at 4.3 grams. The passage from 'Utbi is in Sharma, pp. 158-159.

14. Sharma, p. 143.

15. The material on the later Ghaznavids is drawn from C. E. Bosworth, *The Later Ghaznavids: Splendour and Decay*, New York, 1977.

16. For the reconstructed boundaries of the reduced Ghaznavid state, circa 1100 A.D. see the map on page one, *The Later Ghaznavids*. . . . These territories included Makran, Baluchistan, Sind and Panjab, (i.e. the valley of the Indus) Zamindawar, the territory around Bust at the confluence of the Helmand River, Zabulistan, around Ghazna, and the Kabul and Peshawar valleys.

Indus barely fifty miles to the east of the great river from the sea to Chalandhar (and excluding Kashmir.)

For nearly one hundred years the later Ghaznavid (or Yamini) dynasty sustained this position until, in the fourth decade of the twelfth century, it lost Ghazna to another group of invading Turkomans. Nonetheless the citadel of Ghaznavid power at Lahore remained intact until the incursions of Muhammad of Ghur in 1186 A.D.

Unfortunately, for this phase of the Ghaznavid kingdom, the earlier narrative sources falter. Instead we must rely upon scattered references in histories of other regimes, upon epigraphs, coins and archaeological remains, and a body of surviving epic poetry. Reconstruction of a precise, detailed political narrative is not possible, given the lack of dates provided in these sources. However, we can be reasonably sure that the style, the religious appeal and the patterns of aggressive, long-distance raiding into Hindu territory continued well into the twelfth century. Lahore remained a redoubt and a base for *ghazis* and professional soldiers launched into the *jihad* against pagan India. After carefully scrutinizing the evidence for the later Ghaznavids, C. E. Bosworth stresses the continuing importance of the frontier in Hind especially after the loss of Khurasan and western Afghanistan:[17]

> India . . . was left as the special war-ground of the Ghaznavids. The raids of the middle and later sultans are singularly ill-documented from the Islamic side, and the allusiveness and chronological vagueness of the native Indian sources here provide no complementary dimension of source material; but it is clear that pressure was substantially maintained on the Indian princes, who nevertheless resisted fiercely and were never really overwhelmed by Ghaznavid arms. Hence, although the temple treasures of India continued to be brought back to Ghazna for the beautification of palaces and gardens there, although the flow of bullion continued to keep the economy of the Ghaznavid empire buoyant and its currency of high quality, there were no major gains of territory beyond the eastern fringes of the Panjab and. . . . the Ganges-Jumna doab . . .

Slaves, treasure, ransom and tribute, were the tangible benefits of the *jihad* in Hind. The intensity of emotion engendered by this concept is revealed in the great Kufic poetic epigraphy recently discovered on the marble slabs placed on the walls of the palace of M'asud III b. Ibrahim (1099-115). Composing in the heroic style and metre of the *Shah-Nama* the unknown author celebrates Mahmud as "the protagonist of the Islamic religion and also as an Iranian warrior-hero."[18]

Occasional literary references support Bosworth's view. For example the chronicler Ibn al Athir, writing in Iraq, described three major campaigns mounted by the later Ghaznavid Sultans, but did not record any of the plunder taken in these victories.[19] M'asudi Sadi, the Ghaznavid poet, has

17. *Ibid.*, p. 4.
18. *Ibid.*, p. 89.
19. *Ibid.*, p. 62.

addressed two of his poems to the victories of Prince Mahmud, son of Sultan Ibrahim (1059-1099) in a raid carried out between 1068-1070 A.D. upon Agra, then held by what has been identified as Gopala of the Rastrakuta dynasty. When the Muslim forces took the fortress at Agra after a stubborn defense, the ruler of Agra and several other local rulers (presumably his vassals) also submitted and "brought rich presents of treasure and elephants for Mahmud."[20] In these and a number of other known instances, the mechanism of long-distance raiding, plundering and exaction of tribute sustained a substantial flow of treasure in the eleventh and the twelfth centuries into Lahore and on to Ghazna on the reverse side of the Hindu Kush.

Movement beyond Ghazna of at least part of the Indian treasure took place by means of varied, innumerable exchanges and transfers of the normal trade, gift, currency and remittance operations of the medieval Islamic economy. (One could argue that some part of the plunder found its way back to Hind in the same fashion). Nevertheless as we have observed earlier, plundering, dishoarding and the westward movement of treasure was not a simple one-step process. That is, the accumulation of Indian bullion at Ghazna and Lahore was often an invitation to further assaults from the west and the subsequent carriage of the great treasure of these Muslim capitals further west. For example, the victorious Seljuk armies of Sultan Sanjar, in the course of intervening in a Ghaznavid succession dispute, entered the never before captured city of Ghazna in 1117 A.D. During forty days' occupation of the city, the looting soldiery stripped silver plate from the walls of the palaces and dug up silver irrigation ducts from the Persian gardens. Sanjar himself carried off most of the great Ghaznavid treasure "amongst which is mentioned five crowns, seventeen gold and silver thrones and 1,300 settings of precious metals and jewels."[21] The portable wealth of Ghazna was unquestionably, in its greater part, the wealth of Hind forcibly seized and moved out once again into the channels of medieval Islamic economic exchange.

In 1191 A.D. the aggressive ruler of an obscure Muslim dynasty, the Shansabani of Ghur, after taking Ghazna and Lahore moved into the North Indian plain to confront a coalition of Rajput rulers at the first battle of Tarain, on the outskirts of Delhi. Recovering from an initial setback, Muizzuddin Muhammad bin Sam returned to win a second set-piece battle at Tarain. Thereafter, the Sultan and his principal Turkish slave general Qutbuddin Aibak (later to become the first Sultan of Delhi), spent the next fifteen years (1192-1206 A.D.) in a series of bloody battles, raids and sieges to achieve their strategic objective: occupation of the major cities of the Indo-Gangetic plain up to the borders of Bengal, and defeat of the leading Rajput kings of the region.[22] In 1194 A.D. the combined armies of Sultan Muizzuddin from Ghazni and Qutb-ud-din Aibak's forces from Delhi defeated and killed the

20. *Ibid.*, p. 66.
21. *Ibid.*, p. 97.
22. A. B. M. Habibullah, *The Foundation of Muslim Rule in India* (2nd rev. ed., Dacca, 1961), pp. 53-86 supplies the most comprehensive account of these events.

last Gahadavala Rajput ruler in a battle which opened the entire eastern Gangetic valley to Muslim occupation. During the follow-up operations Sultan Muizzuddin occupied Banaras (Varanasi), the great Hindu holy city. He "destroyed approximately one thousand idol temples and loaded four thousand camels with nothing but the most valuable goods: jewels, gold and precious silken cloth." Again if we assume as a rough guide that an average camel load might be approximately 225 kg. Muizzuddin's plunder returned to Ghazni from this one city may have reached a total of 900,000 kg. or 900 metric tons, much, if not most of which, was likely to be gold.[23]

As the Muslim forces occupied Kanauj, Banaras, Muthra, Lakhnavauti, Delhi, and smaller towns in this extended war, vast amounts of precious metals, as well as other valuables, flowed northwest to the conqueror's capital. Such systematic stripping of the temple, royal, and private hoards built up and preserved as sacred trusts over generations must have caused a large-scale transfer of specie from India to the eastern Islamic world.

After the death of Sultan Muizzuddin in 1206 A.D. partial resolution of a bitter power struggle left Sultan Iltutmish by 1220 A.D. ruler of Delhi and dominant member of the most powerful Turkish slave generals surviving their master and ruler. By a strange reversal, the second Sultan of Delhi found himself in the 1220's A.D. one of the few surviving Muslim rulers who had not been overrun by the Mongols under Chingiz Khan. The Turks in North India soon came under devastating raids mounted by the infidel Mongols from the Hindu Kush against Hindustan.[24] Between 1240 and 1260, the Mongols raided virtually annually into the Panjab. Thereafter, until the 1390's, intermittent Mongol massive raids in force failed to break through the Delhi Sultanate defensive zone along the Ravi and Beas rivers, the easternmost branches of the Indus. The heavy mailed cavalry of the Turks could usually turn back even the largest Mongol forces. Hardest hit in this frontier conflict were the cities and inhabitants of the eastern Panjab: Uch, Multan and Lahore. For example in 1241 A.D. Mongol armies occupied Lahore, plundered the city and slaughtered most of the populace. In 1246 A.D. the inhabitants of Lahore and those of Multan each offered an indemnity of 100,000 gold *dinars* to the Mongol command.[25] By the early 1250's the Mongols occupied these major cities of the Punjab and threatened to move on Delhi. In conformity with their general reliance on mobility the Mongols pulled down the walled fortifications of Multan and Lahore rather than utilize an urban base or allow others to use them. The Delhi Sultanate response, when the cities were retaken, as under Sultan Balban in 1267, was

23. Ferishta (Muhammad Qasim Hindushah Astarabadi) *Tarikh-i Ferishta*, (Lucknow, 1867), p. 61. The estimate for a camel's capacity comes from the *Encyclopaedia Britannica* essay "Draft Animals" in which 225 kg. is the low figure for the Indian dromedary with the possibility of loads rising to 295 kg. for steady travel.
24. Cf. Habibullah, p. 210.
25. *Ibid.*, p. 214. Each gold '*dinar*' at this early period was very likely the Delhi Sultanate gold *dinar* or proto-*tanka* of 70.6 grains, Habibullah, p. 285 and Nelson Wright, pp. 15-16 for a sequence of gold *dinars* struck by Shamsuddin Iltumish. (1210-1275 A.D.)

to refortify, to repopulate, and to rearm the cities as defensive strongpoints.[26]

The frontier line remained relatively static at either the Ravi or the Beas throughout the 14th century. Every few years a Mongol incursion in force carried the war on. The great objective of the Mongols was clearly to break through to Delhi and its ever accumulating population and wealth—a feat which was nearly accomplished in 1298 A.D. The Mongol general Qutlugh Khvaja, leaving Transoxania with 200,000 horsemen, headed straight for Delhi which was soon inundated with refugees. Sultan Alauddin Khilji managed to beat off the Mongol assault in a large-scale cavalry battle outside the walls of Delhi. The next year Alauddin held firm through an encirclement of Delhi by a second large Mongol army. After two months the Mongols simply retired rather than face another disastrous battle with Alauddin.[27]

In the absence of even the sketchiest of data, the true economic effect of the Mongol raids on North India is impossible to determine. It is, however, reasonably certain that minor and major incursions sustained a flow of plundered precious metals toward the northwest. Even when driven back by the Delhi Sultanate cavalry, the raiding Mongols usually had completed wide-ranging devastation in both urban places and in the countryside. One contemporary Indo-Muslim chronicler records between *circa* 1240 and 1380 A.D. Mongol armies made twelve major incursions. Their predatory practices were not confined to cities or temples but to the burning and looting of the agrarian countryside as well.[28]

This long drawn-out conflict ended finally with the sedentary settlement of various Mongol groups in Iran and Transoxania who acted as a buffer to the more predatory, mobile raiders of the steppe. The advent of Amir Timur Gurgan, the Central Asian Conqueror, was the final, disastrous episode in this struggle. Timur, at a war council of his sons and leading nobles, decided to lead a holy war in Hindustan, that land of wealth and unbelievers, whose current Muslim rulers were far too lax in suppressing "the filth of infidelity and polytheism." On reaching Delhi with his army, Timur defeated the Sultan of Delhi in the open field and occupied the great imperial capital. A clash between the Turkish soldiery of Timur and the urban crowd of Delhi led to a wholesale sack and massacre. As Muhammad Bihamad Khani, another contemporary historian phrased it:

> They (Timur's soldiers) sacked that great city which had been the seat of the great Sultans and maliks, for more than two hundred years. All the people, high and low, young and old, were imprisoned and put to martyrdom. They (the Mughals) dug out all the treasures and after a few days left Delhi for their homeland.[29]

26. Yaha bin Ahmad bin 'Abdullah Sirhindi, *The Tārīkh-i Mubārakshāhī*, trans. K. K. Basu, (Baroda, 1932), p. 38.

27. K. S. Lal, *History of the Khaljīs* (Bombay, 1967), pp. 132-145.

28. Yahya bin Admad, *Tarikh-i Mubarakshahi*, pp. 28-29, 31, 36, 38, 40-41, 51, 64n, 70-71, 72, 132, 140-41. Additional raids have been also noted by other Muslim historians.

29. Muhammad Bihamad Khani, *Tarikh-i-Muhammadi*, trans. by Muhammad Zaki (Aligarh, 1972), p. 93. This early 15th century chronicle is essentially devoted to the history of Kalpi, a

En route to Samarkhand after a successful (and extraordinarily bloody) campaign, Timur rode at the head of his troops followed by a heavily loaded baggage train carrying the proceeds of the imperial treasure of Delhi as well as that of infidel forts and towns of Hindustan. As his army moved northwest along the Himalayan foothills toward Kashmir, Timur assaulted, virtually on a daily basis, those Hindu chiefs and their followers in that region. When, after a month the army had won twenty pitched battles and seven assaults on mountain fortresses, Timur's *beks* suggested that he might not need to engage in the close fighting in person. Timur "replied that the War for the Faith had two supreme advantages. One was that it gave the warrior eternal merit, a guarantee of Paradise immediately in the world to come. The other was that it also gained for the warrior the treasures of the present world. As Timur hoped to enjoy both advantages, so he intended to justify his claim to them."[30]

Timur's remarks suggest a significant, but as yet unexplored, difference in cultural attitudes toward treasure, money and its usage existing between Muslim and Hindu rulers. For example, Al 'Utbi reports the purported speech of the 11th century Hindu Shahi ruler Jayapal whose move for peace with Amir Subuktigin of Ghazai had been rejected. In response Jayapal said:

> The greedy desire to plunder their property and make them prisoners was the only obstacle to peace. In the path of peace stood the (Mussalman's) greed and temptation to plunder their goods and elephants and to imprison them. But they would cheat this greed by destroying their property and thrusting themselves into the flames.[31]

One readily discernible distinction lies in the relative significance given to coin as a symbol of legitimate rulership. By the 11th century A.D. every Muslim Sultan, at the time of his accession to the throne, issued new coinage. The latter bore his throne titles, date of accession, place of issue and any other message which he wished to include within the limited space of two coin faces. Thereafter, upon acquisition of a city or territory the Sultan automatically included his name in the formula provided for that purpose within the regular Friday prayers in the great congregational mosque of the city. As a corollary he usually issued a commemorative minting of coin stamped with the date and place of victory. One criterion for popular judgement of a ruler seems to have been his reputation as a moneyer whose mint could issue reliable, high quality coins (usually trimetallic, see below) to sustain the economic life of the cities and to meet the needs of the state. Obviously this intense concern with coinage meant that Muslim rulers operated mints manufacturing large quantities of coin. They made every effort to

---

small Indo-Muslim successor state, which emerged after Timur's invasion temporarily destroyed the central structure of the Sultanate.

30. Hookham, p. 199.

31. Sharma, p. 129. The last sentence refers to the Rajput or Kshatriya practice of *jauhar*, the mass killing of dependents by a doomed garrison and a suicidal assault on the enemy wearing saffron robes.

put this into circulation, thus enhancing the volume and velocity of coin and the efficiency of the monetary system. The general Indo-Muslim state policy of demanding taxes paid in money (usually coin of the dynasty), also pulled precious metals from their repositories for conversion to coin by money changers acting as intermediaries for the royal mints.

On the Hindu or Rajput side coin had far less ritual significance. The striking of new coin played no part in the elaborate medieval Hindu sacrament of coronation.[32] Nor did the issue of commemorative coins usually mark new conquests as in the Muslim case. While Rajput kings did coin to meet economic needs, they often simply replicated traditional motifs without bothering to stamp their own titles and dates on the coins.[33] As for the prizes of conquest and domination, Rajput rulers were more intent upon the symbolic submission of a rival ruler who became a vassal (*mahāsāmanta* or *sāmanta*) of the victor. The rewards of victory were not so much treasure nor material tribute, but the obligation to provide troops and to attend in person the court of the dominant raja. Gopal ascribes this inclination to "the working of the peculiar Indian doctrine of *dharmavijaya*. This discouraged the annexation of a conquered territory but recommended the acceptance of subordination by the conquered king. . . . This tendency is very old and deep-rooted in the Indian tradition."[34] Moreover medieval Hindu works on polity stressed a classification of conquest ranging from "righteous conquest" (i.e., *dharmavijaya*) to "greedy conquest" or plundering extending to that form most condemned "demonaic conquest" which resulted in the destruction of a dynasty and the annexation of the former kingdom. Within these terms of reference the denunciation of Muslim "greed" by Jayapala becomes intelligible, expressing as it does a widely divergent view of warfare, of diplomacy and status honor. We can also discern behind this a different understanding of the role of the ruler and his state in the economy and its monetary system.

Centralization, resource flows, and market demand at Delhi

Gradually, in the 13th century, despite the drain of the Mongol defense on the western frontier and occasional Rajput counterattacks, successive rulers at Delhi built up their power and resources. By the commencement of the 14th century A.D., the sultanate of Delhi (with an ancillary regime in western Bengal) was clearly the dominant power in North India. Through inmigration, conversion, and natural increase the city of Delhi had become a large, secure imperial capital inhabited by a sizeable Muslim population—a center of Islamic culture and civilization, as well as a citadel in a strange land,

32. Cf. Ronald Inden, "Ritual, Authority and Cyclic Time in Hindu Kingship" in J. F. Richards, ed., *Kingship and Authority in South Asia*, (Madison, South Asia Center, 1978).
33. Cf. Madowell, *op. cit.*
34. Lallanjii Gopal, "Sāmanta—Its Varying Significance in Ancient India," *Journal of the Royal Asiatic Society* (1963), pp. 21-28, 29.

to rival the older urban centers of medieval Islam. Slowly, not withstanding the Mongol catastrophe, imperial Delhi was by 1300 A.D. restoring its connections with the rest of the Muslim world.

Over time, successive Indo-Muslim dynasties created a centralized state structure in the Persian tradition whose task was to mobilize human and material resources for the ongoing armed struggle against both Mongol and Hindu infidels. In the hinterland of Delhi, especially the so-called *doab* between the Ganges and Jumana rivers, the state evolved a workable system of taxation by means of the land tax paid in cash by Hindu chiefs and other intermediaries (after a period of violent initial pacification). This vital process fed and supplied the basic needs of the military and civilian urban populace of Delhi.[35] Elsewhere in North India, in Rajasthan, in Gujarat, Central India, northern Bihar, the familiar process of raiding, plundering and tribute taking by Sultanate armies went on. The treasure of centuries moved from the vaults of Rajput rulers and aristocrats to Delhi in increasing quantities. Perhaps the best known and most spectacular such raid, in terms of distance and the prize generated, was that of Alauddin Khilji's long march from Delhi to the Maratha Deccan and his surprise assault on the kingdom of Devagiri. In this instance, a combination of luck and audacity allowed him to extort a war indemnity from the Maratha ruler (and an annual tribute) of 600 *man* of gold (7,680 kg.), and 1000 *man* of silver (12,800 kg.) in addition to large amounts of precious stones and other articles.[36] Alauddin's wealth obtained in this single Deccan raid was sufficient to buy that political support necessary to consolidate his rulership after the murder of his uncle, Jalaluddin. However, the proceeds of steady small-scale punitive campaigns in the Delhi hinterland or in neighboring provinces directed against stubborn local chiefs and rajas, most unrecorded by the chroniclers, enhanced the revenues of the Sultanate as much as occasional raids into the Deccan or into Gujarat.[37]

The concentrated wealth of the Sultanate supported a well managed, well

---

35. For an indication of the sophistication of this system see a translation of the first four chapters of the "Dastur-al 'Aml fi Ilm al-Hisab," a treatise on the Sultanate's tax system written in 1358 by Haji Abdul Hamid. This appears in the *Medieval India Quarterly*, I, pp. 59-69.

The standard monograph on the subject, now rather dated, is W. H. Moreland, *The Agrarian System of Moslem India*, (Cambridge, 1929). Others include Ram Prasad Tripathi, *Some Aspects of Muslim Administration*, (Allahabad, 1956), and I. H. Qureshi, *The Administration of the Sultanate of Delhi*, (Karachi. 4th rev. ed., 1958). However, the most lucid statement of the main features of the Delhi agrarian tax/tribute system can be found in Irfan Habib, "The Social Distribution of Landed Property in Pre-British India," *Enquiry*, New Series, II, (1965), pp. 21-73. The evolved 14th century system which formulated its assessments and made its collections in coin "established a regime that was in some profound respects different from the old (i.e. Rajput). The Sultans achieved power that was, in terms of both territorial extent and centralization, unprecedented. . . . Their principal achievements lay in a great systematization of agrarian exploitation, and an immense concentration of the resources so obtained . . ." p. 45.

36. Ferishta, I. 96 gives the figures. Hinz, p. 22 calculates the Delhi 14th century *man* of 40 *ser* to be 12.824 kg.

37. Cf. M. S. Ahluwalia, "The Sultanate's Penetration into Rajasthan and Central India—a Study Based on Epigraphic Evidence," *Proceedings* Indian History Congress (1966), pp. 144-153.

conceived coinage. Sultan Iltutmish (1220-1235) introduced a basic silver standard and coin, the *tanka*, equivalent in standard to one indigenous *tola* containing 96 *ratis* (jeweler's measures) or 172.8 grains (11.09 grams). The Arabic legend *(kalimah)* on one face bearing the Caliph's titles as well as the Sultan's titles and dates were borrowed from the *dinar*. Devanagari inscriptions remained on the other side. From this basic coin emerged by the 1260's a large gold coin *(tanka)* of the same weight valued at a ratio of 1 to 10 which apparently was the mint ratio till the end of the 14th century. By 1300 or thereabouts bilingual coins had disappeared. Smaller denominations included low grade billon coins in regular sequences and copper coins for everyday commerce (the lowest of which, a half *fals*, was valued at 384 per silver *tanka*).[38]

An efficient state-run system of twelve mints (by mid-14th century) issued coins in extremely large numbers. Until severe factional disputes after the death of Sultan Firuz Shah Tughlaq in 1388 A.D., no discernible debasement of the coinage occurred.[39] Thus, one may suggest that the Indo-Muslim state of the 14th century utilized much (although certainly far from all) of its incoming specie for a coinage impressive both in quality and volume. The resulting increase in monetary liquidity and velocity reflected the economic strength of the state—especially in those monetized sectors determined by state action such as payment demanded of intermediaries for the land or harvest tax, in coin, or tribute from rajas. At the same time, intensified monetary circulation based upon a reliable, high quality coinage added to the strength of the metropolitan economy. We can see this especially in the pull of the North Indian market for at least one essential import.

The horse trade

As Digby reminds us in a recent monograph, strong war horses were essential for the Sultanate since its main battle force depended on heavy cavalry. For reasons of climate, or forage (far from perfectly understood), serviceable riding horses could only be raised in Kutch or Panjab in the western and northwestern borders of the subcontinent or in the north-east towards Assam.[40] These lighter India-bred horses were considered only a secondary option for the Sultanate cavalry, who preferred strong, reasonably fast horses to bear

---

38. See Habibullah, pp. 284-294 for one of the most lucid explanations of the Sultanate currency. The standard work is H. N. Wright, *The Sultans of Delhi: Their Coinage and Metrology* (Delhi, 1936).

39. For the operation of the Delhi mint see V. S. Agarwals, "Dravya Pariksa of Thakkuru Pheru," *The Indian Numismatic Chronicle*, III (1969), 100-114. Thakkuru Pheru was the mintmaster of the Delhi mint for three 14th century Sultans. See Deyell's essay in this volume for one attempt to estimate output. The number of Sultanate coins extant and their provenance is far more extensive than that of the preceding Rajput dynasties. Cf. H. De S. Shortt, "A Bull-And-Horseman Hoard From India," *The Numismatic Chronicle*, XVI, 6th series (1956), 313-325.

40. Simon Digby, *War-Horse and Elephant in the Delhi Sultanate* (Karachi, 1971), pp. 26-28.

the weight of chain-mailed, helmeted riders, their weapons, and similar armor for the animal.[41] Horses with these qualities of power and speed could be imported in the north-east from the Himalayan *terai* (the so-called *tamghan* horses), from Iran (the *bahari* or sea-borne horses), from the Arabian peninsula, Egypt or Syria (sometimes referred to as Shami or Syrian horses), and from Central Asia (the *Tatari* or *bālādastī* or "high-land" horses).

All four types had to be brought by sea or land over expensive long distance routes to the burgeoning market at Delhi, whose merchants were forced to pay for these imports in cash. Apart from a limited demand for return of high value luxury goods, it is doubtful that other payment would suffice. The combination of a discriminating demand, of large and growing purchasing power, a real strategic military need, and a reliable currency brought horses to Delhi by the thousand. Prices and profits were appropriately high: in the mid-14th century an exceptional Central Asian war horse would fetch 500 silver *tankas*, an ordinary Central Asian battle horse or a first-class north-western Panjab battle horse, 100 silver *tankas*. Then, as now, Arabian or Persian horses fit for racing by nobles or princes commanded as much as 1000 to 4000 *tankas*.[42] By contrast the price of a good quality pack mule was 4 to 5 *tankas;* that of a sexually attractive male or female slave 20 to 40 *tankas;* and of an adult male slave 10 to 15 *tankas*.[43] Island trading towns and their hinterlands in the Persian Gulf, such as Kish or the coastal trading towns along the South Arabian coast, came to depend for their prosperity largely on the Indian horse trade.[44]

Overland, the reach of the Delhi market and its ancillary demand at Bengal's western capital was equally impressive. From sizeable horse depots in towns along the southern edge of the Tibetan plateau and the city of Talifu in Chinese Yunnan, merchants convoyed large strings of *tamghan* battle horses down the difficult Himalayan mountain tracks for sale at the Muslim capital at Lakhnavauti. Some of these horses then made the further trek up country into Hindustan to meet the seemingly inexhaustible demand there. Minhajuddin, the early 13th century historian, recalls a visit with a friend of his who years before accompanied Muhammad Bakhtiyar, the conqueror of Lakhnavauti, on an inane military expedition up the Brahmaputra River which terminated in "the open country of Tibet," and at a large border city therein. Becoming interested in the region, Minhaj found that according to local knowledge in Lakhnavauti the latter city called Karampatan (or other

---

41. The Royal Ontario Museum in Toronto, Canada has an excellent collection of representative pieces of Mughal armor.
42. Digby, p. 37.
43. *Ibid.*, p. 38.
44. According to Serjeant, most of the trade was centered at Zufar which could provide fodder for the horses collected there. At least one Arab writer suggests that the profits provided money for agricultural operations in the area. All this was ruined when the Portuguese took over much of the sea-borne trade in the 15th century. See. R. B. Serjeant, *The Portuguese Off the South Arabian Coast* (Hadramī Chronicles, Oxford, 1963), p. 167, note B.

variants) was a major collection point for "every day, at daybreak, in the cattle-market of that city about one thousand five hundred horses are sold; and all the *tangahan* horses which reach the Lakhnavauti country they bring from that place. The route by which they come . . . is well known."[45] Minhaj did not specify the medium of return payment although the most likely possibilities are either Bengal silver coins or highly valuable and specialized goods such as the slave-eunuchs produced in large numbers in the region.

Marco Polo, on the other hand, refers to large strong horses carried from what has been identified as Yunnan by an organized system of imports into Bengal.[46] Both in his travels in the northeast, in the mountainous tracts of the upper reaches of the Brahmaputra river, and in the neighboring Chinese provinces, including Yunnan, as well as his extensive tour of India's western neighbors, especially the Persian Gulf, Marco Polo commented on the horse trade. He was impressed with the scale and the profits of this exchange. In his brief description of Anin (Yunnan) he writes:

Anin is a province towards the east, the people of which are subject to the Great Kaan and are Idolators. They live by cattle and tillage, and have a peculiar language. The women wear on the legs and arms bracelets of gold and silver of great value, and the men wear such as are even yet more costly. They have plenty of horses which they sell in great numbers to the Indians, making a great profit thereby.[47]

Similarly, in describing the Arabian port of Kalhat, he comments: "For, as I have told you before, the number of horses exported from this and the other cities (of South Arabia) is to India yearly is something astonishing."[48]

As we might expect the overland horse trade from the northwest and Central Asia was more far-reaching and probably greater in volume than that of the difficult routes of the eastern Himalayas. During the longish intervals between mass Mongol invasion of North India, the horse traders seem to have carried on their business without much interruption. One of the principal (and surprising) source for battle horses for Delhi was the area, now Russian, under control of the Mongol Golden Horde. At the city of Azof on the estuary of the Don, Ibn Batuta enjoyed the hospitality of the Mongol Amir Tuluktumur. The ubiquitous traveller commented that "horses in this country are exceedingly numerous and their price is negligible. A good horse costs fifty to sixty of their *dirhams* which equals one *dirham* of our money or therabouts."[49] Moreover, according to Ibn Batuta, these horses were exported to North India in droves of about six thousand animals by traders owning one to two hundred animals, who personally accompanied their investment. The greatest attrition in their stock along the journey occurred in the desert areas

---

45. Minhaj-ud-Din, I, 568-69.
46. Marco Polo, *The Book of Ser Marco Polo*, trans. and ed. Henry Yule, and Henri Cordier, 2 vols., 3rd. rev. ed., (New York, 1903), II, 119.
47. *Ibid.*
48. *Ibid.*, p. 450 Cf. Digby, p. 43. Digby comments that "Marco Polo is generally remarkably accurate with regard to the patterns of Asian trade."
49. Ibn Batuta, Travels, trans. and ed. by H. A. R. Gibb, II, 478-479.

of Sind, devoid of forage. Here the Sultan of Delhi imposed one import tariff and a further duty at Multan on those horses surviving. The *Travels* further state that:

> In spite of this, there remains a handsome profit for the traders in these horses, for they sell the cheapest of them in the land of India for a hundred silver *dinars* (the exchange value of which in Moroccan gold is twenty-five *dinars*) and often sell them for twice or three times as much. The good horses are worth five hundred (silver) *dinars* or more. The people of India do not buy them for (their qualities) running or racing, because they themselves wear coats of mail in battle and they cover their horses with armour, and what they prize in these horses is strength and length of pace.[50]

Indirect confirmation of the trade connection between Delhi and the Golden Horde domains derives from the evidence of surviving Delhi Sultanate coins found in hoards near the Volga. According to A. A. Bykov, "a new appearance of India coins in Europe falls in the 14th century, when dirhams of the Golden Horde were widely spread in the east and south districts of the east European plain. Now they were not the silver ones, but golden, struck by the Sultans of Delhi-Khiljis and Tughluqs."[51] A total of seventeen finds of coins from Kazan, Astrakhan, and Perm show a registration of forty-five golden *tankas* struck by two Khijlji and three Tughluq Sultans, all appearing in the known circulation area of Golden Horde silver coinages. Most of the coins are isolated pieces (save for one hoard of 18 golden *tankas*) but as Bykov points out, the golden *tankas* of the Sultans, weighing nearly eleven grams, possessed an enormous value in relation to the silver *dirhams* of the Mongols which only weighed on the average 1.3 to 1.4 grams.[52]

The necessarily sketchy evidence concerning the North India horse trade suggests the formation of a new or at least a greatly enhanced market demand for horses at the imperial Muslim capital of Hindustan by the 14th century. Further, that this exchange of animals for coined money constituted a substantial reverse flow of both gold and silver from India in the early Muslim period. So lucrative was this trade in desperately needed war material (which drove the price up) that it seems to have made its impact felt on several intermediary trading regions in the Indian Ocean and in the Brahmaputra valley as well as on the economies of the actual horse breeding zones in Iran, on the steppe of Central Asia, on the Tibetan plain and possibly even in southwestern China. Especially in the steppe areas this new demand must have coincided with and competed against the Chinese import system of the period which depended upon the export of tea to the Mongol and Turkish horse nomads to pay for Chinese battle horses.[53]

50. Ibid., p. 479.
51. A. A. Bykov, "Finds of Indian Medieval Coins in East Europe," *The Journal of the Numismatic Society of India*, XXVII (1965), 146-156, 151-152.
52. Ibid., p. 154.
53. See S. Jagchid and C. R. Bawden, "Some Notes on the Horse Policy of the Yuan Dynasty," *Central Asiatic Journal*, X (1965), 246-268. Also Morris Rossabi, "The Tea and Horse Trade with Inner Asia During the Ming," *Journal of Asian History*, IV (1970), 136-168.

Cultural and economic interaction with the wider Muslim world

As has frequently been observed, thirteenth-century Delhi (like Mamluk Damascus, Aleppo and Cairo) served as a sanctuary for demoralized refugees fleeing the Mongols. By the next century, as the Islamization of the Mongols restored the unity and interconnections of the greater Islamic world, Delhi grown in stature and prosperity became an impressive vital urban center for the reenergized high culture of Islam.[54] Ibn Batuta's experiences in a cosmopolitan Delhi in the mid-fourteenth century testify to the linkages of a city no longer a frontier outpost, but the proud capital of a hardconquered empire in the Indian east.[55]

The community of Orthodox Indian Islam in this early period, like that of other distant Muslim regions, depended for doctrinal guidance, reinvigorated piety, and communal solidarity upon regular connections with the great shrines of the Muslim world—especially through the annual pilgrimage to Mecca. Every year thousands upon thousands of pilgrims from India travelled on the Haj to fulfill their once-in-a-lifetime obligation by circumambulating the Ka'aba. As was the approved custom, the pilgrims usually carried goods with them to trade in a peddling fashion in order to meet their travel expenses. The cumulative scale of this pilgrim exchange was a major form of goods exchange in the Muslim world. However, the primary economic impact of this practice, impossible to quantify at any given time, was probably to benefit those regions along the established pilgrim routes while reducing somewhat the net export surplus of India usually expressed in its specie import surpluses.

Mecca and Medina and the surrounding regions of the Hijaz profited directly and in large measure survived on the expenditures of the pilgrims. In addition, the Sharif of the holy cities was the recipient of more direct largess from pious rulers and wealthy individuals elsewhere. Presentations from North Indian rulers formed a consequential and recurrent part of this support for the holiest shrines of Islam. For example, in 1409 A.D., local Arab historians of Mecca noted that Sultan Ghiyasuddin Abu'l-Muzaffar Azam Shah, ruler of Bengal, sent a personal emissary with money and letters for the Sharif of Mecca and funds for the needy of the town. The Sultan's agent also brought with him funds for three projects: the first was to construct a theological school or *madrasa* and supervise its initial operation. After obtaining the approval of the Shaikh, the Bengal emissary, Yaqut, had the *madrasa* constructed and opened it soon thereafter with four schools, one for each of the four recognized legal traditions enrolling a total of sixty subsidized

---

54. Cf. Ira Marvin Lapidus, *Muslim Cities in the Later Middle Ages* (Cambridge, Mass. 1967), pp. 1-25.

55. Even more effective testimony for the integration of Delhi into the wider Islamic world is the number and variety of informants with personal knowledge of India available in Damascus to Al'Umari when he set about compiling the Indian chapters of his vast encyclopedia. Cf. Siddiqi and Ahmad, p. 6.

students. The entire project, including an ongoing endowment supported by income from purchased lands and a water reservoir, cost 12,000 gold Egyptian *misqals* (just over 66 kg. of gold).[56] For the second project, Yaqut, acting under his rulers' instructions, opened a similar *madrasa* at Medina, although the precise cost is not given by the local historians.[57]

However, the Bengali emissary was less successful in his third task. The Sultan had sent an even larger sum to repair the stream of Arafa in Mecca (presumably for irrigation). The Sharif of Mecca seized as much as 30,000 gold *misqals* (140.4 kg.) of the money in order to repair other streams and reservoirs instead.[58] However, Sultan Azam Shah's successors seem to have continued the practice of establishing institutions and endowments at the holy shrines. In 1429-30 A.D. Sultan Jalaluddin Firuz Shah financed the construction of a large *madrasa* at Medina.[59]

A continuing outflow of funds from India also went to lesser shrines and even to well-known individuals in other countries noted for their piety, sanctity or scholarship. Ibn Batuta made a special effort to visit the grave of the celebrated Sufi saint, Abu Ishaq al-Kazaruni (963-1035), located just two days travel from Shiraz in southern Iran at the town of Kazarun. Upon arival, he stayed with the hundred or more devotees of the Shaikh and members of the order founded by him. Abu Ishaq, who had come to be "highly venerated by the people of India and China," was especially noted for his protection of travellers by sea. When in danger of foundering or of pirate attack, such travellers made vows to the saint and put in writing a pledge of a sum of money:

> Then, when they come safely to land, the servitors of the hospice go on board the ship, take the inventory, and exact (the amount of) his vow from each person who has pledged himself. There is not a ship that comes from China or from India but has thousands of *dinars* in it (vowed to the saint) and the agents on behalf of the intendant of the hospice come to take delivery of that sum.[60]

Despite the unusual fiscal arrangements of the Kazarun shrine, its ability to attract funds from pious Muslims in India was not at all unique. The

---

56. Hinz, p. 4 giving the weight of 4.68 grams per Egyptian *misgal*.

57. Ziauddin Desai, "Some New Data Regarding the Pre-Mughal Muslim Rulers of Bengal," *Islamic Culture*, XXXII (1958), 199-200. Desai has located and utilized material from several Arab historians including Ibn-i Hajar 'Asqalani (1372-1449), author of a major work in history and jurisprudence; Al-Fasi (1373-1429), author of a work on the history and topography of Mecca; and Al-Maqrizi (1364-1442), compiler of a biographical dictionary of noted figures of his time. A further examination of Al-Fasi's work, in particular, might well reveal data on other royal donations.

58. Ibid., p. 200.

59. Ibid., p. 204.

60. Ibn Batuta *Travels*, Gibb trans. II, 320. Frequently, when poor mendicants came to the shrine for alms, they were given a written order from the intendant, sealed with Abu Ishaq's devise, which ordered anyone who had made a vow to the Shaikh to let him give that amount to the bearer. These charitable bills of exchange might even reach India.

sanctity of shrines and the piety and learning of individual scholars was well known and eagerly supported even at great distances in the medieval Islamic world. For example, Desai records the case of gifts sent in 1430 by Sultan Jalaluddin Firuz Shah, an exceptionally devoted Muslim from Bengal, to Shaikh Alauddin Bukhari, an Indian scholar who resided alternately in Cairo and Damascus, and for a group of other scholars in those cities.[61]

Finally, the whole question of remittances sent homeward by foreign migrants and home retirement with Indian funds is one that like so many others of this type can never be satisfactorily analyzed. We can be reasonably certain, however, that large sums of gold and silver flowed to the various Islamic countries of the Mediterranean, to Iran and to Central Asia when individuals either remitted part of their funds obtained in Hind or returned after service in India. Obviously some Muslim migrants to North India cut all connections with their home country and never returned. On the other hand, some came to India in an effort to make and to return with a fortune. Minhajuddin records the good fortune of a scholar from Firoz Koh who attracted the favorable attention of Sultan Ghiyasuddin of Bengal and was able to leave for his home with ten thousand gold and silver *tankas* in presents from the Sultan and his courtiers.[62]

The largess of Sultan Muhammad b. Tughlaq (1326-1351) toward foreigners (especially noblemen and scholars) was legendary. The Sultan's generosity even extended to alien merchants—not necessarily Muslim. In 1338 A.D. six Venetian merchants, hearing rumours of the odd behavior of the "signor del Delhi" formed a *compagnia*, and left Venice for a lengthy trading journey to Delhi. Surviving notarial documents in Venice, originating in a 1344 inheritance dispute over the estate of one of the traders who died en route, supply some corroborative evidence for the scale of Muhammad bin Tughlaq's liberality. After arriving by sea in the Crimea (carrying cloth to be sold in Russia), the merchants marched overland through Mongol territories via Astrakhan, Urjenc, Bukhara, Ghazni and eventually arrived at Delhi. From the Sultan the company received a gift of 200,000 *bezants* (either Byzantine coins, or a unit of account used for silver coins). From this sum 20,000 *bezants* met the customs demanded at Delhi (along with an additional 2,000 for the "scribe of said customs house") and "from 9,000 to 10,000 *bezants* were given away to those who had paid us great honor, that is the barons of the Lord." The amount actually carried away from Delhi by the *compagnia* amounted to 60,000 *bezants* in either gold or silver coin (10,000 to repay expenses for each partner); 102,000 *bezants* jointly invested in pearls; and 28,000 *bezants* invested in Mongol gold coins. When the party reached

---

61. Desai, p. 204.
62. Minhajuddin, *Tabakat-i Nasiri*, Raverty trans. I, 583-584. Al-'Umari records an interview with Shaikh Abu Bakr bin Khallal al-Bizzi as-Sufi in which the Shaikh recalled: "The Sultan (i.e. Muhammad b. Tughlaq) sent a party of which I was one with three lac *mithqals* of gold (300,000 *mithqals* or 210 kg. 4.3g) to Transoxiana to distribute one lac among the scholars, one lac among the poor in alms and the remaining one lac for purchasing goods for the Sultan." (Al-'Umari, p. 49) The party was also charged with finding the renowned scholar, Shaikh

Urjenc on the return journey, the pearls were sold for a profit returning 17,000 *bezants* plus to each including the heirs of the deceased trader.[63]

Extravagent but consistent with other aspects of Muhammad b. Tughlaq's theatrical style of rule, the Sultan's policy objective was clearly to enlarge the resident body of foreign migrants in Delhi and to bind them to him by ties of clientship, and to enhance his status and that of his frontier regime in the wider world of Islamic civilization. Other Indo-Muslim rulers at Delhi and elsewhere also patronized foreign migrants and visitors in a manner, which although not as munificent perhaps, did move substantial wealth from India.

Conclusion

That illustrative or anecdotal evidence pertinent to the question of monetary flows for the early Muslim period in North India at least sketches the outline of two simultaneous processes. One is unquestionably the annual process of plunder, forcible dishoarding of accumulated indigenous treasure by Muslim and Mongol invaders. During the more than two centuries under consideration this raiding process pushed large amounts of gold and silver out of North India to Ghazni, to Bokhara and west to Samarkhand and beyond into Central Asia. Whether coined or uncoined the precious metals of North India were an end product of one of the most bitter conflicts in human history.

At the same time, once the invaders had become established in Hindustan and controlled the riverine cities of the Indo-Gangetic plain, the flow of precious metals seized by their armies was diverted to the new capital at Delhi. In the northwest, in the Panjab, the Mongols continued to raid and to carry off whatever valuables they could seize to Central Asia. But the plunder and tribute of Gujarat, of Mithila (northern Bihar) of Rajasthan and central India came to the coffers of Delhi. This massive inflow of specie which financed and made possible the coinage of the Sultanate, assisted successive centralizing regimes by the liquidity and velocity imparted to the monetary system. Plunder, tribute and regular taxation (in the central districts of Hindustan) engendered wealth which could be used to pay for various imports—especially battle horses of the requisite qualities. The market at Delhi was as much a creation of conquest as the building of the city. By the 14th century surplus funds enabled the Sultans of Delhi and Lakhnavauti to provide subventions for the most central institutions of the greater Muslim world, and for prosperous Indo-Muslims to go as pilgrims on the Haj or to send contributions to shrines outside Inida. Ultimately, as with Ghazna earlier, a Central Asian Conqueror seized Delhi and brutally uprooted the accumulated wealth and talent of the city.

Burhanuddin As-Sagharji of Samarkhand and giving him 40,000 *tankas* to induce him to travel to North India. *Ibid.*

63. The notarial documents are translated in full in Robert S. Lopez and Irving W. Raymond, *Medieval Trade in the Mediterranean World*, (New York, 1955), pp. 281-289. See also, for a more extended discussion Robert S. Lopez, "European Merchants in the Medieval Indies: The Evidence of Commercial Documents" in *The Journal of Economic History*, (1943) pp. 164-184. I am grateful to my colleage Manfred Raschke for this reference.

7

The China connection: Problems of silver supply in medieval Bengal

JOHN DEYELL

The Ghorid conquest of Hindustan as far east as Gauda (Bengal) in the late 12th century A.D., was accompanied by the attempted establishment of a common North Indian currency. Islamic coinage norms, the gold *dinar*, silver *dirhem* and copper *falus* common in Iraq and the west, had been somewhat modified during the slow advance into the Indian cultural sphere in Seistan, Afghanistan and the Punjab. The basic coinage of the Ghaznavid, Shahi and Rajput dynasties of N. W. India in the 9th-12th centuries was the billon (base silver) "bull and horseman" *dramma* of 3.3 g and variable fineness of alloy. This circulated, in India, with a gold coin of 4.1 g of .650-.980 fineness, and in Afghanistan with both a gold coin of 8.7 g and approximately .950 fineness, and a silver coin of 5 g and .950 fineness. The compromise settled on by the early Turkish sultans of Delhi was a currency of four coins, a gold *tanka* of 11 g and .950 fineness, a silver *tanka* of 11 g and .980 fineness, a billon *dehliwal* or *jital* of 3.6 g and .050-.160 fineness, and copper coins of 2.3-4.5 g pure metal.[1]

Both gold and copper coins remained rare in Delhi throughout the 13th century, and the de facto coinage was limited to the pure silver *tankas* and alloy *jitals*. Reliable statistical data as to the relative or absolute quantities of these coins in circulation is not available. This is primarily due, strangely enough, to the overabundance of surviving or "recaptured" coinage: in the aftermath of the annual monsoon, immense numbers of the *tankas* and *jitals* are unearthed. In the face of this intimidating quantity, the numismatic literature has long ignored hoard data for Delhi Sultanate coinage, being largely content with publishing unusual dates, mints, denominations and rulers. The converse of this lack of objective statistical material is the judgment born of the experience of numismatists handling large numbers of the coins over a period of time. Subjective analysis might reveal a profile of annual quantity of coins minted as in *Graph One*.

---

1. Sources of coin weights: A. Cunningham, *Coins of Medieval India*, Delhi, 1967 (reprint); V. A. Smith, *Catalogue of Coins in the Indian Museum, Calcutta, Volume I*; H. N. Wright, *The Sultans of Delhi, Their Coinage and Metrology*, Delhi, 1936.

Sources of coin assays: Wright, op. cit.; S. K. Maity; *Journal of the Numismatic Society of India*, Vol. XXII (1960), p. 270, p. 272, p. 275.

The volume of billon coins in the early 13th century remained fairly constant and was equal to the combined output of the Rajput kingdoms displaced by the sultanate. The issues of broad silver *tankas* were sporadic, those of Iltutmish (A.D. 1210-1235) being scarce and of the other early sovereigns, rare. The assumption of regular and more prolific issues of silver *tankas* following the death of Iltutmish was accompanied by a fall in the quantity of billon coins issued, as well as a drop in the silver content of these billon coins. Clearly the supply of coinage silver was fairly fixed, such that the initiation of silver coinage was made at the expense of silver in the form of billon coins.

The accompanying *Map One* of South and South-east Asian precious metal resources, shows the primary sources of silver available to the Delhi sultanate. During the century prior to A.D. 1294 these would necessarily have been within the immediate sphere of influence of the Turkish ruling groups: the Ganges basin, the Punjab and Afghanistan. The mines at Zawar in Rajasthan are excluded since they were not extensively exploited until the 15th century A.D.[2] The ports of the Gujarat and Bengal coasts in this period were beyond Turkish political influence; the supply of bullion from these trade sources is problematical; not immediately demonstrable. Rather, the major silver mines of the Panjhir and Wakkhan in northeast Afghanistan, which were worked by the Samanid and Ghaznavid dynasties, in succession, are the likely candidates for a reliable but limited silver supply for the Indian sultanate. The immense quantities of bullion mined in the 10th century which flowed north along the Samanid-Varangian trade routes to the Baltic were no more;[3] while not exhausted, the Panjhir mines were definitely more limited in their output.[4] Further, by A.D. 1222 the mines had passed into the control of the Mongols;[5] this became the primary source of silver for the Ghazni *dirham* issues of Chengiz Khan.[6] Hence rather than working a directly controlled precious metal resource, the Delhi sultans were reduced by mid-century to the reminting of foreign silver: *dirhams* which passed by way of trade and booty through the Punjab and Sind buffer states into Hindustan, in addition to those coins derived from the Hindu littoral states.

The effect of this phenomenon of a steady but curtailed silver supply is clear in *Graph One*, as discussed above. Since the bulk of the sultanate's coins were minted at the "Delhi" mint, the gateway of Hindustan, the effect of this stationary bullion stock was largely felt throughout north-west India alone. A secondary mint at Lakhnauti in Bengal was regularly operated by the sultans

2. J. Tod, *Annals and Antiquities of Rajasthan*, London, 1920, Vol. I, p. 321.
3. Michael Mitchiner, *The Multiple Dirhems of Medieval Afghanistan*, p. 7 ff.
4. Hamd-Allah Mustawfi, *Nuzhat-al-Qulub*, tr. G. Le Strange, London, 1919, p. 193.
5. Md. Habib, "The Mongol Invasion of Ajam," in *A Comprehensive History of India, Vol. V, The Delhi Sultanate*, Bombay, 1957 (reprint), p. 80.
6. Stephen Album attributes the Chingiz issues to Ghazni, and presumes the maintenance of the Panjhir silver supply as consistent with the Mongol policy to spare artisans and exploit established resources of wealth.

and their governors, and this in turn may have placed a slight further strain on available silver.

The pre-Islamic coinage experience of this eastern province was in strong contrast to that of Hindustan: with the exception of very rare gold *dinaras* and silver *drammas* during the Pala and Chandra dynasties respectively,[7] the sole recognizeable currency of Bengal was the cowry shell, until the Muslim invasion (the use of metal ingots in trade is indeterminate).[8] The primary coin introduced by the Turks in Bengal was the silver *tanka*, and this initially circulated in restricted quantities comparable to the contemporary Delhi silver issue (see *Graph Two*). The modest issue of Lakhnauti silver *tankas* by the Turkish governors of the Delhi sultans was one facet of the penetration of an alien Central Asian culture into the area, and it seems reasonable to attribute a common source of silver to the initial Delhi and Lakhnauti mint issues. The subdued scale of the eastern mint activity would not unduly disturb silver availability in the western capital, Delhi, especially if the conquest of the province had netted appreciable quantities of silver booty.

The scarcity of bullion in the Delhi sultanate was sharply terminated by the military campaigns of Alauddin Khalji in the Deccan (1295-1311). The war booty and tribute from the expeditions against Devagiri and Warangel brought great quantities of treasure into the possession of the Delhi rulers,[9] which they could, and did, release as specie when required. *Graph One* shows the dramatic impact of this treasure on the quantity of gold, silver and billon coins minted after A.D. 1295. Gold, hitherto rare, became a circulating coin; the silver *tankas* became inordinately common; an unprecedented issue of billon minor coins was made. Immediate effects of this coinage plenty would logically appear to be a general inflation of prices and an emigration of currency to provinces rich in relatively cheap products (pre-eminently Bengal). The former conjecture seems to be supported by the rigid and rigorously-applied system of market controls instituted by Alauddin in order to regulate prices.[10] The latter would appear to the supported by the *Graph Two* curve which shows a sharp increase in coinage issue in the reign of Ruknuddin Kaikaus of Bengal. A.D. 1291-1301, rising steadily throughout the first half of the 14th century.

However caution is required in both instances. It seems likely that the economy of the 13th century Delhi sultanate was under-monetized; that is, the use of coinage was restricted both by custom and coin scarcity. Considerable economic transactions, such as the payment of revenue to government and endowed institutions, and provision of services throughout the society,

---

7. F. A. Khan, *Excavations at Manamati*, Dacca, Dept. of Archaeology Govt. of Pakistan, 1963, re: Chandra silver coins.
8. Minhaj-us-Siraj, *Tabaqat-i-Nasiri*, in Elliot and Dowson, *The History of India as Told by Its own Historians*, Delhi 1967, (reprint), Vol. II, p. 308.
9. Habib, op. cit., pp. 322-3 (Devagiri), 334-5 (Gujarat), 409-410 (Warangel) 416-417 (Madurai).
10. Ibid., pp. 372-391.

were both accounted and carried out in kind, without reference to monetary value.[11] In a mixed barter/money economy, drastic expansion of the money supply would first of all result in increased use of currency in transactions. In this the result may have been increased penetration of money use into traditionally non-monetized relationships.[12]

Secondly the ascending curve of Bengali silver coin production anticipated the Khalji military conquests by some four years; the initial impetus is thus clearly independent of developments in Hindustan. This point is most important, and will be returned to shortly.

It is not at all certain that the opening of the south to Delhi influence provided a reliable long-term supply of silver bullion at least for the North Indian sultanate. *Map One* pinpoints the prolific precious metal resources of South India, the famous Mysore and Kolar gold fields. Indeed, from the demise of the Mauryan rule ca. 200 B.C. until the Muslim conquest, the major coinage metals of the south had been, with a few notable exceptions, limited to gold and copper, and in the Deccan, gold and lead. Such silver as was captured by Alauddin was not locally produced, but came to South India over the centuries by way of sea trade; the great numbers of Roman *denarii* unearthed are testimony to this process.[13] In this sense the successive Muslim penetrations, capped by the raid on Ma'abar in Alauddin Khalji's reign (A.D. 1311), brought rather short-term redistributions of silver resources northward. With the revolt of the southern provinces from Delhi control which resulted in the foundation of the Ma'abar sultanate (1336) and Bahmani Sultanate (1347), tribute collections from the south and the Deccan ceased altogether.

This was not immediately reflected in the billon coin curve of *Graph One* from A.D. 1300 to 1388, where an approximately static output of billon coins in immense quantities is indicated. It must be remembered though that attrition has a significant effect on a static volume of coinage until an equilibrium with silver supplies (from whatever source) is reached (note the silver fineness chart in *Graph One*).

This analysis highlights a vital point, that irregardless of the fiscal health of a treasury, a major currency could not be floated without a regular supply of metal deriving either from internal production or external trade. That the Tughluq treasury held formidable stocks of gold well into the period of decline following Firoz Shah's death is known from descriptions of the booty

---

11. In interpreting epigraphs of the period it is necessary to distinguish between money of account and money itself. Coin names often referred to an evaluation of goods tendered, rather than a sum of actual money.

12. W. H. Moreland, *The Agrarian System of Muslim India*, Allahabad, 1929, p. 38. Moreland feels that the new regimen by which agents of the sultan assessed and collected revenue in cash was a novelty. Presumably the monetization of an in-kind relationship was an integral part of displacing the intermediary classes.

13. P. L. Gupta, *Roman Coins from Andhra Pradesh*, Hyderabad, 1965 p. 47.

Problems of silver supply in medieval Bengal

taken by Timur-i-lang in the sack of Delhi in A.D. 1399.[14] This notwithstanding, the production of gold *tankas* declined drastically in the later 14th century. The progressive weakening of political control over the Deccan, signalled by the foundation of the Bahmani sultanate in mid-century, diverted the direct transmission of current gold production away from the north. Conversely, the Deccani sultan Alauddin Bahman Shah (1347-1359) was unable to establish a silver currency until the seventh year of his reign.[15] The pattern of trade was not sufficiently developed along north/south lines to assure the Bahmani sultans of a silver supply. Only through the influence of maritime trade could this be corrected, later in his reign (A.D. 1354).

The progress of the silver coinage of the Delhi sultanate during the 14th century has been gone into in some detail, the more readily to appreciate, and explain, the concurrent experience of the independent sultanate of Bengal during the same century. Unlike the former state's coinage, the hoard findings of Bengal coinage are regularly and well reported in numismatic literature, and have been gathered into a *Corpus* complete to 1960.[16] These findings are presented in *Graph Two*. Copper coins are limited to the rare issues of late sultans, and are ignored. Gold coins, while issued by most Bengal sultans, are only known from one or two specimens of each type, and are statistically of no importance. Certain cautions are advisable in interpreting this graph,[17] but for our purposes it compares well with the *Graph One* of Delhi coins.

14. E. Thomas, *The Chronicles of the Pathan Kings of Delhi*, Delhi, 1967 (reprint), p. 171. f.n. 1.
15. The earliest known date of silver *tanka* is 755 A.H. Md. Abdul Walikhan, *Bahmani Coins in the A. P. G. Museum, Hyderabad*, Hyderabad, 1964, p. 10, no. 3.
16. Abdul Karim, *Corpus of the Muslim Coins of Bengal (Down to A. D. 1538)*, Dacca, 1960.
17. Findings of buried coins, called hoards or *trouvailles*, provide our best knowledge of the composition of the circulating coinage at the time of accumulation of the hoard. For statistical purposes the ideal hoard would be a true and random cross-section of coins in circulation when the coins were gathered together; this is presumed to be coincident with the time of burial and is signalled by the latest date on coins in the hoard. Factors of selectivity in the making up of the accumulation spoil this randomness, for instance if the owner of the coins chose only high value coins for preservation, or only new coins, or only coins of a favourite type. Aesthetic factors could significantly influence selection. Likewise, the modern reporting of the hoard composition is similarly subject to distorting selectivity; only the unknown, the unusual, or the coins of specific metals or periods might be reported in the literature. This will affect the validity of statistical compilations based on these reported hoards.
  A second source of coin data, collections public and private, are to be avoided as much as possible. Statistical data collected from their holdings will be of limited value since an extreme degree of selection is normally implicit in collection holdings, including an upper limit on the number of coins of each identifiable type held (usually one or two). The lack of validity of statistical extrapolations from this source was explored in the article "Numismatic Methodology in the Estimation of Mughal Coinage Output," *Indian Economic and Social History Review*, Vol. XIII, July-Sept. 1976, pp. 393-403.
  The *Graph Two* is an abstract of the coins published in Karim, op. cit. He reports both museum and hoard holdings; in fact, every specimen of Bengali sultanate coinage ever reported. The resulting graph must be treated with caution for many reasons, one of which is that the

Rigorous consideration of the graph would require analysis of the coinage production as a function of the rate of re-coinage, the differential of metal stocks involved in import/export, the differential of metal stocks involved in industrial activities, the differential of metal stocks involved in savings/ stockholding, and the rate of production of local metals mines, modified by the constant factor of inadvertent loss. Mathematical models for this analysis do exist.[18] The crudity of our base graph is such that statistical operations are meaningless in the face of the probable margin of error. We can remark however, that due to the total absence of indigenous sources of silver in Bengal (note *Map One*), minting (from bullion), export, industrial uses (such as jewellery fabrication), savings (private hoarding and government stockpiling), and attrition losses in circulation, were all directly dependent on the rate of importation of silver. Presuming that these factors all responded in commensurate and direct proportion to the change in the rate of silver inflow, then *Graph Two* is a rough indicator of the history of silver importations to Bengal in the period ca. 1200-1500 A.D., subject to certain provisos.[19] Positive

---

museum holdings cause a variable deviation from the actual curve. To correct this one should exclude museum specimens; they unfortunately make up the bulk of Karim's sample and one must assume that their numbers relate to relative abundance. Secondly, the hoards reported were not buried at regular intervals, so that there is a distorting effect from the attrition due to circulation losses within each hoard, which is not corrected by any regularity of period of hoard dates. Thirdly, the numbers of coins reported are not really sufficient for statistical certainty; the percentage error is higher than ± 100% in some cases. A striking example of this error is the rise in *Graph Two* in A. H. 823: fifty-four of these coins come from a single hoard in which they are the latest date present. Obviously a block of current mint production was included in the hoard prior to burial, in a quantity that bore no relation to the random sample. A check mark shows the revised total for that year if the 54 are excluded.

Finally, the specific specimen-per-year tally caused the exclusion of numbers of undated coins which could not be distributed evenly over known minting dates: the lack of date itself could be a characteristic of the mint product of certain unrepresented minting years. To lessen this error, a five-year totalizing bar chart was superimposed on *Graph One*. Further, an auxiliary graph which averaged the total of published surviving specimens, with and without dates, across the known mint years of each reign, was plotted. This also was related to a five-year totalizing bar chart, showing a fair correspondence (not published here).

Rather unfortunately, until sufficient primary research is undertaken, the *Graph Two* is the only chart of Bengali minting history available, and must suffice for this study. It should be remarked, to balance the *caveats*, that the curve is fairly congruent with a subjectively constructed hologram deriving from the author's experience with numbers of the coins in commercial channels. The graph may be taken as trustworthy as long as a sophisticated degree of exactitude is not expected. It represents, albeit haltingly, the relative production quantities of the silver coinage.

18. For example C. C. Patterson, "Silver Stocks and Losses in Ancient and Medieval Times," *The Economic History Review*, Vol. XXV (1972), No. 2, pp. 205-235.

19. The most serious impediment to this analysis is the re-coinage factor. It was the habit of *sarrafs* (bankers and moneychagers) in medieval Bengal to discount old coins to allow for loss of metal through wear and tear. The practice was widespread, distinctive of Bengal, and was probably adhered to by the government in accepting coinage in settlement of revenue demands. The money-changers "cancelled" old silver *tankas* by defacing them with chisel marks and

slope on the curve (a rising graph) is indicative of precipitate influx of silver. A neutral slope on the curve (a horizontal graph) is indicative of a rate of influx sufficient to cover the inadvertent loss factor. A slight negative slope on the curve (a falling graph) corresponding to a drop in volume of less than 50% in 35 years, indicates no inflow of silver and normal attrition due to inadvertent losses (based on an attrition rate of 0.02 yr. $^{-1}$x total stock).[20] A precipitate negative slope on the curve, greater than a loss of 50 percent in 35 years, indicates a net outflow of silver bullion from Bengal during that period.

Applying these observations to *Graph Two*'s upper hologram, and ignoring for the moment the peaks occurring exactly at the boundary of succeeding reigns, as well as the period in which Muhammad Tughluq of Delhi held Bengal (1325-1338), for which information was not as readily available to the compiler, the net silver bullion import/export position of Bengal appears as follows:

A.D. *1218-1290*   Exhaustion of original stock and sporadic importations of bullion on a limited scale not sufficient to maintain the currency in circulation.

1291-1357   Very large net inflow of bullion on a regular basis and on a greater scale than in the previous century.

1358-1366   Net outflow of bullion from Bengal.

1367-1390   Moderate, regular importation of bullion sufficient to offset attrition losses.

1390-1415   Second period of heavy inflow of bullion, subject to sharp fluctuations year-to-year; a disturbed supply.

---

distinctive punches. Current coin passed at full value; older coins of the current reign were carefully "shroff-marked" so as not to obscure the issuing sovereign's name, and passed at an initial discount (1-3%); coins of previous reigns were probably demonetized and passed as bullion. This practice encouraged the recoinage of older coins; in fact this was profitable to the government as the mint levied a surcharge on bullion coined, above and beyond mint expenses. This was the Mughal system, and observation of Bengali coinage indicates the system originated there.

The statistical difficulty is that re-coinage is in theory the one factor not dependent on bullion supply, since the stock of coins was in existence and any percentage could be re-coined annually. This would be especially true on the accession of a new king, when the coinage of the previous reign would be discounted *en masse*. While the graph does not confirm this phenomenon before the mid 15th century A.D., the accessions of Barbak (1459), Hussein (1493), Nusrat (1519) and Mahmud (1532) are marked by heavy mintages in the first regnal year. As expected, some percentage of this was due to recoinage, but it also seems certain that this re-coinage did not involve the major portion of the existing circulating stock. This is clear from hoard composition, where coins of the accession year are found with earlier coins, and do not form a *terminus ante quem* for each hoard of the late reigns. Furthermore, the coin inscriptions themselves suggest a direct relationship of accessional and exceptionally large coin issues to major bullion inflows. This is discussed in following pages.

20. C. C. Patterson, op. cit., p. 220. The higher value was chosen due to the peculiar conditions in Bengal conducive to outright loss.

1416-1429 Second large net outflow of bullion.
1430-1492 Moderate inflow, somewhat sporadic, sufficient to maintain the volume of currency.
1493-1533 Steadily increasing inflow of silver on a very regular basis.

Whatever the alternate interpretations of this most tentative data, it seems justifiable to conclude that coinage production, probably independent of internal factors and responding to external conditions of bullion supply, peaked in the periods bracketing the years 1357, 1415 and 1493 A.D. *Graph One* shows that the initial period of inflow coincides somewhat with the decline of silver coinage in the Delhi Sultanate, and may have contributed to it by drawing off silver from the Deccan stocks at Delhi. We have already seen however that this decline is accountable by the maintenance of a billon coinage at Delhi, so that we must look elsewhere for at least part of this inflow to Bengal. The 1415 importation occurred after Timur's invasion and local political decline had caused the collapse of Delhi currency and finances and the disturbance of trade routes to western Asia; in this instance the entire inflow must be sought from a source other than Hindustan. The 1493 inflow coincides not only with the somewhat restored and stable economic conditions in Delhi, but also with the period of maximum prosperity of the Malwa and Gujarat sultanates and the kingdom of Mewar. The former two had extensive silver currencies, and the latter control of (relatively) profitable argentiferous lead mines. The period also merges into the arrival of the Portuguese in the east, and all these factors must be considered in seeking a source for Bengali silver in the latest influx.

Although appropriate historical records that could throw light on this problem of bullion supply in the 13th and 14th centuries are lacking, the inscriptions on the Bengali *tankas* go far towards solving the problem. Amongst the initial issues of Ruknuddin Kaikaus is a type bearing the inscription, "... struck at ... Lakhnauti from the land tax (*kharāj*) of Banga in the year 690 [A.D. 1291]."[21] Banga, from which our term Bengal derives, was the geographical name of the delta region and south-eastern Bengal. At this period it was ruled by the independent Hindu dynasty which had lost Gaura to the Muslims.[22] We conclude that a military confrontation left the sultan in possession of silver booty or tribute from which his initial coinage was struck. This would have been drawn from an accumulated stock, since there are no known silver resources in Bengal, and the people of Banga did not use a silver coinage. Banga thus was merely a transmission point for bullion movement. Sea trade was active at this time, with Arab traders in residence at Chatisgaon (Chittagong).[23] Hence any point on the Bay of Bengal or beyond may have

21. Karim, op. cit., p. 25.
22. J. N. Sarkar, *The History of Bengal*, Vol. II, Dacca, 1948, p. 23.
23. Md. A. Rahim, *Social and Cultural History of Bengal*, pp. 37-47.

been the source of this silver. Likewise, the Hindu states of east Bengal, Tripura and Manipur were candidates for transmission by land.

The expansion of the Bengal sultanate into Satgaon in the south-west on the Orissan frontier in A.H. 708 (A.D. 1305) under Shamsuddin Firoz[24] was not reflected in a rise in quantity of coins minted; we may presume a lack of silver available for tribute in this quarter. Permanent expansion towards the south east, with the conquest of Sonargaon and subsequent independent coinage issues by Mubarak Shah (A.D. 1338-1349), on the other hand, was accompanied by rises in the graph, confirming the availability of silver in this coastal area. Military campaigns against the north brought in stores of silver bullion: Sikandar Shah's initial coinage includes an issue of A.D. 1358 from ... *mulk chawalistan 'urf 'arsah kamru* ..., "the land of rice, alias the province of Kamarupa."[25] It appears that the Hindu kingdoms of the Brahmaputra valley were possessed of bullion supplies: the field of potential supply thus expands to include Tibet in the north and Assam in the north east. This is confirmed by the very large accessional issue of Hussein Shah in A.H. 899; on some coins the inscription reads "... conqueror of Kamarupa, Kamata, Jajnagar and Orissa ..."[26] Apparently a goodly proportion of this issue derives from tribute or captured silver. We have mentioned as unlikely that this originated to the southwest (Jajnagar and Orissa). Orissan chronicles, and the temple records of Puri, show that monetary transactions in that kingdom involved cowry shells and gold ingots exclusively.[27] Therefore Hussein's silver must have come from Kamata and Kamarupa, which had earlier provided Sikandar. Since neither area is a silver producer, the transhipment of silver must have been continuous throughout the period, at least to the extent of making up the earlier losses.

It is important to establish the *direction* of bullion travel, since the Hindu kingdoms on the frontiers of the sultanate soon established their own silver currencies: Tripura (east of Sonargaon) in A.D. 1464, Ahom (Assam) in 1543, Koch Kingdom (Kamata) in 1555, Kachar (Meghalaya) 1583, and Jaintia (west of Manipur) in 1670. The presumption has been that these peripheral kingdoms derived their silver from Bengal. While this may have been the case in the Mughal period (post 1576), the above discussion shows the probability that prior to the early 16th century, the historical direction of bullion flow was just the reverse.[28]

24. Karim, op. cit., pp. 28-29.
25. Ibid., pp. 160-161.
26. Ibid., pp. 177-8.
27. And hence the agonies of British-period historians attempting to convert revenue sums from the palm-leaf records, listed in gold *markas*, into rupees. See N. K. Sahu, ed., *A History of Orissa*, pp. 144-145 f.n. 341.
28. The point of view that silver money gravitated from the Bengal sultanate to peripheral kingdoms is based on the hoards of Bengali silver *tankas* unearthed in these latter territories. Inspection of hoard composition permits an alternate interpretation. The great Koch Behar find of 13,500 *tankas*, reported in 1864, contained issues of Balban (1266-1287) to Ghiyasuddin Azam

In order to establish the reliability of this analysis, three critical elements have to be identified: an ultimate source of this bullion, a feasible route of transmission from the source to Bengal, and an agent of transmission.

The portion of *Map One* to the east of India shows the known sources of gold and silver in east and south-east Asia. The gold sources, marked with black squares, are both placer (alluvial) workings, mainly along river banks, and lode (vein or outcropping) workings, mainly in underground mines and surface pits. The silver sources, marked by white squares, are restricted to lode workings; native silver cannot withstand oxidation due to weathering. It is not likely that every known modern or formerly worked location is charted; in Yunnan especially it has been remarked that one is never far from where gold is mined. Rather, the distribution of deposits in different areas is suggested. Apart from a concentration of gold sources in the Tibet-Szechuan-Yunnan borderlands, this element occurs throughout South East Asia. It will be noticed that the occurrence of silver is not so widespread, confined in fact to Yunnan and the northern Shan States. In the medieval context, excepting Siberia, Manchuria, Hunan and Japan, there were no alternate Asiatic sources of silver, and only Yunnan was within practical distance of Bengal. It is to Yunnanese sources that we must direct our attention.

The metallurgy of precious metal extraction and refinement was known in both the Pyu kingdom of Pagan (Burma) and the Tai kingdom of Nan-Chao (Yunnan) in the 9th century A.D. Gold ore was mined in the Nmai Hka mountains,[29] and alluvial gold dredged from the sands of the Irrawaddy River, both in Burma.[30] Concentration of the ore was by hand-picking, while placer sands were washed on ripple boards.[31] Refinement was effected by treatment with mercury; the amalgam was heated to evaporation point, leaving pure gold.[32] The mercury was widely available from cinnabar deposits. Similar activity was reported form Nan-Chou, both as to lode and placer workings.

(1389-1410). (Note by R. J. Mitra, *Proceedings of the Asiatic Society of Bengal*, No. 4, 1864, pp. 480-484; corrected and expanded by Edward Thomas, "The Initial Coinage of Bengal," *Journal of the Asiatic Society of Bengal*, Vol. XXVI, No. 1, 1867, pp. 1-73). The latest coin date in the hoard was A. H. 799 (A.D. 1397). Both the Kamrup and Ahom chronicles (*Kamrupar Buranji:* N. N. Acharyya, *The History of Medieval Assam*, Gauhati, 1966, pp. 170-171; *Ahom Buranji:* Calcutta, 1930, tr. G. C. Barua, pp. 49-51.) refer to Kamata/Ahom disagreements which drew an invasion by Ghiyasuddin which was ultimately defeated. The kings involved were Shudang of the Ahoms, 1397-1407 A.D., and Sukranta of Kamata, 1400-1415 A.D. The engagement must therefore have taken place in the period 1400-1407, which postdates the latest date of the Koch Behar hoard. The treasure was likely to have been part of the expeditionary fund of the Bengal army, lost during the chaos of defeat. Similar stories can be told of other sultanate hoards beyond the Muslim frontiers.

29. *Man-shu* of Fan Ch'o, c. 865 A.D. 8v., 29v., 32v. Quoted by G. H. Luce and Pe Maung Tin, "Burma Down to the Fall of Pagan," *Journal of the Burma Research Society*, Vol. XXIX, No. 111, 1939, p. 270.

30. Ibid., 8v., 9v, 29v, etc.

31. J. Coggin Brown, "The Mines and Mineral Resources of Yunnan," *Memoirs of the Geological Survey of India*, Vol. XLVII, 1923, p. 149.

32. Ibid., p. 155.

Direct evidence of silver mining and smelting is unavailable, but it is known that the Pyu had a coinage of sorts, silver bars "shaped like a half moon."[33] These, from surviving specimens, are known to have been without inscription, passing for metal value. Similar silver bars akin to later Chinese "sycee," in the shape of circular lumps with sunken centers, circulated in Nan-Chao (of which more below) during this period. Primitive refining of silver depended on open hearth furnaces; the remains of these small scale industries are found throughout the Shan states and Yunnan.[34] Fortunately for the early inhabitants of the area, the predominant silver ore was galena, an argentiferous lead sulphide. Roasting in an open furnace oxidised some of the sulphides into sulphates; a higher temperature caused the oxides to react on the remaining sulphides to produce the metallic lead-silver. A simple cupellation process, widely known to ancient metallurgists, took off the lead as a litharge deposit on a stirring ladle; the remainder was fairly pure silver. The technology was sufficient for the local ores; had sulphates such as anglesite been the sole available ore, a reduction process using blast furnaces would have been necessary. This latter technology was not known at the time, and there is no evidence of its use in the Shan States/Yunnan region prior to 1500.[35]

The knowledge of both gold and silver mining and smelting was widespread in Yunnan during the period of our study, and not necessarily held only by the high cultural groups: there are innumerable references to the "Gold Teeth," an Austric-speaking aboriginal group which was noteworthy for the gold and silver cappings or dentures sported by its members.[36] Whether this precious metal was obtained in trade, war, or by industrial effort is uncertain, but the latter is not impossible considering the tribe's relative isolation and the wide extent of the custom.

Once it is established that silver production was both feasible and actual in eastern Burma and Southwest China by the 13th century A.D., routes of trade and communication over which silver resources could pass westward must be investigated. In particular we must identify the land and water routes which connected south-western China with the frontier areas of Bengal which served as entry-points for this metals movement: Kamarupa in the north-east, Tripura/Sylhet in the east and Chittagong/Arakan in the south-east. It is manifest that land routes exclusively were associated with Kamarupa and Tripura, while a combination of land and sea routes communicated with Arakan. In assessing the relative importance of the various routes it seems to

---

33. G. H. Luce and Pe Maung Tin, op. cit., p. 270.

34. Watson and Fedden, *Journal of the Salween Surveying Expedition, 1868*, p. 10 and p. 39, quoted by R. C. Temple, "Notes on Currency and Coinage Among the Burmese," *Indian Antiquary*, July, 1928, pp. 129-130.

35. T. D. Latouche and J. Coggin Brown, "The Silver-Lead Mines of Bawdwin, Northern Shan States," *Records of the Geological Survey of India*, Vol. XXXVII, Part 3, 1908-09, pp. 245-6.

36. Luce, op. cit., p. 270; Padmeswar Gogai, *The Tai and Tai Kingdom*, Gauhati, 1968, p. 167; Marco Polo (tr. Henry Yule) Chapter L. "Zardandan," para. 2.

me important to avoid the over-emphasis on ocean carriage which is common in the literature on the period. Both modes of transportation, land and sea, were used to advantage in the 13th and 14th centuries in South and East Asia. Unfortunately an imbalance in historical resources makes it difficult to establish perspective vis-à-vis the respective utility of these transport modes. While Chinese and Portuguese sources give detailed accounts of oceanic trade in the Bay of Bengal late in this period (Ma Huan and Tome Pires), no such exhaustive chronicles seem to be available for the overland trade of Tibet and Burma during the same era. Each age and geographical area has its distinctive, predominating conditions. While sea trade was a major channel for the transmission of money and/or bullion into medieval Bengal, it cannot be considered *a priori* that it was the exclusive, or even dominant, means of supply.

Previous to the exploitation of Tyrolean, Japanese and Peruvian silver, the stock of available silver in the medieval world was much smaller, and its value relative to non-metallic goods commensurately greater than in either the ancient or modern periods. Patterson has made this observation of the "European and Muslim worlds" ca. 800 A.D., estimating the per capita quantity of silver as being 15 grams, versus 100 grams in the Roman world and 164 grams in the 20th century U.S.A.[37] This is probably true of Muslim India ca. 1200-1500; a selection of items priced in terms of silver mass during the time of Alauddin Khalji (ca. 1305) and Akbar (ca. 1595) in Hindustan, shows an absolute rise in prices (i.e. drop in silver value).[38] In other words, a lesser quantity of silver was capable of supporting an equivalent volume of trade in the earlier period.

The high value of the exchange medium helped to make land transit a competitive, if not superior, means of trade communication in this period. Merchants made transport-mode decisions based on capacity of carrier, speed of carriage, and security of cargo. The vast mercantilist silver bullion shipments of the 17th-18th centuries were most readily transported by sea, especially when sea routes were secured to the transporting nations and alternate routes via land were inordinately long and risky. There was no equivalent security of sea in the earlier period; the lesser quantities of bullion required in trade were equally portable by pack horse train as by ship; and whenever conditions of peace enforced by armed authority prevailed, land travel became relatively more attractive. This is specifically true in a consideration of bullion and coinage movement from the South China Sea to the Bay of Bengal; in this instance given tenable conditions, transit by land was

37. C. C. Patterson, op. cit., p. 216.

38. In 1305, Delhi: rice 5 *jital/man* = 0.11 g Ar/Kg rice (1 *jital* = 0.29 g Ar; 1 *man* = 13.1 Kg)
1595, Agra: rice 95 *dam/man* = 1.05 g Ar/Kg rice (1 *dam* = 0.28 g Ar; 1 *man* = 25.2 Kg)
1304, Delhi: ghee 1 *jital*/2 1/2 *sers* = 0.13 g Ar/Kg ghee (1 *ser* = 0.87 Kg)
1595, Agra: ghee 105 *dam/man* = 1.17 g Ar/Kg ghee
1305, Delhi: horse 65-120 *tanka* ea. = 715 - 1320 g Ar/ea. horse
1595, Agra: horse 10-30 *mohurs* ea. = 1065-3195 g Ar/ea horse (1 *mohur* = 9.4 rupees)

competitive with the water mode. Sufficient cause for merchant caution is demonstrable. The fate of Muhammad bin Tughluq's embassy to China has been related to us by his ambassador Ibn Batuta: the carriers, Chinese junks, were shipwrecked while departing the Malabar port of Calicut.[39]

In the particular case of bullion shipments from Yunnan, maritime routes had to meet very strong competition from overland transportation. As noted, the quantities of bullion were modest prior to 1500: relatively valuable shipments weighed little and could be carried by animal power as readily as by flotation. As late as the 20th century, observers in Yunnan noted the everpresent mule trains used in metal shipments in the area.[40] This mode of transport is eminently well suited to negotiating narrow tracks in mountainous terrain, living off the fodder to be had along the trail. It is still a dominant method of goods carriage in Nepal.

Secondly, a Mekong-Gulf of Tonkin sea route to Bengal involved roughly 5300 km of navigation by country boats and ocean going vessels, versus some 2000 km of land and sea travel by the route via Pegu and Chittagong, or only 1200 km of overland travel via Burma. Even given the swift descent of the Mekong (and ignoring the barriers of rapids), this route would have taken an excessive length of time, and only been practicable during those seasons when favourable winds encouraged sea travel. More to the point, as will be discussed later, the nature of trade was such that local increments were as important as the long range commerce. Major commercial expeditions out of south China in the Yuan and early Ming dynasties were supervised by government and well-regulated; further, they did not originate in Annam itself.[41] At any rate trading in metals, precious or otherwise, was prohibited.[42] The tendency from Annam would therefore have been coastal trade with South-East Asia, where silver especially would be quickly absorbed in those gold-rich and silver-poor markets. In contrast, the mechanism of silver exchange in the Burma land trade, to be discussed below, favoured transmission to the markets of eastern India.

As to the issue of security of transit, the relative perils of storm and pirates at sea are difficult to gauge against landslides and bandits on land. When discipline was imposed along the trails of Burma and Yunnan, as happened during the later 13th-15th centuries by the Mao-Shan empire, the balance of economy and security would seem to have been more evenly distributed between the maritime and terrestrial transit modes.

The roads from China to India through this area were well travelled from the earliest times. They were three in number, all radiating from the city of Yung Chang Fu (see *Map One*). This place was the road terminal for the routes from north China via the Yangtse valley; from south China via the

39. H. Yule, *Cathay and the Way Thither*, Vol. II, p. 419.
40. Coggin Brown, ". . . Yunnan," op. cit., p. 153.
41. Yuan Shih, tr. H. F. Schurmann, *Economic Structure of the Yuan Dynasty*, Chapter XIII, pp. 230 ff. (94. 24a9-27a2).
42. Ibid., p. 227 (Item 14).

West River; and from Tonkin via the Red River.[43] The main stages of the ancient paved roads within Yunnan, starting from Yunnan Fu, were Yunnan-hsien, Tali Fu, and Yung Chang Fu. The roads west from this last staging point were in operation by the 7th century A.D.[44] and may have been in use as early as the first millenium B.C.[45] The divergent routes passed via Upper, North Central and Middle Burma to equivalent entry points in Bengal proper. The first went from Yung Chang to Momien, crossed the Irrawaddy to Mogaung, went north through the Hukong Valley, across passes in the Patkai Range, to the upper Brahmaputra valley. This was the eastern frontier of Kamarupa. The second route followed the Shweli River, crossing the Irrawaddy at Tagaung, followed the Chindwin River north and crossed via the Imole pass to Manipur. This was the eastern approach to Bangala via Tripura. The third route embarked on the Irrawaddy at Tagaung, Ava or Pagan, then passed from Prome over the Arakan Range to Arakan. A variation of this went directly from Pagan to Arakan via the Aeng pass. This gave access to either a land route northwards to Chatisgaon, or embarkation on the coastal trading boats to Bangala.[46] Numerous combinations of these routes were possible using different stretches of the Irrawaddy for navigation. A separate apparently well-travelled and renowned route led from the region of the upper Yangtse-Mekong-Shalween rivers through Tibet. Passes led through Bhutan and Nepal into Kamarupa and Hindustan respectively. The Tsangpo-Indus route led to Central Asia via either Bactria or Khotan. As in all overland travel in the medieval period, while embassies and individuals might journey from one end of a road to the other, it was not necessary for merchants to do so. Goods passed through a series of hands in an entrepot trade, responding not simply to long-distance demands but the complex patterns of local exchange.

It is appropriate to inquire into the political conditions which reigned over this area during the period of this study.[47] Yunnan in A.D. 1200 was under the government of the Tai kingdom of Nan-Chao, with the capital at Tali. The Tai were a people vaguely related racially to the Han Chinese, and possessed of a culture significantly more advanced than the Lawas, Palaungs and Kachins of the area. They were literate, keeping historical records in a

---

43. Coggin Brown, ". . . Yunnan," op. cit., pp. 19-24.
44. According to the geography of Kua-ti Chin' (642 A.D.), quoted by H. B. Sarkar, "Bengal and Her Overland Routes in India and Beyond," *Journal of the Asiatic Society of Calcutta*, Vol. XVI, 1974, pp. 110 ff.
45. Ibid., p. 104. Also Lallanji Gopal, *The Economic Life of Northern India, ca.* A.D. *700-1200*, Delhi, 1965, pp. 109-110.
46. Sarkar, op. cit., p. 111. Gopal, op. cit., pp. 109-110. S. E. Peal, "Note on the Old Burmese Route over the Patkai via Nongyang," *Journal of the Asiatic Soicety of Bengal*, No. 2, 1897, pp. 69-82. R. B. Pemberton, *Report on the Eastern Frontier of British India*, p. 41 and p. 107.
47. The historical survey following is distilled from Gogoi, op. cit.; M. Carthew, "The History of the Thai in Yunnan," *Journal of the Siam Society* Vol. XL, Pt. 1, pp. 1-38; N. Elias, *Introductory Sketch of the History of the Shans in Upper Burma and West Yunnan*. Calcutta, 1876.

script of Indian derivation, and administering their state through ministerial government. The officers of the realm were assigned land revenue and the standing army was raised by levy. Taxes were payable in grain and in specie. This latter consisted of undifferentiated silver alloy ingots.

Beyond the borders of Nan-Chou was a loosely associated grouping of Tai states from Sukhotai (central Thailand), Lannatai (Chieng Mai), Laos (Vieng Chan), Monei (Southern Shan States), Mong Mao (Northern Shan States), Tagoung (Upper Burma), Male (on the Chindwin) to the Ahom (upper Assam). These had been peopled by successive waves of Tai emigrants forced out of Yunnan by Chinese pressure as early as the 6th century B.C. Culturally related but mutually jealous, they were independent of Nan-Chou. Under Hso Hkan Hpa (1220-1250), the armies of Mong-Mao had brought to submission every Tai state in a broad swath from Chieng Mai in northern Thailand to Assam, creating vassals of Manipur, Tripura and the Nagas in India, and overthrowing the northern provinces of Pagan.

Nan-Chou itself was unable to affect this movement, for relations with the Chinese were deteriorating. Nominally subject to the Chinese emperor, Nan-Chou as a frontier state had maintained cordial diplomatic and trade relations with the Sung dynasty. The conquest of China by the Mongols brought the Tais into conflict with the new overlords. Successive military expeditions by the Mongols in 1245 and 1252/53 eventually reduced the capital of Tali and the state became a Mongol province. This induced a great migration south and west into neighbouring Tai states, and confirmed Mong Mao on the Shweli River as predominant center of the Tai race. Elements of the Nan-chao governing classes found their way into many of the Tai principalities, and the consequent spread of administrative skills was noticeable in later principalities as wide ranging as the Ahom kingdom in Assam and Ayudhya on the Gulf of Siam (founded 1350).

The Tais (or Shans, as they are called in Burma) avoided direct confrontation with the Chinese, who began with an initial foray in 1277 the invasion of Pagan. Campaigns from Yunnan in 1283 and 1287 destroyed the Pagan kingdom, demonstrating, incidentally, the full capacity of the road system. The Shans abided their time, until in 1301, freed from any Burmese threat, they defeated the last Chinese expedition to attempt the permanent occupation of Burma. From this date until the 1440s the Tai states were essentially free from hostilities with the Chinese. The decline of the Yuan dynasty in the mid 14th century even permitted the Tais to temporarily regain control of Yunnan as far as Tali. This period was thus one essentially of tense peace on the major lines of contact with both the Ming Chinese and the Ava Burmese. This is not to say of course that there was any political unity or tranquility in the modern sense within the Tai dominions. Rather there existed that state of intertwined cooperation and competition, of alternating amity and rivalry in the commercial and political spheres, not unlike that obtaining in contemporary Italy. The unity underlying dynastic jealousies and the extent of intercourse between Tai states was demonstrated by the marital alliances with

Burmese Shans undertaken by the Ahom rulers in Assam centuries after settlement in that country. An anecdote is told of a jewelled plaque forming part of the royal vestments of the Ahom king, which, when inspected in 1912, was found to bear legends placing its origin in Chieng Mai (northern Thailand) in A.D. 1408.[48]

Of course it is not necessary to insist that the Pagan dynasty preceding Shan dominance, or the Ava dynasty following, were not capable of providing the essential link in this trade network. However it is enough for our purposes to demonstrate that the Tai or Mao-Shan kingdoms were a sufficient bridge between the south Chinese and eastern Indian worlds; if others functioned, all the better.

Among the strengths of the Tai or Mao-Shan states was their control and exploitation of gold and silver resources. The Hsi Wi Chronicle relates the 'gold and silver fields' located in the environs of Mong Mao in the time of the conqueror Hso-Hkan-Hpa (1220-80).[49] *Map One* confirms the known silver-bearing sites in the Mong-Mao region, most notably at Bawdwin. One species of ingot which circulated amongst the Tais was known as "Shan shell-money" to the British. This "money," which was reported circulating in Thailand and among the Lao States, was traced to the Bawdwin mines, where it was found to be the natural end product of cupellation refining.[50] The Bawdwin workings are of unknown antiquity, one estimate placing their discovery in the 10th century.[51] The mine site, a series of outcroppings of galena, exhibits signs of continuous working over a great period of time, with roads, stone bridges, settlements, fortifications and burial grounds in addition to the industrial remains. The hills for several miles around are permanently denuded of vegetation and badly eroded, the result of centuries of sulphurous fumes from open ore roasting furnaces.

This was the main silver source of the Shans; in fact the very name "Bawdwingyi" means "Great Silver Mines." It remained in Tai hands until the early 15th century. An inscription in Chinese found at the site records the start of operations by the Ming dynasty in 1412 A.D.[52] It was worked until 1858 by the Chinese, reopened this century by the British and became the greatest silver mine in the British Empire. It is difficult to estimate the amount of silver that might have been recovered annually by the Tais in the 14th and Chinese in the 15th centuries. The production in 1973 was 21,306 kg of silver;[53] in 1827, 18,000 kg;[54] and the average production for the 450

48. E. Gait, *A History of Assam*, 3rd ed. Calcutta, 1963, f.n. p. 75.
49. Cited by Gogoi, p. 184.
50. R. Temple, "Notes on Currency and Coinage among the Burmese," *Indian Antiquary*, May, 1928, pp. 91-92.
51. C. H. Henniker, *Mining Journal*, Vol. LXXIX, p. 52.
52. Coggin Brown, "Geology and Ore Deposits of the Bawdwin Mines," *Records of the Geological Survey of India*, Vol. XLVIII (1918), p. 124.
53. U.S. Bureau of Mines, *1973 Minerals Yearbook*, Chapter on Burma.
54. J. Crawfurd, *Journal of an Embassy to the Court of Ava*, London, 1829, p. 444. His estimate was £120,000.

years of Chinese occupation (estimated from the visible slag heaps) was 3,300 kg annually.[55] Another site closer to the Shweli shows signs of early working, but had not been thoroughly surveyed by the time this information was compiled. Assuming a more dilatory exploitation during the earlier period, the Shans may have extracted 1000 kg of silver annually from the two mines. This would have been equivalent to roughly 100,000 *tankas* per year of Bengali money.

It has been established that this silver circulated among the Tais as ingots or "shell money." The use of a non-differentiated "money" is important to the question at hand, since unmarked bars, as opposed to minted coins, were a superior exchange item in long-distance trade. Unlike the coins of nearby Bengal, the ingots could not be distinguished as to year of issue; hence could not be discounted due to age. Every time they passed to new owners a test and weighing was necessary. This anonymity was a positive virtue when the silver crossed several administrative zones, since it did not have to be exchanged to be made current specie, nor did it require recoining as it passed beyond the boundaries of the issuer. It was equally acceptable everywhere it passed; in fact an excellent traveller.

The question still remains however as to the pattern of circulation of this silver. There are a number of reasons to presume that this money moved westward out of Burma. The Bengali coinage, once established in silver, required a constant replenishment with fresh silver stocks to maintain the volume of currency in circulation. This was necessary to counteract the inexorable attrition losses. Demand for silver in Bengal was therefore constant and insatiable; this was the logical result of a successful silver exchange medium. We have Marco Polo's witness from his journey to Pagan ca. 1277-1281 in the service of Kublai Khan, that the gold-silver exchange rate for equivalent weights was 8:1 at Yunna Fu, 6:1 at Tali Fu, and 5:1 at Yung Chang Fu.[56] These cities were in east, central and western Yunnan. The cheapness of gold in the west was not due, as Yule would have it, to the naivety of the inhabitants, but to the plentiful supply of gold in the locality (see *Map One*). The effect on trade would be to discourage the transport of silver into Yunnan from Burma, as it would progressively lose its purchasing power in terms of gold. Indeed, just the opposite would be the expected result; in fact Polo comments that the favourable rate ". . . induces merchants to go thither carrying a large supply of silver to change among the people . . . they make immense profits by their exchange business in that country [Burma]"[57] In this case it is clear that Burmese silver did not find its way east, at least at the end of the 13th century; and in addition we must allow for the volume of Yunnanese silver flowing west speculatively.

55. A. B. Calhoun, "The Bawdwin Mine, in Burma," *Engineering and Mining Journal Press*, Vol. 113, No. 25 (June 24, 1922), p. 1089. He estimates 1,000,000 tons of ore at 43 oz. Ar per ton.
56. *The Book of Ser Marco Polo*, tr. H. Yule, Vol. II, pp. 53, 62, 70.
57. Ibid., p. 70.

Correlating the Shan history with the hypothetical calendar of Bengali silver imports/exports, we find some congruence of possible supply with demand, notably the loss of Bawdwin to the Ming Chinese in the early 15th century, contemporary with the second large outflow of bullion from the sultanate ca. 1416-1428. The effect of maritime trade necessarily complicates this correspondence. The establishment of a silver currency by the Bahmani sultan in the 1350's, alluded to earlier, with bullion derived from coastal ports, may have been a sufficient draw on the maritime supply to occasion the first net outflow from Bengal in the same decade. In this analysis of course the peripheral kingdoms on the Bengal frontier were in a sense land "ports," with the significant difference that they themselves had a capacity for the consumption of silver which had a buffering effect between source of supply and the ultimate consumer.

To summarize, there was in the Yunnan/Burma border area a known source of silver; there were established routes for communication between this area and the frontiers of Bengal; there was an agent of transmission, the Tai peoples, who maintained a loose system of contact throughout the period and over many of these routes; and finally an economic inducement for the silver to find its way, albeit through many intermediaries, to Bengal. The analysis has been inductively pursued, as indeed is inevitable considering the lack of historical resources which can be addressed to these monetary problems. The chronicles of the Shan, Ahom, Kamarupa, Manipur and Tripura peoples fight shy of any mention of non-military affairs across borders; it was not their intent to portray, nor could the chroniclers have recognized, economic or monetary trends. The present analysis has established both sufficient and necessary conditions for the existence of bullion movements during the period of the late 13th – early 16th centuries, from the Shan state or Yunnan to the Bengal sultanate. What is lacking is witness, and any means of comparison of quantities of silver involved, relative to other sources such as the sea commerce so well chronicled by Pires. The exact process of exchange, the amount of bullion involved, the mechanisms and products of trade which would provide the flesh for this analytical skeleton are still to be researched.

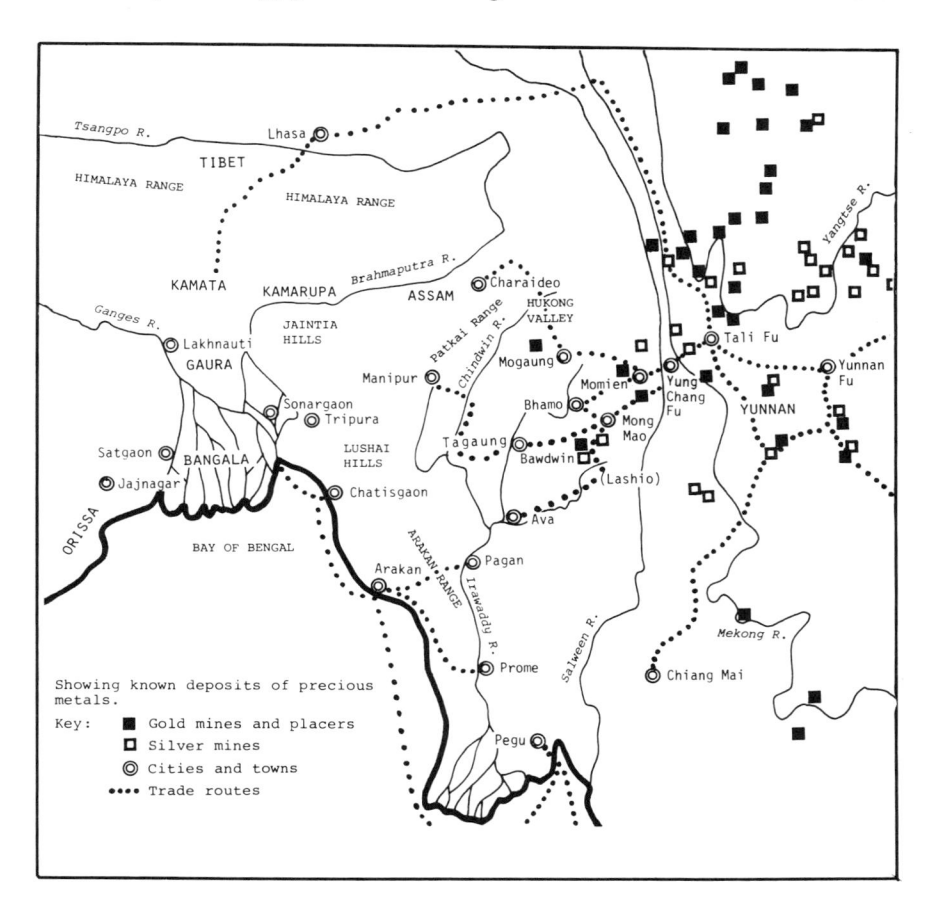

MAP ONE
Bengal and Yunnan
Ca. A.D. 1200–1500

GRAPH ONE

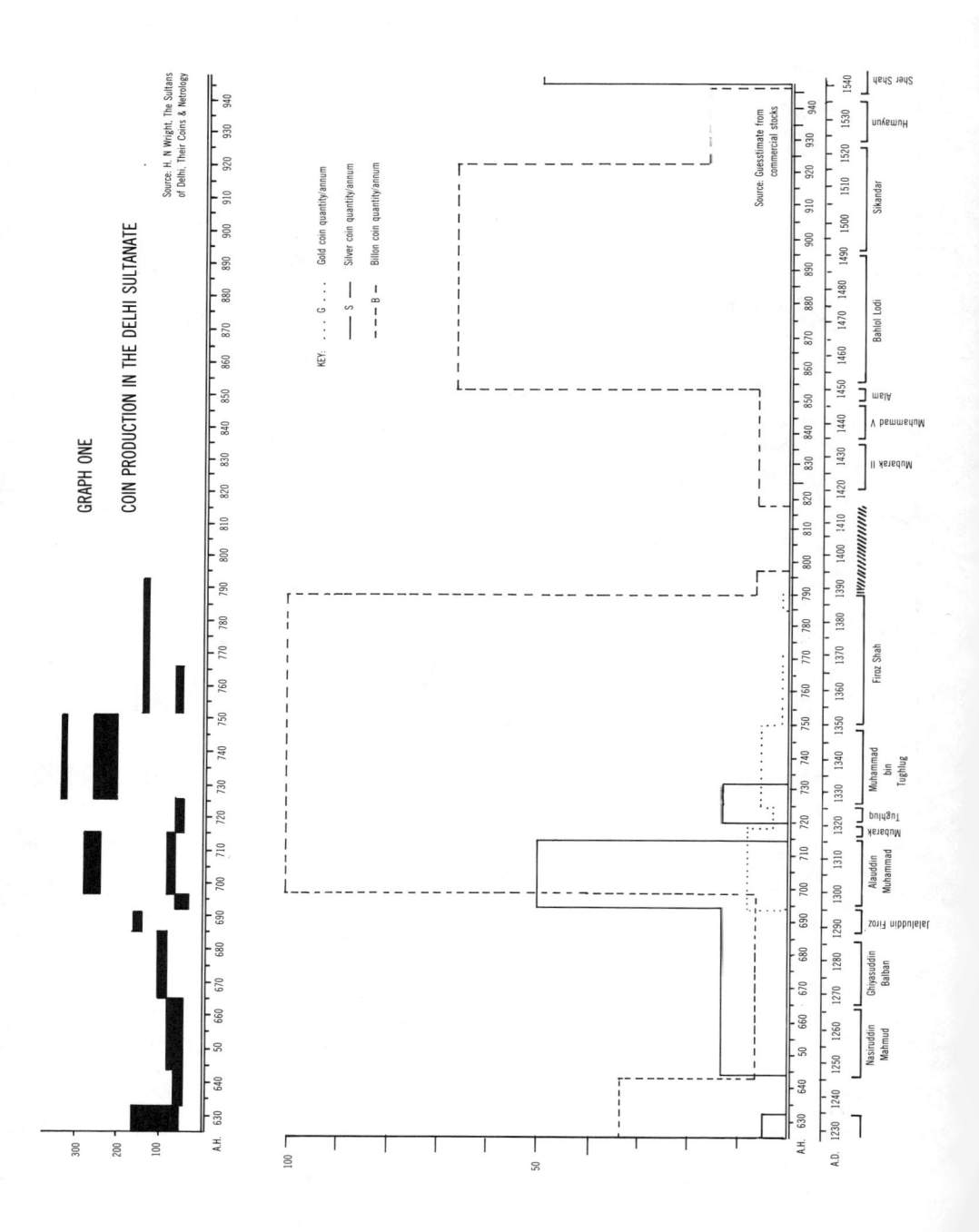

COIN PRODUCTION IN THE DELHI SULTANATE

Source: H. N Wright, The Sultans of Delhi, Their Coins & Metrology

KEY:
..... G ..... Gold coin quantity/annum
——— S ——— Silver coin quantity/annum
– – – B – – – Billon coin quantity/annum

Source: Guesstimate from commercial stocks

226

227

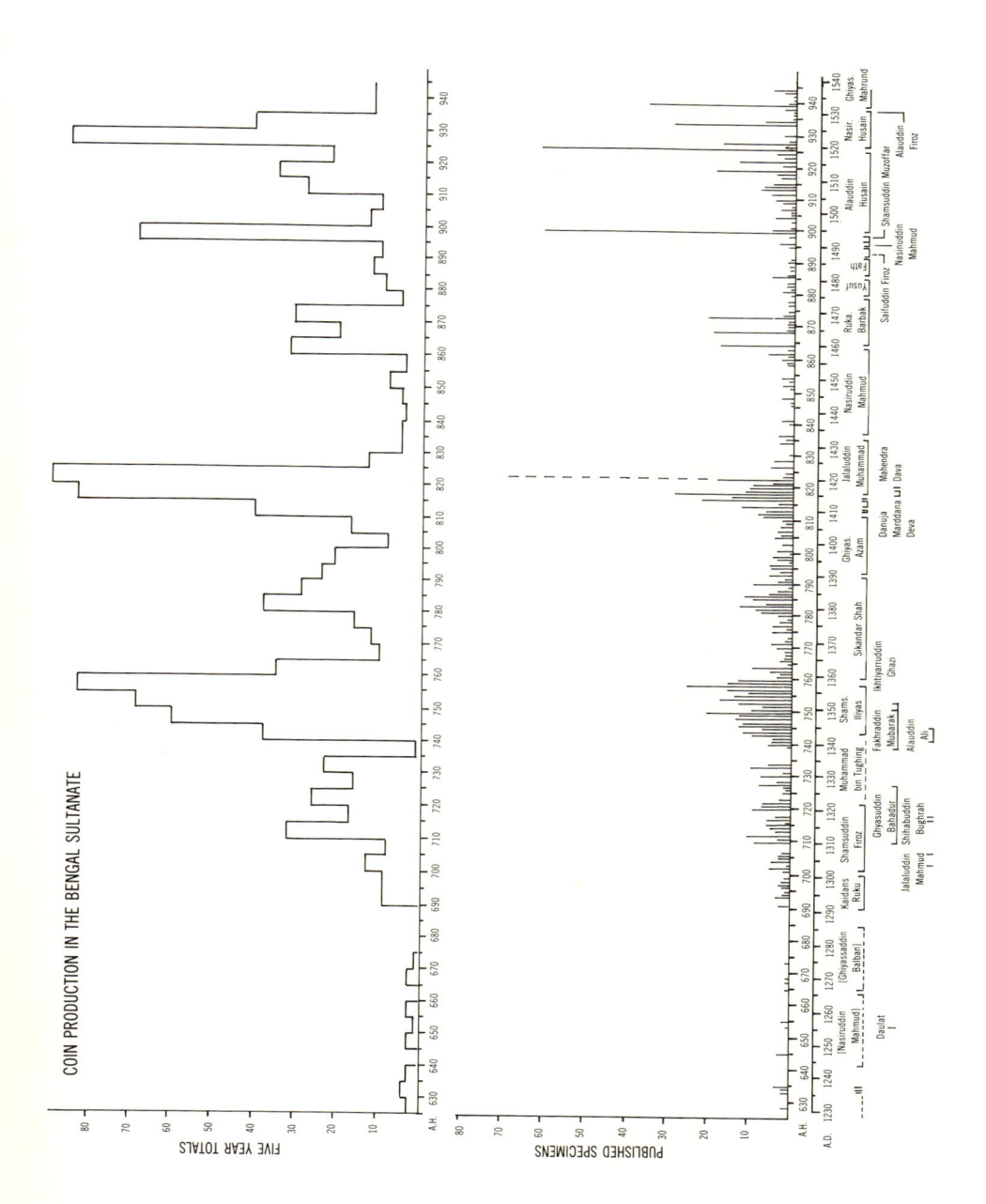

Part II

Monetary expansion and intensified demand for metals

8

Africa and the wider monetary world, 1250-1850

PHILIP D. CURTIN

Until recently, far too many historians thought of Africa—at least sub-Saharan Africa—as a non-monetized zone of "subsistence" economies, a region whose only monetary role in the world economy was that of supplying metals to be coined into Guineas—or other coins with names less indicative of their origin. Recent research by historians and anthropologists has begun to change the picture. African economies were more monetized than outsiders realized, and African household units were not nearly as self-subsistent as they appeared to be; but Africa was nevertheless somewhat less monetized than Europe or Asia. At any rate between 1250 and 1850 A.D., a good deal of the total production of goods and services in sub-Saharan Africa was distributed to the ultimate consumers through market transactions. The regional markets within Africa were less closely integrated than were those in the most developed parts of the Afro-Eurasian landmass. They were semi-isolated from other markets overseas or over deserts, and the comparative isolation lasted into the nineteenth century. This accounts, incidentally, for the extra half-century of data supplied for Africa, in contrast to the other contributions to this volume.

But sub-Saharan Africa's comparative isolation began to be lifted progressively beginning a little before 1000 A.D. In the West, camel caravans began to reach a new intensity of circulation. In the East, maritime contact southward along the east African coast had existed since about the second century B.C., but the rise of Fatimid Egypt diverted some of the Mediterranean-to-India traffic from the Persian Gulf to the Red Sea and the currents of trade up and down the east African coast began to build toward the peak of intensity they were to reach in the fifteenth century.

From the eleventh or twelfth centuries, then, three streams of gold flowed out of Africa in increasing quantities—one northward across the Sahara, mainly to Morocco but sometimes northeastward toward Egypt as well, one from Ethiopia to Egypt, and a third from present-day Zimbabwe to the coast near Sofala and thence overseas. Other forms of money began to reach Africa from overseas. Copper and copper alloys of European origin were a large part of the return cargoes across the Sahara. Once in west Africa these metals were partly monetized and partly used for decorative purposes, including their

231

incorporation in famous works of art like the Ife and Benin "bronzes" of southwestern Nigeria. On the east coast, copper flowed *from*, not *to* Africa, but the counter-current toward the African coast was partly in cowrie shells, carried up the Red Sea to Egypt, transshipped by way of Venice to the Maghrib and then across the Sahara where they were monetized in the region of the Niger bend and some distance up and down the Niger to the southeast and southwest—perhaps as far south as the Gulf of Guinea at some points.

The first phase of integration into the wider monetary world ended about 1500 with the beginning of European maritime contact on all the coasts of Africa. On the east coast during the sixteenth century, the Portuguese naval presence reduced the flow of trade, including monetary metals, and it also redirected the currents of trade. East African trading towns before 1500 had been mainly dependent on the Persian Gulf, and contact with India was largely mediated by Gulf ports. Portuguese posts like Mozambique and Sofala, however, were dependent on Goa; and their contact with other parts of Asia, as with Europe, was through Goa. The shift in trade destinations was therefore from the Persian Gulf to south India, not from the Persian Gulf to Europe. Little if any of the copper or gold exported from southeast Africa ever made its way to Europe. In the west, the substitution of Portuguese caravels for the trans-Saharan caravans was never total, and the seaborne gold exports found their destination from 1500 where they had found it before—in the general Mediterranean monetary system—but a larger part of the total now entered by way of Portugal and a smaller part by way of Morocco.

Important as maritime contact must have been, the economic impact on Africa was not yet very great. African foreign trade was only a very small part of total trade in any case. The African monetary systems sometimes used imported currencies, such as cowries and copper, but they rarely used gold. Gold shipments therefore had an influence on the economy of the mining region only, and monetary imports affected only the regions that used cowries or copper. Most of sub-Saharan Africa used neither. The most important currency over the largest geographical area was cloth, and cloth continued to be dominant until the beginning of the colonial period, even in competition with an expanding cowrie zone, a zone of gold currency centered on the Gold Coast, an enlarged copper zone in southeastern Nigeria, Cameroon, and the lower Congo basin.

Cloth was, of course, imported from overseas as well as produced locally in Africa, but imported cloth was rarely monetized. Nor was the African cloth that reached Europe or India used for monetary purposes. This meant that African monetary systems were economically isolated from the worldwide and interrelated monetary system based on metals or cowries. Marion Johnson of Birmingham University, who has already done the most important research on gold and cowrie currencies in West Africa, is now working on cloth currencies as well. When her research is published, it should be possible to say something more about the relations of cloth currencies to one another, and of any of them to metallic or shell currencies.

Africa and the wider monetary world, 1250-1850

The fact that currency zones were often discrete and only indirectly related to one another tended to increase the ambiguity where a particular product could serve as money or as a commodity for direct consumption. Before the time when coined silver was used as money, it was imported into the western Sudan as a convenient raw material for the African jewelers. Copper and brass were imported as bars, basins, and in many other shapes. Sometimes these were used in the form received, sometimes they were reformed to suit the local taste, and sometimes they were reshaped into one of the conventional units that circulated as money. Goods destined for monetary use were also shipped through zones where they were not money. Cowries were money in India, but not to the Europeans who shipped them around the Cape to European ports, nor to those in the African trade. Even within Africa, Senegambia (not in the cowrie zone) imported substantial quantities in the seventeenth and eighteenth centuries for shipment inland, to be monetized some 800 kms. further east on the upper Niger.

During this second phase of monetary relations between Africa and the outside world, those relations were far from static. The cowrie area expanded steadily from its beachhead at the Niger bend, and from another centered in southern Dahomey. Copper currency areas only expanded in central Africa, while the silver dollar made its entrance in the eighteenth century across the whole belt from the Red Sea and Ethiopian highlands to the Atlantic at Cape Verde. These changes were important, but they were only a prelude to the completely revolutionary changes that came in the second half of the nineteenth century—a third phase, not dealt with here, where the old monetary order was first destroyed and then replaced by one based on the European monetary system of the colonial period.

The monetary flows into and out of Africa in earlier times are most clearly seen if they are separated into the flow of particular goods over particular routes.

Gold—Ethiopia and Southeast Africa

Of the African goldbearing regions, Ethiopia was the earliest to make contact with the wider world. The placer deposits of Enerea, Gojjam, and Demot, still worked in the nineteenth century, were probably the source of gold that entered Egypt by way of the Nile valley as early as the second millenium B.C. In the early centuries A.D., the Ethiopian kingdom of Aksum struck gold coins that passed current in Indian-Ocean trade, and uncoined Ethiopian gold entered the broader patterns of trade by way of the Red Sea. This gold export was large by African standards of the time, but a quantitative estimate is almost impossible on the basis of the current literature. Surviving data tend to be episodic, referring to particular payments or treasure. The Imperial court received 933 kgs of gold as tribute from Gojjam in a specific year of the early sixteenth century: an exaggeration? Enera paid 166 kgs in tribute in the

Table 1. Estimates of Ethiopian gold exports in the nineteenth century

Date	(1) Value in[1] MT$	(2) Gold in Troy Ounces	(3) Equiv. in[2] Pure Gold	(4) MT$ per Ounce of Pure Gold	(5) Gold-Silver Ratio on Exchange	(6) Kgs. of Pure Gold	(7) Gold-Silver Ratio in Pure Gold
1838	40,000	2,000	1,552	25.77	15.4	48	19.4
1840	20,560	1,250	970	21.20	12.4	30	15.9
1842[3]	298,000	17,600	13,658	21.82	12.7	424	16.4
1847	13,900	1,056	819	16.97	9.9	26	12.8
1852	38,125	2,500	1,940	19.65	11.5	60	14.8
1861	38,125	2,500	1,940	19.65	11.5	60	14.8

Source: Pankhurst, *Economic History*, p. 363.
1. Maria Theresa dollars of 23.387 gr. fine silver.
2. The French Scientific expedition responsible for the 1842 data assayed various samples of Ethiopian gold for export and found it to be 77.6 per cent pure gold.
3. Although this estimate is the largest of the group, Pankhurst regards it as the most carefully made and probably the most accurate.

second half of the century, but the emperor Susenyos (1607-32) is said to have had treasure in gold that never exceeded 33 to 50 kgs.[1]

By the nineteenth century, foreign visitors began to arrive at their own estimates, and a number of those were collected by Richard Pankhurst (Table 1).[2] They suggest the same order of magnitude as the earlier episodic figures. If a guess had to be made, the sustained annual average export might have reached 500 kgs for short periods, but was probably somewhat less over the long run.

The Zimbabwe goldfield and the Zambezi valley also supplied gold to world trade at an early date, perhaps as early as the first centuries of the Christian era, but with the most intense growth from the eleventh century to the fifteenth. The main deposits were the river gravels of Zambezi tributaries like the Angwa, Mazoe, Ruenya, and especially the Manyika. The earliest form was placer mining, but reef mining began as well about 1000 A.D. A combination of reef and placer mining built the peak exports of the fifteenth century and economic power for Great Zimbabwe at the same time.[3] Some of the earliest Portuguese visitors estimated that their predecessors exported several thousand kilograms annually (Table 2), but these estimates have to be set aside as a self-serving exaggeration. Most authorities, however, would accept an annual average export of more than 1,000 kgs for the peak periods during the fifteenth century.[4]

The Portuguese seaborne attack on east Africa after 1500 broke the hold of the Arab shippers, but without establishing a Portuguese monopoly. The Arab gold trade was based on the entrepôt town of Kilwa, in what is now southern Tanzania. Kilwa was more important than Sofala, the closest seaport to the goldfields, because the Arabs used the pattern of alternating northeast and southwest monsoons to reach down the coast. They found that they could come as far as Kilwa without difficulty in a single monsoon, but if they went further they risked their return on the alternating wind. They therefore tended to stop near Kilwa, and the gold reached them by a number of different routes combining overland with local maritime trade. When the Portuguese came on the scene, however, they made Mozambique island their principal military and naval base; but they also seized Sofala as their main

---

1. These data were reported in the Ethiopian measure, *oquea* or *waqet*, equal to approximately 33.5 grams. Richard Pankhurst, *An Introduction to the Economic History of Ethiopia from early Times to 1800* (London, Lalibela Huse, 1967), pp. 32, 180, 184-185 and 220.

2. Richard Pankhurst, *Economic History of Ethiopia, 1800-1935* (Addis Ababa, 1968), pp. 363, 400, 412.

3. I. R. Phimister, "Alluvial Gold Mining and Trade in Nineteenth Century South Central Africa," *Journal of African History* (cited hereafter as *JAH*), 15: 445-46 (1974).

4. V. Magalhães-Godinho, *L'économie de l'empire portugais aux xve xvie siècles* (Paris, 1969), pp. 270-1. The citations here and below are to the French edition, which has a later date of publication than the Portuguese edition published under the title *Os descobrimentos e a economia mundial*, 2 vols. (Lisboa, 1963-65). The Portuguese version was written later and contains a number of revisions; it is therefore preferred for most purposes. The data on monetary flows, however, are unchanged.

Table 2. Estimates of gold exports from Southeast Africa mainly through Sofala and based on gold production in Rhodesia and the Zambezi valley

	Exports in kilograms	
	Contemporaneous estimate or accounting	Evaluation of recent scholarship
Arab period before 1500, two estimates of 1502 and 1506 respectively:		
Thome Lopez, Sofala exports	8,500 kg	
Diogo de Alcaçova, Sofala and Angoche exports	7,650 kg	< 1,000[1] kg
1508–09 annual average of exports through Sofala based on a 20-month period		62[2] kg
1512–13 annual average based on eight months of Sofala exports	34 kg[2]	
c. 1512–15 reports of Sofala factory	49 kg[3]	
1518–19 reports of Sofala factory	61 kg[3]	
1585 Exports from Sofala and Quelimane	4 kg[4]	
1591 Exports from Sofala and Quelimane	574 kg[1]	
1610 Exports from Sofala and Quelimane	716 kg[1]	
1667 Exports from Sofala and Quelimane	850 kg[1]	
1750 estimate of Francisco de Mello e Castra for Butua only	1,488 kg[1]	
1758 estimate of Ingacio Caetano Xavier	213 to 255 kg[5]	
1762 anonymous estimate	525 kg[5]	
1806 estimate of A. N. de Barbosa	298 kg[5]	
1806 anonymous estimate for Quelimane only	51 kg[5]	
1820 estimate of Jose Francisco Alves Barbosa for Quelimane only	29 kg[5]	
Nineteenth century and before—based on archaeological evidence in Rhodesia	11 kg[5]	500[6] kg

Table 2. *(continued)*

Notes:
1. Magalhães-Godinho, *Empire Portugais*, pp. 270–71.
2. Axelson, *South-East Africa*, pp. 111–12. These data were mainly reported in *mithqal*, the Arabic measure of weight with a fair degree of local variation. Axelson used the weight of the Sofala *mithqal* at 4.83 gr. while Magalhaes-Godinho and Randles used a *mithqal* at 4.25 gr.
3. Axelson, *South-East Africa*, pp. 123–26.
4. Axelson, *South-East Africa*, p. 150.
5. Randles, *L'empire du Monomotapa*, pp. 113–14.
6. Summers, *Ancient Mining in Rhodesia*, p. 188.

point for the gold trade. Arab shippers could meanwhile participate as well, since the great variety of routes from Kilwa to the goldfields was still available, even though Kilwa itself was destroyed.

The low level of Portuguese-carried gold exports in the sixteenth century (Table 2) is therefore no secure indication that the *total* export fell that far. The Portuguese attempt to interfere with Indian-Ocean trade would certainly have brought it down somewhat from the high fifteenth-century levels, but Great Zimbabwe and other interior kingdoms continued to be prosperous, though the archaeological evidence suggests a decline in reef mining at this period. Nothing but guesswork is possible, but a total export of 500 kgs a year might have been reached in some years, even if that figure could not be sustained as an annual average over time. Then, at the end of the sixteenth century and through the first three quarters of the seventeenth, the Portuguese made a serious military effort to intervene in African affairs at the goldfields and on the routes to the coast. The results are clear on Table 2, but again this is not necessarily evidence that gold exports had increased, only that the Portuguese carried a larger share. The decline shown for the eighteenth and nineteenth centuries, however, is real, a net decline in productivity; first, in the eighteenth century, because of warfare between Great Zimbabwe and the Changamires and then, in the nineteenth, because of the anarchy and warfare that followed in the wake of the Ngoni movement northward. Summers' estimate of 500 kgs a year, on the basis of the archaeological record, checks well with the documentary record for earlier centuries, but it may not have been reached in many years after the middle of the eighteenth century.[5]

West African gold to 1480

The literature on the west African gold trade is far more extensive than it is for either southeast Africa or for Ethiopia, and the west African sources of gold were far more scattered. Many places north of the Gulf of Guinea mined a little gold, but those that mined in quantity and over the long term were three—the Akan goldfield in the hinterland of the Gold Coast (now Ghana), the Bambuk or Bambuhu goldfield between the Falémé and Sénégal Rivers (now in Maili), and the Bouré goldfield on upper tributaries of the Niger (now in Guinea-Conakry). All three supplied gold to the trans-Saharan trade before European mariners appeared off the coast. All three supplied gold for seaborne export afterward. All three supplied *some* gold to the trans-Saharan trade some of the time, even after seaborne export had begun. The problem of evidence is similar to the one posed by Portuguese-Arab rivalry in southeast Africa. The early documentary record is the one left by the Europeans; we know it represents a part of the whole, but not how much.

---

5. Roger Summers, *Ancient Mining in Rhodesia* (Salisbury, 1969), p. 188; W. G. L. Randles, *L'empire du Monomotapa du xve au xixe siècle* (Paris, 1975), pp. 113-114; Phimister, "Alluvial Gold Mining," p. 447.

Table 3. West African gold production in the late pre-colonial and early colonial periods

Period	Annual Average Exports in Kgs.	
	British West Africa	French West Africa
1861–75	844	–
1876–80	922	–
1881–85	664	–
1886–90	886	–
1891–95	990	–
1896–1900	570	208
1901–05	2,070	176
1906–10	7,336	132
1911–20	9,470	82
1921–25	6,381	374

Source: Robert H. Ridgeway, *Summarized Data of Gold Production* (Washington, Bureau of Mines (929), table following p. 63. Economic paper No. 6.

Even before examining the evidence, another kind of guideline comes from even more recent data on gold production. At the eve of the colonial period, production data should show the capacity level of each goldfield, before it was increased by the application of machine technology in this century. These relatively recent production figures can serve as a check on other evidence of gold actually shipped, and as an outer limit for what gold production might have been before maritime contact.

For Bambuhu, production was not much affected by European technology. Gold deposits were widely scattered and uneven in quality. French firms used some gold dredges in the river gravels; but most deposits were not susceptible to machine technology, and production for export has virtually stopped. Estimates based on recorded maritime exports (see Table 8, below) suggest that precolonial production rarely, if ever, exceeded 200 kgs a year. Shipments by sea in a good year—not an annual average, but the capacity of the goldfield when circumstances were most favorable—seem to have run at about 35 kgs in the sixteenth and seventeenth centuries, 60 kgs in the eighteenth, and a little over 100 kgs in the nineteenth.[6] (See Table 8, below)

Bouré gold production is less well documented. Much of it was shipped down the Niger to the desert edge. Reports on coastal export by way of Sierra Leone are sporadic at best. Table 3 shows the exports from French West Africa for the earliest period of French occupation, before European mining technology had been introduced. The range of 100 to 200 kgs a year before

---

6. P. D. Curtin, *Economic Change in Pre-Colonial Africa. Senegambia in the Era of the Slave Trade*, 2 vols. (Madison, 1975), 1: 202; Curtin, "The Lure of Bambuk Gold," *JAH*, 14: 623-31 (1973); J. Meniaud, *Haut-Sénégal-Niger: Géographie économique*, 2 vols. (Paris, 1912), 2: 171-75.

1920 is probably representative of earlier periods as well, while the substantial increase in the early 1920's probably represents the early results of European investment. Labouret estimated fifteenth-century production at no more than 250 kgs a year for Bambuhu and Bouré together, which is as good as any other guess one could make with existing evidence.[7]

The Akan goldfield was not merely better reported; existing reports also indicate that productive capacity did not change very much before the introduction of hard-rock mining toward the end of the nineteenth century. Pacheco Pereira estimated total production at about 1000 kgs in 1505, and the Portuguese appear to have exported nearly that amount in good years during the first two decades of the sixteenth century. Totals exported by sea were about the same in the seventeenth century when both the British and Dutch were heavily involved in the Gold Coast gold trade. They fell during the second half of the eighteenth century but then rose again to the range of 500 to 1000 kgs in a year during the second half of the nineteenth century. The sharp jump between the figure for 1896-1900 and that for 1901-05 in Table 3 represents the beginning of mechanized production about twenty years before the equivalent change in the Bouré field.[8]

Vitorino Magalhães-Godinho, the most careful student of the monetary affairs of the Portuguese empire in the sixteenth century, estimates that total Portuguese exports of gold from all West Africa in the first two decades of the sixteenth century came to about 700 kgs a year. Total exports were hardly more than half again as high in the late nineteenth century, after the trans-Saharan trade had dropped to a mere trickle. Such estimates tell nothing about African consumption in jewelry or for monetary purposes, but they do suggest some kind of outer limit on gold flows across the Sahara before the 1480's at around 1500 kgs a year, with 500 to 1,000 kgs a more probable figure.[9]

This may seem low in comparison with eighteenth-century Brazil or twentieth-century South Africa, but it was an important amount for its influence on Mediterranean currencies. Ronald Messier's recent research on North African coinage of the period from about 1070 to 1250 shows, on the basis of radio-chemical anlaysis, that most of the gold coins struck by Almoravid North Africa, Zirid North Africa, Almoravid Spain and Almohad Morocco were made of coins whose copper content matches the natural copper content of the gold entering the Maghrib from the Sahara at that

---

7. Labouret, Henri, Jean Canu, Jean Fournier, and Georges Bonmarchand, *Le commerce extra-Européene jusqu'aux temps modèrnes* (Paris, 1953), p. 58.

8. Szereszewski, R., *Structural Changes in the Economy of Ghana 1891-1911* (London, 1965) p. 29; Dickson, Kwamina, B., *A Historical Geography of Ghana* (Cambridge, 1969), pp. 188-189.

9. Magalhães-Godinho, *L'empire portugais*, 218. This guesswork is not very precise, but the data on which it is based make it possible to set aside some of the wilder assertions of the past literature, such as R. Mauny's estimate that west African gold production came to 9,000 kgs per year through the whole period from the beginning of the sixteenth century to the end of the nineteenth (*Tableau gégographique de l'ouest Africain au moyen age* (Dakar, 1961) p. 298.)

Table 4. Gold from Guinea imported into Portugal on private account, 1494-1513

Years	Quantity annual average kilograms
1494-96	53
1497-98	182
1505-07	301
1509-10	372
1511-13	277

Source: Magalhães-Godinho, L'économie de l'empire portugais, p. 218.

period. Even the Fatimid coins minted in Egypt appear to have been made of gold from the western Sudan until after 1047, when the Fatimids lost control over Tunisia. Their later coins appear to have used gold from Ethiopia instead.[10]

West African gold—Sixteenth and seventeenth centuries

With the maritime revolution, a variety of time-series on gold exports from West Africa become available, sometimes on an annual basis. The apparent precision of the numbers can, however, be misleading. Gold was comparatively light for its value, easy to hide, and frequently taxed or regulated; hence European statistical returns for gold are probably less accurate than those for other commodities. In the early sixteenth century, Portuguese regulations required that all gold from São Jorge da Mina on the Gold Coast be imported on the royal account and sent directly to the mint. Some gold also came from other parts of Guinea on private account, and it was subject to taxation. Table 4 refers to this private trade in taxable gold. Table 5 gives the available numbers for gold entering the Lisbon mint. The two tables therefore represent two different kinds of gold shipment, so that neither even claims to be a complete return of African gold imported into Portugal. Magalhães-Godinho's estimate of 700 kgs a year in the first two decades of the sixteenth century is markedly higher,[11] but the discrepancy between real and recorded shipments is confirmed by the few surviving records made at Elmina itself. One account covering 27 months in 1529-31 indicates an annual average of 339 kgs of gold exported in contrast to the annual average mint figures of only 212 kgs for the two complete years 1529 and 1531. Again in 1521, recorded departures from Elmina show a total of 502 kgs of gold shipped, while the mint records show only 429 kgs received.[12]

10. R. Messier, "The Almoravids: West African Gold and Gold Currency of the Mediterranean Basin," *Journal of the Economic and Social History of the Orient*, 17: 31-47 (1974).
11. Magalhães-Godinho, *L'empire portugais*, 218.
12. J. Vogt, "Portuguese Gold Trade: An Account Ledger from Elmina, 1529-31," *Transactions of the Historical Society of Ghana*, 14: 93-104 (1973).

Table 5. **Gold-coast gold received by the Lisbon mint**

Year	Quantity-Kilograms of 22 1/8 carat gold
1487	225
1488	225
1489	225
----	---
1494	647
1495	647
1496	647
1497	371
1498	371
1599	371
1500	371
1501	280
----	---
1504	438
1505	438
1506	438
----	---
1511	321
1512	518
1513	414
1514	404
----	---
1517	423
1518	484
----	---
1520	465
1521	429
----	---
1523	300
1524	284
1525 (1st half-year)	206
1526	248
----	---
1528	223
1529	211
1530 (April)	150
1531	213
1532	680
----	---
1534	272
----	---
1540	392
----	---
1543	349
1544	142

Table 5. (*continued*)

Year	Quantity-Kilograms of 22 1/8 carat gold
----	---
1549	168
1550	155
1551	212
1552	123
1553	94
----	---
1555	377
1556	242
----	---
1560	144
1561	145

Source: Magalhães-Godinho, *L'économie de l'empire portugais*, p. 216. Before 1517, John Vogt, *Portuguese Rule on the Gold Coast, 1469–1682* (Athens, Ga., University of Georgia Press, 1979), pp. 217–218. Based on contemporary estimates.

The mint data are nevertheless worth reporting, if only because they reveal a pattern of considerable annual variation and because they may reflect significant trends, if not actual values. The pattern is one of a strong beginning early in the century and sustained until about 1521, falling off to less than half the earlier level through most of the 1520's. Totals picked up again in the 1530's, though not to the earlier peaks, and then fell off once more after 1544. The first of these slumps seems associated with foreign Europeans and Portuguese interlopers. It therefore reflects a weakening of the royal monopoly—not necessarily a drop in the total gold shipped. The royal monopoly, in fact, weakened progressively through the century; some authorities estimate that, by the mid-century, the Dutch were already carrying about half a total annual export of 780 kgs.[13]

But the second half of the sixteenth century also saw a revival of the trans-Saharan trade. The new, Saadian dynasty in Morocco were more and more active in their reach to the south until, in 1591, they sent a military expedition across the desert to capture the Niger bend and destroy the power of the previously dominant empire of Songrai along a large stretch of the desert-fringe. It is clear that the Moroccan presence south of the Sahara helped to speed the flow of gold. What is not so clear is whether this meant a net addition to the total flow of gold to the Mediterranean world, or whether it merely took one part of the previous Portuguese share at a time when the

---

13. Richard Bean, "A Note on the Relative Importance of Slaves and Gold in West African Exports," *JAH* 14: 351-56 (1974).

Table 6. Bosman's estimates of export capacity in the gold trade from the Gold Coast

Carrier	Annual capacity
Dutch West India company	369 kgs.
Royal African company	295 kgs.
Zeeland interlopers	295 kgs.
English interlopers	246 kgs.
Brandenburgers and Danes together	246 kgs.
French and Portuguese together	197 kgs.
Total	1,723 kgs.

Source: William Bosman, *Description of Guinea*, pp. 88–91.

Dutch were taking another part.[14] Only a tentative conclusion is possible, but the strength of Dutch performance by the early seventeenth century, the presence of Portuguese interlopers, and the Moroccan success in the north add up to an impression of something more than the estimated 700 kgs a year carried by the Portuguese at the beginning of the century. A metric ton would be a possible guess.

Some 988 kgs of gold carried by the Dutch alone was a well-known estimate for about 1600.[15] It may be accurate, but, if so, it should be understood as an estimate of capacity when circumstances were favorable, not an annual average. The other famous over-all contemporaneous estimate for the end of the seventeenth century was given by William Bosman (Table 6). This time the figures are explicitly the quantities of gold that could be exported in a good year, with peace in Europe and at sea, and all the paths from the interior open in Africa; and he noted that in recent years, the Dutch West India Company had not carried half the projected quantities, while English interlopers had carried more than twice the amount assigned to them.[16] In fact, as Table 7 shows, the Royal African Company imported only an annual average of 85 kgs in the years 1695-99, not the 295 Bosman projected for them, but the Dutch West Indian Company carried 270 kgs—almost three-quarters of the assigned amount.

14. See Magalhães-Godinho, *L'empire portugais*, pp. 219-226.
15. P. de Marees, *Beschrivinge en Historische Verhael van het Gout Koninchrijck van Guinea 1602* (The Hague, 1915), cited by Kwame K. Daaku, *Trade and Politics of the Gold Coast, 1600-1720* (Oxford, 1970), p. 11; but this estimate was also current in Portuguese circles. See a letter from a Portuguese merchant in Sierra Leone at about this time, quoted in Peter Kup, *A History of Sierra Leone, 1400-1787* (London, 1961), p. 19.
16. William Bosman, *A New and Accurate Description of the Coast of Guinea* (London, 1907), pp. 88-91.

Table 7. Quantities of gold shipped to Europe by:

Dates	Dutch West India Company	Royal African Company	Total
1622–34	766 kg annual average[1]		
1624	309 kg[2]		
1626	518 kg		
1632	560 kg		
1635	528 kg		
1636		203 kg[3]	
1637–44		458 kg annual average[3]	
1638	426 kg		
1639	409 kg		
1640	632 kg		
1641	409 kg		
1642	530 kg		
1643	525 kg		
1644	395 kg		
1645	946 kg		
1646	628 kg		
1647	469 kg		
1648	369 kg		
1653	601 kg		
1655	624 kg		
1668–76	350 kg annual average[4]		
1673	350 kg	17 kg[5]	367
1674	350 kg	61 kg	411
1675	350 kg	94 kg	444
1676	373 kg[6]	117 kg	490
1677	400 kg	260 kg	660
1678	457 kg	38 kg	495
1679	474 kg	194 kg	668
1680	425 kg	132 kg	556
1681	356 kg	191 kg	547
1682	343 kg	179 kg	521
1683	496 kg	197 kg	693
1684	429 kg	159 kg	588
1685	396 kg	354 kg	750
1686	476 kg	180 kg	656
1687	515 kg	249 kg	764
1688	451 kg	303 kg	754
1689	395 kg	196 kg	591
1690	544 kg	12 kg	556
1691	524 kg	205 kg	729
1692	425 kg	146 kg	571

Table 7. (continued)

Dates	Dutch West India Company	Royal African Company	Total
1693	347 kg	58 kg	405
1694	312 kg	121 kg	433
1695	326 kg	165 kg	491
1696	364 kg	19 kg	383
1697	244 kg	88 kg	332
1698	197 kg	137 kg	334
1699	222 kg	17 kg	239
1700	194 kg	–	194
1701	130 kg	38 kg	168
1702	211 kg	–	211
1703	210 kg	7 kg	217
1704	212 kg	14 kg	226
1705	148 kg	10 kg	158
1706	205 kg	–	205
1707	211 kg	43 kg	254
1708	157 kg	22 kg	179
1709	241 kg	86 kg	327
1710	301 kg	80 kg	380
1711	305 kg	–	305
1712	204 kg	22 kg	226
1713	143 kg	1 kg	144
1714	142 kg	–	142
1715	181 kg	–	181
1716	188 kg	–	188
1717	160 kg	–	160
1718	200 kg	–	200
1719	213 kg	–	213
1720	137 kg	–	137
1721	160 kg	–	160
1722	159 kg	–	159
1723	295 kg	–	295
1724	216 kg	–	216
1725	215 kg	–	215
1726	156 kg	–	156
1727	151 kg	–	151
1728	148 kg	–	148
1729	168 kg	–	168
1730	136 kg	–	136
1731	89 kg	–	89

Sources:

1. Kwame Y. Daaku, *Trade and politics on the Gold Coast 1600–1720* (Oxford, 1970), p. 14.
2. Hermann Wätjen, "Zur Geschichte des Tauschhandels an der Goldküste um die

Table 7. (continued)

Mitte des 17. Jahrhundert. Nach Holländischen Quellen." in Dietrich Schafer, *Forschungen und Versuche zur Geschichte des Mittelalters und der Neuzeit* (Jena, 1915) pp. 550–551.
3. R. Porter, "The Crispe family and the African trade in the seventeenth century," *Journal of African History*, 9: 57–77 (1968) p. 64.
4. Johannes Postma, "West African exports and the Dutch West India company, 1675–1731," *Economisch-en Sociaal-Historisch Jaarboek*, 34: 53–74 (1973) pp. 60–69. Each reference number applies to the remainder of that column until interrupted by a further reference superscript.
5. K. G. Davies, *The Royal African company* (London, 1957) p. 360.
6. Postma, "West African exports," p. 71.

But Bosman's figures may be more accurate than they appear to be at first. Looking back in Table 7 to the best sequence of recent years, they seem to be 1687-92. For this six-year period, the English company averaged an export of 216 kgs, while the Dutch company brought out 476 kgs. One was too high and the other too low, but the mean of these annual averages and the mean of Bosman's projections for the two companies were 346 kgs and 322 kgs respectively. While these comparisons provide no certain indication, they *do* suggest that Bosman's estimates could be reasonably accurate for the very best years of the late seventeenth century—and that these years, particularly those in the 1680's, probably represent the highest sustained export in the pre-colonial history of West Africa, at something like 1,700 kgs a year.

Table 7 also hints at the geographical origins of gold within West Africa. Since the Gold Coast was the only gold-shipping part of West Africa where the Dutch traded intensively, all the Dutch exports can be counted as coming from the Akan goldfield. Most of the English-shipped gold also came from the Gold Coast, as did that shipped by the Danes and Brandenburgers, though the English also bought gold where they could.[17] It is possible on this basis to guess that at least 80 percent of the gold shipped came from the Akan goldfield, and this is consonant with Senegambian exports (Table 8), representing the seaborne portion of Bambuhu gold shipments.

These figures represent only the exports from west Africa to Europe. In addition to the trans-Saharan trade to north Africa, some gold was shipped directly from the Gold Coast to India, especially for a brief period between about 1657 to 1663, when the English East India Company was allowed to carry gold direct to its Indian posts. Fragmentary data for these years indicate shipments of 37 kgs in 1657, 77 kgs in 1660 and 97 kgs in 1661. In 1662 and 1663, ships were ordered to the Gold Coast to load gold up to the value of 175

17. K. G. Davies, *Royal African Company* (London, 1957), pp. 350, 360. Johannes Postma, "West African Exports and the Dutch West India Company, 1675-1731," *Economisch-en Sociaal Historisch Jaarboek*, 34: 53-74 (1973).

Table 8. Gold exports from Senegambia by sea, 1687–1850

| | | | | Shipped from | | |
| | | | | Gambia | | |
Year	St. Louis	Gajaaga	Gorée	by French	by English	Total
1687	c 1.7	2.9	–		c 6.7	c 8.4
1693	c 9.1		c 1.2		c 24.5	c 34.8
1698	–	c 6.9	–		–	
1718	–	c 12.3	–		–	
1723		c 12.3				
1730		15.4				
1733		c 24.5				
1737	–	20.1	–			
1740		25.8				
1751		24.5				
1757		c 55.1				
1786	–	2.6	–			
1788	–	3.3	–	–	–	
1820	–	0.3	–	–	–	
1822	–	1.0	–	–	–	
1823	–	–	–	–	70.3	
1826	–	3.94	–	–	–	
1828	–	1.0	–	–	20.7	
1829	–	5.1	–	–	18.9	
1830	–	5.0	–	–	c 14.2	
1832	–	40.3	–	–	22.7	
1833	–	–	–	–	32.3	

Table 8. (continued)

| | | Shipped from | | | | |
| | | | | Gambia | | |
Year	St. Louis	Gajaaga	Gorée	by French	by English	Total
1834	–	2.3	–	–	1.2	
1835	–	–	–	–	8.1	
1836	–	50.2	–	–	35.5	85.7
1837	91.4	0.6	–	–	–	
1838	–	0.8	–	–	8.6	
1839	–	0.9	–	–	9.5	
1840	–	0.4	–	–	9.1	
1841	–	–	–	–	32.1	
1842	–	1.4	–	–	1.7	
1843	–	2.3	–	–	2.2	
1844	–	66.0	–	–	9.7	91.7
1845	86.0	2.3	–	–	5.7	
1846	84.9	2.9	–	–	–	
1847	–	–	–	–	0.8	

Source: P. D. Curtin, *Economic Change*, 2: 62–68.

kgs and 73 kgs respectively.[18] Although these exports to Asia were within the general scale of the Royal African Company's later performance, they were a comparatively small part of west African gold exports and were only a temporary phenomenon—at least as an official and permitted trade. It may well have been that interlopers also engaged in direct trade from west Africa to Asia, but that remains an unknown for the moment.

West African gold—Eighteenth century and after

The reason gold flowed from Africa to Europe is explained simply enough by the fact that Europeans were willing to pay a price high enough to cover the cost of mining and transportation. If necessary, the explanation can be pushed one step further, showing that the European price of gold was sustained by its scarcity, its acceptability as money, or its value for decoration. Since these conditions seem to have been in effect one century after another, it is puzzling to find, in the second half of the eighteenth century, that the export of gold almost ceased, and west Africa may have become a net importer of gold. In 1676, just before the peak of gold exports from the Gold Coast, Europeans could buy gold for a bundle of goods with a prime cost in Europe as little as £1.74 sterling. The prime-cost price of gold remained in the range of £1.50 to £2.50 until the middle of the eighteenth century. Then, in the 1760's, it began to rise till it reached the range of £5.50 to £6.00 during the 1770's.[19]

Nor was this price rise confined to the Gold Coast. The prime-cost value of a *gros* (equal to 3.82 grams) of gold on the Sénégal rose from 3 livres tournois worth of goods in 1745 to 10 livres in the 1780's. There too, the increase began in the 1760's. By the seventies and eighties, merchants reported that it was no longer profitable to buy gold for export.[20]

Something somewhere worked to create a higher economic demand for gold in west Africa. There was no drastic shift of gold prices in Europe at that time. A dramatic increase in north African gold prices might have attracted gold exports across the Sahara, but it seems unlikely that north African demand was strong enough to bear the cost of transportation across the

---

18. Reference to this trade are to be found in the India Office Records, Foreign and Commonwealth Office, London. Original Corespondence, vol. 26, no. 2865, f. 1, 11 January 1661; Court Minutes, vol. 24, f. 213, 11 August 1661; Letter Books (renamed Despatch Books), vol. II, f. 266, November 1657; and ff. 326 and 349 (1660); vol. III, ff. 46, 66, 153, 170, 173, 200-02 (1661-63). These data and references were made available through the courtesy of Dr. K. N. Chauduri of the School of Oriental and African Studies, London.

19. Marion Johnson, "The Ounce Trade in Eighteenth-Century West African Trade," *JAH*, 7: 197-214 (1966), pp. 2-3, 5.

20. "Commerce de la concession du Sénégal, "unsigned, undated report, c. 1745, Archives nationales, Paris, C⁶ 12; J. B. L. Durand, *Voyage au Sénégal*, 2nd ed., 2 vols. (Paris, 1807), 2: 45. Evidence of John Barnes to House of Commons Committee on the African Trade. Accounts and Papers, XXV (635), 26 March 1789, p. 24; Evidence of Capt. Heatley, in Great Britain, Privy Council, *Report of the Lords of the Committee of Council for . . . Trade and Foreign*

Sahara when European demand was not enough to pay for transportation by sea. This leaves the explanation to be sought in west Africa itself, and it has to be an explanation strong enough to reach across the west African gold markets from Saint Louis du Sénégal to Lagos in Nigeria—a distance of some 2000 kms. Some local demand came from jewelry-making, but the historical record mentions no extraordinary change in fashion. The only obvious change, indeed, is the spread of gold as a currency and a standard of value in non-gold-production countries. Akwamu, so placed as to be an intermediary in the gold trade from the Akan region to the coast, began importing gold wherever possible after about 1708. Even further east, after Dahomey captured Allada in 1724 and Whydah in 1727, Dahomey also became a net importer of gold.[21] Neither of these states introduced a widespread gold currency on the Akan model. Both were within the cowrie zone, but they apparently used the gold as a store of wealth and as a form of currency for major transactions. These states together were far too small, however, to create such a wide ranging increase in the price of gold throughout west Africa. If a switch to gold for treasury purposes were very general, that might have been enough to produce the change, but so far we have a hint only—not an explanation.

Even before the major rise of the 1760's and after, the rising African demand brought at least one counter-current in the flow of gold within the Atlantic basin. The Portuguese became net importers of gold into the Gold Coast. This time, it was not the peninsular Portuguese but Brazilian traders in slaves. After the gold strikes in Minas Gerais at the end of the seventeenth century, gold became cheap enough in Brazil to be sent to Africa along with rolls of tobacco for the purchase of slaves. This trade was illegal in Porguguese law, but Dutch records on the Gold Coast suggest its possible size. By Luso-Dutch treaty, Brazilians paid the Dutch at Elmina 10 percent of the value of their cargoes for the right to trade on the *costa da mina*. Their payments between 1718 and 1723 implied that they came to the coast with an annual average cargo of about 40 kgs of gold. They sold the gold for slaves. Some of it remained in Africa, but the Dutch in particular bought some for reexport to Europe. A governor of Brazil estimated from that side of the ocean that as much as 500,000 cruzados in gold was shipped to Africa in the single year 1721, which implies a weight of 478 kgs in gold. This was far more than the Dutch were able to ship to Europe; even 40 kgs was nearly a quarter of the Dutch return to Europe in gold, but at least that much of annual Brazilian gold production could have been diverted to Africa. Some of the

---

Plantations . . . Concerning the Present State of Africa, and Particularly the Trade in Slaves . . . (London, 1789), part 1.

21. Walter Rodney, "Gold and Slaves on the Gold Coast," *Transaction of the Historical Society of Ghana*, 10: 13-28 (1969), p. 17; David Henige and Marion Johnson, "Agaja and the Slave Trade: Another look at the Evidence," *History in Africa*, 3: 57-67 (1976).

Dutch returns, indeed, came in the form of gold cruzados minted in Brazil—up to 10,400 coins in 1722 alone.[22]

Whatever caused the high price of African gold, it declined somewhat at the end of the eighteenth century. Gold shipments resumed from Senegambia (Table 8) and the Gold Coast alike, with Gold Coast exports reaching 31 kgs by about 1805, going higher than 1,500 kgs in 1850.[23] But the general level in the second half of the nineteenth century appears to have been less in an ordinary year—not far from the level attained some four centuries earlier.

Cowries

The use of cowrie shells as currency in India is easy enough to explain. The shells are small, uniform, and comparatively scarce. The only source of *Cyprea moneta,* the true money cowrie, is the Maldive Islands, though other *Cyprea* exist in the Indian Ocean and have sometimes been used as currency as well. They are therefore a cheap and convenient fractional currency, virtually impossible to counterfeit. It is somewhat harder to explain how they came to be used in the western Sudan as early as the mid-fourteenth century, separated by at least 4,000 kms from the nearest cowrie-currency zone. The usual hypothesis is that other, local shells were already in use as currency, which prompted north African merchants who knew of the Indian-Ocean cowrie zone to introduce the exotic shells. The price was low enough, Gresham's law operated, and the local shells disappeared as currency.

That hypothesis will probably never be verifiable, but shell currencies *were* used in other parts of western Africa, notably the *nzimbu* or *Olivancillaria nana* (Lamarck) from Luanda island in northern Angola. When the Portuguese arried in the 1480's, they found the nzimbu, which they called *zimbos,* as the dominant currency for the kingdom of Kongo and its neighborhood, and the shells continued in their monetary role. Meanwhile, a little after 1500, the Portuguese introduced the *Cyprea moneta* from India to west Africa, where cowries may have already been known by way of the desert. In any event, the shells caught on as the main currency for small transactions from the Niger delta westward nearly to the Gold Coast.

Cowries never replaced the zimbos, but they were used in central Africa as well. A steady stream of imports by way of the Mediterranean and around the Cape of Good Hope sustained the other major cowrie zones. In sheer quantity, the flow of cowries was far larger than the flow of gold, silver, copper, or any other monetary metals. Portuguese observers in the middle of the sixteenth century estimated that Portuguese ships were already bringing

22. Henige and Johnson, "Agaja," p. 63; Postma, "West African Exports," p. 33; Count of Sabugosa to King Don João, 18 February 1921, Ordens Résjas, 13. doc. 236. I am indebted to Professor John Russell-Wood of The Johns Hopkins University for the reference to the Brazilian data.

23. Mungo Park, *Travels in Africa,* 2 vols. (London, 1815), 2: cxcii; Ivor Wilks, *Asante in the Nineteenth Century* (Cambridge, 1975), p. 193.

as much as 150 metric tons a year in ballast from India. Marion Johnson estimated a west African import of about 160 metric tons a year through the eighteenth century—not just a capacity estimate this time, but an annual average for the whole period. The rate of import increased again in the early nineteenth century, with 305 tons shipped from Great Britain to the Gold Coast alone in 1853.[24]

Through all this time, cowries remained a comparative stable currency, but then, in the second half of the nineteenth century, the great cowrie inflation began. With the *Cyprea moneta* in short supply, Europeans began bringing in the somewhat larger *Cyprea annulus* directly from Zanzibar and vicinity. They flooded the region that is now southern Nigeria with a sustained import of more than 1,870 metric tons a year over the period 1841-69. The western part of the cowrie zone resisted the new shells, but it too was caught up in the general inflation. In Accra and Whydah where the *Cyprea moneta* was still dominant, the sterling value of cowries dropped from index 100 in 1850 to index 24 by 1895. In Lagos, where the change to *Cyprea annulus* was more complete, the price dropped from index 100 to index 13 over the same period. Colonial governments would no doubt have tried to substitute their own currencies for the local monetary system, but the cowries inflation ended the useful life of cowries as a widespread currency, though the shells are still sometimes used in relatively isolated places.[25]

Copper

Since Africa exported copper from the east coast and imported it on the west, it can be conveniently divided into two zones—of copper surplus and copper deficiency. The copper surplus zone would have included the whole of the southern bulge, supplied by many small deposits in addition to the major copper ores of Shaba or Katanga (Zaire) and the Zambian copper belt—deposits that have recently accounted for about 20 percent of the world copper production entering trade. Further north, near the geographical center of the African continent, the mines at Hofrat en Nahas were formerly of great importance, though they have not been so in recent centuries. Another zone of copper surplus included the Maghrib and stretched into the Sahara. Between these northern and southern zones of copper production was a broad zone of copper deficiency including all of sub-Saharan west Africa and stretching along the sahel through Chad and the Republic of the Sudan to include the horn of Africa and the region northward through

---

24. Marion Johnson, "Cowrie Currencies in West Africa," *JAH*, 11: 17-49 and 331-353 (1970), pp. 16-19.

25. Johnson, "Cowrie Currencies," pp. 18, 23-24, 340 and 343; Paul E. Lovejoy, "Interregional Monetary Flows in the Pre-Colonial Trade of Nigeria," *JAH*, 15: 560-85 (1974). The *C. annulus* is markedly larger than the *C. moneta* with only about 1.12 million shells to the metric ton, compared with about 2.36 million shells to a ton of *C. moneta*. This difference was reflected in the West African prices for each kind of shell.

Egypt.[26] The only west African copper deposits ever exploited in the pre-colonial period were one near Nioro in Mali and another at Azelick in Niger, but neither of these has been in production since the fourteenth century.[27]

The location of the ores had something to do with the movement of copper and copper alloys in trade. Broadly speaking, the Maghrib supplied sub-Saharan west Africa, drawing on metal from Europe. The southern copper zone was self-sufficient and sometimes exported copper northward to Egypt as well as by sea from the east African coast. Movement by sea or overland between the central African copper zone and west Africa appears to have been insignificant.

The quantities of copper passing along any of these routes in early centuries are recorded far less often than gold or silver movements are. W. R. G. Randles looked thoroughly through the literature on trade from southeast Africa. He turned up only one estimate, some 46 metric tons exported from the lower Zambezi in 1762.[28] The size of these shipments indicates that the trade must have been significant in other years as well, but it could have been higher or lower.

The trans-Saharan copper trade to west Africa was very old. Excavations at Igbo Ukwu in southeastern Nigeria turned up some 685 pieces of copper or brassware with radiocarbon dates indicating the ninth and tenth centuries A.D. These dates seem a little early, in terms of the development of the trans-Sahara trade generally, but a modification toward the eleventh and twelfth centuries would still lie within the margin of error for radiocarbon dating. By that time, the movement of copper across the desert was already important; archaeologists recovered more than a metric ton of brass rods from a single caravan site in the Sahara. Well before 1500 A.D. the Ife and Benin "bronzes" were already produced by a sophisticated process of lost-wax brass casting, where the level of technology indicates a long probable background in time. Copper currency was reported in use in the western Sudan by the middle of the fourteenth century, and it appears that the Cross River hinterland near the Cameroon-Nigerian border already used brass or copper currency at the time of the first maritime contact with the Portuguese.[29]

Throughout the southern bulge of Africa, a great variety of different standardized shapes and sizes of copper or brass circulated as equivalents to the minted coin—that is, a piece of metal sufficiently standardized to be suitable for circulation by count rather than weight. Some were in the form of rods, with African-made wire in circulation to facilitate small purchase. Others were in the form of copper or brass stylized bracelets or horseshoes in

---

26. Sundström, Lars, *The Exchange Economy of Pre-Colonial Tropical Africa*, 2nd ed. (London, 1974), pp. 217-23. First ed. title *The Trade of Guinea*.

27. Eugenia W. Herbert, "Aspects of the Use of Copper in Pre-Colonial West Africa," *JAH*, 14: 179-194 (1973), pp. 192-194.

28. Randles, *Monomotapa*, p. 115.

29. Herbert, "Copper in West Africa," passim; Merrick Posnansky, "Brass Casting and its Antecedents in West Africa," *JAH*, 18: 287-300 (1977).

Africa and the wider monetary world, 1250-1850

several different style or sizes. The Europeans quickly adapted these shapes and manufactured their brassware to meet African demand calling them *manilas*, from the Portuguese *manilha*, bracelet.

Even without long time-series, it seems likely that copper was the most important monetary metal imported into Africa in the sixteenth century and perhaps in the seventeenth as well. It is possible, indeed, that such gold-producing regions as Asante used copper or brass coinage before their trade relations with the Europeans helped to convert them to gold. In any event, copper and copper alloys appear to have been the dominant monetary import during the sixteenth century, making way for increasing quantities of cowries in the course of the seventeenth and for silver in the course of the eighteenth and nineteenth.[30] The quantitative evidence, however, is still only spotty. From 1504 to 1507, for example, the Portuguese used 287,800 manilas in the gold trade at São Jorge da Mina.[31] If these were the comparatively large manilas (of 0.6 kgs each) used in the Mina trade a couple of decades later, the copper or brass import for these four years would have been 173 metric tons. The surviving Elmina account books for 1529-31 show 242,000 manilas of this size imported over 27 months, or 325 metric tons a year for this one port. At that period, metals (chiefly brass and copper) accounted for more than half of the total imports into the Gold Coast by sea.[32]

Seventeenth-century sample data suggest a lower proportion of total imports and perhaps lower quantities as well. Even as early as 1600, estimates of the Dutch trade suggest about 20 metric tons of copper and brass as the ordinary level.[33] By 1680-82, at the height of the Royal African Company's trade, copper and brassware came to only 7.7 percent of all the Company's imports into Africa—or an annual average of 41 metric tons, compared to an annual average of 77 tons in cowrie shells.[34] The Royal African Company was only one carrier among many, but its share of the trade makes possible a wild guess at total copper and copper alloy imports into West Africa on the order of 100 metric tons a year, and probably no more than 500 tons at the outside. This is an obvious comparative decline from the 1520's, when one port, albeit a crucial one, could absorb more than 300 tons a year.

Cloth

Cloth, the most common African currency of the period, is also the one least well represented in the literature. While foreign cloths were rarely monetized, their import and use no doubt had some kind of impact on the value

---

30. Herbert, "Copper in West Africa," pp. 192-94: A. J. H. Latham, "Currency, Credit, and Capitalism on the Cross River in the Pre-Colonial Era," *JAH*, 12: 599-605 (1971). Sundström, *Exchange Economy*, pp. 161-178, 241-46.
31. Magalhães-Godinho, *L'empire portugais*, 383.
32. Vogt, "Portuguese Gold Trade," pp. 93-94.
33. Ddaku, *Trade and Politics*, p. 11. For early trade with manilas, see also Alan Ryder, *Benin and the Europeans 1485-1897* (New York, 1969).
34. Calculations based on Davies, *Royal African Company*, pp. 350-75.

Table 9. Gambian imports 1683–88 compared to royal African company's exports to Africa 1680–82

Commodities	(1) Gambian imports 1684–88		Royal African company exports, 1680–82	
Metals and metalware		33.7		22.4
Iron bars	24.9		11.1	
Silver coins	4.2		–	
Copper rods	1.0		4.5	
Brassware	1.7		4.9	
Pewterware	1.9		1.9	
Woolen cloth		1.4		28.1
Perpetumnas	–		13.9	
Red cloth	1.4		–	
Says	–		13.3	
Welsh plains	–		0.9	
Indian textiles		1.6		18.3
Allejaes	–		1.0	
Bafts	–		1.4	
Brawles	–		2.5	
Guinea stuffs	–		1.9	
Long cloths	–		2.8	
Longees	–		1.2	
Nicconees	–		2.3	
Pautkes	–		1.6	
Tapseels	1.6		3.6	
Other textiles		1.0		14.4
Annabasses	–		3.5	
Carpets	–		0.1	
Silesias (linen)	1.0		4.7	
Sheets	–		5.4	
Boyasados	–		0.7	
Fringe	–		–	
Blanchester cloth	–		–	
Cutlery and weapons		8.1		6.4
Firearms	1.5		2.7	
Gunpowder	1.2		3.1	
Knives	–		0.6	
Swords	5.4		–	
Cowrie shells	–	–	7.2	7.2
Brandy	14.1	14.1	–	–
Beads and semiprecious stones		39.9		3.1
Glass beads	13.9		1.8	
Coral	7.0		1.3	
Amber	11.4		–	
Crystal	7.6		–	
Carnelians	–		–	
Total	100.0	100.0	100.0	100.0

Source: Curtin, Economic Change, 1:313.

and role of the local cloths that *were* monetized, but the impact remains obscure. We know something about the more precise make-up of goods imported into Africa from Britain, at least for the period when the Royal African Company was closest in fact to exercising the monopoly it enjoyed in theory, but separate data for Senegambia indicate that some regions in tropical Africa responded with distinct patterns of demand. (Table 9)

It is also possible to follow the import patterns for Senegambia over time (Table 10). Although this region may not be typical, it is striking that metallic imports generally declined over the course of the late seventeenth to the early nineteenth century. Cloth imports increased, and the absolute quantity increase is far higher than the percentage change. Keeping in mind that import totals were a reflex of export totals, the value of Senegambian exports rose from index 100 in 1680's to index 242 in the 1730's, 724 in the 1780's and index 2330 in the 1830's.[35]

This increase in textile imports was far higher than the equivalent increase in cowrie imports, but cloth had too many non-monetary uses to make that observation very significant. It is possible, however, that the vast new cloth imports could have had an inflationary effect on cloth currencies, in some way comparable to the cowrie inflation. If so, it too could have played a role in the destruction of the older African currency system and the substitution of European coinage.

One significant example of monetizing imported cloth did take place in the transitional period of the nineteenth century. One of the most distinctive cloths imported to Africa from India was a heavy, hand-woven cotton, very heavily impregnated with indigo dye. These cloths were manufactured in the region of Madras and Pondichéry and shipped to Africa with increasing frequency after the beginning of the eighteenth century. The British called them blue bafts, the French, *pièces de Guinée*, and they were especially popular in the gum trade of the southern Mauritanian sahel. The currency of account on the adjacent coasts of Senegambia had been iron bars since the early seventeenth century, but, by the 1720's, blue bafts began to replace bars as the currency of account along the Senegal river. That is, the price of gum was expressed as so many blue bafts of the standard size and weight, and other commodities followed. French merchants even kept their accounts in bafts, with the symbol "P." for *pièces* followed by the number of cloths.

In some respects, they were not a very useful currency of account, their value was subject to wild fluctuations based on local disorder, war at sea, disorder in India, or alternative demand from other parts of Asia. They were nevertheless so firmly established by the early nineteenth century that the French colonial government of Senegal decided to monetize them officially. The French first set up quality standards to be enforced by inspectors at the Indian colony of Pondichéry. Cloths passing inspection were marked with a

---

35. Curtin, *Economic Change*, 1: 329.

Table 10. Estimated proportions of certain major imports in Senegambian total imports 1680's, 1730's, and 1830's

Commodities	1680's (per cent)	1730's (per cent)	1830's (per cent)
Metals			
Iron	24.9	21.2	1.5
Silver	4.2	10.2	3.9
Copper	1.0	0.5	–
Brass	1.7	4.8	–
Pewter	1.9	1.1	–
European textiles	2.4	9.7	25.0
Indian textiles	1.6	18.5	33.9
Cutlery and weapons			
Firearms	1.5	5.6	6.0
Gunpowder	1.2	4.0	3.8
Cutlery	5.4	1.0	–
Beads and semiprecious stones	39.9	18.0	8.8
Spirits	14.1	4.8	7.1
Paper	–	0.5	–
Tobacco	–	–	10.0

Source: Curtin, *Economic Change*, 1:318.

special stamp, which was designed to have the same consequences as the coining of silver, copper, or gold. Further Senegalese regulations of 1844 prohibited trade along the Senegal in any cloth that was not officially stamped, and further regulations of 1847 made it illegal to trade with any other currency—specifically outlawing gold and silver coins, even though these still passed current in the French-controlled port of Saint Louis.[36]

The experiment was not a complete success. The Indian inspectors had somewhat different standards of value from those of the Mauritanian consumers. Mauritanians sometimes accepted unstamped cloth, in spite of the law and refused to accept stamped cloth they regarded as inferior. As time passed, and even before the colonial conquest was completed in the 1890's, silver coins won a larger and larger place as a currency rivaling the local cloth currencies. But the *coup de grace* for the blue bafts was the European development of techniques to produce by machine counterfeit cloths that would pass muster with African customers. The result was again inflation, the operation of Gresham's law, and the gradual substitution of cheaper machine-made bafts for the genuine article from India.

36. Curtin, *Economic Change*, 1: 260-64, 268-70.

Value ratios of differing money materials

Unlike the literature on economic history for Europe and Asia, the historical literature on Africa has little to say about value ratios of gold to silver, gold to copper, or silver to cowries. Most writers apparently began with the assumption that the "true" value of, say, gold to silver, was the value on the European market, and that any deviation was simply a reflection of distance and transportation costs. They may have been right for the nineteenth century; but they were probably not right for earlier centuries, and we are very short of data.

Those assembled in Tables 11 through 17 are therefore little more than an initial presentation of what is readily available in the printed literature. In general, they reflect transportation costs and some of the major changes in value—like the west Africa rise in the price of gold in the second half of the eighteenth century. Since little or no silver was ever mined in Africa, one would expect a lower gold-to-silver ratio than was common in Europe or Asia, and the data met that expectation. Magalhães-Godinho calculated a rate of 1 gold to 8.1 silver for the southeast African coast in the sixteenth century.[37] Rates were sometimes lower for the interior of west Africa, but they tended to increase through time, converging with the European price. The nineteenth-century Ethiopian data, for example, show a range of 10 to 13 to one, at a time when the European rate was about 15 to one.[38] (Table 1) That difference may be accountable to the cost of transportation to Europe, but if all the gold was as impure as that assayed by the French expedition of 1842, the ratios would be higher than those prevalent in Europe at that time, and therefore something of a puzzle.

The gold-silver ratios recorded for the interior of west Africa are markedly lower than those for Ethiopia, and this should be accountable to the more difficult access from the coast. (Tables 10, 13 and 14) Especially in Table 14, the increase in the value of cowries or silver against gold as one moved inland from the coast seems to reflect transportation costs in a systematic way, and to nearly the same degree for cowries and silver. A unit of silver would have increased in value by 44 percent between the coast and Katsina, while a unit of cowries would have increased by 42 percent. This, of course, is a measure in gold alone. The picture might be different, if it were possible to measure the changes against a bundle of consumer's goods.

The gold-cowrie ratios may take on more meaning when they are compared to similar ratios for Bengal and the Maldive Islands. Magalhães-Godinho reports that their value was very stable in Bengal between the sixteenth and eighteenth centuries,[39] and it appears to have been very stable

---

37. Magalhães-Godinho, *L'empire portugais*, p. 385.

38. But two alternate calculations are possible using these data. One is the actual exchange of coins for gold, the other based on the amount of pure gold contained in the not-yet-refined metal, based on the assays carried out by the French expedition of 1842.

39. Magalhães-Godinho, *L'empire portugais*, p. 390.

Table 11. Changing value ratios of gold, silver, and cowries Western Sudan

| | Ratio to one unit of gold || Ratio to one unit of silver to cowries |
	Silver	Cowries	
c. 1350 Niger Bend[1]		280	
After c. 1580[1]		700	
c. 1780[2] Katsina	7	548	78
c. 1780–1800[2] Gobir			79
c. 1805[3]	1.6	646	
c. 1820[4] Southern Morocco			90
c. 1820[1] Katsina	7.8	702	90
c. 1900[1] Hausaland		1,447	

Sources:
1. Calculated from data in M. Johnson, "The Cowrie Currencies of West Africa," *Journal of African History*, 11:331, 336 (1970), on the basis of cowries at 880 per kilogram, silver dollars at 24.42 grams each, mithqals at 4.859 gr.
2. Paul E. Lovejoy, "Interregional Monetary Flows in the Precolonial Trade of Nigeria," *Journal of African History*, 15:563–85 (1974), p. 567.
3. Calculated from data of Mungo Park, *Travels in Africa*, 2 vols. (London, 1816), 2:clxxiii.
4. Johnson, "Cowrie Currencies," 11:29.

in the western Sudan as well. Whatever the past situation may have been, however, that stability did not last into the seventeenth and eighteenth centuries. The variance of the annual average price paid by the East India Company in India between 1669 and 1753 (with some years missing from the data) was .49. The variance in the annual sale prices in London was somewhat higher, at .57 for the period 1669-1705, but the London market settled down to a pattern of fluctuation that appears to have been substantially less than this through the remainder of the eighteenth century. (Tables 12 and 13)

Marion Johnson calls attention to the rise on gold-cowrie ratios on the west African coast between the 1720's and about 1780. Part of this change could be accountable to changes in the prices of cowries delivered to London from the Indian Ocean, but the London cowrie price was only about 30 percent less in the 1780's than it had been a century earlier. Marion Johnson shows, however, that cowries held their value, at least against the value of live hens, and suggests that they probably did the same against other food stuffs.

If so, the rise in gold prices was a movement of gold alone while other prices held more or less constant, which makes the puzzle of the rising gold prices even more intriguing.

The stability of cowrie prices is all the more striking, because west Africa imported about 160 metric tons of shells each year through the eighteenth century. Given the fact that cowries have few non-monetary uses, they must either have disappeared into hoards or have been absorbed into the regional economy. Since we know that the cowrie region was expanding at this period, and the economy seems to have been increasingly monetized, it may well be that the stability of cowries in the face of significant imports had the same ultimate cause as the rise in the price of gold. That is, both gold and cowries were in demand in west Africa because progressive monetization of exchange absorbed currency. Hypothetically, then, both demands for currency were a product of deep-seated changes in west Africa away from "subsistence" agriculture and toward marketed production, but a firm answer will have to wait for further research.[40]

Conclusion

No conclusion is possible. The essay is no more than an introduction—depending largely on the careful archival work of K. G. Davies, Marion Johnson, Vitorino Magalhães-Godinho, and Johannes Postma. Without their work, even this starting point would not have been possible.

40. Johnson, "Cowrie Currencies," pp. 347-48.

Table 12. East India company's trade in Cowries, 1669–1753

Year	Quantity imported from India to England (metric tons)	Average annual price paid in India (pounds sterling)	Silver-cowrie ratio, India
1669	32	40	1:226
1670	98	25	1:354
1671	107	32	1:283
1672	14	40	1:223
1673	133	37	1:240
1674	148	40	1:227
1669–1674	Annual average 88.70	Annual average 35	
1680	117	34	1:265
1681	–	–	–
1682	79	31	1:291
1683	207	31	1:287
1684	171	31	1:291
1685	139	35	1:258
1686	95	54	1:167
1687	0	0	–
1688	0	0	–
1680–1686	Annual average 89.81	Annual average 36	
1690	41	32	1:279
1691	0	0	–
1692	0	0	–
1693	226	29	1:313
1690–1693	Annual average 66.90	Annual average 15	
1698	34	35	1:261
1699	5	161	1:56
1700	37	34	1:261
1701	19	50	1:179
1702	30	49	1:183
1703	27	74	1:121
1704	16	70	1:129
1705	3	21	1:424
1698–1705	Annual average 21.49	Annual average 61	
1710	19	96	1:94
1711	18	100	1:90
1712	50	69	1:130
1713	25	48	1:188
1714	35	41	1:221
1715	36	15	1:589
1716	82	42	1:215
1717	9	30	1:302
1718	7	25	1:361
1719	54	45	1:201
1710–1719	Annual average 33.43	Annual average 51	
1720	148	41	1:218

Table 12. (continued)

Year	Quantity imported from India to England (metric tons)	Average annual price paid in India (pounds sterling)	Silver-cowrie ratio, India
1721	65	39	1:229
1722	213	53	1:171
1723	152	48	1:185
1724	105	40	1:222
1725	115	42	1:215
1726	122	45	1:199
1727	181	54	1:166
1728	143	43	1:211
1729	90	43	1:211
1720–1729	Annual average 133.50	Annual average 45	
1730	4	41	1:222
1731	61	47	1:193
1732	80	47	1:192
1733	88	40	1:223
1730–1733	Annual average 58.27	Annual average 43	
1740	50	41	1:217
1741	92	41	1:220
1742	133	39	1:229
1743	83	25	1:358
1744	102	41	1:217
1745	124	41	1:218
1746	42	41	1:218
1747	26	38	1:235
1740–1747	Annual average 81.38	Annual average 33	
1669–1753		Annual average 73.25	
Standard deviation		22.14	
Coefficient of variation		.49	

Source: India Office Records, Foreign and Commonwealth Office, Accountant General's Department, General Commerce Journals and General Ledger Books. I am most grateful to Professor K. N. Chaudhuri, for providing a computer print out of the data he had assembled on this subject.

Table 13. East India and Royal African company trade in Britain in Cowries

| Year | East India company ||| Royal African company |||
	Quantity sold in metric tons	Price in pounds sterling per metric ton	Gold-cowrie ratio	Quantity sold in metric tons	Price in pounds sterling per metric ton	Gold-cowrie ratio
1669	9	103	1:1332			
1670	121	84	1:1622			
1671	38	63	1:2153			
1672	87	60	1:2304			
1673						
1674	254	69	1:1986	12	76[1]	1:1807
1675				9	79	1:1730
1676				7	78	1:1743
1677						
1678						
1679						
1680	117	92	1:1487	8	67	1:2047
1681				62	82	1:1672
1682	79	113	1:1213	67	105	1:1307
1683	155	73	1:1885	83	106	1:1287
1684	185	59	1:2312	53	80	1:1713
1685	3	64	1:2132	115	84	1:1621
1686	94	78	1:1754			
1687	32	90	1:1523			
1688	151	68	1:2012	29	89	1:1535
1689				17	89	1:1538
1690	7	60	1:2279	17	90	1:1525
1691	58	59	1:2309	8	94	1:1458

Table 13. (Continued)

	East India company			Royal African company		
	Quantity sold in metric tons	Price in pounds sterling per metric ton	Gold-cowrie ratio	Quantity sold in metric tons	Price in pounds sterling per metric ton	Gold-cowrie ratio
1692	169	63	1:2156	24	98	1:1388
1693	75	63	1:2171	38	87	1:1563
1694				29	72	1:1896
1695				19	89	1:1538
1696				22	138	1:991
1697				4	148	1:920
1698	28	285	1:479	5	281	1:487
1699	5	161	1:848			
1700	42	167	1:820			
1701	37	193	1:708	16	184	1:744
1702	12	142	1:961	16	217	1:629
1703	22	94	1:1455	16	133	1:1025
1704	19	104	1:1320	20	102	1:1342
1705	3	21	1:6451			
Annual mean				111.17		
Standard deviation				51.26		
Coefficient of variation				.46		

Source: EIC data from India Office Records, Foreign and Commonwealth Office, London, Accountant General's Department, General Commerce Journals and General Ledger Books. I am most indebted to Professor K. N. Chaudhuri of the School of Oriental and African Studies for a computer printout of the data he assembled. Royal African Company data from K. G. Davies, The Royal African Company, p. 357. Prices in either case are the total volume for the year divided by the total price paid or received.

Table 14. Prices of Cowries, F.O.B. British Ports, 1714–1736

Year	Annual price (pounds per metric ton)	Prices over periods of time	Annual implied gold-cowrie ratio	Gold-cowrie ratios over periods
1714–1730		price range £157–197		ratio implied 870–690
1720	176[2]		774	
1721	176		—	
1722	98		1388	
1723	118		1156	
1724	138		991	
1725	154		889	
1726	130		1051	
1727	110		1239	
1728	86		1588	
1720–1729		annual average £ 119		908
1730	131		1043	
1731				
1732	122		1119	
1733	140*		977	
1734	138*		991	
1735	138*		991	
1736	138*		991	
1737	94		1458	
1738	90		1512	
1739	110		1239	
1730–1739		annual average £ 110		1032
1751	69		1982	
1752	79		1734	

266

Table 13. (continued)

| | East India company ||| Royal African company |||
Year	Quantity sold in metric tons	Price in pounds sterling per metric ton	Gold-cowrie ratio	Quantity sold in metric tons	Price in pounds sterling per metric ton	Gold-cowrie ratio
1753		81			1692	
1754		81			1692	
1755		79			1734	
1756		79			1734	
1757		74			1850	
1758						
1759		77			1779	
1760		77			1779	
1751–1760			annual average £ 70			160

Mean 110.66
Coefficient of variation .33

Sources: References apply to all data below in the same column until interrupted by another reference superscript number.
1. M. Johnson, "The Cowrie Currencies of West Africa," *Journal of African History*.
2. Royal African Company and successors, Invoice Books Outward, Public Record Office, London, T 70/921, T 70/558–575, T 70/926–27. Data marked with an * are estimated prices for inventory purposes, not actual invoices. Since these are prices that happened to turn up on invoices, they are not averages for the year and are judged to be less accurate than Davies' data for the earlier period.

Table 15. Changing value ratios of gold to Cowries Western Sudan and West African coast

	(1) Approximate date	(2) Western Sudan	(3) West African coast	(4) Great Britain
	1500	100	300	–
Before	1728	600	600	1100 (mean of 1720–28)
After	1728	600	875 (Dahomey)	1100
	1780	730	1200	2300
	1800–16	730	940	1400
	1820	730	1200	3000
	1850	910	1400	1300

Sources: Cols. 1 and 2, M. Johnson, "Cowrie Currencies," 11:344. Col. 3 of table 9.

Table 16. Value ratios of gold, silver, and cowries about 1850 at differing places in Western Sudan

	Gold-silver ratio	Gold-cowrie ratio	Silver-cowrie ratio
Hausaland[1]		940	
Timbuktu[1]	6.4	890	140
Jenne[1]		940	
Kaarta[1]	5.9	590	100
Bambuhu (not in the cowrie zone)[2]	5.9		

Sources:
1. Johnson, "Cowrie Currencies," 11:343.
2. L. L. C. Faidherbe, "Notice sur la colonie du Sénégal," *Annuaire du Sénégal et Dependences,* 1858, pp. 71–144, p. 111.

Table 17. Value ratios of cowries to gold and silver along the route from the Gold Coast to Hausaland about 1820

Successive points from the coast inland	Gold-silver ratio	Gold-cowrie ratio
Coast	c. 14	1200
Salaga	12.6	1000
Yendi		980
Zongho		930
Nikki		890
Niger at Boussa		800
Katsina	7.8	700

Source: M. Johnson, "Cowrie Currencies," 11:31 and 34.

9

The role of international monetary and metal movements in Ottoman monetary history 1300-1750

HALIL SAHILLIOĞLU

The Ottoman Empire, situated at the meeting place of East-West trade and ruling major centers of commercial exchange between the two continents, was accordingly affected by international movements of money and precious metals. These effects can be observed at every stage of its history. The purpose of this essay is to examine such effects on the monetary history of the Ottoman Empire.

Silver monometallism during the epoch of the empire's foundation

The foundations of the Ottoman Empire were laid at the beginning of the 14th Century. Changes in the distribution of power were accompanied by a change in the pattern of trading routes within the Mediterranean Basin. The Mamluks, after seizing power in Egypt and Syria from the Ayyubids (1250 A.D.) stopped, at the battle of *Ayn Djalut* (1260 A.D.), the westward expansion of the Mongols. Somewhat later, they helped to terminate Crusader power in the Levant by conquering the Castle of *Akka* (Acre) (1291 A.D.). These developments helped create a common interest between the Mongols, the Crusaders and the trading city states of Italy. By the late thirteenth century the most intensively used east-west trade routes had shifted north to Anatolia and the Black Sea instead of through the more southerly route terminating in Syria.

Byzantine recovered Istanbul in 1261 from the Latins who had been able to capture it in 1204, while the Mongols settled in Anatolia (Battle of Kösedağ, 1243). These events hastened the break up of the Sultanate of the Anatolian Seldjuks. However, so long as the Mongol Ilkhanid Empire in Iran continued to exist, Western trade passing through Anatolia and the Black Sea could flourish under the imposed Mongol Peace. The collapse of the Ilkhans in 1335 resulted in the emergence of numerous small states (Beyliks) in Anatolia. This hindered trade proceeding along the northern routes. Inferior coins used in Anatolia during this period suggest the consequences of declining trade, if only in a very general way.

In Anatolia, the minting of gold coins came to an end with the demise of the Seldjuks and the Ilkhans. Because Byzantium controlled the Bosporus and the trade passing through it she was able to remain the only state in the north pursuing a monometal gold-money policy. The Byzantine gold money the "Perpero" however, was no longer as strong a currency as it had been during the Byzantine period of effloresence.[1]

Referring to the immediate years after the collapse of the Ilkhan Empire, Ibn Fadullah al Omari[2] and the Moroccan traveler Ibn Battutah[3] relate that the local lords or Beys of Anatolia minted silver coins, smaller in size than Mamluk *dirhams,* but having a high purchasing power, thus suggesting that Anatolia was a region of low prices. If the lowered price level may be attributed to failure of demand, it may be suggested that there was a real decline in trade. We must also remember that gold and large silver coins are usually minted in response to economic or commercial needs, hence their absence may indicate that no such needs were felt.[4]

Western economic historians offer various explanations for apparent economic stagnation in Europe in the 14th and 15th centuries. The rapid economic shifts occurring in the Near and the Middle East probably should be added to the list of explanations already suggested. During this period of stagnation in the West, none of the petty states or *Beyliks* in Anatolia minted gold money or large silver coins. Instead they confined themselves to the minting of small silver coins, generally referred to as *akçe* in Turkish.[5] In use until 1470, the *akçe* was a coin of about 1.152 grams. It might be

Abbreviations: BBA (Başbakanlik Arişivi, Istanbul: The Archives of the Prime Ministry in Istanbul), AN., (Archives Nationales, Paris), AFF. ETR. (Affaires Etrangères).

1. "... una moneta d'oro che s'appellano perperi, i quali sono di lega di carati si ne sono li 6 carati per oncia, e lo rimanete della lega infino in 24 carati si ne sono li 6 carati d'argento fine e li 7 di rame per ogni oncia." F. B. Pegolotti, *La Pratica della Mercatura,* Ed., by A. Evans (Cambridge, 1936), p. 40.
2. Ibn Fadl Allah al-Omeri: *Mesâlik al-absâr fi Memâlik al-amsâr,* Lib. of Ayasofya, Mnsc. No. 3416, f⁰ 98-115.
3. C. Défermery et B. R. Sanginetti: *Voyages d'ibn Battoutah,* Vol. 2, pp. 252-352.
4. F. Braudel et F. C. Spooner: "Les Métaux monétaires et l'économie du xvi. siècle" Relazioni del Cogresso Internazionale di Scienze Storiche, Roma 1955, Vol. IV, Storia Maderna, Firenze, 1955, pp. 233-264.
5. According to Pegolotti's account of commercial life in Anatolia and neighbouring islands, large silver coins were being minted outside of Byzantium. Cyprus had *Bisante bianco* equivalent to 7 akçes in *Alaiyye (Candolore)* (Pegolotti, id. p. 92). Rhodes *gigliati* weighed 1/57th of the *marchio di Rodi* (id. p. 103). Tana's silver bars called *sommo* weighed 202 *aspri (akçe)* (Ibid. p. 25). In Tebriz gold coins in circulation were named *cassinini (Hasene)* (Ibid. p. 27). *Aspro (akçe)* were minted in Tana, Rhodes, *Candolore (Alâiye)* and Trabzon. The Ottomans by the end of this period had silver coins in circulation which weighed 2.30 grams (a twopiece) and 5.76 grams (a fivepiece). They were minted during the first years of Orhan Bey's reign. In these years Ilkhanids still flourished and Anatolia's economy continued to be favourable. But after 1335, such coins disappeared from circulation. Numismatic catalogues do not describe such issues (I. Galib: *Takvim-i meskûkât-i Osmaniyye,* Istanbul 1311 (1895); Halil Ethem, *Meskûkât-i Osmaniyye Kataloğu,* Istanbul 1334 (1915); Ahmed Tevhid, *Türkistan Hâkanlari* . . . *Anadolu*

appropriate to consider Aydĭnoğlu—who were involved in Mediterranean trade—as an exception in this regard; it was known that he had minted imitation Venitian ducats but had terminated this practice in the face of Venetian protest.[6]

At the beginning of the early Modern Age, enhanced economic development experienced in Europe encouraged minting of new large silver coins. The Ottomans, following suit, in 1470 introduced into circulation a large silver coin of 10.14 grams called *"Muhammed Hânî."*[7] Because of its size, this coin was referred to as *"âkçe-i Büzürk"* or the Large *âkçe*. It was also known by a third name, the *Gümüş Sultânîye*, inspired by the first gold coin the Ottomans had minted, the *Sultânî*. The new coin was thus referred to as *Gümüş Sultânîye* (Silver Sultânî) or its Arabic equivalent "Sultânîyye fiddiyye."

Up to the reign of Mohammed the Conqueror, Ottoman currency in circulation could be reminted in accordance with the "old *akçe* prohibition" which made it possible to remint old coins by paying minting royalties and fees. Under Mohammad's predecessors the weight and silver content of Ottoman coins, even when reminted, had remained largely unchanged. However, during the five occasions when old coins were reminted under the reign of Mohammed the Conqueror, the weight of the *âkçe* was reduced by a grain (approximately 48 milligrams) each time. Thus, the *âkçe* which weighed 1.182 grams in 1425 had been reduced to 0.768 grams at the time of the last *refonte* coinciding with the death year of the Conqueror.[8]

Although the *âkçe* was the official money of the Empire, the state did not interfere with the circulation of foreign gold and silver coins in its domain. Under metallic money systems the value of coins derive from the value of the precious metal they contain. Any loss of value in a coin is a function of the differences in the price of the same metal in different markets. Thus, if a government established a fixed exchange rate for a particular foreign coin, this would be tantamount to favoring it generally. For otherwise, foreign coins would find their value according to the price of the precious metal they contained.

Tavâif-i Mülûki . . . Sikkeleri Katalogu, Istanbul 1321 (1903) . . .). *Grossi*, the large Venetian coins disappeared from circulation between 1330-1350 (F. C. Lane: "Salaires et régime alimentaire des marins au début du xıve siècle," *Annales*, E. S. C., 18e année, No. 1 (1936). p. 135. It is true that three Turcoman states, Saruhan, Mentese, and Aydınoğulları, minted large silver coin (*gigliati*) as described by Schlumberger (in his *Numismatique de l'Orient latin*) as late as the mid-fourteenth century. Occasional gold coins from this period have also appeared. After 1350, however, these larger issues dwindle and disappear in favor of the smaller coin issues in silver.

6. W. Heyd, *Histoire du commerce de Levant*, Vol. I. (Leipzig 1936), p. 544.

7. R. Anhegger and H. Inalcik: *Kanun-nâme-i Sultânî ber-mucib-i Örf-i Osmanî*, Türk Tarih Kurumu, Ankara 1956; and N. Beldiceanu: *Les Actes des Premiers Sultans . . . I. Actes de Mehmed II et de Bayezid II . . .*, (Paris, 1960), Document No. 15.

8. H. Sahillioğlu, "Bir Mültezim Zimem Defterine Göre 15. Yüzyıl sonunda Osmanli Darphane Mukataalari," Mint taxfarms in late 15th century according to the register of tax farmers debts *Iktisat Fakultesi Mecmuasi*, Istanbul, Vol. 23, No. 1-2 (1962-63), pp. 145-218.

From the vague and brief accounts of the Ottoman chroniclers it can be generally inferred that a gold shortage existed, at least in Anatolia. The chronicles reporting on the marriage of the heir-apparent Yıldırım Bayezid, son of Murad, to the daughter of Germiyanoğlu Yakup Bey, describe the presents which the bride brought with her as consisting of horses and gold embroidered white cloth of *Denizli*. They proceed to add apologetically "in those days, gold and silver were scarce." The Beys of Osman on the other hand, they add, brought trays of gold as presents.[9] Similarly, considerable difficulties were experienced in collecting gold when there was talk of paying ransom in gold to free Yıldırım Bayezid who had fallen captive to Temur the Lame.[10]

During the founding stages of the Ottoman State, many active gold and silver mines came into Ottoman possession as a result of their conquests in Rumelia. They took possession of Novobrdo (1440), Kratovo (1490), and Srebernice (1455) mines along with the Sidrekapsa mines in Halkidikia near Salonica which produced lead with gold in it. There were also many mines of secondary importance in the Balkans. During the 14th and 15th Centuries these mines were generally operated by entrepreneurs from Dubrovnik who shipped most gold and silver directly to the Adriatic ports.[11] The Ottomans attempted to direct these metals to Ottoman mints by transferring these tax farms to other entrepreneurs. The Ottoman state established a mint beside nearly every mine. The places where the mines were to send their product was also regulated. Contrary to what Tadić has assumed, these mines did not close down once they came under Ottoman possession, but continued to function through the 16th Century and some till the third decade of the 17th Century. Ottoman Balkan workings finally closed down in the face of the abundance of American silver and increasing costs of production, but did resume production at the beginning of the 18th Century.

An examination of the *Kaun-nâme-i Sultânî* reveals the routes through which these metals reached Istanbul as well as their official purchase price in major Ottoman cities.[12] These Balkan mines offered one source of precious metals to meet the monetary needs of the imperial economy. Another source of money for circulation was gold coins of various foreign origins which the Ottomans welcomed before they began to mint gold coins themselves. The most favored of these gold coins was the "flöri" or the "frengi flöri," the levantine version of *Florin* or *Fiorino*. This, however, was not the Florentine coin but usually the Venetian *ducat*. More generally, *flori* was used as a

---

9. Neşrī, *Kitâb-i Cihân-mumâ*, published by F. Babinger, p. 55.
10. Id. p. 99. The ransom was 90,000 gold coins.
11. J. Tadić, "Les Premiers Eléments du Capitalisme dans les Balkans du 14e au 17e Siècle," *Fourth Int. Conf. of Economic History*, (Bloomington, 1968), ed. by F. C. Lane, pp. 69-77.
12. The purchasing price of silver by the mints varied according to the location as follows: 281 in Novo Brdo, 283 in Siroz (Serrai), 285 *akçes* in Istanbul, Bursa, Ayasluk and Konya. These were the prices of 100 *dirhams* (302.2 grams) of pure silver.

generic name for all gold coins in a fashion similar to *"dinar"* and *"Hasene."* The origin was typically specified by adding an adjective. For example, when Florentine gold was to be identified, it was referred to as "Floriyyen Efrentiyyen" or "Efrenti Flöri."

Ottoman bimetalism

The upsurge of trade and production in European cities brought new efforts to mint large silver coins and introduce them into circulation. Exposed to similar conditions, we would expect the Ottomans to have behaved in a similar fashion. And indeed, in an imperial letter (*Nâme-i Hümâyûn*) which Mohammed the Conqueror sent to the Sultan of Egypt and the Sherif of Mecca, after having completed the conquest of Istanbul, he relates that he had minted new gold coins from the booty, samples of which he is sending to them as a gift and souvenir.[13] Examples of Venetian, Genoese or other foreign gold coins minted by the Turks do exist in museums or numismatic catalogues. But since they are imitations and were intended to pass as such they are sometimes difficult to identify. Nevertheless, travellers' accounts and the presence of extremely large numbers of ducats listed in the official treasury accounts confirm the Ottoman practice of issuing imitation European coinage.[14] We may surmise that the minting of these imitation coins coincided with a protracted state of war between Venice and the Ottoman Empire (1465-1479).

In general tax farms called "Tahvil" by the Ottomans were leased for a period of three years. The "Tahvil" of 15 Safer 883 A. H. (January 12, 1479) establishing the first mint concession *Daru'd-darb-i hasene-i Sultânî-i cedîd* or, the new mint for the minting of gold *Sultânîs* to make Ottoman gold coins, was issued within two weeks of the peace treaty with Venice. Establishment of an Ottoman coinage thus placed the empire under a bimetallist system.

The Court Records of Bursa (Kadi Sicili No. 3A) which began to be kept during the last three years of the reign of Mohammed the Conqueror, testify to lively trade. During this time, it was possible to meet traders from Venice and Genoa and also from India, Iran and Egypt in Bursa. Bursa merchants themselves were active in the lands of the Mamluk, Basrah, Iran, Hejaz, India, Crimea and many cities of Rumelia. Some had settled in these lands to carry on business and commercial relations. Surviving documents indicate that gold coins of Italian, Hungarian and Egyptian origin were also in circulation in the regions surrounding the Sea of Marmara.

The Court Records of Bursa contain in notarial records references to the form of coins of various origins. Our study of the internal treasury of the Emperor also enables us to determine both the kind and the quantity of gold

---

13. Feridun Bey: *Münşeâtü's-selâtîn*, Vol. I. p. 240.
14. Sahillioğlu, Id.

coins held by the Sultan and the merchants and the share of each kind in total holdings. (See Tables 1, 2, and 3.) An inventory taken in 1481 of Mohammed the Conqueror's treasury at his death reveals that there were 2,308,200 gold coins, worth 104,537,200 *âkçes*. When the holdings in silver were added, the total value of precious coins reached 346,827,030 *âkçes*. Thus, while silver *âkçes* constituted 69.38 percent of the holdings, gold coins came only to 30.14 percent. Among the gold coins, *Efrenjiye (ducats)* were the most numerous, 1,600,000 in all, accounting for 69.31 percent of all the gold pieces and 21.22 percent of the total value held by the treasury. Second were the 340,000 *Esrefiyyes* (Ashrafiah, Mamluk coins) constituting 14.73 percent of the gold coinage and 4.11 of total value. *Sultânî*, the newly-introduced Ottoman gold coin, took third rank (150,000 pieces) followed by the Engürüsiye of Hungary (70,000 pieces). (See Table no. 1.)

That the *Efrenjiyye* accounted for most of the gold coin holdings of the treasury may be attributed to the probability that some of these were in fact minted in Istanbul. The *Sultâniyyes* minted by the Ottomans needed in principle to match the standards of the Venetian *Ducat* both in weight and in metallic content. According to standards defined in the *Kanûn-nâme*, 129 ducats should weight 100 *misqals*. A *misqal* weighed 4.608 grams, hence a single coin weighed 3.572 grams. This is only 13 milligrams more than the 3.559 grams we obtain when we convert the standard Venetian and Ottoman measures to grams.[15]

15. Anhegger and Inalcik, I., Beldicéanu, I. document No. 1. The Ottoman *misqal* Ottomans used, more or less, until 1650s was 4.608 grams. Three sources of evidence exist: 1) The law code of Mehmet the Conqueror ordered that 129 *Sultânîs* had to be minted from 100 *misqals* of gold. The Ottoman *Sultânî* had the form of ducats that were minted in the Ottoman mints. 2) a ducat was 3.559 grams. A *Sultânî* could not be heavier than this. If calculated with that *misqal*, one *Sultanî* should weigh (100 × 4.608):129 = 3.572 grams. This shows that an Ottoman *Sultânî* made from light *misqal* was heavier than a *ducat* by 13 milligrams. On the other hand, if we calculate with a *misqal* of 4.810 grams used after 1650, one *Sultânî* weighs 3.729 grams. Consequently, a difference of 3.729-3.559 = 0.170 grams appears and this in turn prepares the way for the Law of Gresham to operate, i.e., the ducat and *Sultânî* were exchanged at the same rate. 3) During the reign of Süleyman the Magnificent the Ottomans replaced the Hashimî *dirham* by their unit of weight in Egypt. (See: Gaston Viet: *Memoire d'un Bourgeoise de Caire*, items *dirham* and *zirâ* in the index.) When the Ottomans changed the *dirham* they used, for some reason that is not clear. The Egyptians however insisted on using the *dirham* introduced by Ottomans. They called this the *Misrî* (Egyptian) *dirham* and named the new *dirham* that was then used in Istanbul *Rumî*—*pertaining to the Byzantine Greeks*—*dirham*. A *misqal* was 1.5 dirhams. Thus, the old Ottoman *dirham* which was also called *Misrî dirham* due to its usage in Egypt weighed 4.608:1.5 = 3.072 grams. The new *dirham* in Istanbul, on the other hand, weighed 3.207 grams. In 1690s the Sultan ordered the mint in Egypt to produce at the same standards as the mint in Istanbul had maintained. He commanded the same mint to stamp 100 *Şerifi* with 115 *Misrî dirhams* which were equal to 110 *Rumî dirhams*.

115 × 3.072 = 353.28
110 × 3.207 = 352.77
———
0.51

The difference thus calculated is negligible.

The silver *âkçe* remained the major indigenous coin in circulation between the date that the Ottoman Empire adopted bimetallism until the third quarter of the 16th Century. Nonetheless, throughout this period of nearly a century a relative abundance of gold money can be observed. Gold coins circulating in the Ottoman territories were mostly *Efrenjis* from Western Europe, Hungarian coins from Central Europe and coins of African origin coming to the Empire from trade with Egypt and North Africa. The primary source for Ottoman gold coinage continued to be that gold mined in the Balkan peninsula. Foreign coins, on the other hand, often provided additional gold from which Ottoman coins were minted. The mines in Anatolia no longer contributed significantly to the gold stocks of the Empire.

Inventories of the Ottoman treasury taken at different times show both the quantity and the percentage share of the gold holdings (appendix table n: 4). The following table shows the relative value of those various gold coins expressed in *âkçe* up to the time of Suleiman the Magnificent:

Sultan		Sultânîyye	Efrenjiyye (Ducat)	Engürüsiyye	Eşrefiyye
Fâtih	(1476)	6.60%	70.40%	2.81%	13.66%
Bayezid II	(1st)	29.54%	66.22%	—	4.24%
	(2nd)	26.39%	69.35%	—	4.24%
Selim I.	(1518)	7.82%	74.81%	5.91%	9.22%

The holdings of the internal Treasury of the Ottoman Sultans continue to show a low proportionate holding of indigenous Sultaniyye coins throughout the 15th Century. Ottoman coins in this vault approach one-third of the total only under Bayezid the First. Instead, *ducats* from Europe retain a predominant place in the treasury—rising to as high as three-quarters in the early years of the 16th century. Mamluk gold coins (*dinar*) rarely exceed ten percent of the contents. However, these figures are somewhat misleading in that it is difficult to assume that they reflect general circulation patterns. A better source is the coin proportions found in the budget of Suleiman the Magnificent from the year 1523-24 A.D.:

Sultânîyye	*Efrenjiyee*	*Engürüsiyye*	*Esrefiyye*
47.5%	30.22%	4.15%	14.12%

We have also examined the Inheritance Records of the religious courts in Bursa at the end of the 15th Century. These records indicate both the quantity and the value of the money in private holding, and thus provide some basis for comparison with the above table.

Record No.	Years	Akce	Sultânî	Efrenci	Engürüsi	Esrefî
1A	1462 A.D.	93.39%	—	6.60%	—	—
2A	1467	98.38%	—	1.62%	—	—
6A	1487-1488	65.87%	0.03%	10.56%	—	23.35%
12A	1497-1499	44.55%	—	38.27%	—	17.16%
13A	1499-1500	52.57%	4.44%	31.24%	1.20%	10.46%
31A	1510-1512	37.78%	—	8.98%	1.01%	52.05%
21A	1513	48.26%	0.32%	4.60%	0.02%	46.79%

Although foreign silver coins circulated during this early period, they did not add up to significant amounts. The records of the religious courts in Bursa refer to Iranian silver coins in debt suits. However, in these cases, usually, both the debtor and the creditor were foreign subjects. The action of the Court was confined to either the issuing or the recognition of documents employed as bills of exchange, (*Nakl-i Şehâdet* or *Kitâb-ül-Kadi*) which certified the claims of the creditor.[16] Iranian coins mentioned include those minted by the Timurids, Shahruh, Kılıç Bey and Hasan Sultan. These coins referred to as *dirhams* were identified either by the name of the ruler who had minted them or the name of the Iranian city where they had been minted.

Bursa court records also document law suits involving Bursa merchants who traveled to other Islamic lands to trade. The documents refer to trading capital carried by the merchants and list restrictions imposed by the Ottoman on the export of precious metals and coins. Often traders caught violating these regulations suffered confiscation of their assets. Most Ottoman restrictions on monetary outflows were directed toward the east. Traders caught smuggling silver bars, old *âkçe*, or precious goods who appeared in the Bursa court were either themselves from Central Asia or farther East or were Ottoman traders destined for eastern Emporia. Restrictions on the export of silver or the production of goods from silver coincide with Ottoman old *âkçe* Prohibitions or *refonte*. During the reign of Mohammed the Conqueror, as will be recalled, the weight of the *âkçe* was reduced with each refonte. Those possessing old *âkçe*, during this time, found it more profitable not to turn their old *âkçe* in only to receive new but devalued *âkçe*, but preferred to sell them in outside markets, melt them into bars or make them into silver braids. Each *refonte*, was therefore, accompanied by a set of new monetary regulations. Ottomas officials tried to prevent smuggling of silver out of the country and to control the activities of silversmiths and braidmakers.[17]

16. H. Sahillioğlu, "Bursa Kadı Sicillerinde İç ve Dış ödemelerde 'Kitâbü'l-kadi' ve 'Süftece'ler," *Kitâbü'l-kadi* (letter of the court) and *Süftece* (Exchange letter) in the internal and external payments according to the registers of Bursa Shar Court *Turkiye iktisad Tarihi Kongresi Semineri*, 8-10, June 1973, (Ankara, 1975), pp. 103-144.

17. Anhegger and Inalcik, I. Beldicéanu, I. documents No. 2, 5, and 13. Scarcity of gold in the middle of 16th century and especially in 1552 led to a ban on gold embroidery except for

Enforcement was achieved through specially appointed officers called *Yasakçı* or *Gümüş Yasakçısı*. While these regulations did remain in effect over long periods of time, enforcement seems to have been lax except around the time of a new *refonte*.[18]

The ascension of Suleiman the Magnificent to the throne, (or possibly earlier during the conquests of his father Selim I), at a time when gold was supreme and relatively abundant, marks the beginning of a new area. With the conquest of Istanbul, Anatolia and Rumelia were united. The Ottoman State had now become an Empire. These developments paved the way to a new era of monetary and metal movements.

Ottoman territorial expansion which continued during the 16th Century did not impose monetary union, however. Ottoman policy aimed at non-interference with local customs and traditions. Only those laws and practices that the new subjects had found disagreeable under their former rulers were repealed. This, it was anticipated, would facilitate the acceptance of Ottoman rule. Within the various conquered provinces the populace was allowed to use its own local currencies temporarily or permanently. This had been true even in Anatolia. The Ottomans did mint new coins bearing the name of the Sultan, but the weights and metallic content of the coins remained unchanged. For example, the coin issued in Egypt and Syria continued to be the same unit circulated by the Mamluks. Known as *nısıf fıdda* in Arabic, it is referred to as *medin* or *medain* in western sources and called first *kıt'a* and later by its Persian equivalent *pâre* by the Ottomans. It is in fact this latter name that later came to denote the Turkish word for money, *para*.[19] *Pâre*, which eventually replaced the *âkçe*, were minted not only in Syria but also in Diyarbekir and even to the north of Erzurum in the mint of Canca (now Gümüşhâne).

---

cloth embroidered for palace usage (see the *firmān* in Ahkam Defteri in Topkapi Palace Library, Koğuşlar No. 888, f⁰ 273). At the end of the 16th century silver wire-drawing was prohibited. In the 17th century silver wire-drawing was allowed only in the city of Istanbul. But at the end of this century Istanbul, Bursa and Selanik (Thesaloniki) all had silver wire-drawing factories (*simkeşhâne*). During the monetary reform of 1640, silver wire-drawing was totally prohibited for several years (BBA. Mihimme 89, *firmān* No: 181, sent to Kadi of Bursa in 27. IV. 1643).

18. In 1498 (February) Mehmed Çelebi, a merchant from Bursa, delivered 200,000 *akçe* minted in the name of Sultan Bayezid Han to Hadji Musa of Amasya and left one of his slaves (Dede) with him to go to Persia for trading. (See Bursa Court Registers no: 14 A, f⁰ 196r.)

19. *Pâre*, originally the smaller of the two silver coins minted by the Mameluk al-malik al-Muayyad Shaykh Mahmudi during the monetary reform and put into circulation in 1415 May 6th. At 14 and seven carats it was one and one-half of the *dirham shar'* in weight. Since the smaller coin's weight was 1/2 *dirham* the Egyptians named it *nısıf* (half) or *nısıf fiddah* (one half silver). From the name of the Sultan al-Muayyed the Europeans reproduced the words Medain . . . The Ottomans used first the Arabic word *kıt'a* and later its Persian synonym *pâre* (piece) to distinguish it from the Ottoman silver coin, the *akçe*. In reference to its origin the Syrians used the word *misriyyah* (Egyptian) in the singular and *masari* in the plural.

According to the *Kânûn-nâme* of Egypt, the *pâre* weighed 1.224 grams and contained 1.050 grams of pure silver, i.e., 84% silver and 16% copper.[20] During the 17th and 18th Centuries, the exchange rate between the *pâre* and the *âkçe* went up from 1.2 to 2.0 and then to 3.0. Apparently as changes were made in weight and the silver content of the *âkçe*, changes were also made in the *pâre*, but the ratios were different. Unfortunately, we are unable to document the dates and the content of all the changes. The monetary system in Ottoman Egypt also retained a gold money system. Until late into the 18th Century, Cairo, Damascus and Aleppo prepared their budgets in pâre, but also quoted the hasene (gold coin) equivalents of each item.

On the European side, Budin prepared its budgets using both the Ottoman *âkçe* and a local copper unit called *penz*. There being no Ottoman mint in Budin, local coins, Hungarian gold coins and *groschen* from Central Europe were used in circulation. In the Crimea, a different *âkçe* known as "Kefevî *âkçe*," (the *âkçe* of Kaffa) was minted.

In North Africa in the Ottoman provinces of Tripoli, Tunisia and Algeria different silver coins were minted in the name of the Sultan. Various gold and silver coins were also minted in the Yemen. That the gold content of coins minted in major cities varied suggests that not even within a province could monetary unity be achieved.

Among the various coins minted and circulated in various parts of the Empire, that of greatest importance next to the *âkçe* and the *pâre* was the *şâhî*. Originally a Persian coin, it was issued at the mints of Basrah, Baghdad, Van, Gümüşhane, Diyarbekir and Aleppo. The *şâhî* was minted exclusively for export, but was used extensively within the Ottoman economy. Until the third decade of the 17th Century, Ottoman officials made several attempts to ban internal use of the *şâhî* and to terminate its issue. Nevertheless, the coin continued to be used widely as a local currency.

In addition to the many coins already cited, silver *lârî* (*larin* in Western usage) and *mohammedî* were used for the Indian Ocean points past Basrah. The *lârî* consisted of a wire pounded or hammered in the middle, then folded. The *mohammedî* was minted into coin. Both were struck only for export purposes.[21] At first the *lârî* and *mohammedî* were struck in the hope that minting royalties and fees could be obtained from the silver that was being smuggled to Iran and the Gulf in any case. But, their minting was at times terminated because such activity was viewed as facilitating the outflow of silver from Ottoman territories.

Until the death of Suleiman the Magnificent, the Ottomans did use gold for external trade. Like its contemporaries, however, the Ottoman State

---

20. Ö. L. Barkan, *Kanunlar* I., (Istanbul, 1945), p. 386.
21. BBA. Mühimme No. 3, *firmān* No: 616, The *firmān* sent to the *Beglerberg* of Baghdad in 12.4.967 A.H./3.I.1560 A.D. For the *lârî*, or *larin* in European languages, see Tavernier; *Les Six Voyages* . . ., Vol. I., (Paris 1713), pp. 168 and 303. For a drawing of this coin see F. P. Bonneville, *Traité des monnaies d'or et d'argent*, (Paris 1806), p. 286.

behaved in a mercantilist spirit and tried to prevent an outflow of gold and silver. It therefore favored barter whereby a merchant would take goods out and return with other goods in exchange.[22] The exporting of certain goods perceived to be strategic were banned. Among these were copper used in making of cannon, horses, leather, armaments and grains and raw materials such as cotton and cotton thread.[23]

The Ottoman Empire covered huge tracts of territory. Within it were what might be called monetary areas, the *akçe* area (Anatolia), the *pâre* area (Egypt), the *şâhî* area (Persia-Iraq), and the *penz* area (Balkan). In each area, different factors and choices operated on the gold and the silver market. For example, the same foreign coin was valued differently in each region. The conditions of the silver market were also radically altered after the introduction of American silver into the markets. Generalizations made about the Ottoman monetary system, for this reason, may often be misleading.[24]

An important development occurred during the reign of Suleiman the Magnificent and continued into later periods. Ottoman financial officials tried to ensure that income from the provinces conquered by Selim I be paid in gold to the treasury in Istanbul. Egypt's annual payments were nearly 600,000 gold coins. Payments sent collectively from Aleppo, Damascus and Diyarbekir also approximate this sum. Income from Egypt became the private income of the Sultan and went into the Internal Treasury in order to be spent on items

---

22. The merchants who brought silk to Bursa were obliged to convert their proceeds into commodities. However, a merchant managed to smuggle 8.25 *okkas* (10.138 kilograms) of silver, partly in Ottoman (*osmanî*) and partly in Muscovite (*masko*) akçe (BBA., Kepeci No: 63, *Ahkâm-i mâliyye*, p. 277, September 1553). In another *firmân* sent to the Erzurum *Beglerbeg* in July 1560, the *Beglerbeg* was informed that the Persian merchants were exporting their earnings in silver coins and that nothing had been done to stop this. (BBA. Mühimme No: 3, *firmân* No. 1341).

23. Commodities forbidden to be sold to the infidels were: wheat, powder, arms, horses, cotton, cotton thread, lead, wax, Morocco leather, tallow, coarse sole leather, tanned sheepskin, sheepskin and tar (BBA. *Mühimme* No: 66, *firmân* No: 6, 997/1588-89).

24. Depending on local supply and demand and local preference for particular coins, exchange rates varied in various cities of the Empire. Often unaware of this fact, European travellers did incur losses. Some of the coins they had brought with them, though highly regarded in some regions, were not so in others.

"L'argent dont se doit charger sont les piastres d'espagne qui passent pour un écu à Marseille, à Alexandrette, à Alep et à Bagdad et non Bassora où l'on Pèse fort exactement. Il y a 7 a 8% à gagner à Alep et à Bagdad et 10 à 12% à Bassora, ce qui s'entend de toutes les piastres d'Espagne qui sont de poids et non celles du Perou, qui sont de mauvaise alloy. Il en est du même de nos Louis d'argent à Bassora, où il y a 10% à gagner.

"Pour ce qui regard l'or, il y a un notable profit à faire sur le vieu sequin de Venise et ceux de la Hongrie. On nous offrit à Lyon à six livres trois sols, que nous ne prime point, faut de scavoir ce que valoient en Turpuie. Cependant nous les eussions mis à sept livres dix sols à Alep et à Bassora."

"Cette connaissance de monnoyes est d'une telle importance . . . qu'il y en a pour s'en estre bien servis, ont sauvé plus que la dépence de leur voyage d'Europe en Bassora . . ." M. de Bourge: *Relation du voyage de monseigneur l'evesque de Beryte vicaire apostolique du royaume de la Cochinchine par la Turquie, la Perse* . . ., (Paris, 1683), p. 42.

not covered in the state budget. By thus injecting large quantities of gold into the center of the political system, the usual constraints in government spending were eased. This transfer of gold also affected the ratio of gold and silver in circulation.

African gold minted into *sultânîs* constituted a significant portion of the gold sent from Egypt.[25] Returned to circulation through government spending, Egyptian coin remained in the redistributive cycle via mercantile activity in the southern provinces. But these gold coins returned as taxes from the provinces the next year to the treasury.[26] All these gold flows facilitated internal trade. Since the gold coming from Egypt was minted there in the name of the Ottoman Sultan, it is impossible to calculate the share of gold with African origin in the annual payments.

A *Rûznâmçe* (journal) of 1547-1548 shows a carryover of 86,889,845 *âkçes* from the official imperial budget of the previous year. This sum consisted of the following:

Kind	Amount	Value in *Akces*	Percent of Total Value
Âkçe		36,021,027	41.46
Sultânîyye	662,944	37,680,802	43.37
Efrenciyye	197,207	11,240,799	12.93
Efrenciyye (Halves)	11	319	—
Engürüsiyye	25,625	1,409,375	1.62
Sakiz (Chios)	2,732	150,560	0.17
Kurona	337	16,176	0.02
Gold Bars (23,153 kgs.)	7,537	326,619	0.38
Silver Bars (34,151 kgs.)	11,117	44,468	0.05

Total revenue for the same year was 198,887,294 *âkçes*. Aleppo, Damascus and Diyarbekir together had sent in a total of 489,401 gold pieces. The revenue from Egypt, going to the internal treasury, was not included in the budget and total expenditure was 111,997,449 *âkçes*.[27]

---

25. Naturally, part of this gold returned to Egypt via domestic trade, and another part travelled from Europe as a result of foreign trade. But the major portion, according to the *Kanun-nâme* of Egypt (See, Barkan, *Kanunlar I.*, p. 386), came from Takrur. In a document dated August 1746, the persons who transported *tibr* (gold dust) from Sudan to Egypt, were called *cellâbe* or *gallabah* (drovers, transporters), and those who collected gold from the market and sold it to the mint after converting them into bars were called *muvrid* (*emprovidea*). (See BBA. Cevdet, Darphane No: 169).

26. At the end of each financial year when provinces were obliged to send their excess revenues to Istanbul, they distributed their holdings of silver coins to the merchants and artisans and obtained gold coins in return. This process of coin exchange was called *tebdil-i hasene*.

27. Ö. L. Barkan, "H. 954-955 (1547-1548) Mali yilina ait bir Osmanli bütçesi," *Iktisat Fakültesi Mecmuasi*, Vol. 19, No. 1-4 (1960).

If the current holding of the treasury can be assumed to reflect the general configuration of money in circulation, we may infer that the *âkçe* constituted nearly half (40 percent) of the coins in current use if not actually even more. The treasury valued gold coins as a more secure means of storing value. It is interesting to note the relative abundance of Ottoman gold coins. Gold coins from Sakız (Chios) can be found in records dating throughout the 16th Century. The origin of the *kurona* (Crown) is unknown, although it is likely that it came from Central Europe.

During periods when there was an abundance of gold in circulation, Ottoman gold and silver coins did not suffer a notable decline in value. Around 1565, the weight of the *âkçe* was adjusted downwards to a weight of 0.683 grams. Similarly, the weight of the gold coin was reduced to 3.544 grams in 1552 and to 3.517 grams in 1565. (A ducat weighed 3.426 grams in 1526.)

The advent of American silver to the East

The Ottoman gave the name of *guruş* (derived from the German *groschen*) to the large silver coins, weighing 9.5 *dirhams* (29.184 grams), which the Western European merchants imported in increasing numbers to the Empire. This unit, so far as we know, first appears in treasury accounts from Hungary. The treasury accounts from Budin reveal that a *guruş* was worth 100 *penz* or 50 Ottoman *âkçes* in 1554.[28] The coin is frequently mentioned in official documents of Belgrade, Timişoara, Transylvania and Valachia thereafter. Central Europe was the initial contact zone for trade with the West. These coins flowed into the remainder of the Empire from the Balkan provinces. At the beginning of the 17th Century, the Ottomans referred to Spanish *guruş* frequently as *kebir* (Large or Big) as well as *tam* or *tamâm guruş*. The term *riyal* (real) also employed furnished the name of some later Arabic currencies. The French word used to denote the same coin, *piastre*, on the other hand, has not had a lasting influence.

Until the end of the 16th Century, the Ottomans did not distinguish between large silver coins of varying size and weight, be they of Western European or American origin. They identified all simply as *guruş*. For example, in the budget of 1582-1583, entries for *guruş* valued at 55, 48, 44, 40, and 39 *âkçes* appear without giving each kind a different name.[29]

Identifying *guruş* by different names such as the Spanish *real* and the Dutch *aslanli* or *esedi* (on account of the picture of a lion it bore) is a 17th Century phenomenon. The Ottoman public, generally unfamiliar with the Latin Alphabet, was unconcerned about the origin of these coins. Some

---

28. L. Fekete: *Die Siyaqat-Shcrift in der Türkischen Finanzverwaltung*, Budapest 1955, pp. 206-279.

29. This year 522,228 gold coins were received from Egypt. On the other hand, there were revenues of 29,764,012 *akçes* from Aleppo, 7,323,408 *akçes* from Damascus and 1,542,420 *akçes* from Timişoara of which the value in gold is not mentioned. (BBA., *Maliyeden Müdevver* No. 893, The Budget of A.D. 1582-1583).

Spanish *reals* obviously weighed less and had a lower silver content. That there were actually Mexican or Peruvian *reals* and not *guruş* minted in Sevilla was not important.

The *guruş*, first gaining wide usage in Budin in 1554 and subsequently in the Balkan Peninsula, attained official recognition by the Ottoman Treasury in 1570. These silver coins became an accepted currency throughout the Empire in the last three decades of the 16th Century.[30] Nevertheless, the silver *guruş* did not occupy a major place in the Ottoman currency until 1584. As late as the fiscal year 1582-1583, *guruş* accounted for only 2.5 percent of coin received by the Treasury in revenue in contrast to gold coins which accounted for 20 percent. (Annex, Table No. 5.) It is only after the 1584 devaluation that these Spanish silver coins circulated more plentifully.

The rise of silver

The influx of *guruş* to Ottoman markets soon began to affect money circulation in this region. Difficulties began to be experienced in the trade of the Eastern Mediterranean, which had so far operated on about a million pieces of gold.[31] The demands and supplies of goods and precious metals were now reaching new magnitudes.

The *Defterdars* (Directors of Public Finance) of Aleppo and Diyarbekir found it more and more difficult to send surplus revenue of their provinces to Istanbul in gold, as they had formerly done. Previously, in common with the *Defterdars* of Damascus and Cairo, they had to convert tax revenues to gold at the official exchange rate by taking their collections to local gold dealers and merchants. However, because European merchants no longer exported gold it became increasingly difficult to exchange other money for gold. Delays of payment to the imperial center resulted. These officers began to ask for authorization to send their coin to nearby cities and villages in order that it might be traded for gold.[32]

30. According to the day-book of the treasury in 27 August 1569, 790 *sultânîyyes* (Ottoman gold coins) priced at 60 *akçe* and another 100 priced at 59 *akçe* with 282 *Efrenciyyes* (European gold coins) and 108 *guruş* (silver large coins) priced at 40 *akçe* entered the treasury. A week later a quantity of the *guruş* valued at 40, 20 and 10 *akçes* had also entered. The latter two must have been halves and quarters respectively. See BBA. Kepeci No. 1786, *Rûznamce* (day-book), p. 4, 14/3/977 A. H.

31. Over a million gold coins collected; and approximately the same amount spent. If we keep in mind that the treasury was the sole source of state distribution and collection, one may hypothesize that this amount constituted the money in circulation. Put differently, the volume of domestic trade was financed with these one million gold coins. Meanwhile, the exportation of gold for imports and the amount that was kept for savings was balanced by the import of gold through exports.

32. An inflow of abundant silver caused the gold entries to diminish. As a result, in Aleppo and Diyarbekir difficulties were experienced in finding gold to send to Istanbul. In 1577, May the 6th, the *Defterdar* of Diyarbekir could find only 150,000 *filoris* (gold coins) and requested permission to provide 40,000 gold coins from Aleppo at the exchange rate of 1 *filori* to 91 *akçes* (See BBA. *Mühimme* No. 30, *Firman* No. 254). In 1572 July 22nd, silver money was sent to be

In the Balkans, particularly in provinces near the border, tax collectors encountered difficulties in finding *âkçes*. They complained that the people did not have any money but the *penz* and the *guruş*.[33] The level of demand for silver on the part of the Iranian merchants coming to Aleppo in 1576 was extremely high. They collected all the *guruş* they could find at a rate 6-7 *âkçe* higher than the going rate and smuggled them in large quantities out of the country to Iran.[34] *Guruşes* of proper value were followed by counterfeited ones without any substantial public protest. This suggests that in the Near East the demand for silver was intense. The ratio of the price of gold to silver in Europe was in the neighborhood of 1/13,[35] but had reached a value of 1/10 in the Near East.[36] For this reason it was natural that *guruş* should be exported from Europe to the Ottoman lands. In addition to this long-distance movement of silver from Europe to Iran, there existed a parallel movement of both Ottoman and foreign specie to Iran carried out by local merchants. Whereas at one time the Anatolian merchants had purchased gold for their trade in Aleppo, they now took with them raw silver and various kinds of silver coins.[37] We can follow the eastward drainage of silver for commercial purposes from both the European and Anatolian Ottoman provinces. In order to prevent the flight of precious metals the Ottoman authorities insti-

---

exchanged for gold coins to Tarablus, Hama, Ayntab, Antakya and Sermin. Due to the Cyprus war no gold arrived from Europe to Aleppo and the *tebdil-i Hasene* became very difficult there (BBA. *Mühimme* No. 19, *firman* No. 443). Similar "lack of gold" complaints continued for twenty to thirty more years.

33. On October 12th, 1573, in a *firmān* sent to the *beglerbeg* of Budun (Buda) it was stated that the vineyards were destroyed because *guruş* was demanded from the people who were using *penz*. (BBA. *Mühimme* No. 22, *firmān* No. 653). In 1579 a new copper money called *patka*, valued at 2 *penz*, came from Transylvania and circulated in the *sançaks* of Buda, Seçen (Szécsény) and Hatvan. In 1698 two kinds of *guruş* named *murcin* and *ducat* were in circulation in Ioannina and Albania (valued respectively at 120 and 110 *akçes*) (See BBA. Cevdet, *Darphane* No. 1652).

34. BBA. Maliyeden Müdevver No. 7534, *Ahkâm-i Mâliyye Defteri*, p. 369, the *firmān* sent to the *defterdar* of Aleppo in 12.2 984 A. H./19. Dec. 1576 A.D.

35. F. G. Spooner, *L'économie Mondiale et la Frappe Monétaire en France 1493-1680*, (Paris, 1956), p. 9, and C. M. Cipolla, *Mouvements Monétaires dans l'Etat de Milan, 1581-1700*, (Paris, 1952), p. 51.

36. During various monetary reforms the ratio of precious metal to the weight of the coins changed as follows:

Year	Akçe (Silver in grams)	Sultânî (Gold in grams)	Price of Gold (in akçes)	Ratio
1550	0.731	3.559	60	1/12.32
1565	0.681	3.554	60	/11.52
1586	0.384	3.517	120	/13.10
1600	0.323	3.517	120	/11.01
1618	0.307	3.517	120	/10.47

37. See the same *firmān* in note 36.

tuted strict regulations enforced by check points installed in the Dardenelles and the Bosphorus to prevent passage of silver ore and coins.[38] These measures had a deleterious effect on the flow of trade. For example, the inhibited flow of *sof* (Mohair) from Ankara to Istanbul meant that Istanbul suffered shortages as a result.[39] In addition to these commercial shifts the Ottoman Empire suffered other unwelcome effects from the increase in production of the American silver mines. As a result of the new discoveries in America, many Ottoman mines producing both silver and gold no longer profitable were closed, together with the mints which were attached to them.[40]

When silver had been relatively less abundant, small mines, with the help of certain tax immunities and compulsory labor services, operated in various parts of Ottoman Rumelia. However, the sudden abundance and cheapness of American silver meant that operation of these mines was no longer economically practicable. Mining entrepreneurs found it more convenient to meet their official obligations for silver ore by purchasing *guruş* on the market and simply melting them down.[41]

Similarly, workers met their obligation for compulsory service to the mines by cash payments in *guruş*. Thus a great number of mines closed. The mints dependent on their production were obliged to close. Mint masters also resorted to the same expedient of obtaining silver by purchasing *guruş* and melting them down for reissue. However, the standard of these reissued coins tended to be inferior. It was impossible for the mint officers to reproduce the same amount of *âkçe* at the official rate from silver obtained in melting down *guruş*. They therefore had two options: They could either mint *âkçe* at the standard of the *guruş*, or else mint a smaller *âkçe* with a lesser silver content. In theory, *âkçe* were minted only from pure silver whereas in the *guruş* coins weighing 9.5 *dirhems*, a portion of 1 *dirhem* was composed of copper. (note: $Dirhem = 3.072$ gm.) The exchange rate for a *guruş* coin of 9.5 *dirhem* composed of pure silver was thus 80 *âkçes*. The Ottoman fiscal authorities considered a high rate of exchange for the *guruş* desirable in order

---

38. BBA. *Mühimme* No. 51, *firmān* dated Receb 991 A. H. (August 1538 A.D.) sent to the *Kadi* of Geliboli. In another *firmān* (no. 217) in the same *Mühimme*, sent to the *emīn* (intendent) of Istanbul mint it was stated that leaving aside the regular *akçe* which ought to weigh 100 *dirhams* per 450 *akçes*, even those weighing 100 *dirhams* per 470 and 480 *akçes* were hard to find. He was ordered to take new measures in order to prevent the flight of these coins to the Asian side.

39. BBA. *Mühimme* No. 78, *firmān* No. 1628.

40. Under Murad III (1574-1595) 42 mints, Mehmed III (1595-1603) 25 mints, Ahmed I (1603-1617) 20 mints and Murad IV (1623-1640) only 10 mints operated intermittently. A 1595 order sent to the *kadis* of the districts containing mines and mints stated that "Precious metals, in your districts, are not being extracted as before, some of the mines are on the edge of collapsing and some of them are not operating at all." (See BBA. *Mühimme* No. 83, *firmān* No. 229.)

41. See for the case of the mine of Ohri (Ohrid), BBA. Mâliyeden Müdevver No. 18142 *Ahkâm defteri*, p. 138, the *firmān* dated 1.10.1007 A. H./27.IV.1599 A.D. and sent to the *kadi* of Okhrida.

to attract silver to the Ottoman domains. However, following the 1584 devaluation the silver content of the 9.5 *dirhems* coin was only 8.5 *dirhems*, hence its actual value was only 66 *âkçes*. For this reason it would have been more rational to have either lowered the rate of exchange for *guruş* or adjust the value of the *âkçe*.[42] The mints which produced *âkçe* from melting down *guruş* did so without official sanction. The government, stating its opposition to the practice of debasement of the coinage, one by one closed the remaining mints which were operating in this fashion. The government, however, did tolerate the practice of arbitrage and made no attempt to stop it. Pure *âkçe* removed from the market either found their way to foreign countries or were melted for use in the jewelry trade or in the production of silver thread at the *simkeşhâne*.[43] To counter this development, new regulations were announced. In addition, the practice of clipping coins became increasingly common. Some coins even had been cut into two, three or four pieces.[44] This practice led to a further increase in the price of the *guruş* and gold coins resulting in the necessity for successive devaluation of the *âkçe* or attempted stabilizations of the currency after 1584 in 1600, 1618, 1624 and 1641 A.D.

Following the devaluation of the *âkçe* and the currency adjustments between 1584-86 all of the Empire's mints aside from the most important centers such as Cairo and Istanbul closed. Following the currency adjustments of 1584-86, the ratio of the price of silver to the price of gold in the Ottoman lands approached for a time the ratio established in Europe. However, after achieving for a time a ratio of 1: 13 after the devaluation of 1600

---

42. In 1600, of the *guruş* in circulation, *riyal guruş* contained only 72 and *hançerli guruş* (*Grosch* of dagger or sword) only 74 *akçes* weight of silver. However, their prices were 80 *akçes* in Ankara. *Ankara shar' courts registers*, No. 7, p. 194. The *firmān* sent to the *kadi* of Ankara in 7.4.1009 A. H./25.X.1600 A.D.

43. See BBA. *Mühimme* No. 3, Firmān No. 329, dated 13.2.985 A. H./21.V.1577 A.D. Gilding stirrups and trappings of horses was prohibited in Egypt because of the waste in gold. Gilding had also been forbidden several times. For example, gilding *üsküf* (Tall caps, formerly worn by the officers of the Janissaries) was banned in Istanbul in 1.V.1577 (BBA. DVN. 100). The annalist Âlî of Gelibolu estimated the waste of gold in gilding of about 200,000 to 300,000 gold coins in a year (Âlî; Nasihatü'l Mülûk, *Library of Süleymaniyye the manuscript of Fâtih* No. 3225, f⁰ 133°). The waste in silver and in gold in the factories of silver wire (the *simkeşhâne*) was higher. Therefore, the silver wire factories of Istanbul, Bursa and Selanik (Tessaloniki) were long closed. Later the Istanbul factory reopened; the others remained closed. (BBA. *Mühimme* No. 69, *firmān* No. 89, 3.5.1000 A. H./3.III.1592 A.D.)

44. *Bursa Court Registers* No. 113 A, *firmān* dated 15.7.989 A. H./16.VIII. 1591 A.D. Clipping silver from the edges of the coins after the devaluation of 1586 became a great industry. The money-changers while visiting inns and soup kitchens (*imâret*), were buying pure and heavy *akçes*, 50 to 55 of them for a gold coin, and sold 70 for a gold coin after clipping, meanwhile the official rate was 60 *akçes* for a gold coin. (BBA. *Mühimme* No. 28, *firmān* No. 7, 1576.) Merchants of Saruhan (a district in south-West Anatolia) brought loads of clipped *akçes* from Aleppo and put them into circulation in their province (BBA. *Mühimme* No. 40, *firmān* No. 581, dated September 1579).

and those following it, the ratio again found its way back to 1: 11, a level similar to that which had prevailed earlier.[45]

The *şâhî*, a silver Iranian coin, played an important role in Ottoman currency adjustments. This coin, weighing one *misqal* or 4.608 gm., although originally Iranian, was minted in the Ottoman lands as well. The mints of Bagdad, Tabriz, Basra, Van, and Diyarbekir all produced *şâhîs*. Following the devaluation carried out in the mid-sixteenth century by the Safavids, the Ottoman authorities permitted the Bagdad mint to produce similar 20 *kirats* (3.840 gm.) coin (February 1566). The value of this coin in *akçes* was maintained at a high level. Officially valued in excess of 7 *akçe*, its silver content was somewhat less than the equivalent silver contained in 7 *akçe* pieces. Therefore, we find a constant flow of silver from Anatolia to Bagdad for the purpose of minting *şâhî* coins. In similar fashion, the excess silver resulting from the profits in minting *şâhî* coins escaped from Bagdad to Iran. The exchange rate of the *şâhî* had been established in terms of *pâre* rather than *akçe*. Therefore, because in this instance the Ottomans did not follow the principle of parity, (maintenance of the same currency measure for the establishment of the relative rates of various coins) constant arbitrage forced the flow of silver towards the place of its highest value (Greshams law). Had the silver content of the *şâhî* been maintained at the exact equivalent of 7 *akçes* this flow of precious metals would never have occurred, nor would it have been necessary to adjust the value of other coins to make up for the loss. For unless the government took steps to adjust the rates of various coins, the money changers, counterfeiters, and guilds took matters into their own hands and clipped the corners of the undervalued *akçe*.

Many mints, hoping to profit from the coining of *şâhîs* from *guruş* undertook active production. A private mint even opened at Trablus-i Sam (Tripoli of Syria) which had not previously possessed a mint, but soon afterwards it was closed.[46] Fierce competition arose between the mints of Diyarbekir, Aleppo, and Bagdad for the production of *şâhî* from *guruş*. Because Aleppo was the gateway to the East for Western commerce other mints applied to the central government for permission to obtain their silver stocks from Aleppo. The master of the Aleppo mint tried to conceal from the central government the profits which he was realizing from the coining of *şâhîs*. This officer tried to prevent official permission to the other mints for *şâhî* coinage. He accused the other mint masters of smuggling the *şâhîs* which they had produced to Iran.[47]

---

45. See footnote 36.
46. Khalil Sahilî (H. Sahillioğlu), "The Monetary System in the Arab Countries under Ottomans (in Arabic)," *Faculty of Arts Journal*, University of Jordan (Amman), Vol. II., May 1971.
47. Id. and BBA. *Mühimme* No. 27, *firmân* 377, the *firmân* sent to the Beglerbeg of Aleppo in 12.9.987 A. H./16.XII.1575 A.D. A peasant claiming that he was only transporting the load to Diyarbekir, was caught near the fortress in Aleppo with 40,000 *dirhem* (123 kg.) of pure silver.

Developments between 1600-1640

Official failure to establish parity between relative rates for coins of various origins and denominations lead to devaluation of the *âkçe* on four separate occasions. At the same time most of the smaller mints were closed, accused of producing coins of inferior quality and standard. Because most of the mints besides Istanbul and Cairo were now out of operation, the Ottoman Empire became increasingly dependent on foreign coin to meet its circulation demands. As a result, the coins of foreign origin came to be accepted among the populace along side the locally produced coins. Many of the provincial *defterdars* even kept their accounts in terms of the now farmiliar Dutch (*esedî*) or Spanish (*riyal*) *guruş* in their budgets and daily accounts.[48] They sent tax receipts to the central treasury in the form of Dutch or Spanish *guruş*, but not with the local gold coin (*altun*) nor with the *âkçe*.[49] Just as counterfeiters had produced *akçe*, *şâhî*, and *pâre* coins during earlier periods of monetary crisis, so now counterfeit *guruş* began to be produced.[50]

The period 1640-1687

After 1640 most of the mints of the Ottoman lands remained inactive. Even those in Istanbul were either entirely closed or only sporadically produced coin. For this reason, there was a constant shortage of small coin in circulation, and those coins actually in circulation were either worn out from use, or sub-standard. The needs of the Ottoman Empire for small coins for circulation were met in large measure by the sub-standard coin produced in Europe for the Turkish market. One such coin was the "Louis de cinq sous" produced in the South of France. When first introduced in the Ottoman realm these coins (Turkish *sumn*) were exchanged at the rate of 1/8 of a Spanish *riyal* and enjoyed some popular acceptance. However, many of these coins were so debased and substandard that their circulation gave rise to protests as a result of which they were banned from the market. (*Sumn* of the proper standard, however, did continue in circulation for a time.) The effect of this scandal was greater in Europe than in Turkey itself.[51] For, as a result of their preoccupation with the Campaign of Crete and major wars with

48. In the budget of Baghdad in 1057 A. H./1647 A.D. the amounts were shown in *riyal* (*real*) *guruş*. Those of Diyarbakir of the same year and of Cyprus from 1643 and Aleppo from 1644 were also expressed in the same *guruş*. On the other hand, the budgets of Bassorah at the end of the XVII Century, of Taralblus (in 1670), and Damascus (in 1672) used *esedî guruş* (*guruş* of Lyon). Surplus revenues from these provinces were sent to the capital in *riyal* in the former and in *esedî* in the latter.

49. The funds of Aleppo, Damascus, Egypt and Diyarbakir were calculated in *pâre* and the equivalent calculated in gold. Surplus income sent to the capital consisted of gold. When gold in circulation diminished, this surplus was sent in *guruş*.

50. In 1580, in Ohri (Ohrid), the jeweler Gorgi was minting counterfeit *akçe*, *guruş* and gold coins (BBA. *Mühimme* No. 39, *firmān* No. 68).

51. J. B. Tavernier: *Nouvelle relation de l'interieure du Sarrail du Grand Seigneur*, Paris 1675, p. 49.

Austria and Poland the Ottomans apparently had little time to concern themselves with problems of currency adjustment. In most cases, unless counterfeiting reached intolerable proportions, the government let the production of coin take its own course without interfering. In general, official policy when Ottoman mints did operate was to reduce the weight of the *akçe* coins minted, since it was impossible under the circumstances to produce *akçe* of the old standard. This practice amounted to a de-facto devaluation of the *akçe*.

Tri-metallism (gold-silver-copper) during the period 1687-1691

From the time of Murad Bey (1352) the Ottomans produced copper coins for use in small purchases. It was a near token money since its metallic value often diverged from its exchange value. These copper coins circulated under a variety of names such as *fels, fulus, pul,* and *mankur*.[52] Because the face value of Ottoman coppers was usually beneath their metallic value, they were used only in small purchases. Circulation under these conditions can be looked on as a kind of hidden tax. Distribution of copper coin in the market was known as *salgun* (obligatory imposition). The larger of two copper coins weighed one *dirhem* (3.072 gm.) and the smaller coin was one-third of it.

At the end of the 16th Century, the exchange value of the larger copper coin was 8 per *akçe* and the smaller coin circulated at the rate of 24 to the *akçe*. Following the *akçe* devaluation in 1584, the largest copper circulated at the rate of 4 to the *akçe*.[53] In 1687, the *mankur*, two of which weighed one *dirham* and exchanged for one *akçe*, was introduced. But soon, an order was given that this *mankur* should be exchanged for one *akçe*, and no limit was put on the quantity to be produced.[54] At this point, a tri-metallic monetary system prevailed in the Ottoman empire.

The underlying cause for these short term monetary reforms should be sought in the desperate fiscal straits of the state. Pressures from failure at the seige of Vienna, unpaid back-wages of the army, and distribution of an accession largesse (*cülûs bahşişi*) amounting to a full year's wages to all

---

52. H. Sahillioğlu, "Fâtih' in son yillarinda bakír para basílmasí ve dağítílmasí ile ilgili belgeler," (documents concerning the coinage and distribution of copper coins during the last years of the conqueror) *Belgelerle Türk Iktisat Tarihi Dergisi,* Istanbul, No. 6 (March 1968), pp. 72-75.

53. In 1604, in Galata, copper coins (*füls* or *mankur*) were distributed to the tradesmen according to the rule "one for two *pul* (supposedly small copper coins) four for one *akçe*." These weighed one *dirham* as previously. Because the *akçe* was reissued, its silver content declined by 50%, the copper coin also had to be adjusted at the same ratio. But this adjustment was not made by weight but by value. *Mankur*, previously at 1/8 *akçe* or one *pul* was now worth 2/8 or 1/4 *akçe*, or two *puls*. The adjusted *akçe* was now half the value of the old *akçe* which was worth 4 *mankurs*. This change retained a habit of more than one century, of exchanging one *akçe* for 8 *puls*.

54. Fındfklı Mehmed Ağa, *Silahdar Tarihi,* Vol. II., p. 306; *Râşid, Tarih-i* Râşid, Vol. II, p. 61.

officials paid cash salaries following the deposition of Mehmed IV in favor of Süleyman II, made such measures inevitable.[55]

Before his deposition, Mehmed IV had been considering monetary reforms. He engaged the services of a Venetian convert named Cerrah Mustafa to construct machines for use in cutting coins (*rakkas pendulum balançoire*).[56] These coin presses were first put to use during an inflationary period. As a result of the silver inflation the copper value of the coins was greatly in excess of their face value. For example, one *okka* (400 *dirhems*) of copper (1.288 kg.) was worth 110 *âkçes* in silver (apart from the losses due to oxidization and the minting of coins.) Thus the mint realized a profit of 625 *âkçe* from coining one *okka* of copper. For this profit to be realized it was necessary that a sufficient quantity of copper be put into circulation. From the date when the coining machines were fully introduced, a quantity of 80,000 *mankurs* could be produced in a single day and put into circulation. In the new mint house with additional machinery 300,000 *mankurs* and starting with 21.VIII.1686 620,000 *mankurs* were produced.[57] It is probable that these figures do not include the fee for the use of the coin press.

When in 1687 the price of copper reached the same level as the price of silver (the copper *mankur* and the silver *akçe* exchanged at the same rate), production of copper coins by means of the new coin presses was an extremely profitable enterprise. Consequently, a flourishing counterfeit coin industry sprang up in Rumelia. Europe, too, following suit, sent secretly produced copper coins into Ottoman circulation by the ship load.[58] The inflationary experience took three years but was stopped before its dispersal. Disturbances occurred due to the refusal of the central treasury to accept tax payments in the form of the copper *mankur* coins it had itself issued. Finally, the treasury compromised by agreeing that 1/3 of all payments would be accepted in copper coin, the other 2/3 to be paid in either silver *guruş* or gold *altun* coins. Tables No. 6 and 7 show that in the year 1101-1102 A. H./1690-91 A.D. payments were made to the treasury in copper coin (*mankur*) as opposed to other coin at a ratio of approximately 1/3. In other words, between 32-39 percent of all payments were made in terms of the copper coin. During the period of the copper inflation in the Ottoman Tri-metallism experience, a new Ottoman silver *guruş* was also introduced into circulation. In another expedient the treasury melted down unused gold objects belonging to it and issued a new gold coin (the *şerifi*).

---

55. Ibid.
56. BBA. Ibnülemin. *Darphane-Maden* No. 96.
57. Ibid. BBA. Mâliyeden Müdevver No. 3427, *Kuyûd-i Mühimmât Defteri* pp. 323 and 524.
58. BBA. *Mühimme* No. 100, *firmān* No. 453. In this *firmān* the *Kadis* of the Aegean and Marmara ports were warned against the French who were trying to bring in counterfeit *mankurs*. The French ambassador Castagnère de Chateauneuf writes in a letter dated 25.VIII.1690 that Sieur Fabre loaded *mankurs* to a ship in Marseilles. AN. Aff. Etr. Bi. 381.

The period following 1691

When, after 1691, the Ottomans renewed production of the new silver *guruş* and gold coins, they had achieved a near parity with the widely current coin of European manufactured standard. We noted earlier that the first Ottoman gold pieces were struck at a standard equivalent to the Venetian *ducat*. Similarly the new Ottoman gold coin, the *şerifi*, matched the Hungarian gold coin.[59] The *guruş* modelled on the *esedi*, its Dutch counterpart, omitted only the lion on its reverse side. The Ottomans made even greater use of the *zolota*, originally a Polish coin, 2/3 the weight of the *guruş*, and later reduced in value to 3/4 its weight.[60] According to the reports of the French ambassador in Istanbul at the time, the *zolota* was produced in Holland as well.[61]

In 1696 Sultan Mustafa II considered new money reforms. By enforcing a monetary union within the empire he could stop circulation of all foreign coin. This move would end the profits of foreign nations who used the Near East as the market for a flourishing trade in counterfeit coin.[62] However, the outbreak of war prevented realization of his plans.

Differences between the metal markets and minting costs in Egypt and Istanbul made it difficult to issue gold coins at the same standard in Alexandria and Istanbul. The gold coins of the Istanbul mint were issued at a higher standard while the Egyptian mints obtained permission to issue gold coins of a lower standard. The difference in standard amounted to about 1/115. However, by issuing gold coin of a standard well below the allowed 1/115 difference, Egypt managed to attract the gold coin of Istanbul as well. The gold standard was changed year after year and reissued to circulation under a new name. In 1703 it was called *tuğrali*, in 1713 *zencirli*, in 1716 *fındık*, and in 1729 *zer-i mahbub*. But on each occasion, the difference between the standard of the Egyptian and Istanbul minted gold coins continued and arbitrage thus acted to equalize the difference.[63] Low standard coins from the Egyptian mint found their way to Istanbul, while Istanbul-minted gold coins flowed towards Egypt.

Mustafa II hoped to attract circulating gold coin to the Istanbul mint. He was especially concerned to prevent circulation of the Venetian gold *ducat* or

---

59. The Ottoman *şerefi* gold and the Hungarian gold coin was equivalent to 270 good *akçe* or 360 sunk *akçes*. The Venetian gold coin was equivalent to 300 good or 400 sunk *akçes*. Henceforth, Ottoman gold coins were introduced into circulation under different names and standards. All were modelled after Central European gold coins.

60. The *zolota* was equivalent to 80 *akçes* until 1717 and 90 after this date. On the other hand, the *guruş* was always 120 good (160 sunk) *akçes*. *Zolota* was first worth 2/3 and later 3/4 of a *guruş*. When minted to the standard of 2/3, half pieces of it were also put into circulation.

61. AN. Aff. Etr. Bi. 381. The letters of the French ambassador C. de Chateauneuf and evaluations of some Frenchmen who were trading in Istanbul are worth citing. "According to the "Copie de l'avis de Sieur Maynard" the *isolotte (zolota)* was an alloyed German coin exported to the Levant.

62. BBA. Maliyeden Müdevver No. 3419, *Kuyûd-i Mühimmât Defteri* p. 551.

63. BBA. *Mühimme* No. 110, pp. 94-95. The *firmân* sent to Egypt at the end of 1108 (1701 mid-May) is one of many examples which describe similar situations.

yaldız. The latter was nearly pure gold while the Istanbul *şerifi* contained a small quantity of silver and some copper. Anyone bringing 100 gold *ducats* to the Ottoman treasury received 102 *şerifi* gold pieces in exchange.[64] In riposte, the Venetians began to mint a separate *ducat* for the Near East Market.[65]

As can be seen from the tables showing the sums entering and expended by the treasury during the last two years of the five year tri-metallic period, apart from the *frengi* gold piece all of the coins listed were of local provenance. Even the *esedî*, *guruş* and *zolota* were now produced locally. Among the several types of coins entering the treasury in the year 1110 H./1698 A.D. a small quantity of Hungarian gold coins was included (see table 6, 56 coins marked "M" for Magir.)

The term *mamul hasene* is a term applied to the Ottoman gold coins. In fact it was a fictitious money of account artificially fixed at the rate of 270 *âkçes* in much the same way as the *livre* was used in France as a money of account. Included in the *mamul hasene* figures for the year 1110H. are a quantity of 202 and 142 *eşrefi-i cedid* (new *eşrefî*) that is the Ottoman gold pieces with the Imperial Seal (*tuğra*). Included in the *esedî guruş* figure for the same year are coins labelled as *atik* (old *esedî*) (table 6.) Starting with the 18th century one comes across *guruş* of European origin in increasing numbers recognized under various names. Most of these are of central European provenance.

An important new development was the increasing use of metal of Anatolian origin in the minting of gold and silver coins. The mines at *Gümuşhâne* and *Keban* and their immediate surroundings constituted an important part of the raw materials used by the mints. During the years 1730-1740, every year between 2 and 4 million *dirhams* of pure silver were produced at the mine in Keban. In 1730, 5,722 kg. of silver were produced while in 1736 the level had risen to 12,771 kg. At the same time, in the year 1730, 15,484 kg. of silver were produced at the Ergani mine while in 1735 this increased to 975 kg. The gold production at this mine for the year 1730 amounted to 1,976 kg. At the mine of Espiye in the year 1737 a quantity of 41,574 kg. silver and 144 kg. gold was produced. Although great efforts were made to make the levels of production of Rumelian mines reach what they had been in the sixteenth century, they failed to equal even the levels of metal production in Anatolia at the time.

---

64. BBA. Mâliyeden Müdevver, No. 3462, *Kuyûd-i Berevat . . . Defteri*, p. 131.

65. Holland and England traded in the Near East with false *guruş*. France, in 1685, minted false *sequins* (Venetian *ducat* known as Yaldız in the Levant) outside France and blamed the Venetians (AN. Aff. Etr. Bi. 235, *Proposition du débit . . . des sequins qui se batte à la Monnoye d'Orange*). These coins were called "Ibrahim Çelebi" in Istanbul according to the Annals of Silahdar. But the coins which were minted in Venice itself and put into circulation in the Levant in 1696 were given the name "Rouspi" (Prostitute) rather cynically (AN. Aff. Etf. Bi. 1046, the letter dated 11.III.1696, f⁰ 348).

In summary, it can be said that the Ottoman lands remained as an important stage in the path of East-West trade, and played a significant role in the money system and changes in the values of currencies through time. Throughout the course of internal and external monetary history one observes the effect of Gresham's law. The Ottoman Empire succeeded in establishing one of the most forceful states in world history. But despite this, its currency never achieved the status of a universal standard as had the money of both the early Islamic States and the Byzantine Empire. The money of the Venetian Empire on the other hand managed to achieve the status of a "pilot" currency even within the boundaries of the Ottoman Empire.

Translated by Prof. Dr. Ilter Turan and Dr. Rhoads Murphey

Figure ... Imperial Treasury ... which were transported to the new treasury building under the care of Mehmed Paşa (after **A.H.** 881/A.D. 1476)

Coin type	Number of coins	ER*	Akçes	Percentage a^1	b^2	c^3
Gold						
Efrenijyye (Venetian g.m.)	1,600,000	46	73,600,000	70.40	69.31	21.22
Eshrefiyye (Egyptian g.m.)	300,000	42	14,280,000	13.66	14.73	4.11
Sultânî (Ottoman g.m.)	100,000	46	6,900,000	6.60	6.49	1.98
Mixed						
Eshrefiyye	30,000					
Engurusiyye (Hungarian)	70,000	42	2,940,000	2.81	3.03	0.84
Total	2,100,000					
Sultânî	50,000					
Eshrefiyye	10,000					
Total	2,160,000					
Hasene-i Haşt (?)	148,200	46	6,817,200	6.51	6.42	1.96
Total	2,308,200		104,537,200	99.98	99.98	30.14
Silver						
Muhammed Hânî (great silver coins)	163,983	10	1,639,830	0.47		
Akçes (small)	240,650,000	1	240,650,000		69.38	
Total			346,827,030			99.96

*Exchange rate
[1] The ratio of the value of the kinds of gold coins to the total value of the gold holdings of the treasury
[2] The ratio of the number of kinds of gold coffer to the total number of gold coins
[3] The ratio of each kind of currency to the value of the total holdings of the treasury

293

Table 2. The ratio of gold and silver money in circulation in late 15th and early 16th centuries according to inheritance registers of Bursa.[a]

Regis-ter No.	Date	Total No.	Containing specie No.	Containing specie %	Silver Total shares No.	Silver Total shares %	Gold Total shares No.	Gold Total shares %	Sultânî No.	Sultânî %	Esrafiyye No.	Esrafiyye %	Efrenjiyye No.	Efrenjiyye %	Engürüsiyye No.	Engürüsiyye %
1 A 1462.	III–1462. XII	164	66	40.24	64	39.24	8	4.87	–	–	–	–	8	4.87	–	–
2 A 1467.	VIII–1467. X	395	124	31.38	121	30.63	10	2.53	–	–	–	–	10	2.53	–	–
6 A 1487.	V–1488. XII	450	154	34.22	145	32.22	31	6.88	2	0.44	27	6.60	23	5.11	–	–
12 A 1497.	III–1499. V	745	217	29.12	208	27.91	29	4.16	–	–	19	2.55	26	3.48	–	–
13 A 1499.	X–1500. VI	250	53	21.20	48	19.20	23	9.20	2	0.80	15	6.00	20	8.00	3	1.20
165 A 1501.	X–1503. X						85		3		42		36		4	
31 A 1510.	I–1512. II	210	38	18.09	21	10.00	26	12.38	–	–	25	11.90	3	1.42	4	1.90
22 A 1513.	VIII–1513. IX	286	78	27.27	67	23.42	43	15.43	3	1.05	40	13.98	16	5.59	2	0.69

[a]Based on the number of inheritances containing money

Table 3. The relative importance of gold and silver money in circulation in the late 15th and early 16th century according to the inheritance registers of Bursa (based on the akçe value of kinds of specie in inheritances)

Register No.	Total value	Akçes Quantity	%	Gold coins Value (akçes)	%	Sultani Quantity	Value (akçe)	%	% of total gold coins	Eshrafiyye Quantity	Value (akçe)	%	% of total gold coins	Efrenjiyye Quantity	Value (akçes)	%	% of total gold coins	Engürüsiyye Quantity	Value (akçe)	%	% of total gold coins
1 A	360,906	337,083	93.4	23,823	6.6	–	–	–	–	–	–	–	–	584	23,823	6.6	100.0	–	–	–	–
2 A	179,724	176,531	98.4	3,893	1.6	–	–	–	–	–	–	–	–	91	3,893	1.6	100.0	–	–	–	–
6 A	639,588	421,355	65.9	217,157	34.1	4	194	0.1	0.1	3,176	149,367	23.4	68.8	1,386	67,594	10.6	31.1	–	–	–	–
12 A	1,451,773	646,786	44.6	804,987	55.4	–	–	–	–	4,795	249,252	17.2	31.0	10,300	555,735	38.3	69.0	–	–	–	–
13 A	198,044	104,113	52.6	93,799	47.4	163	8,802	4.4	9.4	399	20,718	10.5	22.1	1,151	61,887	31.2	66.0	46	2,392	1.2	2.5
165 A	–	–	–	821,954	–	709	37,617	–	4.6	9,636	483,275	–	58.8	4,936	279,684	–	34.0	405	21,372	–	2.6
31 A	74,548	28,164	37.8	46,260	62.2	–	–	–	–	816	38,808	52.1	83.9	120	6,695	9.0	14.5	17	717	1.0	1.6
22 A	463,948	223,911	48.3	240,065	51.7	28	1,490	0.3	0.6	3,805	217,089	46.8	90.4	391	21,380	4.6	8.9	2	106	0.0	0.1

295

Two inventories made during Bayezid II's reign (1481–1512)

Table 4. Inventories of the internal treasury

Coins	1st	%	2nd	%
Sultânî	556,000	29.54	496,900	26.39
Efrenci	1,246,880	66.22	1,305,900	69.35
Eshrefi	80,000	4.24	80,000	4.24
Total	1,288,880	100.00	1,882,800	99.98

Table 4a. Inventories of the internal treasury

Inventory made under Selim the first the 6.I.1518

	Quantity	Value Akçes	%[a]	%[b]
A. Silver				
Akçes (osmani)	9,880,325	9,880,325		7.31
Akçes (Halebî, of Aleppo)	19,119,850	38,239,700		28.30
Gümüs (silver) sultaniye	10,000	100,000		0.07
Akçe of Mekka	113,875	113,875		0.08
Silver ingot, dirhems	5,885	23,540		0.01
B. Gold	1,582,336	86,747,621		64.21
1. Sultâniyye	120,325	6,782,875	7.82	5.02
2. Efrenjiyye	1,180,496	64,927,280	74.85	48.05
3. Engürüsiyye (Hungarians)	98,563	5,125,276	5.90	3.79
4. Eşrefiyye (Egyptians)	160,000	8,000,000	9.22	5.92
5. Mağribî eşrefiyye (North Africans)	115	5,750	0.006	0.004
6. Eşrefiyye of Mekka	8,175	408,850	0.47	0.30
7. Eshrefiyye (Pâre)	11,233	561,650	0.65	0.41
8. Takyanos (Antique gold)	75	4,500	0.005	0.003
9. Golden akçes	532	10,560	0.01	0.007
10. Golden pâre	23,022	920,880	1.06	0.68
Total		135,105,061	99.99	99.98

[a]The ratio of each kind of gold coin to the total amount of gold in the treasury.
[b]The ratio of each coin to the total value of the treasury.

297

Table 5. The kinds of money entering the treasury according to the Budget of 990–991 (1582–1583). The carryover to the next one and their ratios to the total revenue on one hand and to the total carryover on the other.

	The Revenue			Carryover		
	Quantity	Value in akçes	%	Quantity	Value in akçes	%
A. *Silver coins*						
Akçes		182,746,348	58.25		18,953,015	52.41
B. *Gold coins*	**1,090,352**	**64,361,561**	**20.51**	**235,281**	**13,870,234**	**38.35**
Sultânî and Efrenji						
Efrenjiyye and Sultaniyye	1,074,159	63,440,916	20.22	229,919	13,565,221	37.51
Sakiz (Chios g.m.)	300	17,000	0.00	300	17,000	0.05
Engürüsiye (Hungarians)	15,056	858,192	0.27	5,040	287,280	0.79
Debased coins (Nakis)	794	43,670	0.02	–	–	–
Korona (Crowns) and Tarablus (Tripolitans)	21	1,050	0.00	–	–	–
Selimî (G.c. of Selim II.)	1	35	0.00	1	35	0.00
Half coins (Nisif)	1	29	0.00	1	29	0.00
Hasene-i nukre (1/4 ?)	20	269	0.00	20	269	0.00
C. *Gold and Silver bars,*						
Guruṣh (Piastres) and Ṣâhis		**61,723,966**	**19.67**		**3,342,641**	**9.24**
Gold bars 11.535 grs		172,335	0.05	Grs.1,468.5	32,426	0.09
Silver bars 82.754 grs		121,194	0.04	Grs. 418	612	0.00
Gurush (Piasters)	196,159	7,693,180	2.45	66,869	2,607,891	7.21
706 coins of 50 akçes						
430 coins of 48 akçes						

Table 5. (*continued*)

	The Revenue			Carryover		
	Quantity	Value in akçes	%	Quantity	Value in akçes	%
3,635 coins of 44 akçes						
13,168 coins of 40 akçes						
178,220 coins of 39 akçes						
Şahis	7,663,712	53,669,752	17.11	90,601 × 7 akçes	634,207	1.75
7,616,175 × 7						
47,537 × 7.5						
Pâre-i Halebî (of Aleppo)	459	688	0.00	459	688	0.00
Penz (Small)	102,778 × 0.5	51,389	0.02	102,778 × 0.5	51,389	0.17
Penz (Large)	7,830 × 1.5	11,745	0.00	7,830 × 1.5	11,745	0.00
Silver chains and bit		3,683	0.00	–	3,683	
D. *Difference in exchange rates of gold coins and piasters*		**4,912,497**	1.57	–	–	–
Total		**313,744,372**			**36,165,890**	

Table 6. Entries into the treasury (revenue) 1101–1105, 1110 (1689–1694, 1698)

Years A.H.	Efrenciyye	Mamul Hasene	P. of Lion	Zloty	Pâre	Akçe	Mankur	Total
A. Amount of kinds of money coming into the treasury								
1101	214,345	119,715	138,593.5	–	37,179,365	90,000	146,010,396	–
1102	270,227	274,792	262,812	–	64,353,867	9,552,700	192,408,621	–
1103	281,984	288,874	185,413	266,731.5	10,853,077	7,371,723	141,946	–
1104[a]	53,555	192,300	107,859.5	89,037	3,329,726	176	–	–
1105	144,309	506,900	461,834.5	44,625.5	6,600,204	389	–	–
1110[b]	7,400F	202,142EC	844,319.5A	1,418,895.5	4,068,710	691	–	–
	56M	211,177M						
B. Value of money expressed in akçes								
1101	64,303,500	32,323,050	16,631,220	–	111,538,095	90,000	146,010,396	370,896,261
1102	81,068,100	74,193,840	31,537,440	–	193,061,601	9,552,700	192,408,621	581,822,302
1103	112,793,600	103,994,640	29,669,280	28,451,440	43,412,308	7,371,723	141,946	325,834,937
1104[a]	21,422,000	69,228,000	17,257,520	9,497,280	13,318,904	176	–	130,723,880
1105	57,723,600	182,484,000	73,893,520	4,760,240	24,400,816	389	–	343,262,565
1110[b]	2,960,000	80,856,800	135,091,120	151,348,880	16,274,840	60,024	–	462,635,444
	20,160	76,023,720						
C. Percentage of value to the sum of moneys entering the treasury								
1101	17.37	8.71	4.48	–	30.17	0.03	39.37	100.00
1102	13.93	12.75	5.42	–	33.18	1.64	33.07	100.00
1103	34.62	31.92	9.11	8.73	13.32	2.26	0.04	100.00
1104[a]	16.38	52.96	13.20	7.27	10.19	–	–	100.00

300

Table 6. (*continued*)

Years A.H.	Efrenciyye	Mamul Hasene	P. of Lion	Zloty	Pâre	Akçe	Mankur	Total
1105	16.81	53.16	21.52	1.38	7.11	–	–	100.00
1110[b]	0.64	17.48	29.20	32.71	3.52	0.01	–	100.00
	0.004	16.43						

[a]Three months.
[b]Ten months.

Table 7. Outflows from the treasury (expenditures) 1101–1105, 1689–1694

Years A.H.	Efrenciyye	Mamul Hasene	P. of Lion	Zloty	Pâre	Akçe	Mankur	Total
A. Different moneys used in payments
1101	176,011	177,881	182,776.5	–	68,190,404	1,700,235	156,953,117	–
1102	259,649	279,992	276,662.5	–	16,558,360	58,490	171,014,199	–
1103	297,829	303,167	247,210.5	155,803.5	10,226,009	10,844,416	–	–
1104[a]	43,741	82,523	55,535	21,967.5	1,779,599	6,484	–	–
1105	256,767	425,931	483,194	42,855.5	8,484,119	828	–	–
B. The value of payments expressed in akçes
1101	52,803,300	48,027,870	21,933,180	–	204,761,212	1,700,235	156,953,117	486,178,914
1102	77,894,700	75,597,840	33,194,700	–	139,675,080	58,490	171,014,199	497,435,009
1103	119,131,600	109,140,120	39,353,680	16,619,040	40,904,036	10,844,416	–	336,191,892
1104[a]	17,496,400	29,708,280	8,885,680	2,343,200	1,118,336	6,484	–	65,558,380
1105	102,706,800	153,335,160	77,311,040	4,571,280	33,936,476	828	–	371,861,584
C. Ratio of value to the sum of expenditures
1101	10.86	9.88	4.51	–	42.12	0.35	32.28	100.00
1102	15.66	15.20	6.67	–	28.08	0.01	34.38	100.00
1103	35.44	32.46	11.77	4.94	12.17	3.22	–	100.00
1104[a]	26.69	45.32	13.55	3.57	10.86	0.01	–	100.00
1105	27.62	41.23	20.79	1.23	9.13	–	–	100.00

[a]Three months.

Ottoman monetary history 1300-1750

Table 8. Changes in the weight of Akçe 1326-1340

Years	Coins/100 dirham	Weight gr
1326	266	1.152
1431	260	1.181
1460	330	0.931
1480	400	0.768
1492	420	0.731
1565	450	0.681
1586	800	0.384
1600	950	0.323
1618	1,000	0.307
1624	1,000	0.307
1640	1,000	0.307
1659	1,250	0.257
1666	1,400	0.229
1685	1,250	0.257
1688	1,700	0.188
1696	1,900	0.169
1697	1,800	0.178
1699	2,300	0.139
1705	1,900	0.169
1740	–	0.180
		0.135

Table 9. Changes in the weight and silver content of Pâre 1524-1740

Years	Grade %	Weight of 1000 coins (*dirhams*)	Gross Weight gr.	Silver content gr.
1524	84	400	1.224	1.050
1564	70	340	1.054	0.731
1685	70	240	0.769	0.539
1687	70	230	0.738	0.516
1696	70	220	0.705	0.491
1705	68	200	0.641	0.436
1720	60	181	0.580	0.349
1740	60	175	0.561	0.336

Figure I: Rates of exchange, akçes 1640-1740.

1) Rate of exchange: ducat in akçes
2) Rate of exchange: Şerifi (Sharifi, Ottoman gold money) in akçes
3) Number of akçes per 10 dirhams (30.72 g) silver.
4) Price of wheat.

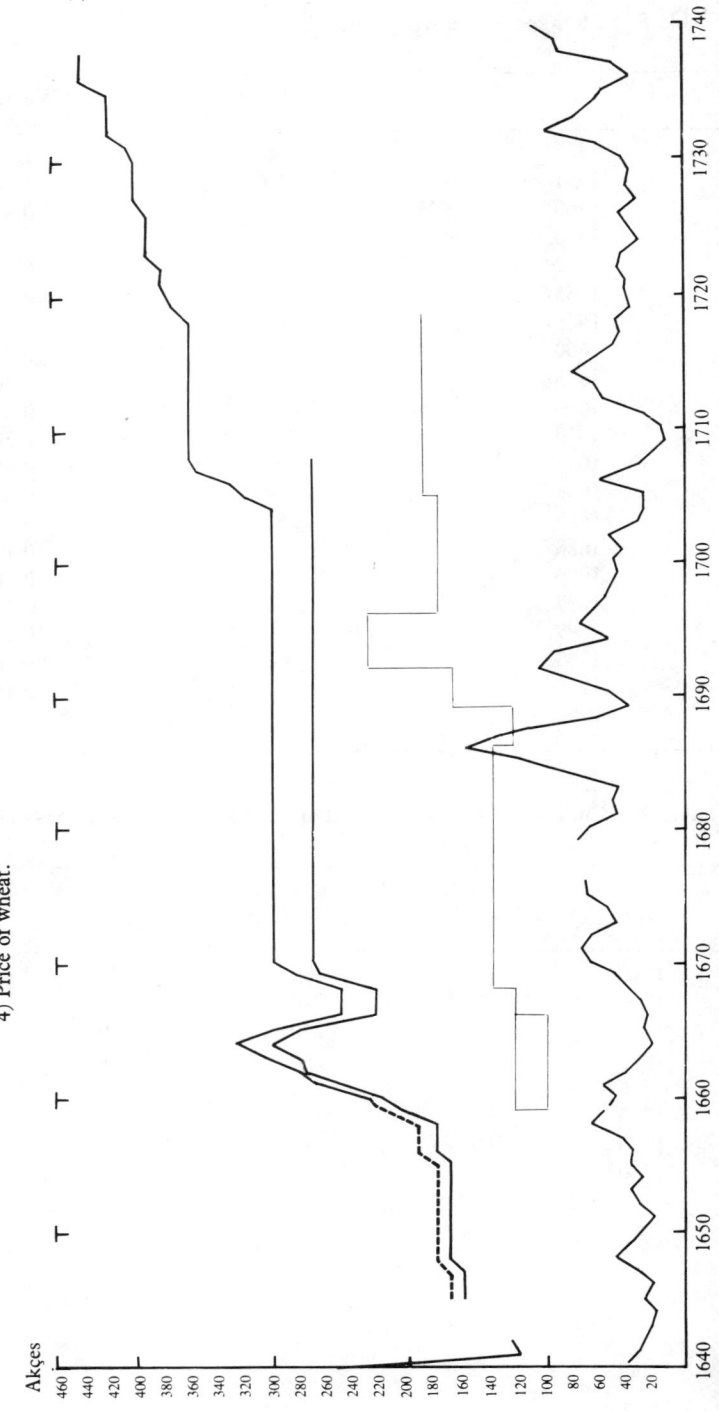

10

Gold and silver exchanges between Egypt and Sudan, 16th-18th centuries

TERENCE WALZ

The impact of gold mined in the Western Sudan on the medieval Egyptian economy—and by extension on the entire Middle East—has in recent years been likened to the effect of American gold on the economy of 16th-century Spain and Europe. Consequently, the arrival in the late 15th century of the Portuguese on what became known as the Gold Coast has been seen as catastrophic for the trans-Saharan gold trade as their arrival in India was for the Middle Eastern spice trade in general. None of these parallels is particularly satisfying. On the one hand it has become increasingly evident that the trans-Saharan gold trade functioned independently of European activities along the west coast of Africa, and on the other that the impact of the Portuguese in India on the Middle East has been exaggerated.[1] As we are concerned with the modern period, attention must also be focused on gold sources in the Eastern Sudan and Ethiopia which were opened or became available to Egypt for the first time.

Since the movement of precious metals into and out of Europe is marked by deliberate secrecy and obscurity,[2] it is not surprising to know that similar movements between Africa and the Middle East are also obscured. Nonetheless certain patterns can be traced for the period 1500-1750 and they will be etched below in order of the intrinsic value of the metal concerned.

Coinage in Ottoman Egypt

First, however, a preliminary word about coins in Ottoman Egypt. Unlike the Iranians, the Turks succumbed to Islamic tradition by issuing a series of gold coins. Until the end of the 17th century, the only Ottoman gold coin minted and in use in Egypt was the *sharīfī muhammadī*, known colloquially

---

1. See Halil Inalcik, "The Ottoman Economic Mind and Aspects of the Ottoman Economy," in M. A. Cook, ed., *Studies in the Economic History of the Middle East* (London, 1970), p. 212; Roger Owen, "The Middle East in the 18th Century—An 'Islamic' Society in Decline?", *Review of Middle East Studies*, 1 (1975), pp. 101-12.
2. Thus gold has been called "a silent factor", quite appropriately for this study. See Frank C. Spooner, *The International Economy and Monetary Movements in France, 1493-1725* (Cambridge, Mass., 1972), p. 26.

as the *ashrafī*, which weighed 3.448 grams with a fineness of 996, thereby resembling the Venetian ducat (known as the *bunduqī*) and the famous *ashrafī* of Sultan Barsbay. In 1680 the Venetian coin enjoyed a higher value than did the *sharīfī* (105 *nisf fidda* versus 95) in the Cairo market, and in the closing decades of that century its value continued to rise due to the apparently decreasing gold content of the Egyptian-minted coin. The *sharīfī* ceased to be minted in 1697 and was replaced by the *tughralī*, weighing 3.542 grams and equivalent, once more, to the ducat. However, the *tughralī* rapidly depreciated, though it continued to circulate until mid-1730's. For the weights of these coins as minted in Istanbul and years in which they were introduced, see the contribution by Dr. Sahillioğlu.

In 1707 the Egyptian mint was ordered to issue a new gold coin, the *zingirlī* ("chainies"), which had been introduced ten years earlier in Istanbul and whose weight in Egypt appears to have been 3.508 grams. It was valued at 107 *nisf* while the venerable ducat stood at 115 *nisf*. According to Raymond, the best authority on coins and coinage in Ottoman Egypt,[3] one million *zingirlīs* with a reduced gold content of 3.36 grams were minted in 1725.

That year a new coin, the *findiqlī*, appeared; half and double *findiqlīs* were also minted. Also modeled after the ducat, its fineness was 23 *qīrāts* and was valued on a par with the Venetian coin at 134 *nisf*. By 1730 its value had risen to 180 *nisf*, along with the ducat. As with most Egyptian gold coins, debasing and clipping soon took place and by the end of the 18th century the *findiqlī* contained 23 percent less gold than did the original 1725 issue. It should be noted here, as Dr. Sahillioğlu points out in his article, that Egyptian minted coins were allowed to have less gold content that Istanbul mints.

Orders were received by the Egyptian authorities to mint a new type of gold coin in 1736 which was called the *zar mahbūb* or *mahbūb* ("beloveds"). It weighed a quarter less than the ducat (2.598 grams versus 3.494) and was valued at 110 *nisf*. Half and quarter *mahbūbs* were also struck. This occurred at a time of general worldwide shortage of gold and silver. The *mahbūb* widely circulated throughout the remainder of the 18th century and is often mentioned in local Cairo court records. In fact, it was the only gold coin minted in Egyptian after 1754. Its weight, however, inevitably dropped and by 1798, when the French occupied Egypt, the *mahbūb* weighed only 71 percent of its original weight. But because it was one of the few gold coins circulating in Cairo, its market value continuously rose. During the three

---

3. André Raymond, *Artisans et commerçants au Caire au XVIIIe siècle*, 2 vols. (Damascus, 1973-74), I, 30. We have relied heavily on this work, especially Chapter 1, "Les Monnaies," I, pp. 17-52; for other works dealing with Egyptian currencies, Stanford J. Shaw, *Ottoman Egypt in the Age of the French Revolution* (Cambridge, Mass., 1966); E. W. Lane, *Manners and Customs of the Modern Egyptians*, Everyman Edition (New York, 1963), p. 579; Samuel Bernard, "Mémoire sur les monnaies d'Egypte," *Description de l'Egypte, État moderne*, 2nd ed., vol. XVI (Paris, 1825), pp. 267-506.

years (or 33 months) of the French occupation, 261,727 *maḥbūbs* were issued by the imperial mint at Cairo.[4]

The basic silver coin was a paper-thin para or *nisf fiḍḍa*, resembling the Turkish asper. Originally it weighed 1.289 grams, or less than half the traditional weight of the traditional silver *dirham* (2.975 grams), but it began to depreciate almost immediately. By 1750, the *nisf* weighed 0.342 grams and by 1798, 0.079 grams. The debasement of the *nisf fiḍḍa* throughout the whole Ottoman period in Egypt is a central historical and monetary theme. The only other silver coin minted in Egypt during this period was the short-lived piaster (*ghirsh*) which appeared during the reign of Alī bey al-Kabīr (1769-72). It weighed 15 grams and was valued at 40 *nisf*. When Alī bey was deposed, the coin was withdrawn from circulation.

Finally, a copper coin, the *jadīd*, also circulated. Its weight approached the *dirham* but varied greatly. It was valued at 8 per *nisf fiḍḍa* during the 17th century, falling to 10 and 12 per *nisf* in the 18th. With the great depreciation of the *nisf*, its value at the end of the century was negligible.

The preponderance of high-value locally minted gold coins coupled with an unstable lesser-value silver coin resulted typically in the wide acceptance of imported foreign coins. Foreign silver coins in fact became the most commonly used coins in Egypt. Beginning in the 17th century, they included the Dutch rixdale (called *Abū Kalb* or *asadī*, therefore referring to the liothaler, which was actually lighter in weight) and though heavily debased by copper — up to 50 percent of their metallic content — the Egyptians trusted them as much as they trusted the Spanish dollar which had a higher silver content. The appearance of false rixdales in the 18th century undermined public confidence and they disappeared from circulation around 1730.

The Spanish (Sevillian or Mexican) dollar supplanted the rixdale; foreign consuls at the end of the 17th century noted their wide acceptance in Egypt. Known as the *riyāl Abū Madfaʿ* or more commonly *riyāl Abū Masht* because its pillars resembled cannons or combs, vast numbers were brought into Egypt from Europe. Between 1690 and 1720, Marseille exported 300,000 to one million to Egypt annually. They also came in from other parts of the Ottoman empire and Europe. Public acceptance of them is partially explained by their adoption among coffee and textile merchants trading with Jidda. Spanish dollars were overvalued in Egypt and European merchants found they could earn from 15 to 20 percent profit by exporting them there. According to Raymond, fewer Spanish *riyāls* were exported after 1730.

The German thaler, known locally as *riyāl Abū Tāqa* and by the Europeans as *pataque*, circulated in Egypt in the 17th century, becoming more widely accepted after 1703. Though superior in silver content to the Spanish dollar, it was valued on a par with it. By the middle of the 18th century, the thaler had ousted the Spanish dollar generally, a development also related to the thaler's circulation in the Hijaz (and India). By 1769, it had

---

4. Bernard, p. 441.

become the basic silver coin in use in Egypt, finally being elevated to the position of an account money valued nominally at 90 *nisf*.

European gold coins rarely circulated because Egyptians tended to suspect all gold coins (their own being so unreliable), because gold coins were scarce in Europe, and because gold was cheaper in Egypt than in Europe, at least at the end of the 18th century.[5]

There are references in local court records to gold coins from Morocco (*dīnār ismā'īlī; dhahab maghribī; sharīfī maghribī*) and to silver coins from the Hijaz (*ghirsh hijāzī*). Raymond believes the Hijaz coins are too few to indicate a currency flow. On the contrary Egypt exported great quantities of gold and silver coins, especially silver, to the Hijaz to pay for coffee and Indian textile imports. In turn, almost all Arabian imports from India were paid for by specie. Thus much of the imbalance in trade between Egypt and the Hijaz, paid for in silver and gold specie, seems to have ended up in Indian markets.[6]

Finally, we must deal with curiously persistent reports that gold and silver coins from the Western Sudan circulated in Egypt. The medieval historian Ibn Ḥajar is said to report the circulation of a *dīnār takrūrī* during the Mamluk period; al-'Aynī mentions a *dirham takrūrī*.[7] Even as late as the early 19th century it is claimed that a gold coin minted in northern Dahomey, the so-called Nikki *mithqāl*, found its way to Egypt.[8] Such references would appear to refer to gold blanks fashioned into coin shapes for easier transportation rather than to mint issues;[9] the silver coin reference will be discussed further on. No coins of West African origin are mentioned in Ottoman-period court registers we have surveyed and their import into Egypt, if indeed such coins ever really existed, seems dubious.

On the other hand, gold dust from the Western Sudan is mentioned in these registers under the name *dhahab takrūrī* or the more correct *tibr min dhahab takrūrī*. At the end of the 18th century, it seems to have had a gold purity of between 89 percent and 93 percent, necessitating nonetheless at

---

5. *Ibid.*, p. 438.

6. Raymond, I, pp. 136-7; "The balance of trade with India is carried back in Spanish, Venetian and German coins, and occasionally a few pearls," *Milburn's Oriental Commerce*, 2 vols. (London, 1813), I, p. 89; for the 17th century, Jean-Baptiste Tavernier, *The Six Travels of John Baptiste Tavernier, Baron of Aubonne*, 2 parts in 1 (London, 1684), part 2, pp. 18-21.

7. Eliyahu Ashtor, *Les métaux précieux et la balance des payements du Proche-Orient à la basse époque* (Paris, 1971), p. 29; on the *dirham*, see below.

8. Marion Johnson, "The Nineteenth-Century Gold 'Mithqal' in West and North Africa," *Journal of African History*, IX, 4 (1968), pp. 552-3.

9. They would thus resemble the famous "bald dinars" of the Sudan mentioned by al-Bakri, Leo Africanus and O. Dapper; see Jean Devisse, "Routes de commerce et échanges en Afrique occidentale en relation avec la Mediterranée," *Revue d'histoire économique et sociale*, L (1972), p. 374, note 208; Ronald A. Messier, "The Almoravids/West African Gold and the Gold Currency of the Mediterranean Basin," *Journal of the Economic and Social History of the Orient*, XVII, 1 (1974), p. 36-7; A. G. Hopkins believes they were minted: *An Economic History of West Africa*, 2nd ed. (London, 1975), p. 68. These pieces were to be found at Jenne, Timbuktu, Tadamakka and Tadla.

least two stages of refining.[10] In the early medieval period, gold from the Western Sudan is said to have had a purity of 92.2 percent.[11]

Gold of Takrūr: the legend

Egypt's formal discovery of the gold of Takrūr, as Arab geographers termed that region of sub-Saharan Africa stretching from as east as Dār Fūr (present-day Sudan) to as west as Senegal, but classically the area of the Niger Bend, dates to the 14th century and the pilgrimage of Mansa Mūsā, the ruler of Mali. He arrived in Egypt in 1324 on his way to the Hijaz with eighty or one hundred camel loads of gold, depending on the source.[12] If a camel load is weighed at three *qintārs* (kantars or quintals), as was usual, then one hundred camel loads is equivalent to 12.2 metric tons at the canonical weight of the *mithqāl*.[13] This may be compared to 43 metric tons of gold brought to Spain during the height of gold shipments from America, 1551-60.[14] The

---

10. Bernard, pp. 429-30; he notes, however, that a total of 2,919 *dirhams* of gold were reduced in the refining process to 2,602.51 *dirhams* of pure gold, indicating an average purity of 89 percent. On the stages of refining, *ibid.*, pp. 476-7.

11. Messier, *loc. cit.*

12. Among countless versions of this pilgrimage, E. W. Bovill, *The Golden Trade of the Moors*, 2nd ed. (London, 1968), pp. 85-8; Nehemia Levtzion, *Ancient Ghana and Mali* (London, 1973), p. 66; Ashtor, *Les métaux*, pp. 18-19; Hasanein Rabie, *The Financial System of Egypt A. H. 564-741/A.D. 1169-1341* (London, 1972), pp. 191-2; Charles de la Roncière, *La découverte de l'Afrique*, 3 vols. (Cairo, 1924-7), I, pp. 115-16 (who refers to only 48 camel loads).

13. *A note on weights.* Precious metals were weighed by *dirham*, *mithqāl* (1 1/2 *dirhams*, or 72 barley grains), *wiqīya* (ounce, or 12 *dirhams*), *ratl* (pound, or 144 *dirhams*, or 96 *mithqāls*) and *qintār* (100 *ratls*, or 9,600 *mithqāls*). As is well known, the actual weight of these units in grams or kilos varied from time to time and place to place. For the medieval period in Egypt, for instance, Popper adopted a *dirham* weighing 3.186 grams, a *wiqīya* of 38.23 grams, a *ratl* of 458.78 grams and a *mithqāl* of 4.25 grams, which was considered the canonical weight of the *mithqāl*. According to Dr. Sahillioglu, whose findings are based on Turkish records, the Ottomans weighed the *dirham* at 3.072 grams, the *mithqāl* at 4.608 grams, the *wiqīya* at 36.86 grams, the *ratl* at 442.4 grams and the *qintār* at 44.24 kilos. We have followed the weights given in Bernard, "Notice sur les poids arabes anciens et modernes," *Description de l'Egypte, État moderne*, 2nd ed., XVI (Paris, 1825), pp. 73-106, whose findings were fractionally different. Accordingly:

dirham	=	3.088 grams
mithqāl	=	4.63 grams
wiqīya	=	37.06 grams
ratl	=	444.7 grams
qintār	=	44.47 kilos

See William Popper, *Egypt and Syria under the Circassian Sultans/1382-1468 A. D./Systematic Notes to Ibn Taghrī Birdī's Chronicles of Egypt*, University of California Publications in Semitic Philology, XVI (Berkeley and Los Angeles, 1957), pp. 39-40; Raymond, *Artisans*, p. lvii; Lane, *Manners*, p. 579. When Marion Johnson read a draft of this paper, she strenuously warned against using a uniform weight for the *dirham* and *mithqāl*. Were the sources available, it would indeed be valuable to check against variations over time; in their absence, Mrs. Johnson's warning merits attention (personal communication with the author). For her discussion of weights of the *mithqāl* in North Africa, see her article, "Mithqāl," pp. 547-52.

14. Spooner, p. 13.

ruler and his entourage were well received by the sultan of Egypt but badly treated by local merchants who, according to a well-known story by al-ʿUmarī, tended to overcharge their black African guests, and Mansa Mūsā returned to Mali with a host of Egyptian creditors. Once there they could not have missed the opportunity to redeem their credit notes in gold dust and thus began, it is believed, the period of regular gold shipments from West Africa to Egypt.[15] The total annual trans-Saharan gold trade to the whole of North Africa, including Egypt, has been calculated by Mauny at 9 tons a year; this figure has subsequently been revised downward to between half and one ton per year in the late medieval period (see Dr. Curtin's contribution).[16]

Although statistics on gold imports from the Western Sudan are virtually non-existent, there is considerable disagreement as to when these imports subsided. The concern here is not so much the impact of "Sudanese" gold on the international economy as the flow of gold to Egypt itself. Udovitch, using Darrag as his source, has dated the decline to "the latter part of the 14th century."[17] If so, then the gold of Takrūr would have been so short-lived as to not be worth discussing. On the basis of European accounts in the middle of the 15th century, Ashtor has conjectured "the supply of gold . . . although much diminished, should have been sufficient until the second half of the 15th century, when the Portuguese share grew considerably."[18] Yet in a more recent work he states that the light-weight *dīnār* minted by Sultan Barsbay in 1425 already reflected a "diminished" supply of Takrūr gold as it was then being siphoned off by Italians as well as Portuguese.[19] European activity at this date in both North and West Africa is certainly overstated. Ashtor's argument may be questioned on two grounds, that the Mamluk sultan's new monetary policy was motivated by entirely different economic considerations, unrelated to the supply of gold,[20] and second, that the Portuguese gold

---

15. As per Rabie, *loc. cit;* Ashtor dates gold imports to the Fatimid period, some four centuries earlier: Eliyahu Ashtor, *A Social and Economic History of the Near East in the Middle Ages*, (Berkeley, Los Angeles and London, 1976), pp. 80-1, 195, 291-2.

16. Raymond Mauny, *Tableau géographique de l'ouest africain au moyen âge* (Dakar, 1961), 301 as quoted in Hopkins, *Economic History*, p. 46; for an earlier revision of Mauny's figures, Vitorino Magalhaes-Godinho, *L'economie de l'empire portugais aux XVe et XVIe siécle* (Paris, 1969), p. 119.

17. Robert Lopez, Harry Miskimin and Abraham Udovitch, "England to Egypt, 1350-1500; Long-term Trends and Long-distance Trade," in *Studies in the Economic History of the Middle East*, p. 126; Ahmad Darrag, *L'Egypte sous le règne de Barsbay 825-841/1422-1438* (Damascus, 1961), p. 92; we have been unable to follow up the references in Suhbi Labib, *Handelsgeschichte Aegyptens im Spätmittelalter* (Wiesbaden, 1965); assuredly Udovitch's 14th century is an error for 15th.

18. Eliyahu Ashtor, "Recent Research on Levantine Trade," *Journal of European Economic History*, II, 1 (1973), p. 200.

19. *Social and Economic History*, p. 324.

20. Jere L. Bacharach, "The Dinar versus the Ducat," *Int. J. Middle East Stud.*, 4 (1973), pp. 77-96.

Gold and silver exchanges between Egypt and Sudan

imports from Elmina, their principal West African trading post, were not significant until the last two decades of the 15th century.[21]

In fact Portuguese agents in Alexandria in 1511 were writing Lisbon that two caravans came each year from Takrūr carrying "great quantities" of gold, and on this and other European archival information, Braudel argues in the English edition of his *La Mediterranée* that the Turkish conquest of Egypt in 1517 actually gave the Ottomans access to Takrūr gold and the means to maintain a bi-metallic monetary system. He dates the end of the "age of Sudanese gold" to the middle of the 16th century, the reference being to the impact of Takrūr gold on the international economy.[22] Yet it is pertinent to note here that Egypt annually sent 600,000 gold coins to Istanbul during much of the 16th century as its yearly remittance to the Ottoman rulers, of which Sudanese gold must have played a significant if unknown part of Egypt's total precious metal resources.[23] Thus recent historical research from the Mediterranean end has pushed forward the demise of the elusive trans-Saharan gold trade of Egypt by two centuries. But that date is also premature, since gold still flowed from the Western Sudan to North Africa until the end of the 18th century.

Gold of Takrūr—16th-18th centuries

In order to put the Takrūr gold trade in proper perspective, two aspects may be pointed out at the outset. First, it operated essentially as an adjunct of the royal pilgrimage tradition in West Africa, in which sultans strove to display their largesse, wealth and position, thereby winning international recognition, while individual pilgrims who accompanied them needed to carry pocket money to sustain themselves while on the *ḥājj*. Within this framework, enterprising merchants in northeast Africa sought to acquire gold in Takrūr by taking advantage of already organized caravans that would have been too expensive for them to finance by themselves. Second, the Ottoman governor of Egypt, presumably on instruction from Istanbul, encouraged the import of gold by exempting it from customs duty.[24]

---

21. Portuguese gold imports from Arguin, their trading post nearest the Saharan caravan routes, amounted to only 20-23 kilos during the last quarter of the 15th century and beginning of the 16th: Godinho, quoted in Levtzion, p. 134; imports from Elmina from 1482 onward averaged 400 kilos a year: Godinho, pp. 210-20; see also Richard Bean, "A Note on the Relative Importance of Slaves and Gold in West African Exports," *J. African History*, XV, 3 (1974), p. 352.

22. Godinho, 122; Fernand Braudel, *The Mediterranean and the Mediterranean World in the Age of Philip II*, tr. by Sian Reynolds, 2 vols. (New York, 1972-73), I, pp. 182, 466-7; II, p. 668; for additional European archival data, Spooner, p. 14.

23. Stanford J. Shaw, *The Financial and Administrative Organization and Development of Ottoman Egypt, 1517-1798* (Princeton, 1962), p. 284; see also Dr. Sahillioğlu's contribution in this volume.

24. Comte d'Estève, "Mémoire sur les finances de l'Egypte," *Description de l'Egypte, État moderne*, I (Paris, 1809), p. 338; Halil Inalcik, "Capital in the Ottoman Empire," *J. Eco. History*, 1-2 (1969), p. 135. Gold and silver coins were also exempt.

While the pilgrimage of royal personages from the Western Sudan can be followed in chronicles, both Egyptian and Sudanese, thereby giving indications of probable years of gold movements, the absence of official customs statistics does not permit those movements to be quantified. In this unhappy quandary, recourse may be made to the religious court archives of Cairo which contain documents pertaining to 17th-century pilgrims and merchants who either brought gold with them from Takrūr or who were trading privately in gold from Takrūr. They provide the data for a hypothetical model of gold imports during the Ottoman period in Egypt. It could be argued that the individual trade is gold, though never equalling the vast sum brought by Mansa Mūsā in the 14th century, which is improbably large anyway, was a primary means by which Egypt obtained its pure gold; the other was the importation of European gold coins. The fact that West African sultans arrived in Cairo during much of the Ottoman period suggests the existence of a gold trade then as much as it did for the medieval period.

Pilgrimages

Beginning with Mansa Mūsā's *ḥājj* in 1324 and until 1513, four years before the Ottoman conquest, at least 16 pilgrim caravans arrived from Takrūr. On the basis of available records this amounts to an average of one caravan every ten-and-a-half years. The first half of the 15th century was particularly active, with caravans arriving in 1432, 1433-35, 1439, 1443, 1445, 1446 and 1454, some described by chroniclers as "important" or bringing "many slaves and [much] gold dust.[25] Askiyā Muhammad, the Songhai sultan who performed the *ḥājj* in 1495-97, must also have brought significant amounts of gold, as will be mentioned further on. These 190 years are taken as the "classical" period of Egypto-Takrūr relations, the one in which the wealth of Sudanese monarchs and their subjects not only was materially manifested in the Middle East but also firmly imprinted on the Middle Eastern mind. During this period the pilgrimage tradition was strongly maintained by monarchs in the Niger Bend, rulers of the successive kingdoms of Mali and Songhai whose territory bordered on the auriferous regions

---

25. The list, which is incomplete, has been compiled from the following sources: Joseph M. Cuoq, *Recueil des sources arabes concernant l'Afrique occidentale du VIIIe au XIV siècle* (Paris, 1975), pp. 390-2; Jean Claude Garcin, *Un centre musulman dans la Haute Egypte médiévale: Qus* (Cairo, 1976), p. 426; Auguste Cherboneau, *Essai sur la littérature arabe au Soudan* (Constantine, 1855), p. 4; 'Umar al-Naqar, *The Pilgrimage Tradition in West Africa* (Khartoum, 1972), p. 35; Jacques Jomier, *La mahmal et la caravane égyptienne des pèlerins de la Mecque (XIIIe-XXe siècle)* (Cairo, 1953), pp. 77-8, 85-6; H. T. Norris, *The Tuaregs/Their Islamic Legacy and Its Diffusion in the Sahel* (Warminister, Eng., 1975), pp. 41, 60; Muhammad Ahmad Ibn Iyās, *Journal d'un bourgeois du Caire*, tr. Gaston Wiet, 2 vols. (Paris, 1955-60), I, pp. 85, 321, 324; E. M. Sartain, *Jalāl al-Dīn al-Suyūṭī*, 2 vols. (Cambridge, 1975), II, p. 100; the author first broached this subject in a paper delivered at the Conference on the Economic History of the Central Savanna (Kano, 1976), entitled "Trade between Egypt and *bilād al-Takrūr* in the XVIIIth Century."

of Bambuk (Bambuhu) and Buré and whose markets were frequented by traders from newly-opened mines in present-day Ghana, beginning in the mid-15th century.[26]

The end of the 15th century and beginning of the 16th century saw the development of strong states to the east of Mali-Songhai, and the pilgrimage tradition shifted eastward to the emerging states of Bornu and Agades. Sultans of both began to organize caravans to Mecca, travelling the well-known routes through the desert to the oases of Libya and Cairo. Between the *ḥajj* of Mai Idrīs of Bornu in 1574 and the *ḥajj* of Mai Dunāma 'Alī in 1728, 12 royal personages passed through Cairo, averaging one pilgrimage every 13 years.[27] Yet while the medieval-period pilgrimages were often mentioned by contemporary Egyptian historians, the latter-period pilgrimages were virtually unnoticed. References to them, however, are found in West African historical sources and in travelogues of foreign visitors to Cairo. The Turkish traveller, Evliya Čelebī, for instance, talked to Mai 'Alī 'Umar of Bornu during his fourth or fifth pilgrimage in 1670, concluding that "the sultans of Bornu come every year" bringing gold dust and slaves. De Maillet, the French consul, encountered his two sons in 1707, remarking that they had been "most generous during their passage through Cairo.[28]

With the exception of the pilgrimages of Mansa Mūsā and Askiya Muḥammad there are no references in the sources to the amount of gold brought by pilgrimage caravans. At best chroniclers note them as "great quantities" or "small quantities". Nor did "great quantities" only accompany the arrival of Sudanic rulers. The Egyptian historian al-Ishāqī, for instance, met the shaikh of the Takrūr caravan during the *ḥajj* of 1028/1619, noting upon his death outside Cairo that he had left an estate of "much wealth and gold dust" which an amir tried to confiscate.[29] The Italian traveller Seguezzi learned of a company of pilgrims in 1635 and remarked they had brought so much gold that its value dipped in the Cairo market[30]—a

26. On these mines, Hopkins, *Economic History*, p. 46; Philip D. Curtin, "The Lure of Bambuk Gold," *J. Afr. History*, XIV, 4 *(1973)*, pp. 623-31; Ivor Wilks, *The Northern Factor in Ashanti History* (Legon, 1961).

27. For these pilgrimages, Naqar, pp. 31-3; Norris, pp. 79, 94-6; Terence Walz, *Trade between Egypt and Bilād as-Sūdān (1700-1820)* (Cairo, 1978), p. 18, note; Y. Urvoy, *Historie de l'empire du Bornou* (Paris, 1949), p. 83; and his "Chroniques d'Agadès," *J. Soc. Africanistes*, IV (1934), pp. 160-71.

28. Tomasz Habraszewski, "Kanuri—Language and People—in the Travel-Book *(Siyahatename)* of Evliya Čelebī," *Africana Bulletin* (Warsaw), 6 (1967), p. 65; C. Beccari, ed., *Rerum Aethiopicarum Scriptores Occidentales*, XIV (Rome, 1914), p. 384. De Maillet indicates the scope of their generosity—and of their father's—by mentioning that they had established hostels in Cairo, Mecca and Medina for Bornu pilgrims. That they should have done so means that they must have brought slaves or gold dust to purchase the buildings.

29. Muḥammad 'Abd al-Mu'ṭī al-Ishāqī, *Kitāb akhbār al-awwal* (Cairo, 1315 AH), p. 169.

30. Santo Seguezzi, "Estat des revenus d'Aegypte par le Sieur . . ." in Auguste Courbe, comp., *Relations veritables et curieuses de l'isle de Madagascar et du Brèsil . . . [et] trois relations d'Egypte et une du Royaume de Perse* (Paris, 1651), pp. 98-8.

story incidentally which parallels the local tradition of Mansa Mūsā's passage. Neither of these reports jives with known, dates of royal pilgrimages.

In addition, therefore, to royal pilgrimages, smaller goldladen *ḥajj* caravans from western Africa also transited the Sahara. According to late 18th-century information, when relations between Egypt and Takrūr had reached a low, they came every three years and were composed of a hundred members or so.[31] Such numbers may be taken as an absolute minimum. Individual pilgrims equipped themselves with small quantities of dust to pay their travelling expenses and to purchase goods in Cairo and the Hijaz which were then sold profitably upon their return home. This aspect of the *ḥajj* has in fact been documented.[32] Court records from Cairo dating to the 16th and 17th centuries suggest that pilgrims imported sums varying from 24 to 300 *mithqāls* (111 grams to 1.39 kilos). In one instance, a party of 19 pilgrims, all identified by the place-name *nisba* of al-Takrūrī, had stored 1,463 *mithqāls* (6.77 kilos) of gold in Cairo while they went to Mecca on pilgrimage.[33]

On the basis of such evidence a hypothetical model might be projected to indicate the average amounts of gold imported by "royal" and private pilgrimage caravans. It may be supposed that each pilgrim tried to bring at least one *ṣurra* or purse of 100 *mithqāls* (463 grams) and possibly two, in which case a "small" pilgrim party of 100 people would have brought between 10,000 and 20,000 *mithqāls* (from 46.3 to 92.6 kilos) and a larger party of 1,000 pilgrims would have averaged 694 kilos. Sultan Askiyā Muhammad is said to have spent 100,000 *mithqāls* for traveling expenses from West Africa to the Holy Cities, 100,000 for charities in the Hijaz, and a final 100,000 *mithqāls* for goods purchased in Cairo (850 kilos would, therefore, have been dispensed en route and in Egypt at the canonical weight of the *mithqāl*).[34] If larger or royal *ḥajj* caravans passed through Egypt every 13 years and "small" caravans came every three years, then Egypt may have received 9,022 kilos from larger caravans during the period 1574 and 1728 and an

---

31. Lucas in *Proceedings of the Association for Promoting the Discovery of the Interior Parts of Africa*, 2 vols. (London, 1810), I, p. 191.

32. Prax, "Commerce de l'Algerie avec la Mecque et le Soudan," *Revue de l'Orient*, 2e serie, V (1840), pp. 337-48; VI (1849), pp. 1-17. He estimated the pilgrimage trade between Algeria and Mecca at about 2 million francs annually.

33. The archive of the old religious courts of Cairo (*al-maḥākim al-sharʿiyya*) was housed during the period of my researches (1970-76) in the Land Registration Office (Maṣlaḥat al-shahr al-ʿaqārī), Cairo. They are bound in registers according to name of court; pages are numbered as is each document. Abbrevations of court records used in this study are as follows: Maḥkama al-Qismat al-ʿAskariyya (Askar); Maḥkama al-Qismat al-ʿArabiyya (Arab); Maḥkama al-Ṣāliḥiyya al-Najmīyya (SN); Miscellaneous Papers (Dashṭ) (D); Maḥkama al-Bāb al-ʿĀlī (BA); Arab 41, p. 404, no. 534 (1061/1651); al-Ḥājj ʿAlī Abū Bakr al-Takrūrī al-Wankārī brought 200 *mithqāls*: BA 20, p. 25, no. 114 (970/1562); al-Ḥājj Aḥmad ʿAlī Abū Ḥāmid al-Maghribī al-Waḥadānī (?) brought 4 purses: Arab 31, p. 251, no. 359 (991/1583)

34. ʿAbd al-Rahman Sadi, *Tarīkh al-Sūdān*, trans. O. Houdas (Paris, 1900), French translation pp. 119-20; Arabic text, p. 73. The translator interpreted the Arabic *dhahaban* as "pieces of gold"; these may have been the "bald dinars" mentioned in geographical texts and travelogues, but more probably gold dust particles measured in *mithqāls*.

average of 3,562.5 kilos from smaller caravans, making an "annual" import of 81.7 kilos from the pilgrimage traffic alone.

According to Raymond one million *zingirlīs* were minted in 1725 with a reduced gold content of 3.36 grams; the hypothetical Takrūr supply would have met only 2.4 percent of the mint's needs that year. By way of comment on the attempt to calculate imports without reliable statistics, it should be noted that the "Maghribī" caravan that arrived in Cairo in 1799 brought 2,919 *dirhams* of gold dust, or only 9.01 kilos. After refining it was reduced to 2,602 *dirhams* (8.03 kilos), sufficient to mint 3,100 gold coins (*mahbūbs*) at a gold content fixed by the French at 2.592 grams. During the period of the French occupation, the annual mint output was 95,173 *mahbūbs* of which the gold brought by pilgrims contributed 3 percent.[35] Bernard noted the *peu d'activité* of the mint during those years, yet even in periods of more active production, it would seem that by the late 18th century Takrūr supplied only a fraction of local needs. In any event, *ḥājj* caravans would have included numerous indigent pilgrims who would have brought no gold at all, and thus the question of "annual" gold imports becomes even more problematic.

Pilgrims sold their gold to private merchants in Cairo, not to the mint, since they offered better prices than did the Government.[36] Market transactions were observed by several Europeans and described with mystical embellishment. Gold dust was enclosed in leather sacks, of interesting tooling and adorned by "African" amulets often in the shape of snakes and tortoises. Such was the honesty of the sellers that the sacks were not weighed. Consul De Maillet noted that pilgrims would not accept payment for their dust in equivalent gold coin, commenting that in doing so "they would have committed a very great sin." At the end of the 18th century, Bernard, too, noted a reluctance to accept gold for gold, and believed their peculiar behavior reflected some idosyncratic Sudanese religious practice.[37] In fact pilgrims were probably motivated by sound economic thought, realizing that locally-minted gold coins were not accepted currency in the Hijaz, whence they were all bound.

Private gold trade

Whereas in the medieval period, Timbuktu, and Jenne had been important gold market towns in the Western Sudan, the shift of power eastwardly created markets for gold in the new centers of Katsina, Birni N'gazargamu

---

35. Bernard, "Monnaies," pp. 429, 441.
36. *Ibid.*, p. 426.
37. Abbé le Mascrier, *Description de l'Egypte . . . composée sur les mémoires de M. de Maillet*, 2 parts in 1 (Paris, 1735), p. 196* (hereafter cited De Maillet, *Description*); Bernard, "Monnaies," 428. De Maillet's account is repeated in *Le troisième voyage du Sieur Lucas* (Paris, 1744) and Richard Pococke, *A Description of the East and Some Other Countries*, 2 vols., 3 parts (London, 1743-8).

and Agades. It is now apparent that the development of trade routes between cities in northern Nigeria and the powerful Ashanti state in modern day Ghana meant that rulers, merchants and prospective pilgrims in Bornu and Agades had access to gold from the Ashanti and Lobi mines as well as the older sources that fed into the market at Timbuktu. No gold is mined in the Lake Chad area.

The emerging gold trade in this region proved of paramount importance to the Agades rulers as the state developed in the 16th and 17th centuries. The founding state "charter", which probably dates to the end of the 15th century when Agades paid tribute to the Songhai sultans, states that the tax collected on caravans should be given to "the men who guard the road which is situated between Egypt and Timbuktu because it is a major route, used by all [engaged] in the transport of Egyptian cloth and the gold from Timbuktu. These men are your power and strength."[38] In his recent study of the Tuareg, Norris concludes that the region of Aïr, which surrounds the Agades capital, became in fact increasingly identified as "Takrūr" by the Muslim East, thereby inheriting the fame of Takrūr as a region of gold, while at the same time it was drawing close to Egypt. Mamluks fleeting Cairo sought refuge in Agades.[39] Indeed, as evidence shown below will make clear, the road from Cairo to Timbuktu and other gold markets in the Western Sudan was until the 18th century through Murzuq and Agades, and it was along this route that merchants from Cairo travelled to obtain gold.

The independent attempts of those merchants to obtain Takrūr gold can also be followed in the Cairo court records. The usual pattern was for merchants to invest camel loads of goods with travelling merchants (*jallāba*, or *tujjār saffār*) headed for Fazzān, Agades or Timbuktu. Sometimes they sent their own agents, as in the case of Ibrāhīm Madkūr "al-Jallāb," a Cairene merchant, who dispatched his freed slave and agent, one al-Ḥājj 'Anbar 'Abd Allāh al-Ḥabashī, to *bilād al-Takrūr* to obtain gold. He returned in 1678 with 38 purses, each purse containing 100 *mithqāls* (3,800 *mithqāls* or 17.6 kilos). In 1690, his nephew al-Ḥājj Aḥmad Madkūr Māzin "al-Jallāb" sent seven camel loads of beads to Fazzān worth 21,000 *niṣf fiḍḍa*. (A statement indicating the gold return on his investment, however, could not be located.) In March of that year another merchant, Aḥmad Jūrbajī "al-Jazzār," dispatched 8 camel loads of merchandise (2 loads of cloth [*qumāsh*], 2 loads of silk, 4 loads of cowries) worth 53,560 *niṣf*, and in June, 1691, his heirs formally received their shares of the return investment, now reckoned in "gold dust from *bilād as-Sūdān*" of which 1,125 *mithqāls* (5.2 kilos) were valued at 135,000 *niṣf fiḍḍa*.[40] In 1690, moreover, one of the Cairo judges heard the case of al-Khawāja Muḥammad "Shalabī" Jamāl

38. Norris, *Tuaregs*, p. 55; also translated by R. Palmer in his *Bornu, Sahara and Sudan* (London, 1936), p. 56; and by Urvoy, "Chroniques d'Agadès," p. 155.
39. Norris, p. 51.
40. Askar 74, pp. 292-6 (1089/1678); Askar 83, p. 319, no. 420 (1101/1690); Askar 84, p. 58 (1102/1690); Askar 84, p. 425, no. 650 (1102/1691).

al-Dīn, a merchant in Indian textiles, who testified he had entrusted four camel loads of silks and Indian satin brocades with one al-Ḥājj 'Abd al-Salām Aḥmad al-Maghribī al-Fāsī to take to *bilād K-nā* (probably Kashnā, the Arabic spelling of Katsina) which had been exchanged there for 1,606 *mithqāls* of gold (7.4 kilos). His complaint concerned the parcelling of that amount between himself and the travelling merchant.[41] Thus during the year 1690 at least 12.6 kilos of gold from Takrūr materialized through private merchant channels.

The largest amount of imported Takrūr gold we found in the Cairo archives is noted in the estate papers of al-Ḥājj Muḥammad 'Abd al-Raḥman al-Ghariyānī dated 1724. Though Libyan, his family was settled in Cairo. According to the estate *ḥisba*, Muḥammad traveled to Timbuktu in 1715, traded in salt and gold, and died there about 1722. His brother 'Abd Allāh left Cairo for Timbuktu to settle the estate and returned with 7,103 *mithqāls* (32.9 kilos) for his brother's family. The document also discloses that the cost of a return journey to Timbuktu amounted to 2,272 *mithqāls* (10.5 kilos) and that while passing through Agades, 'Abd Allāh was required to pay a tax of 333 *mithqāls* (1.5 kilos).[42]

The series of late 17th-century archival references to gold imports suggests that the gold trade between Egypt and Takrūr was more significant than is generally appreciated. This belief is encouraged by a contemporary report, albeit grossly exaggerated, written by De Maillet in 1692. In a *mémoire* on the state of Egypt he states that between 1,000 and 1,200 *qinṭārs* of gold were imported annually from the Sudan. This figure amounts to between 44 and 53 metric tons.[43] While De Maillet's claim should be treated with extreme caution, it may nonetheless be set against the background of numerous reports that considerable amounts of gold were circulating in various parts of Libya at this time—circumstantial evidence so to speak of Takrūr gold movements in an Egyptian direction—and that newer, heretofore untapped sources, were being developed in the Eastern Sudan and Ethiopia.

Gold in Libya

Libyan evidence is as unsatisfactory as Egyptian data, though it has the merit of being more contemporary and more persistent. It concerns the three key regions of the country through which trans-Saharan caravan traffic passed:

41. BA 175, p. 279, no. 1087 (1101/1690).
42. D 234, p. 613 (1136/1724); the Agades reference is *lil-dawla bi-nāḥiya Aghadidh*. 'Abd Allāh al-Ghariyānī seems to have become one of the eminent members of the wealthy North African community in Cairo.
43. "Mémoire sur le gouvernement d'Egypte," (Archives du Ministère des Affaires Étrangère, Paris), Mémoires et Documents, Egypte, I, p. 20; see also Raymond, I, p. 48. Curiously enough, De Maillet's great interest in the affairs of Ethiopia and Sinnar makes him one of the few reliable early 18th-century sources.

the southern territory of Fazzān, the eastern oasis of Awjila, and the western oasis of Ghadāmas.

The earliest annual tribute fixed by the mid-16th century Ottoman governors of Tripoli on the independent rulers of Fazzān amounted to 1,140 *mithqāls* of gold (5.28 Kilos).[44] The energetic 17th-century governor, Muḥammed al-Saqizlī (1620-40) moved to control the trade in gold which was transiting Fazzān and being siphoned off via Awjila to Egypt. His overtures to the Bornu sultans are well known.[45] In 1636 Fazzān was newly beseiged and an annual tribute of 4,000 *mithqāls*, half to be paid in gold dust (2,000 *mithqāls* = 9.25 kilos), was established.[46] An expedition against Awjila four years later, however, even better illustrates the direction of trade. According to Libyan chroniclers, whose statements must also be treated cautiously, Turkish troops plundered 12 camel loads of gold (1.6 metric tons), the ruler of Awjila was ransomed 28,000 *mithqāls* (129.6 kilos) and an annual tribute was levied amounting to 15,000 *mithqāls* (69.4 kilos), almost four times the Fazzān tribute. Ibn Ghalbūn observed that the gold plundered in Awjila was sufficient to provide for the minting of a new gold coin in Tripoli, called the Saqizlīyya after the governor, from that time until the rule of Khalīl pasha (1706-09).[47] There is no way to check the veracity of such reports, but two conclusions may be drawn from them: significant quantities of gold were passing through Libya in the 17th century; and the important movement of gold was not northward, in the direction of Tripoli, but northeastward, in the direction of Cairo. The enormous tribute levied on Awjila was no doubt designed to stop this movement.

Throughout the 17th and 18th centuries European observers in Libya refer to the import of gold: Consul Le Maire (1686); Delalaude (1698); Savary and l'Abbé Raynal (mid-18th century); Consul de Lancy (1766) and Consul Frazer (1767).[48] De Lancy believed Ghadāmas was the emporium for gold from both Timbuktu and Agades, the value of the former being twice that of the latter. Ghadāmas' trade connections were with both Tripoli and Tunis. But Venture de Paradis argued twenty years later (1788) that because the pasha at Tripoli fixed the price of gold, most Sudanese imports went to Cairo where an official price was not applied.[49]

44. John Lavers, "The Trans-Saharan Trade, ca. 1500-1800: A Survey of Sources," paper presented at the Conference on the Economic History of the Central Savanna, Kano, p. 7

45. B. G. Martin, "Kanen, Bornu and the Fazzan: Notes on the Political History of a Trade Route," *J. Afr. History*, X, (1969), pp. 15-27; Lavers, p. 9.

46. G. A. Krause, "Zur Geschichte von Fesan und Tripoli in Afrika," *Zeitschrift der Gessellschaft für Erdkunde zu Berlin*, Band 13 (1878), p. 365; Ettore Rossi, *La Cronaca araba Tripolina di Ibn Ġalbūn (sec. XVIII)* (Bologna, 1936), p. 90; C. L. Feraud, *Annales tripolitaines* (Tunis and Paris, 1927), p. 100.

47. Feraud, pp. 102-04. Gold-laden camels continued to pop up in Libyan history. The Arabic history of Fazzān used by Krause mentions 15 camel loads of gold plundered by Turkish troops in Murzuq ca. 1682-3 (Krause, p. 367).

48. Lavers, pp. 13-16, and appendix.

49. "Notions sur le royaume de Fezzan," *Bull. Soc. Géog. Paris*, 2ème série, IV (1835), p. 193.

Cairo court records, in addition to those already referred to, indicate that gold was indeed seeping in through Libya. Merchants of Awjila who plied the desert route between Cairo and Murzuq were entrusted with purses of *mithqāls* by Awjila notables. The local *qāḍī*, for example, entrusted a purse worth 212 *mithqāls* (981 grams) with one al-Ḥājj 'Īsā al-Awjilī "al-Qaysi," no doubt to purchase Cairo merchandise for him. Inventories of other Awjila merchants reveal the presence of gold dust among their belongings.[50] Thus despite strenous efforts, Tripoli authorities were never able to stem the flow of gold eastward, at least not until the 19th century.[51]

Gold from Ethiopia and the Eastern Sudan

The Turkish conquest of Sawākin, the Red Sea port of Sudan, and Massawa, the Ethiopian outlet, in 1556-7 brought the Ottomans in contact with gold sources in eastern Sudan and Ethiopia. Ethiopian gold was collected in a number of provinces, in Gojjam, Damot and the so-called Shanquella country south of the Blue Nile, taken to markets in Damot and Gondar and eventually to Massawa and Sinnār. The gold wealth of Ethiopia was almost as powerful a legend in Europe as Takrūr gold.[52]

Yet very little is known about gold imported directly by Egypt from Ethiopia. Most Ethiopian gold flowing into the Red Sea trade network almost certainly ended up in Jidda. Nevertheless, the Ethiopian emperor sent Muslim agents to Egypt at the end of the 17th century to purchase goods in Cairo. De Maillet states they brought 500,000 *ecus* of gold dust, civet and ambergris each time they came.[53]

More regular direct relations probably occurred each time the Ethiopians required a new abuna. In 1745, for instance, an Ethiopian party was given 450 ounces of gold (16.7 kilos) by the emperor to finance such negotiations.[54] A certain curious analogy exists between the gold trade of Egypt with

50. Askar 144, p. 288, no. 416 (1149/1737); Walz, *Trade*, p. 90.
51. Thus a series of reports on the gold trade of Libya are available for the period 1820-60: P. B., "Lettere di un viaggiatore . . .", *Biblioteca italiana ossia Giornale di letteratura, scienze ed arti* (Milan), XVIII (1820), p. 251; Jacopo Gräberg di Hemsö, "Prospetto del commercio di Tripoli d'Africca . . .," *Antologia* (Florence), part 1, XXVII, no. 81 (1827), p. 91; France, Ministère de l'intérieur, *Annales du commerce extérieur, Afrique*, II, 6 (dated 1848); *ibid.*, no. 11 (1861); Praz, "Commerce," p. 344; E. Pellissier de Reynaud, "Le règence de Tripoli," *Revue des Deux Mondes*, 2ème série, nouvelle periode, XII (1855), p. 43, note 2; on the other hand, there are virtually no 19th-century Egyptian sources on the gold trade from Libya or western Africa, despite Muḥammad 'Alī's intense search for gold.
52. See Richard Pankhust, *An Introduction to the Economic History of Ethiopia* (Lalibela House, Rochester?, 1961), pp. 224-7.
53. *Ibid.*, pp. 303, 344-5; Europeans believed Egypt obtained gold from Ethiopia "*a copia*": Spooner, p. 39; on Ethiopian gold obtained through Sawākin, Godinho, p. 293.
54. Pankhurst, p. 335; in the 19th century 500 kilos of gold were exported annually from Ethiopia to India and Egypt: Richard Pankhurst, "Indian Trade with Ethiopia, the Gulf of Aden and the Horn of Africa in the Nineteenth and Early Twentieth Centuries," *Cahiers d'études africaines*, 55 (1974), p. 472; no further breakdown of Ethiopian gold exports is known.

Ethiopia and the Western Sudan, both functioning in tandem with religious obligations.

Ethiopian gold brought to markets in Sinnār was one source for merchants in the Eastern Sudan. Another source was alluvial gold collected in the Shaybūn region of the Nūbā Mountains in southern Kurdufān. D'Anania's *Universale fabrica*, published in 1573, refers to merchants from Cairo who travelled to Uri, the capital of the Tunjur kingdom in the area of what became Dār Fūr, to exchange arms for gold,[55] but according to Bell Shaybūn gold was not exploited before the 17th century. He relates its development to the tribute paid by the Nūbā chiefs to the sultans of Sinnār, beginning with Bādī II (Abū Diqn) who reigned 1664-81.[56] The private movement of Shaybūn gold both east and west of Kurdufān after that date is the likely explanation of why gold "in small quantities" often occurs in descriptions of Egypt's caravan trade with both Sinnār and Dār Fūr.[57] Dār Fūr has no gold deposits.[58] The conquest of Kurdufān by Dār Fūr in the early 1780's gave that kingdom direct access to Shaybūn gold,[59] some of which may then have flowed northward to Egypt.

Gold in ingots, dust, rings and coin passed as currency in 18th-century Sinnār. Both Bruce and Burckhardt, who reached the kingdom in its declining years, indicate that the value of gold increased proportionately as it neared Red Sea ports.[60] Merchants travelling the roads between eastern Sudan and Egypt may have avoided transporting gold for security reasons, but as we shall see later the price in Cairo was not high enough to make the trade profitable. Court records dating to the first half of the 18th century

---

55. Dierk Lange, "L'intérieur de l'Afrique Occidentale d'après Giovanni Lorenzo Anania (XVIe siècle)", *Cahiers d'histoire mondiale*, XIV, 2 (1972), pp. 343-5.

56. G. W. Bell, "Shaibun gold," *Sudan Notes and Records*, XX (1937), pp. 127-9; R. S. O'Fahey and J. L. Spaulding, *Kingdoms of the Sudan* (London, 1974), pp. 57, 64-5, 111

57. Maillet, *Description*, p. 198*; S. Sauneron, "Une description des Djellabs datant du milieu du XVIIIe siècle," in his "Villes et legendes d'Egypte," *Bull. Institut Francais d'Archéologie Orientale*, LXVII (1969), p. 145; Paolo Revelli, "Il viaggio in Oriente di Vitaliano Donati (1759-1762)," *Cosmos*, XII (1894-6), p. 324; Carsten Niebuhr, *Travels through Arabia*, 2 vols. (Edinburgh, 1792), I, p. 100; Truguet, "Mémoire sur l'Egypte," in (AE) Memoires et Documents, Egypte, I, p. 71 *verso*; P. S. Girard, "Mémoire sur l'agriculture, l'industrie et le commerce de l'Egypte," *Description de l'Egypte, État moderne*, II (Paris, 1813), p. 638 (with reference to imports from Sinnār).

58. W. G. Browne, *Travels in Africa, Egypt and Syria from the Years 1792 to 1798* (London, 1799), p. 268; O'Fahey and Spaulding, *Kingdoms*, p. 111.

59. Gustav Nachtigal, *Sahara and Sudan, IV: Wadai and Dār Fūr*, tr. by Allan G. B. Fisher and Humphrey J. Fisher (London, 1971), p. 296, note; from this source the Dār Fūr sultan obtained gold to bestow on notables; a gift of 80 *mithqāls* (370 grams) is mentioned in R. S. O'Fahey and Jay Spaulding, "A Sultanic Gift," unpublished paper, xerox, kindly sent to the author by Dr. O'Fahey.

60. James Bruce, *Travels to Discover the Source of the Nile in the Years 1768, 1769, 1770, 1771, 1772 ad 1773*, 5 vols. (Edinburgh, 1790), IV, p. 486; John Lewis Burckhardt, *Travels in Nubia*, 2nd ed. (London, 1822), p. 277; this was also true in western Africa as gold neared the Mediterranean: Johnson, "Mithqāl", p. 556.

provide rare instances of *jallāba* bringing gold with them from Sinnār.[61] The sums, however, are not significant.

Silver and copper: imports from the Sudan

There are only small silver deposits in the Western and Eastern Sudan and in Ethiopia; they have scarcely been mined before the present century.[62] Nonetheless, the medieval Egyptian historian, al-'Aynī, claimed that a silver "Takrūrī *dirham*" circulated in Egypt in the early 15th century, though for reasons not disclosed it was later withdrawn from circulation.[63] The question of whether any coins, gold or silver, were minted in West Africa until modern times is open to debate. If al-'Aynī's remark is thought trustworthy, it may indeed be wondered how the Egyptians could recognize a "Takrūrī *dirham*" if it were not minted.[64]

On the other hand, copper mines in the Sudan are well documented. The mines around Takadda, near the Aïr Massif, produced copper for the Bornu trade; copper mines at Ḥufrat an-Nuḥās in southern Dār Fūr are known to have been exploited in the 18th century.[65] Devisse has suggested that the great demand for copper in 15th-century Egypt provided the incentive for an important copper export trade from Takrūr to Egypt. He argues that the existence of crude copper lingots among the goods of a lost caravan dating to the 11th or 12th century is proof that the Western Sudan exported copper to the Mediterranean while it imported only worked or finished copper items.[66]

Evidence collected from the Egyptian end, based on Ottoman-period travellers' accounts, official reports and local court records, consistently supports the more traditional view that copper moved southward from the Mediterranean to the Sudan. Exports included copper bars, sheets, wire and scrap.[67]

Monetary developments in the Sudan: Silver and copper imports from Egypt

The export of silver and copper into the Sudan during the 16th through 18th centuries is tied to the development of currency forms. The development is

---

61. Shaʿbān Ṣāliḥ al-Maghribī died in Sinnār in 1719 leaving 55 *wiqīya* (2.04 kilos), of which only 16 *wiqīya* were brought to Cairo; al-Ḥājj Sulaymān Muṣṭafā al-Baghdādī travelled to Sinnār in the early 1740's, bringing back 14 *wiqīya* (519 grams): Askar 118, p. 303, no. 448 (1133/1719); SN 519, p. 290, no. 595 (1156/1743).

62. Hopkins, *Economic History*, p. 44.

63. The passage was copied by al-Jawharī al-Ṣayrafī in his *Nuzhat al-nufūs wa'l-abdān fī tawārīkh ahl al-zamān*, ed. Ḥasan Ḥabashī, III (Cairo, 1974), p. 24; the author is most grateful to Dr. Bacharach for sending him this reference which was copied from the Dār al-Kutub manuscript of al-'Aynī's work.

64. Dr. Bacharach raised this point in a personal communication.

65. Ibn Baṭṭūṭa, in Bovill, p. 96; Hopkins, p. 44; on the Dār Fūr mines, Browne, p. 26.

66. Devisse, p. 361; on export of copper from Mali to Egypt, p. 363.

67. Walz, *Trade*, pp. 41-2

particularly evident in the Eastern Sudan, quite possibly because of its geographical proximity to the Islamic heartland but also because of an increasingly closer economic tie with the Egyptian and Red Sea trading zones. While the use of cowries as money was a characteristic feature of the monetary system in the Western Sudan, the "cowrie currency zone" seems not to have extended across the Sudanic belt east of Bornu.[68] Rather, the system that developed in that part of the Sudan closest to Egypt used a confusing number of trade articles as money, including cotton strips, beads, millet, sorghum, iron rings, tin bars and imported specie. Of primary interest are developments in the Sinnār region.

The Sinnār monetary system was apparently modeled on the traditional three-tiered Islamic system of gold, silver and copper. Copper, however, proved far too valuable to be used in small transactions, so iron rings, cotton strips and handsful of millet and sorghum were substituted in towns and villages where copper was scarce. The travel account of Theodore Krump, who provides a rare glimpse into the Sinnār state at the beginning of the 18th century, discloses that Egyptian gold coins and European gold and silver coins were already circulating among the foreign merchant community in the larger Sinnār towns. They included *ashrafīs*[69] and possibly *tughralīs*, "guilders" (probably liothalers), "florins" (rather ducats) and Spanish dollars, the latter just finding acceptance among tradesmen. By 1720 Egyptian *zingirlīs* were also circulating.[70]

Gold dust in weights of *mithqāls* and ounces (*wiqīya*) were used in the largest transactions; according to Spaulding, who subscribes to the archaic economy school of thought, the sultan controlled the distribution of gold and regulated its exchange rate vis-à-vis other currencies.[71] The state had also acquired the technology to mint coins and during Krump's tour in 1701 the local mint was turning out small silver coins which were those probably known at a later date as *maḥallaqāt*. Though tariffed at 45 per Spanish dollar, they traded for 60 and were equivalent, therefore, to the Egyptian *niṣf fiḍḍa* which also traded at 60 per dollar.[72] *Maḥallaqāt* from this period were seen by Brocchi, the 19th-century traveller, who drew a sketch of them: some exhibit a simple grill-work pattern; others a sort of chain.[73]

Krump also mentions the circulation of a smaller coin, made of iron, which was valued at 12, 16 or 20 per *maḥallaq*. These may be identified with a

68. On the term, Marion Johnson, "The Cowrie Currencies of West Africa, Part 1," *J. Afr. History*, XI, 1 (1970), pp. 32-7; Hopkins, *Economic History*, pp. 67-9.

69. The "gold hoard" found by Brocchi contained gold coins minted in 1603; most dated from the reign of Sultan Murād IV (1623): G. B. Brocchi, *Giornale delle osservazioni fatte ne' viaggi in Egitto, nella Siria e nelle Nubia*, 5 vols. (Bassano, 1843), V, p. 475.

70. They were included in the belongings of Shaʻbān Ṣāliḥ al-Maghribī, Askar 118, p. 303, no. 448 (1133/1719). For other currencies, see the reference to Krump in footnote 72.

71. O'Fahey and Spaulding, *Kingdoms*, p. 55.

72. Theodore Krump, *Hoher und Fruchtbarer Palm-Baum des Heiligen Evangelij* (Augsburg, 1710), pp. 287-88; the author is indebted to Dr. Spaulding for his translation of the travelogue.

73. *Giornale*, V, tavola 7, figure 3 (endpages).

crescent-shape iron piece known as *hashshāsha* which circulated in early 19th-century Kurdufān and southern Dār Fūr.[74] They appear to have been equivalents of the Egyptian copper coin (the *jadīd*) which was valued at from 8 to 12 per *niṣf fiḍḍa*.

Between Krump's visit and the arrival of Bruce in 1771, copper had evidently become much more available as a result of trade with Egypt. Copper had replaced the silver content of the *maḥallaqāt*, which Bruce described as "a very bad copper coin." Moreover, the Sinnār government had begun minting a heavier silver coin, called *ghirsh* ("crush" in Bruce) or *riyāl Sinnārī* in Cairo court records. Bruce gives the exchange rate at 20 *maḥallaqāt* per *ghirsh*, a rate which also corresponds to court record information.[75] It may be assumed that an inflation in prices accompanied the transformation of iron pieces into copper and copper *maḥallaqāt* into silver *maḥallaqāt*.

The *maḥallaqāt* and *ghirsh* were pegged to the gold ounce; in Bruce's time, the tariff was 20 *maḥallaqāt* per *ghirsh*, 12 *ghirsh* per *mithqāls*, 4 *mithqāls* per gold ounce. The exchange rate of 4 *mithqāls* per *wiqīya* is a local variation of the traditional exchange rate in the Muslim world of 8 *mithqāls* per *wiqīa*. This rate was still in effect when Burckhardt reached Sinnār in 1813-14.[76]

If the exchange rate current in Sinnār is compared to the exchange rate current in Egypt, it might be expected that merchants would have found it profitable to export gold northward. Very simply, if the ounce in Sinnār were equivalent to 4 *mithqāls* or 48 locally-minted silver *ghirsh*, merchants trading into Egypt could count on exchanging twice as many Sinnār *ghirsh* for one Spanish dollar according to the accepted exchange rate in Cairo,[77] or what would have been equivalent to 24 dollars per ounce. Yet the gold ounce in Cairo during the 1770's was valued at only 16 Spanish dollars,[78] a price which importers must have thought disadvantageous. Moreover, the Sinnār ounce, contrary to Bruce's claim, was smaller than the Egyptian ounce, being equal to 10 "drams" (*dirhams* = 30.88 grams) instead of 12 (37.05 grams);[79]

74. Krump, p. 289; Muḥammad 'Umar al-Tūnisī, *Voyage au Darfour* (Paris, 1845), p. 318; Arabic text, *Tashḥīdh al-adhhān bi-sīra al-'arab wa's-Sūdān*, ed. Khalīl 'Asākir and Muṣṭafā Mus'ad (Cairo, 1965), pp. 301-02; Ignatius Pallme, *Travels in Kordofan* (London, 1844), p. 303.

75. Bruce, *Travels*, IV, p. 406; see also Brocchi, V, p. 476; SN 520 (bis), p. 46, no. 101 (1165/1752): the defendant testifies he paid silver *ghurūsh* "each worth 20 *niṣf*" to a travelling merchant in Dongola the previous year: for *riyāl Sinnārī*, SN 519, p. 290, no. 595 (1156/1743).

76. *Travels in Nubia*, pp. 216-17.

77. According to one court case, 800 *riyāls* "current [or minted] in *bilād as-Sūdān*" (*riyāl min mu 'āmala bilād as-Sūdān*) was equal to 400 "Egyptian *riyāls*"; D 245, p. 256 (1150/1737).

78. Askar 194, p. 330, no. 308 (1188/1774): 1 *mithqāl* at 2 *riyāls*.

79. *Travels*, IV, p. 486; however, he does mention another ounce, which he terms "Atareys", weighing 12 grams, equivalent to the Egyptian ounce: *ibid.*, p. 487. In the 19th century, Brocchi believed the ounce weighed 320 "grains" (the *mithqāl* weighing 80 grains) which would have meant the Sinnār ounce weighed 55 percent the usual Egyptian ounce, providing that barley grains were used in Sinnār. It is not surprising that Pallme wrote "the oock-ckah [*wiqīya*] of Kordofan is rather heavier than that of Sennaar." *Travels in Kordofan*, p. 291.

consequently the Sinnār ounce should have earned only 13 1/3 Spanish dollars in Cairo. Far higher prices were available to gold merchants in Red Sea ports.[80]

At the same time, it was profitable to export silver specie to Sinnār, the exchange rate working in favor of merchants coming from Egypt, and the movement of silver coin up the Nile Valley probably increased during the latter half of the 18th century. Spanish dollars began to appear in Dār Fūr at the end of the century,[81] and in the year 1796, approximately 72,000 dollars were carried back from Egypt by caravan to that kingdom and to Sinnār.[82] In fact, by the early 19th century, Burckhardt found that dollar so "common" in Sinnār that it had apparently replaced the locally-minted silver *ghirsh* in the country's monetary system.[83]

Conclusion

An attempt had been made to relate the trade in gold between Egypt and various parts of sub-Saharan Africa to political and religious movements in those regions. It is believed that by placing the gold trade in the framework of the pilgrimage tradition in West Africa or the abuna-needs of Ethiopia that a sounder basis for understanding that trade can be provided. If Takrūr gold flooded northeast Africa during the 14th-16th centuries, then the mechanism which triggered the flood should also have continued to function in some fashion in the 16th-18th centuries. It is believed that Egypt continued to receive gold from Takrūr up to the beginning of the 18th century, and possibly as late as 1730.

The drying up of the Takrūr source is not only related to the collapse of the pilgrimage tradition in West Africa, but also the development of strong local economies there with increasing ties to states along the West Africa coast. It is possible that the entrenched position of Libyan merchants in Sahelian states diverted more and more of the gold trade away from Egypt and towards Tripoli. At the same time, Egypt's trade with the Sudan was undergoing a reorientation, away from western Sudan and toward eastern regions. In the course of the 18th century, Dār Fūr emerged as Egypt's most active trading partner, supplying her with requisite quantities of slaves, ivory, feathers, tamarind, camels and hides, but not gold. The fact that the

---

80. Bruce mentions the ounce selling for 22 *riyāls* at Mocha; IV, p. 487; Burckhardt quotes the ounce at 12 dollars at Sinnār; 16 at Shandī; 20 at Sawākin; 22 at Jidda: *Travels*, p. 277.

81. al-Tūnisī, *Darfour*, p. 316; *Tashḥīdh*, p. 298.

82. Walz, "The Trade between Egypt and *Billād as-Sūdān*, 1700-1820," Ph.D. thesis (Boston University), p. 74, based on a statistical analysis of the Sudan trade in 1796, as per Girard, "Mémoire," pp. 632-8. The movement of Austrian dollars from Tripoli to Fazzān and possibly further south was noted in 1790 by Beaufoy: R. Hallet, ed., *Records of the African Association, 1788-1831* (London, 1964), p. 93.

83. *Travels*, pp. 216-17: all exchange rates are in dollars for ounces.

price of gold was usually high in Cairo at the end of the century merely indicates the idiosyncracies of the French occupation and not the continued high demand for silver. Egypt's trade with Europe and with Jidda, which operated on the import of European specie, tied her ever closer to the European monetary system. The monetary reforms carried out by Muḥammad 'Alī in 1834 pegged the local currency to the Austrian dollar, a move which simply capped what had long been apparent.

Appendix

Notes on the value of gold, silver and copper in the 17th and 18th centuries

Gold

The value of the gold *mithqāl* is difficult to follow in local records. Its value is usually rendered in local or foreign coin, both of whose values fluctuated greatly from year to year. To control these fluctuations, André Raymond devised a table from which the real or constant value of the *nisf fiḍḍa*, the highly unreliable small silver coin, may be determined.[84] Values of of the *mithqāl* are given below according to current and constant *nisf*.

According to our information, which extends only from 1640 to 1798, the *mithqāl* (4.63 grams) held steady until 1678; between the months June and September, the price of gold rose 9 percent. Raymond would relate this to the period of intense inflation which took place 1677-78 generally. The year 1690 was another period of inflation, after which the price of gold seems to have declined. In 1691 it stood at 110 *nisf* per *mithqāl*; by 1719, it had declined in real terms to 97.75 *nisf*.

The period 1740-80 was a prosperous one; at the beginning gold was valued at 117.6 *nisf*, a result of several economic crises during the years 1721-5, 1731-5, and towards the end of the period, it had dropped to only 99 *nisf*. When the French arrived in 1798, they found the price of gold cheap; we have no information on its value in the years immediately before their arrival. They pegged the Spanish dollar at 150 *nisf*, in accordance with its current value. The *mithqāl* was selling at 3 3/4 Spanish dollars in 1799, not quite twice the rate that prevailed in 1774. This undoubtedly reflects an intensified French demand for gold.

TABLE

Date	Number of *mithqāls*	Other weight	Value as declared	current *nisf*	constant *nisf*
1640	1		2 *riyāls*	66 (?)	98.3
1678 (June)	1			90	94.5
1678 (Sept)	1			100	105
1691	1			120	110
1719	(8)	1 ounce	11 1/2 *ghirsh* (920 *nisf*?)	115	97.75
1737	212		407 *riyāls* (36,630 *nisf*)	173	117.6

mithqāl value in terms of

84. *Artisans*, p. liv.

TABLE (*Continued*)

				mithqāl value in terms of	
Date	Number of *mithqāls*	Other weight	Value as declared	current *nisf*	constant *nisf*
1743	1		1 *findiqlī* (180 *nisf*)	180	117
1750	1			170	103.7
1774	1		2 *riyāls*	180	99
1798	65	97 *dirhams*	244 *riyāls* (36,600 *nisf*)	563	180

Sources:

Arab 36, p. 379, no. 608 (1094/1640); Arab 58, p. 662, no. 113 (1089/1678); Askar 74, pp. 292-6 (1089/1678); Askar 84, p. 425, no. 650 (1102/1691); Askar 118, p. 303, no. 448 (1133/1719); Askar 144, p. 288, no. 416 (1149/1737); SN 519, p. 290, no. 595 (1156/1743); Askar 161, p. 293, no. 376 (1164/1750); Askar 194, p. 330, no. 308 (1188/1774); Bernard, "Monnaies," p. 427. For abbreviations of citations, see footnotes 3 and 33 above.

Silver

Even fewer statistics have been collected on the price of silver, although in fact it is often mentioned in inventories when determining the price of silver anklets, earrings and other personal jewelry. Silver was valued at 6 *nisf fidda* per *dirham* for the period 1741-56, dropping to 5 *nisf* per *dirham* during 1764-71 and as low as 4.86 per *dirham* in 1774.

According to Raymond, the gold-silver ratio remained stable at 1:15 throughout the period 1685 to 1796, despite the occasional shortage of silver; this ratio is, therefore, comparable to the European ratio in the early 18th century. Our figures for the middle and latter part of the 18th century, calculated in constant currency terms on the basis of *dirhams* of gold versus *dirhams* of silver, suggest a gold-silver ratio of 1:10.8 around 1740, of 1:11.3 around 1775 and 1:21.3 during the period of the French occupation (1798-1801). The former ratios appear to correspond to gold-silver ratios typical of the Middle East while the later reflects the unusual French demand for gold.

Sources: Other than those indicated in the previous note, Askar 50, pp. 129-30 (1154/1741); Askar 161, p. 293, no. 376 (1164/1756); Askar 185, p.

404, no. 669 (1185/1771); Raymond, *Artisans*, I, p. 47 n. 4; Spooner, p. 21. For abbreviations of citations, see footnotes 2, 3, and 33 above.

Copper

While neither tabulating the price of gold or silver, Raymond provides a table on copper prices. Averaging 20 *nisf* per *ratl*, they remained steady throughout most of the 17th century. A drop in price began around 1690, reaching 14 *nisf* in 1730, rising to 16 *nisf* at mid-century and dropping again to 9.7 and 8.3 *nisf* per *ratl* in 1788 and 1792. By the end of the century it reached 12 *nisf*.

Source: *Artisans*, Table 19, I, Raymond p. 75.

11

Silver mines and Sung coins—A monetary history of medieval and modern Japan in international perspective

KOZO YAMAMURA and TETSUO KAMIKI

Introduction

The goal of this essay is to describe and analyze the monetary system of medieval and early modern Japan, from the 13th through the 18th centuries. We will focus our attention primarily on those aspects of the monetary system which were affected, directly or indirectly, by the flows of precious metals and copper coins between Japan and her trading partners: her East Asian neighbors and then, from the 16th century, the Western traders who came to Japan.

The first section of this essay provides a brief description of the monetary system and the production of gold, silver, and copper before the 13th century. This description is intended only to provide the background necessary for pursuing our examination of the historical development of the 1200-1750 A.D. period. The second section contains a description of the essential aspects of the changes in the monetary system from the 13th to the mid-16th centuries, with an analysis of the reasons for and effects of the large inflow of Chinese copper coins to Japan and the outflow of gold to China during these centuries. We will offer speculations and hypotheses concerning the quantitative magnitude of the copper coins and gold traded between China and Japan. We will also be concerned with the relative exchange value of gold to silver and gold to copper that prevailed between these two East Asian trading partners. The main subjects of the third section, which covers the mid-16th to the 18th centuries are: the large outflow of silver from Japan to China, and the roles which the Western traders played in the outflow of precious metals. We will also present data on the magnitude of the silver outflow and on the changes in the prices of gold and silver.

Before proceeding, a few paragraphs must be added lest readers' expectations of what we will be able to accomplish are unduly raised. The monetary history of preindustrial Japan is one of the most important areas of historical research remaining which has yet to receive the careful attention of scholars.

To date, only a dozen or so Japanese historians have published on subjects broadly relating to the monetary history of Tokugawa Japan, and the number of scholars interested in the monetary history of the pre-1600 period is even fewer. Published works on the subject in Western languages number no more than two dozen, even including those articles which are only marginally useful to the study of the monetary history of Japan.

There are two major reasons for the neglect suffered by this important aspect of Japanese economic history. One is that most Japanese historians specializing in the premodern period are primarily interested in political and institutional history, and many have adopted the Marxist framework of analysis. This means that their primary interest in the economic aspects of premodern history is, in most instances, confined to providing the necessary ingredients for explaining political changes and conflicts among various classes of individuals within society.[1] Thus, monetary history as defined by most Western economic historians continues to be neglected. The earlier works of Kobata and a few of his contemporaries, and recent works on the Tokugawa period (1600-1867) by Shimbo, Iwahashi, and several other scholars are notable exceptions.[2]

Despite these obstacles, we can ill afford to continue to neglect the monetary history of pre-modern Japan, especially if we are to obtain an understanding of the economic history of premodern Japan and to increase our knowledge of the position which Japan occupied within the international market in East Asia and in the world. The goal of this paper, therefore, is to take a modest step towards a monetary history of premodern Japan which may be useful to those interested in the economic history of Japan as well as to those students of monetary history who may find the Japanese experience useful in their efforts to improve and refine their own research.

The monetary system and the output of gold, silver, and copper before the 13th century

In the middle of the 7th century, a new government was established in Japan by a group which succeeded in subduing its rivals to attain a political and military hegemoney.[3] This hegemon, who came to be called an emperor, fashioned what is known as the Ritsuryō government after the model provided by T'ang China, and it was under the aegis of this government that gold, silver,

---

1. For discussions of the effects which the Marxist framework of analysis has had on the analyses of the economic history of the pre-1868 period, see: Kozo Yamamura, "Introduction to the Workship papers on the Economic and Institutional History of Medieval Japan," *The Journal of Japanese Studies*, Vol. 1, No. 2 (Spring 1975), pp. 225-267; and K. Yamamura, "Toward a Reexamination of the Economic History of Tokugawa Japan, 1600-1867," *The Journal of Economic History*, Vol. 33, No. 3 (1973), pp. 509-546.

2. The most useful of these works are included in the Bibliography appended to this essay.

3. The economy of the Ritsuryō period is described and analysed in Kozo Yamamura, "The Decline of the Ritsuryō System: Hpotheses on Economic and Institutional Change," *The Journal of Japanese Studies*, Vol. 1, No. 1 (Autumn, 1974), pp. 3-37.

and copper were produced and made into coins to be used in making tax payments, in gift-giving, and as media of exchange.

We will begin with a discussion of gold, the most precious of these metals. The first record of the production of gold in Japan is dated 701 and notes that in that year the government dispatched an official to Mutsu, the northeastern part of the main island of Japan, to begin producing gold.[4] The attempt was obviously successful. In 749, gold in the amount of 900 *ryō* (approximately 11,250 grams) was submitted to the government by the provincial magistrate of Mutsu.[5] The magistrate, we should note, was said to have been a descendant of the king of Koguryo (on the Korean Peninsula) which, if true, may explain why he possessed advanced knowledge of gold production.[6] Mutsu continued to remain the principal source of gold until the 16th century.[7]

Gold was used for three purposes. First, it served as gifts to the Imperial household, nobles, temples, and foreign (Chinese and Korean) emissaries. Gold used in gift-giving usually consisted of gold dust (alluvial gold) contained in a small cloth bag. (We know that from the early 14th century on, each bag generally contained ten *ryō*, i.e., 168.75 grams of gold.)[8] The second use made of the gold was in gilding and plating the statues of Buddha and other ceremonial paraphernalia relating to the Buddhist religion. And, lastly, gold was used as a means of paying for the goods brought by the Chinese and Korean ships to Japan, beginning in the 9th century. All of Japan's trading with her Asian neighbors was conducted by the government and payments were made in gold dust, silk batting, and raw silk. After the 10th century when trade with Sung and Southern Sung China (960-1279) was carried on, the trade balance was settled principally in gold. Though we are unable to establish the magnitude of this trade, the extant records note that gold was in chronically short supply in Japan.[9] We are thus led to assume that while trade increased and required more gold, the supply of gold lagged behind demand due to the limitations imposed by the technology of the

---

4. *Shoku Nihongi* included in Saeki Ariyoshi *ed.*, *Kōtei zōho rikkokushi* (A revised and expanded compilation of six historical sources) Vol. 3, (Tokyo: Asahi Shinbunsha, 1940), p. 18.

5. *Ibid.*, pp. 356-7.

6. See: Imai Keiichi, "Kudara no Konikishi Kyofuku" (The king of Koguryo Kyŏng Bok) (Kyoto: Sōgeisha, 1965).

7. Until the mid-16th century when new mining technology became available (as will be discussed in the text), gold produced in Japan was limited to alluvial gold. Kobata Atsushi, *Nihon Kōzanshi* (A study of the history of Japanese mines), (Tokyo; Iwanami Shoten, 1968).

8. Kobata Atsushi, *Kaitei zōhō Nihon kahei ryūtsūshi* (A revised and expanded history of the uses of money in Japan) (Tokyo: Tōkō Shoin, 1943), pp. 282-302.

9. For example, because of a shortage of gold the Japanese government in 988 was unable to settle its trade balance with Sung merchants. See: an entry of the thirteenth day, seventh month, 988 in *Gonki* (a diary written by a high-ranked noble, Fujiwara Yukinari) in Yano Tarō *ed.*, *Shiryō taisei* (A compilation of historical sources), Vol. 35, (Tokyo: Naigai Shoseki, 1939), p. 40.

time.[10] We can, however, be certain that the government exerted direct control over the production of gold and that gold was the most important product which Japan then exported.

Though it is known that the Ritsuryō government minted gold coins once in 760, it is most unlikely that these were used as a medium of exchange. The quantity minted was presumably extremely small—only thirty-one such gold coins have been discovered to date—and, given the types and quantities of exchange then carried out in Japan, gold coins could hardly have been useful in daily transactions.[11] It is safe, therefore, to conclude that the gold coins were most likely used in gift-giving and that only in the late sixteenth century were the first gold coins minted in Japan to serve as a medium of exchange.[12]

Silver, according to the *Nihon shoki*,[13] was first presented in 675 to the Ritsuryō government by the provincial magistrate of Tsushima—the island province which continued to be the only major producer of silver in Japan until the early 16th century. Tsushima, located halfway between the Korean Peninsula and the island of Kyushu, was the most important link in the transmission of advanced technologies from China and Korea to Japan during this period. It is perhaps correct to assume, therefore, that the technology necessary in mining and smelting silver was first brought to Tsushima from the continent. The silver mines on the island were managed by the magistrate of Tsushima, and the mining operations appear to have been supervised by high-ranking officers of the Dazaifu, the Ritsuryō government's regional office in charge of Kyushu.[14]

Though the extant records are silent on the output of the Tsushima mines, it is known that in one year (not identified) 890 *ryō* (33,375 grams) of silver were included in the dues paid to the government through a provincial

---

10. Though the province of Shimotsuke is known to have produced gold during the mid-9th century, the principal and almost sole source of gold was Mutsu province. However, the veins in Mutsu were exhausted by the end of the 10th century as the result of the continued mining to meet the demands of merchants who came from Kyoto. Mori Katsumi, *Shintei Nisso boekishi* (A revised study of the international trade between Japan and Sung China) 3 vols. (Tokyo: Kokusho Kankōkai, 1975), p. 266.

11. Kobata Atsushi, *Nihon no kahei* (Monies in Japan) (Tokyo: Shibundō, 1958), p. 7.

12. See footnote 41 below.

13. *Nihon shoki* included in Kuroita Katsumi ed., *Shintei zōho kokushi taikei* (A revised and expanded compilation of sources for Japanese history), Part I of Volume 1, (Tokyo: Yoshikawa Kōbunkan, 1951), p. 335.

14. As was the case in gold mining, advanced mining and smelting techniques used in the production of silver were also undoubtedly brought from China and Korea. Using the capacity of silver to resist rapid oxidation, the smelting technique of the period consisted of oxidizing lead, sulphur, and other impurities by fire for about two weeks. Also, according to the extant records, the silver mines in Tsushima had tunnels as long as 1,000 meters by the second half of the 9th century. Vertical shafts too became much deeper after the 10th century, requiring a large amount of labor to drain them. The cost of this labor was defrayed by the Ritsuryō government. Kobata Atsushi, *Nihon kōzanshi no kenkyū*, (A study of the history of Japanese mines) (Tokyo: Iwanami Shoten, 1968), pp. 45-50, and pp. 97-100.

Table 1. The Twelve Imperial Coinages (*Kōchō 12-sen*)

Name of coins[a]	Period of coinage	Diameter of coins (cm)[b]	Weight of coins (gr)[b]
(1) Wadō kaichin	708–760	2.73–2.42	5.25–3.15
(2) Man-nen Tsūhō	760–unknown	2.73–2.42	5.63–3.68
(3) Jingō kaihō	765–unknown	2.58–2.27	4.43–3.00
(4) Ryūhei eihō	796–817	2.64–2.42	3.75–2.78
(5) Fuju shinpō	818–834	2.42–2.27	3.75–3.00
(6) Jōwa shōhō	835–837	2.27–1.91	3.38–2.33
(7) Chōnen taihō	848–859	2.12–1.67	3.00–1.54
(8) Gyō-eki shinpō	859–868	2.12–1.76	2.63–1.50
(9) Jōgan eihō	870–889	2.00–1.76	2.63–2.10
(10) Kanpyō taihō	889–906	1.91–1.52	2.63–2.29
(11) Engi tsūhō	907–957	1.91–1.82	2.63–2.03
(12) Keng en taihō	958–unknown	1.97–1.82	2.51–1.86

[a]All of these coins contain some lead. The last three coinages are known to have contained as much as one-third their weight in lead.

[b]These figures are for the maximum and the minimum of the coins' diameter and weight.

Sources: (1) Ōkurashō ed., *Dai-Nihon kahei-shi*, (Tokyo: Chōyōkai, 1925) Vol. 1, p. 29, pp. 71-79. All nine volumes were reissued in 1969 by Rekishi Toshosha in Tokyo. (2) Kusama Naokata, "Sanka zui" in Takimoto Seiichi ed., *Nihon keizai taiten*, Vol. 39, (Tokyo: Keimeisha, 1929), pp. 32-78.

office. We also know that 300 *ryō* (11,250 grams) of silver were exchanged by the government for the goods which it had received in tax payments in one year during this same period, and that the inhabitants of Tsushima ordinarily paid their dues in silver.[15]

We can hardly begin to speculate on the annual output of silver with such meager evidence. However, we are justified, we believe, in asserting that the output of silver was even smaller than that of gold, which was the principal export product of the period. Supporting this assertion is the fact that Chinese silver is known to have been imported from the late 11th century to the mid-16th century when the output of silver in Japan began to rise sharply as we shall describe later in this essay.

Copper was produced during the Ritsuryō period in much larger quantities than either of these precious metals, and copper coins were minted in fairly large quantities to serve as an increasingly significant medium of exchange. The first major minting of these coins, called *Wadō kaichin*, occurred in 708 and was the first in a series of twelve known major mintings of copper coins which were undertaken by the Ritsuryō government during

---

15. *Engishiki* in Kuroita ed., *op. cit.*, Vol. 26, p. 587, p. 554, and p. 622.

the following two and a half centuries. As shown in Table 1, the last of these mintings was in 958.

We can only speculate on the total amount of copper used in these mintings from the occasional mention of quantitative magnitude in the historical records. From one such description we know that in the fifth minting, which lasted from 813-834, the annual amount of copper used was 5,670 *kan* (21,262.5 kilograms) for the 818-821 period; 3,500 *kan* (13,125.0 kilograms) for the 822-828 period. Though it is most likely that the real output was less than that indicated in these figures because of the constraints imposed by the lagging supply of copper, these data, if taken at face value, indicate that the total output of copper coins during the 818-834 period amounted to 113,180 *kan* or about 424,425 kilograms.[16] Because one *kan* of cooper coins consisted of 1,000 pieces of one *mon* coins, the total output was in excess of 113 million coins. Believing that the supply of copper did in fact place a significant limitation on the minting made in these early decades of the ninth century, the authors of the *Dai-Nihon kaheishi* estimated that the total output of the minting for the entire 818-834 period did not exceed 51,990 *kan* (194,962 kilograms), or 51,990,000 pieces of coin.[17] Since it is evident in the records that shortages of copper were frequent, and that the government was known to have melted down coins issued earlier in order to mint new coins, we are fully justified in placing confidence in a more conservative estimate of the total output.[18] On the other hand, we must not neglect the fact that these mintings contributed to a secular inflation. Though no reliable time series data exist, we can clearly establish for the Kinai region (the seat of the central government) and its environs, that the price level began to show a distinct rising trend from the beginning of the 8th century, reaching a peak during the mid-8th century, and that the trend continued into the first half of the 9th century after a brief reversal during the 760s.[19]

Although we need not elaborate in this essay, we note briefly that on the basis of extant evidence and our analysis of the economies of the Ritsuryō and Kamakura (1185-1333) periods, we believe that a thorough reexamination must be made of the view held by most Japanese historians that the Ritsuryō economy was never monetized. Many scholars stress that coins were minted by the Ritsuryō government basically to "imitate the monetary system of T'ang China" and "not to meet natural demand rising from below."[20] They argue that money was used mostly to meet various tax

16. *Ruijū sandai kyaku*, in *Ibid.*, Vol. 25, p. 418.
17. Okura-shō ed., *Dai Nihon kaheishi* (A history of monies in Japan) (Tokyo: Chōyōkai, 1925), Vol. 1, p. 48.
18. Contemporary sources often note that "old coins have been exhausted" indicating that the government mint suffered from constant shortages of copper. See pp. 419-420 of the source cited in footnote 16.
19. Toyoda Takeshi and Kodama Kōta eds., *Ryūtsūshi* (A history of commerce), Vol. 1, (Tokyo: Yamakawa Shuppan, 1969), pp. 36-38.
20. Sakudō Yōtarō, *Nihon kaheishi gairon* (An outline of the Japanese monetary history), (Tokyo: DaiNihon Kaheishi Kankō-kai, 1970), p. 78.

obligations (including cash payment in lieu of corvée) and only in part to pay the wage labor engaged by the government. Thus, they maintain, "use of cash could hardly be expected when exchanges were conducted using unhulled rice and cloth as media of exchange."[21]

We believe that this widely accepted view seriously underrates the degree of monetization achieved during the Ritsuryō period. To cite only a few facts in support of our view, we can, for example, show that use of wage labor increased from the 8th to the 9th century, replacing corvée labor, and that cash was paid to an increasing number of the wage laborers employed by the government. We can also show that these wage laborers were in fact part-time workers who were primarily cultivators, i.e., they were supplementing their income by earning cash from doing government work. Furthermore, indicating that a labor "market" had developed during this period, wages paid to skilled workers (craftsmen) are known to have been as much as six times those paid to nonskilled workers.[22]

To us, such evidence and the fact that copper coins began to be minted illegally during the 10th century suggest that the economy was not one based on "slave-like labor" which was freely commanded by the government, but was rather an economy which had achieved a considerable degree of monetization and in which the demand for cash, even by cultivators, was steadily rising.[23] On these grounds and others we argue that the degree of monetization achieved during this period was not a reflection of the limited demand for a medium of exchange. Rather it reflected the shortage of copper, shortsighted policies adopted by the government to enforce unrealistic exchange rates between newly minted and old coins, and the decline of the Ritsuryō system itself by the 10th century.[24] If we subscribe to the view widely held by Japanese scholars, we are hard put indeed to explain why a large quantity of Chinese copper coins came to be imported in the 11th century.

21. Hirano Kunio, "Nihon ni okeru kodai kōgyō to shukōgyō," (Mining and handicraft in ancient Japan), in Ishimoda Shō, ed., *Kodaishi kōza*, (Lectures on the ancient history), Vol. 9, (Tokyo: Gakuseisha, 1963), p. 169.
22. Seiki Akira and Aoki Kazuo, "Heijō-kyō" in Ishimoda Shō, ed., *Nihon rekishi kōza* (Lectures in Japanese history), Vol. I, (Tokyo: Kawade Shobō, 1953), pp. 253-258.
23. On economic changes in this period, see: Yamamura, "The Decline of the Ritsuryō System," *op. cit*. Also see: Hosokawa Kameichi, *Jōdai kahei keizaishi* (A monetary history of the ancient period), (Tokyo: Moriyama Shoten, 1934), pp. 186-205 for a description of the appearance of illegal privately minted coins during the second half of the 8th century. Despite the fact that the law of this period called for the death penalty for counterfeiters, the magistrate of the province of Yamato is reported to have apprehended many such offenders. The punishment obviously did not deter them.
24. For example, the Ritsuryō government, facing a shortage of copper coins, reduced the weight and diameter of the coins (as seen in Table I, the coins became lighter and smaller over time) and also attempted to force an exchange rate of one of the smaller and lighter new coins to ten old coins. Of course, the declining Ritsuryō government was unable to enforce this measure. See: Kobata, *Monies in Japan, op. cit.*, pp. 11-14.

The influx of Chinese copper coins in medieval Japan

The Ritsuryō government, a civil government of nobles, was replaced during the closing decades of the 12th century by the Kamakura Bakufu, a military government (*bakufu*) supported by an increasing number of local powers who could be called *samurai* in the broad sense of the term.[25] The Kamakura period, which continued until 1333, was characterized by the steady growth of agricultural productivity, the growth of local markets, and an increased monetization of the economy.[26]

Despite the clearly discernible development of the economy, we have little information concerning gold and silver. The Kamakura Bakufu had defeated the regional power who ruled the gold-producing province of Mutsu and gained direct control over the gold produced in this province. But the Bakufu's acquisition of the mines was undoubtedly a by-product of its successful campaign as no evidence exists indicating that the Bakufu had wanted to acquire Mutsu for its gold.

The existing records show, however, that the Kamakura Bakufu continued to have the Mutsu mines worked and it continuously presented gold to the Imperial Court (which helped to establish the Bakufu's claim for legitimacy). However, the amount of gold presented to the Court was small. For example, 450 *ryō* (5.63 kilograms) was presented in 1186, 500 *ryō* in 1200, and 450 *ryō* in 1285.[27] Though we cannot establish the total amount of gold produced during this period, these gifts to the Court most likely constituted only a small part of the gold which the Bakufu acquired from the Mutsu mines and from mines in the Suruga, Hitachi, and Sado regions which are also known to have begun producing gold during this period.[28]

Gold was, along with sulphur, an important export product in Japan's trade with Sung (960-1126) and Southern Sung (1127-1279). Though no reliable data exist from which to estimate the total amount of gold exported to the Southern Sung, we know that during the mid-13th century the Japanese merchants exported 4,000-5,000 *ryō* (67.5-84.4 kilograms) of gold to Ch'ing-yüan fu in Ningpo. If we add the estimated total amount of gold which was exported to Ch'uan-chou and that taken to China by Sung merchants themselves, the total amount exported during this period could have been as large as 10,000 *ryō* (168.75 kilograms).[29] We cannot ascertain for how many years gold exports to China continued at this high level, but we can be reasonably

---

25. Although its power declined steadily, the "civil" administrative structure of the Ritsuryō government continued to coexist with the "military" administrative structure of the newly established Kamakura Bakufu.

26. For a good description of the Kamakura economy, see: Nagahara Keiji, *Chūsei no shōen shakai* (The *shōen* society of medieval Japan), (Tokyo: Iwanami Shoten, 1968), pp. 156-171.

27. For a list of gifts made to the Imperial Court, see: *Azuma kagami* on pp. 244, 516, and 576, Vol. 32 of the source cited in footnote 13.

28. Kobata, *A revised and Expanded History, op. cit.* p. 268-274.

29. Kato Shigeru, *Tō-Sō jidai ni okeru kin gin no kenkyū*, (A study of gold and silver of the T'and and Sung periods), (Tokyo: Tōyō Bunko, 1926), pp. 557-560.

Silver mines and Sung coins 337

certain that a large quantity of gold was exported to China during the 13th century and the early decades of the 14th. As the export of gold was closely tied to the Japanese import of copper coins from China, we shall return to this trade shortly.

We know little concerning silver. There exists no historical evidence indicating that silver was exported from Japan during the period under discussion. Rather, it is known that during the mid-15th century and the early 16th, the Ming envoys customarily brought gifts of silver from the Ming Emperor to the Ashikaga Shōgun (the hegemon of the new Bakufu which replaced the Kamakura Bakufu during the late 14th century).[30] This seems to suggest that the direction of the flow of silver was from China to Japan until it was reversed during the 16th century.

The most noteworthy development in the monetary history of medieval Japan was a large influx of copper coins from China. This influx occurred because Japan, which no longer produced its own coins after the decline of the Ritsuryō government, was eager to obtain the large quantity of Chinese copper coins needed as a medium of exchange for the expanding market activities within Japan. Thus, when trade between Japan and Sung China increased after the mid-12th century, China became the supplier of this principal medium of exchange.

Most of the Chinese copper coins, it is frequently said, consisted of those minted during the Northern Sung (960-1126) period, and the frequency distribution of the coins unearthed by archaeologists and historians, as presented in Table 2, seems to support this view. As is evident in the table, the highest proportion (84.77 percent) of the coins found to date are Northern Sung coins, and if we add Southern Sung coins to this, Sung coins account for nearly 86 percent of the total number of coins unearthed to date. Though we cannot demonstrate that the high proportion of the Sung coins found in Table 2 accurately reflects the real proportion of coins imported, we feel justified in concluding that, of the total Chinese copper coins imported during this period, 80-85 percent were the Sung coins. We should also remind ourselves that not all of the Sung coins had to have been imported during the Sung period because it is readily conceivable that the coins continued to be imported to Japan even during the centuries following the demise of the Sung dynasty.

The continued importation of these copper coins was the result of Japan's desire to obtain her own medium of exchange and not of the Chinese desire to export them is evident since it is well known that the Chinese, from as early as the T'ang period, attempted to prohibit export of their copper coins. In fact, from the Northern Sung period on, the shortages caused by the large outflow of these coins were seen as a grave problem by the Chinese govern-

---

30. Kobata Atsushi, *Chūsei Nisshi tsūkō bōekishi no kenkyū* (A study of international trade between Japan and China during the medieval period), (Tokyo: Tōkō Shoin, 1941), pp. 392-393.

Table 2. Chinese copper coins unearthed in Japan[a]

Dynastic names	Irita survey[b]	Kobata survey[c]	Total	Percent
Han	20		20	
Sui				
T'ang	47,299	31,008	78,307	8.87
Ch'ien Shu	68		68	
Hou Chin	3		3	
Nan T'ang	445		445	0.05
Hou Han	11		11	
Hou Chou	72		72	
Northern Sung	456,088	292,098	748,186	84.77
Liao	6		6	
Hsi Hsia	5		5	
Southern Sung	8,065	928	8,993	1.02
Chin	1,016		1,016	0.12
Yuan	163	31	194	0.02
Han Ch'en Yu-liang	4		4	
Ming	40,559	3,077	43,636	4.94
Korea	621	128	749	0.08
Others[d]		832	832	0.08
Total	554,608	328,005	882,613	

[a]A small number of Korean coins and privately minted Japanese coins are included in the category "others."

[b]Irita's work includes the Chinese coins which were minted before 1433 and were unearthed in eighteen specific provinces.

[c]The Kobata survey includes the Chinese coins which were minted before the end of the 15th century. There is no overlap between those coins identified by Kobata and by Irita.

[d]This category includes the Japanese coins referred to in a and those coins which could not be identified, due to erosion or chipping.

Sources: Irita Seizo, "Hakkutsu sen ni tsuiteno kōsatsu," (An analysis of the unearthed coins), *Kōkōgaku zasshi*, vol. 20, no. 12 (1930), pp. 820–829. Kobata Atsushi, *Kaitei zōho Nihon kahei ryūtsushi* (A revised and expanded history of the uses of money in Japan) (Tokyo: Toko-shoin, 1943), p. 64.

ments of the subsequent periods. The Chinese sources first name Japan as the principal culprit of this major economic problem in 1171, and a decree issued in 1214 by the Sung government which prohibited the outflow of the coins also cites the Japanese, along with the Koryo (Korean) merchants, as the offenders who constantly endeavored to carry away copper coins.[31]

31. Mori, *A revised study, op. cit.*, pp. 493-494; and Kobata, *Monies in Japan, op. cit.*, pp. 39-40. Also, for a discussion of the prohibition of the reexport of copper coins during the Sung period, see Sogabe Shizuo, *Nisso kin kahei kōryū-shi* (A history of trade in monies between Japan and Chin China), (Tokyo: Hōbunkan, 1949), pp. 75-96.

The Chinese accusations of the Japanese were well-deserved, because it was precisely during this period—from the latter part of the 12th century to the mid-13th—that the Japanese succeeded in importing increasing quantities of Chinese copper coins. The Japanese ships visited not only the designated port of Ch'ing-yüan (Ningpo) but also the coastal towns of Wen-chou and T'ai-chou to trade, i.e., to obtain copper coins. It is said that copper coins totally disappeared from the cities in T'ai-chou within a day after these Japanese traders came in search of them.[32] Though observations made in the Chinese sources cannot be verified, there is little doubt, especially given the facts presented in Table 2, that an enormous number of Chinese copper coins were brought to Japan during the first half of the 13th century.

The trade between China and Japan slackened and was reduced to a virtual standstill during the late 13th century and throughout the 14th. The reasons for this prolonged decline in trade were many. The Southern Sung, which had been attempting to limit trade with Japan, yielded in 1279 to the Mongols who twice attempted to invade Japan during the final decades of the 13th century. The Mongols in turn were repelled to the borderlands in 1368 by Chu Yüan-chang, who established the Ming dynasty. In Japan, the Kamakura Bakufu was growing increasingly unable to control the emerging regional powers by the late 13th century. Further weakened by the heavy burden imposed on it and its vassals by the Mongol invasions, the Bakufu collapsed in 1333, and the nation suffered more than half a century of political instability and military conflicts, until Ashikaga Yoshimitsu was able to solidify the foundation of a new Bakufu during the final decades of the 14th century.

Political stability was finally regained during the late 14th century in both nations and the resumption of trade between them became possible.[33] The Bakufu took the initiative to reopen commerce, as it was eager to replenish its coffers even at the cost of appearing subservient to the Emperor of Ming China. The trade resumed in 1401 consisted both of tribute paid by the Bakufu to the Ming emperor and of private trade which was also permitted under strict conditions imposed by the Chinese. Because of the restrictions imposed by the Ming government, Japanese could send her ships—limited in number, cargo, and size of crew—to China only once a decade after 1465. Thus, during the 150 years between 1401 and 1547, the Japanese were able to

---

32. Sogabe, *op. cit.*, pp. 162-163.
33. Though reduced in quantity, trade between Japan and China continued even during the years in the 1330s between the fall of the Kamakura Bakufu and the establishment of the Ashikaga Bakufu. Although the official relationship between the two nations had been disrupted by the Mongol invasions of 1274 and 1281, the new Yuan government allowed the Japanese to trade. Thus, after 1325, the Japanese sent ships to China. The Japanese ships sent in 1341, for example, were known for the large profits they earned which were used in building the Tenryū temple. Trade with Korea also continued, but it was on a small scale because of the political instability on the Korean peninsula and the difficulties created by the *wakō* (Japanese pirates). For further discussions of Japanese trade with China and Korea during this period, see: Tanaka Takeo, *Wakō to kangō bōeki* (*Wakō* and the tally trade), (Tokyo: Shibundō, 1961).

Table 3. Selected facts concerning the Ming China-Japan "Tally" trade, 1401–1547

	Year of departure to China	Year of arrival in China	Number of ships
1	1401	1401	
2	1402		
3[a]	1403	1403	
4	1404	1404	
5	1405	1405	
6	1406	1407	38 each
7	1408	1408	
8	?	1408	
9	?	1410	
10	1432	1433	5
11	1434	1435	6
12	1453	1453	9
13[b]	1465	1468	3
14	1476	1477	3
15	1483	1484	3
16	1493	1496	3
17[c]	?	1509	1
	1511	1512	3
18[c]	?	1523	3
	1523	1523	1
19	1539	1540	3
20	1547	1549	4

[a]The "tally" trade began in 1403.

[b]Beginning in this year, the Ming government limited the tally trade to once a decade and to no more than three ships per voyage. Also, each ship was limited to a total of 300 persons including crew, merchants, and government officials.

[c]In these years, two major regional powers (*shugo daimyo* Hosokawa and Ouchi) dispatched ships which they sponsored separately.

Source: Tanaka Takeo, "Japan's Relations with Overseas Countries" in J. W. Hall and T. Toyoda, eds., *Japan in the Muromachi Age* (Berkeley: University of California Press, 1977), pp. 168–170.

dispatch their ships only twenty times, and the ships sent after 1403 were required to possess a tally issued by the Ming government.[34] Table 3 presents basic facts known concerning the tribute trade—the so-called *kangō* (tally) trade—of this period. Though such critical facts as the total volume and value of cargo carried in each voyage, the prices of goods shipped, etc., are unavailable, we learn from the table that the Japanese tribute, though it changed in composition over time, consisted mostly of gold, swords, sulphur,

34. *Ibid.*, pp. 61-63.

and other goods, and that the Chinese emperor sent copper coins, silver, and silk goods in return.[35]

As is evident in extant sources and shown in Table 3, private (i.e., non-official) trade was allowed to increase after the first decade of the fifteenth century. On the Japanese side, trade was financed by large merchants in port cities and the major regional powers *(shugo daimyo)*. Once in China, trade was closely controlled by the Ming government and was conducted on a consignment basis by officially designated merchants. Officially set exchange rates (expressed in terms of the Ming note, paper money in the *ch'ao* unit denomination) had to be used in conducting private trade. The rates which remained in effect during the span of this trade were: one *k'uan* of the Ming note to one *k'uan* of copper coins; and one *k'uan* of copper coins to one *liang* of silver.[36] (The Chinese characters used for *k'uan* and the Japanese *kan* were identical, as were those for *liang* and *ryō*, and *wen* and *mon*.)[37] The most important item of "non-official" Japanese export was swords and imports consisted mostly of copper coins. (It is interesting to note also that the Japanese exported a fairly large quantity of copper while importing copper coins. We shall return to comment on this fact in the concluding section.)

Leaving to the last section the crucial question of why Japan did not mint her own coins instead of depending on the imported Chinese copper coins as the principal medium of exchange, let us now turn to examine the relative values of gold, silver, and copper coins prevailing in China and Japan during the 12th through 14th centuries.

Although the data from Chinese sources of this period (presented in Table 4) are limited, what there is indicates that in China from the beginning of the 12th century to the early decades of the 13th, the value of gold rose in terms of copper coins and that one *liang* of gold was worth 30,000-40,000 *wen* of copper coins during most of the 12th century. The value of silver also rose vis-à-vis copper coins during this same period as is shown in Table 5. The data in Tables 4 and 5 suggest that the relative value of gold to silver was generally around 1:12-13, except for a much higher value of silver at the beginning of the 11th century. However, we believe that in Japan the relative value of gold to silver was around 1:5 and declined to 1:3-4 by the end of the 13th century. (We believe that the relative value of gold to silver did not rise during the 14th century, and the relative value during the first half of the 15th century was as shown in Table 10.) Our speculations on the relative values of gold and silver are based primarily on the relative values of gold to

35. For a more detailed description of the tally trade, see Tanaka Takeo, "Japan's Relations with Overseas Countries," in Hall, J. W. and Toyoda, T. eds., *Japan in the Muromachi Age*, pp. 159-178.

36. Tanaka, *Wakō, op. cit.*, p. 129.

37. Throughout the periods which are covered in this essay, the weights of copper coins, that is, of *k'uan* and *kan*, and *wen* and *mon* were identical. Both in China and Japan, one *k'uan* or one *kan* consisted of 1,000 copper coins (1,000 *wen* or 1,000 *mon*). However, for gold and silver, despite the same Chinese ideographs used in both nations, the weights of one *liang* and one *ryō* differed after the 13th century. For the differences in weights, see Tables 4 and 5.

Table 4. Price of gold in terms of copper coins: China[a]

Year	Price of gold in copper coins (in *wen*) I[b]	II[c]	Location
1100	10,000	4,500	Pien-ching
1126	20,000	9,000	"
1126	32,000	14,400	"
1127	35,000	15,750	"
1127	32,000	14,400	Hangchow
1134	30,000	13,500	"
1209	40,000	18,000	"

[a]As the data are extremely scarce, all known quantitative evidence, including that for the same year, is presented.
[b]Price of one *liang* of gold (37.5 grams).
[c]Price of gold weighing 16.875 grams, which is the weight of Japanese one *ryō* of gold. The gold content of the Chinese and Japanese gold coins is assumed to be identical.
Source: Kato Shigeru, *Tō Sō jidai ni okeru kin gin no kenkyū*, (A study on gold and silver of the T'ang and Sung period), (Tokyo: Tōyō Bunko, 1926), Vol. II, p. 473.

Table 5. Price of silver in terms of copper coins: China[a]

Year	Price of silver in copper coins (in *wen*) I[b]	II[c]	Location
1126	1,500	645	Pien-ching
1126	2,200	946	"
1127	2,500	1,075	"
1134	2,300	989	Hangchow
1160	3,000	1,290	Kuang-hsi lu
1198	3,000	1,290	Hangchow
1198	3,300	1,419	"
1225	3,300	1,419	"

[a]As the data are extremely scarce, all known quantitative evidence, including that for the same year, is presented.
[b]Price of one *liang* of silver (37.5 grams).
[c]Price of silver weighing 16.125 grams, which is the weight of Japanese one *ryō* silver coins. The silver content of the Chinese and Japanese silver coins is assumed to be identical.
Source: Kato Shigeru, *op. cit.*, p. 473.

rice, and silver to rice. For example, 3 *koku* of rice are known to have been exchanged for one *ryō* of gold in 1156. This means that if the contemporary observation was accurate in noting that one *koku* of rice was commonly sold for one *kan* (1,000 *mon*) of copper coins, one *ryō* of gold was roughly

Table 6. Output of the Iwami silver mines

Year	Annual (average) output (unit-kilogram)
1581	434.070[a]
1597	298.151[a]
1683–1706	1,348.277[b]
1707–1732	692.761
1733–1758	354.790
1759–1784	465.032
1785–1810	454.666
1811–1836	404.955
1873–1857	159.924

[a]These amounts are *not* output but the amounts which were submitted to Toyotomi Hideyoshi who, by this time, had virtually succeeded in unifying the nation under his rule. If we assume, based on the standard practice of the period, that only one-fourth or at most one-third of the total output was submitted as *de facto* tax to Hideyoshi, the total output could have been three or four times the amount indicated in the table.

[b]The data from this row and those below are the annual average outputs of the period indicated.

Source: Kobata Atsushi, *Nihon kōzan-shi no kenkyū* (A study of the history of Japanese mines), (Tokyo: Iwanami-Shoten, 1968), p. 122, p. 136, and p. 137.

equivalent to 3,000 *mon* of copper coins. Silver, on the other hand, was known to have been traded at one *ryō* to 600 *mon* of copper coins.[38]

Limited though our evidence may be, we can safely assume that the relative value of silver to gold was significantly lower in China than in Japan. And this fact must be the reason for the continued inflow of silver from China to Japan from the Ritsuryō period on. However, the outflow of gold from Japan to China and the influx of copper coins to Japan should be analyzed, we believe, by concentrating on the differentials in gold/copper coin ratios prevailing in China and Japan. During the 12th and 13th centuries, the differential was large: one *ryō* of gold in China could be exchanged for nearly five to six times the amount of copper coins than could be obtained in Japan (See Table 4). A substantial part of this large differential clearly must have been due to the continuing increase in the demand for these coins by the growing and monetizing Japanese economy of the 13th century, and to the declining value of copper coins in China as we have just noted. That is, we argue here that the diverging value of copper coins in China and Japan was the principal factor accounting for the sustained outflow of gold from Japan to China and the observed huge influx of Chinese copper coins to Japan.

Finally, we should note briefly that the relative value of gold to copper coins in Japan was generally stable until the early 15th century and possibly to as late as the 1530s despite the continued huge influx of copper coins in

38. Kobata, *A revised and expanded history, op. cit.*, pp. 309-310.

Table 7. Estimates of outflow of silver from Japan

	Subperiod				
Year	I 1601–94	II 1695–1709	III 1710–13	IV 1714–35	V 1736–1817
Satō estimate[a]	3,911,028.8	299,265.0	352.5	1,481.3	1,833.8
Iwahashi estimate[b]	3,750,000.0	112,500.0	11,250.0	131,250.0	56,250.0
Annual average of Iwahashi estimate	39,893.6	7,500.0	2,812.5	5,966.9	686.0

[a]Satō Chuzaburō's estimate is based on a report made by the Ministry of Finance in 1875 on the supplies of gold and silver coins and on the flows of precious metals. Though long used by earlier scholars, Satō's estimate should be seen only as a crude summary of the historical data included in the report of 1875.

[b]For subperiod I, Iwashashi chose to rely on a 1736 report of the Silver Guild which stated the outflow for the period was about 3.4 million kilograms. Iwashashi, allowing for nonreported outflows, adjusted the figure upward by about 10 percent. For subperiods II to V, Iwashashi adjusted the Satō data to take into account recent estimates on outflows of silver from Tsushima and the Ryukyu islands.

Sources: For the Satō estimate, see: Yamaguchi Kazuo, "Edo jidai ni okeru kingin-ka no aridaka" (The stocks of gold and silver coins during the Tokugawa period), *Keizaigaku Ronshū*, Vol. 28, No. 4 (1954). Iwashashi Masaru, "Tokugawa jidai no kahei sūryō—Satō Chūzaburō sakusei kahei aridaka-hyō no kentō" (The supply of monies in the Tokugawa period—an examination of the tables of stocks of monies, prepared by Satō Chuzaburō), in Umemura Mataji et al ed., *Nihon keizai no hatten* (The development of the Japanese economy), Vol. 1 of the quantitative economic history series, (Tokyo: Nihon Keizai Shinbunsha, 1976), p. 254. Also for the outflow of silver from Tsushima and the Ryukyu islands to Korea, see: Tashiro Kazui, "Tokugawa jidai ni okeru gin yushitsu to kahei aridaka," (The export of silver and stocks of monies during the Tokugawa period), also in the book cited in (2) above.

Japan (see Table 9). Because we have no evidence indicating that the supply of gold and silver in Japan rose appreciably before the beginning of the 16th century, this stability in the relative value of gold to copper coins in Japan over such a long period must be explained in terms of the rapidly increasing demand for copper coins to meet the needs of the growth and monetization which the Japanese economy was undergoing during this period.[39]

39. Here, we refer to the rapid increase in the transactions demand for these coins. The principal medium of exchange shifted from rice (and cloth) to copper coins during the Kamakura period. See: Takeuchi Rizō, ed., *Tochi seidoshi* (A history of the land system), (Tokyo: Yamakawa Shuppan, 1973), pp. 342-352.

Silver mines and Sung coins

Table 8. Estimates of silver transferred to the Bakufu (unit = kilogram)

	Subperiod	Sado Amount transferred	Sado Annual average	Ikuno Amount transferred	Ikuno Annual average	Iwami Amount transferred	Iwami Annual average
I	1601–1694	588,592.5	6,195.0				
II	1695–1709	61,260.0	4,083.8	45,266.3	3,018.8	17,328.8	1155.0
III	1710–1713	15,971.3	3,993.8	7,023.8	1,755.0	2,317.5	581.3
IV	1714–1735	51,438.8	2,340.0	37,672.5	1,713.8	14,028.8	637.5
V	1736–1817	91,537.5	1,117.5	142,695.0	1,740.0	35,996.3	438.8

Note: See text for a discussion of the periodization.
Source: Iwahashi, *op. cit.*, p. 254.

Table 9. Relative value of silver to gold in China (silver per one unit of gold)

Year	Silver
1000	6.3
1126	13.3
1134	13.0
1198	12.1
1375	4.0
1385	5.0
1386	6.0
1397	5.0
1415	10.0[a]
1436	5.0
1481	7.0
1502	9.0
1572	8.0
1620	8.0
1635	10.0
1637–40	13.0

[a]We are unable to explain this departure from the observed trend.
Sources: Kato, *op. cit.*, p. 475. Kobata Atsushi, "Nihon no kin gin gaikoku boeki ni kansuru kenkyū (II)" (A study on the Japanese international trade of gold and silver), *Shigaku Zasshi*, vol. 44, no. 11 (1933), pp. 62–63. Kobata Atsushi, *Kaitei zōho Nihon kahei ryūtsū-shi, op. cit.* p. 365 and pp. 416–417.

Table 10. Relative value of silver to gold in Japan (silver per unit of gold)

Year	Silver	Year	Silver
1434	4.66[a]	1588	9.15
1438	5.74[b]	1589	11.06
1447	4.04[c]	1594	10.34[f]
1540	3.62[d]	1604	10.99[g]
1571	7.37[e]	1609	12.19
1575	10.34	1610	11.84
1579	8.77	1615	11.38
1581	8.92	1620	13.05
1583	9.19	1622	14.00

[a]This price was calculated assuming that the price of gold was 2,750 *mon* in copper coins.
[b-d]These prices were calculated assuming the price of gold during these years was 3,000 *mon*.
[e]The price of gold in 1572 was used.
[f]The price of gold in 1590 was used.
[g]The price of silver in 1605 was used.
Source: See the notes for tables 11 and 12.

Increases in the output of precious metals and the outflow of silver after the mid-16th century

After the *de facto* collapse of the Ashikaga Bakufu in the Ōnin War of the 1460s, waged among the regional military powers, Japan underwent a century of intermittent civil wars. It was not until the first decades of the 17th century, following a major and decisive battle in 1600, that another new Bakufu, the Tokugawa, was firmly established to rule the nation until 1867.

As these fundamental changes were taking place in the political arena, the economy of Japan also underwent significant change during the century and a half of the civil wars and the first decades of the Pax Tokugawa.[40] Because of the policy adopted to create small peasant landholders who were free of tax and rent obligations imposed on them by various levels of local and regional rulers, and because of the rapid growth of commerce benefiting from the gradual unification of the regional and national markets, the tempo of economic growth accelerated during the 1550-1650 period.

These political and economic changes had two important consequences. One was the rapid increase in Japan's capacity to produce precious metals as the result of the efforts made by the warring regional military powers who had both the motivation and the resources to undertake costly mining opera-

---

40. For an examination of the economic growth of this period, see: Kozo Yamamura, "Returns to Unification: The Economic Growth of Japan, 1550-1650," a paper presented to the SSRC conference on the Sengoku period: August 1977, Maui, Hawaii.

tions, and the other was the resumption of the minting of coins by Hideyoshi and then by the Tokugawa Bakufu after a hiatus of more than five centuries since the last of the Twelve Imperial Mintings in the 10th century.[41]

We must also note that it was during the 16th century that Western traders, first the Portuguese and then the Dutch and English, began to trade with the Japanese and in so doing acted as a conduit in exporting Japanese silver to China and Southeast Asia and in importing gold to Japan from China, the Philippines, and elsewhere. Though it was not exported to the West, Japanese silver thus began to play a role in the Western traders' successful efforts to increase their activities in Asia.

The principal factor accounting for the rapid increase of precious metals, mostly silver, was the development of the mining and smelting technology which was brought to Japan from China by way of the Korean Peninsula.[42] The technology was first adopted by a Hakata merchant in working a large silver mine in Iwami, a western province facing the Sea of Japan. It is said that two skilled miners (presumably of Korean origin) employed by the merchant succeeded in 1553 in using the new method brought from Korea to smelt ore from a mine which had been worked since the 1520s. The new process, called *haibukihō* (literally ash-blowing method), consisted of first smelting silver ore with lead to let the lead absorb the silver and then isolating the silver by extracting the lead from the molten mixture by means of "blowing ash" onto the mixture.[43]

As the process was perfected, the output of the Iwami silver mine increased rapidly, and this fact was not lost on people interested in silver mining. No sooner was ore containing silver discovered in 1542 in Ikuno, not far from Iwami, than the *haibuki* process was adopted from Iwami, resulting in a rapid growth of silver in Ikuno. Other mines soon followed the example of these mines and the total output of silver continued to increase. In order to indicate the approximate magnitude of the rate of increase of total output (which is not available), we have presented the output of the Iwami mine for 1581 and for the 1683-1857 period in Table 6. Output must have started to rise much earlier, in the 1530s, because of the new technology adopted, but we have no data on this period. However, our data do show that the peak of silver production was reached during the 1683-1706 period.

Aided by a technological innovation of the mid-16th century, gold output also can be assumed to have risen during this period, though we are unable to provide quantitative data to prove this. The innovation was essentially identical to that for silver mining and consisted of melting lead with pulverized ore which contained gold and then applying the *haibuki* method.[44] The techno-

41. The Tokugawa Bakufu, which took an active role in promoting mining and smelting technology, directly controlled all the silver and gold mines in Japan. After 1601, the Bakufu possessed a virtual monopoly over the gold and silver output. Sakudo Yōtarō, *Kinsei Nihon kaheishi* (A monetary history of Tokugawa Japan), (Tokyo: Kōbundō, 1958), pp. 26-37.
42. Kobata, *A Study of the history of Japanese mines, op. cit.*, p. 26, p. 110, and pp. 154-155.
43. *Ibid.*, p. 51, and p. 155.
44. *Ibid.*, p. 52 and p. 63.

logical capability to excavate and pulverize the ores and to dig and maintain horizontal tunnels were also developed during the early to mid-16th century and were important both for gold and silver mining. These developments were largely by-products of the developments in civil engineering techniques due to the efforts of warring regional powers to build larger and better-fortified castles and moats.[45] It is also possible that the contacts with the Europeans, mostly via the missionaries, provided the Japanese with yet another method of smelting—the "amalgam method"—which began to be adopted during this period.[46]

In examining the international trade of the 16th and the 17th centuries against the background of the increasing output of the precious metals just described, we must divide our discussion into two periods: the period between the mid-16th century and the 1630s when the Tokugawa Bakufu adopted, primarily out of its desire to prevent the further dissemination of Christianity, the well-known policy of "secluding" Japan from international contacts (sakoku); and the period following the adoption of the seclusion policy and lasting to the end of the 18th century.

The Portuguese, the first to come to the shores of Japan, had been engaged in the spice trade in Asia since the early 16th century and had traded with merchants on the Southern coast of China. By the end of the 1530s, they were competing with the Chinese merchants of Canton and were gradually forced to seek new markets and bases of operation farther north along the Chinese coast. Thus, Japan became their next port of call. The Portuguese first arrived in Japan in 1542 (or in 1543) and began full-scale trade following their acquisition of Macao as the main base for their East Asian trade.[47]

The Portuguese were in luck. The Japanese trade with Ming China had just been substantially reduced because of the Japanese pirate activities along the Southern coast of China and because of the subsequent Chinese policy of limiting Japanese access to its ports. This meant that the Portuguese could trade with the eager Japanese who had just lost the services of the Chinese and Ryukyuan ships. The Portuguese could supply the Japanese with the guns, gun powder, raw silk, silk goods, leather, and other goods they were most eager to purchase. The most important among these goods was raw silk which the Portuguese bought in China to be resold to the Japanese. The Japanese paid for all these imports in silver, which the Portuguese used in

---

45. See Yamaura, cited in footnote 40.

46. The "amalgam" method, which most likely was introduced by the Spaniards at the beginning of the 17th century, consisted of: (1) mixing pulverized gold or silver ore with mercury to obtain an amalgam and (2) distilling mercury to obtain gold or silver. The problem faced by the Japanese in adopting this technique was an inadequate supply of mercury. For further descriptions of this method and its adoption in Japan, see: Kobata, *A study of the history of Japanese mines*, p. 64 and Okumura Shōji, *Hinawajū kara kurofune made—Edo jidai gijutsushi* (From the matchlock to the Black ships—A History of technology in Tokugawa Japan), (Tokyo: Iwanami Shoten, 1970), pp. 9-13.

47. Matsuda Kiichi, *Nippo kōshōshi* (A history of the relationship between Japan and Portugal), (Tokyo: Kyōbunkan, 1963), p. 66.

Silver mines and Sung coins

their trade with the Chinese.[48] The quantity traded was by no means small. One source estimates that in performing the intermediary role between China and Japan, Portuguese ships came to Japan in the summer of each year with a load of 1,500-2,000 *pico* (one *pico* = 60 kilograms) of raw silk which was sold for 400,000 to 800,000 *cruzados* of silver.[49] This was about 15,000-30,000 kilograms of silver, since one *cruzado* was the equivalent of one *ryō* of 10 *momme* of silver.

This trade was welcomed by the warring regional lords of Kyushu as it provided them with a lucrative source of additional income which was sorely needed to fill their military coffers. After 1588 Hideyoshi—one of the powerful unifiers of the warring lords—also engaged in active trade with the Portuguese, purchasing raw silk and gold from them and selling it at a handsome profit within Japan.[50]

However, by the turn of the 17th century, the Portuguese faced serious competition from English and Dutch traders in the profitable trade with the Japanese. Though little data exist with which to quantify the magnitude of trade conducted between these newcomers and Japan, it is believed that the total amount of trade was large. Thus, Japan and the West were linked by silver, mostly produced in Iwami and other major mines actively worked during this period.[51]

Then came the seclusion policy of the late 1630s. Under the policy, the Portuguese, along with the Spaniards, were expelled from Japan because of a variety of political reasons along with fears of the foreign religion of Christianity. The English had left Japan in 1623 for political and economic reasons of their own,[52] and thus the expulsion of the Catholic traders left only the Protestant Dutch and the Chinese with whom the Bakufu, mindful of profits which could be earned, was willing to continue to trade through the port of Nagasaki. Thus the access of the outside world to Japanese silver was not cut, but it was now under the strict control of the Bakufu which intended to

48. Kōda Shigetomo, *Nichi-ō tsūkō-shi* (A history of the international trade between Japan and Europe), (Tokyo: Iwanami Shoten, 1943), p. 66.

49. Matsuda, *op. cit.*, p. 12 and Yanai Kenji and Numata Jirō, *Kaigai kōshōshi no shiten*, (A study of Japan's overseas' relationships), Vol. 2, (Tokyo: Nihon Shoseki, 1976), p. 95.

50. Beginning in 1588 Toyotomi Hideyoshi bought raw silk and gold from the Portuguese and sold them within Japan at a large profit. Yanai and Numata, *op. cit.*, p. 88.

51. During this period the Dutch and English are known to have acquired in Japan high grade silver called either *somo* or *soma* silver. This clearly was the *haibuki* silver then being produced at the Iwami mines because these mines were at the time often referred to as the Sama mines. Similarly silver called Seda, Nagites and Tajimon by the Europeans was silver produced in Sado, Nagato, and Tajima. Kobata, *A study of the history of Japanese mines, op. cit.*, pp. 150-151.

52. The English ships of the East India Company first came to the port of Hirado (in Kyushu) in 1613. Establishing a trading house there, the English, who had signed an agreement with the Dutch to share the trade with Japan, engaged in active trade for several years. However, due to the political conflict which soon developed between the Dutch and English and also due to meagre profits which the English were able to earn in trading with the Japanese, the English closed the trading house in 1623 and ceased all trade with Japan. Koda, *A history of the international trade, op. cit.*, pp. 325-326.

monopolize trade, restricting it to Nagasaki and two other ports; Tsushima for trade with Korea, and Satsuma for trade with the Ryukyu islands.[53]

The Bakufu's principal means of payment for the raw silk and other goods which they bought was silver, supplemented by lead and gold. Though Nagasaki, by far the most important of the three "windows," the Dutch and the Chinese carried away over 300,000 *kan* (112,500 kilograms) of silver during the 1640-72 period. As Table 7 shows, the exodus of silver was sharply reduced in 1688 when the export of silver was prohibited by the Bakufu, and by the beginning of the 18th century the export of silver became virtually nil.

Though the amount was not large, gold was known to have been imported from China, the Philippines, and other parts of Asia until the 1620s, after which Japan was seen exporting very small amounts of gold. For example, we can establish that the Dutch were allowed to receive payments in gold after 1668 when the export of silver was prohibited, but the amount which they received could not have exceeded a few hundred kilograms per year, and it is most unlikely that they took even this amount.[54] Thus, it is accurate to say that for the period between the mid-16th century and the first few decades of the next Japan was a net importer of gold.

The foregoing shows that Japan's active role in the East Asian trade and in the international flow of gold and silver effectively ended in 1668. The recent efforts of several scholars to reexamine the significance of trade after this date, largely with Korea and the Ryukyu islands, show that it was not as limited as has been maintained.[55] But their findings do not require a revision of our assessment that the outflow of silver and gold was far smaller during the 18th century than during the 17th century and of the statement that Japan remained in an effective self-imposed exile for nearly two centuries until 1853 when she was forced by Commodore Perry to rejoin the community of trading nations.[56]

Let us now present several estimates of the outflow of silver from Japan during the 17th century. The dates selected in dividing the period in Tables 7

---

53. For information on the trade between Japan and Korea, and Japan and the Ryūkyū Islands, see: Sydney Crawcour, "Notes on Shipping and Trade in Japan and the Ryukyus," *The Journal of Asian Studies*, Vol. XXIII, No. 3 (May 1964); and Robert Sakai, "The Satsuma-Ryukyu Trade and the Tokugawa Seclusion Policy," *The Journal of Asian Studies*, Vol. XXIII, No. 3 (May 1964).

54. According to the estimate made by Iwahashi, the average annual outflow of gold during the 1695-1709 period was 220 kilograms. The annual outflow fell to less than 100 kilograms during the beginning of the 18th century and further fell to less than 10 kilograms after the 1730s. Iwahashi, *op. cit.*, p. 247.

55. The Iwahashi estimates for subperiod III on (in Table 7) include Tashiro's estimates of silver exported from Japan to Korea and the Ryūkyū Islands. Tashiro, "Silver Export," *op. cit.*, pp. 237-239.

56. Excellent English sources on the effect of the "opening" of Japan on the flow of silver and gold are: Hugh Patrick, "External Equilibrium and Internal Convertability: Financial Policy in Meiji Japan," *The Journal of Economic History*, Vol. XXV, No. 2 (1965); and Peter Frost, *The Bakumatsu Currency Crises*, (Cambridge, Mass.: Harvard University Press, 1970).

and 8 follow the practice which is adopted by Japanese scholars. Subperiod I (1601-1695) covers most of the 17th century, for which no official data of outflows exist; and subperiods II through V correspond either to changes in policy on the international trade or to the duration of the reigns of one or two Emperors (the period names changed frequently during the reign of one emperor as well as at the time of the beginning of a new emperor's reign). As is evident, the periodization based on changes in policy and on period names are sufficient for our purposes of demonstrating the magnitude of the outflows and their change over time.

Table 7 clearly shows that, according to both the estimates made by Satō in 1875 and those by Iwahashi in 1976, there was a rapid outflow of silver during the 17th century, but the flow was sharply reduced for subperiods III through V. Even in Iwahashi's estimate for the 18th century which is much higher than that made by Satō, there is a sharp contrast in the magnitude of the outflow between subperiod I and the subsequent subperiods.

Also, the Satō estimate, which is slightly larger than that made by Iwahashi for the 1601-1695 period, may still have been smaller than the real magnitude of silver exported, since his estimate does not include the so-called *haibuki* silver which was also known to have been exported along with the silver coins minted during the Keichō period (1596-1614). Table 8 is added to show that, as estimated by Iwahashi, the output of silver (the total and average annual amounts turned over to the Bakufu) from the largest mines also declined visibly after the first decade of the 18th century.

Despite their limitations, the general reliability of these estimates can be corroborated by contemporary observations. For example, Ralph Fitch, who travelled through Asia during the latter part of the 16th century, reported that up to 600,000 *cruzados* (22,500 kilograms) of Japanese silver were annually brought to China by Portuguese ships. Another source for the 1620s notes that 12,000-15,000 *kan* (45,000-56,250 kilograms) of silver were exported annually from Japan to China, also by Portuguese ships.[57] On the basis of the data presented in Table 7 and these contemporary observations, it may not be too far wide of the mark to estimate that the Portuguese ships carried an annual average of 6,000-10,000 *kan* (22,500-37,500 kilograms) of silver from Japan during the 1560-1600 period. And, if we add another 3,000 *kan* (11,250 kilograms) as a reasonable estimate for the amount of silver carried from Japan by the Chinese and Japanese ships that were continuing a limited scale of trade during this period,[58] then the total annual export would have been approximately 9,000-13,000 *kan* (33,750-48,750 kilograms). That is, the total amount of silver exported from Japan to China during the 1560-1600 period could have been within the range of 1.35-1.95 million kilograms.

For the period between the last years of the 16th century and and

57. Iwao Seiichi, *Shuinsen bōekishi no kenkyū*, (A study of the vermilion-seal trade), (Tokyo: Kobundō, 1958), pp 325-326.
58. Yanai Kenji and Numata Jiro, *op. cit.*, pp. 92-94; and Kobata Atsushi, "Nihon no kin gin gaikoku bōeki no kenkyū," (A study of gold and silver in the international trade of Japan), *Shigaku Zasshi*, Vol. 44, No. 11 (1933), p. 74.

beginning of the effective enforcement of the seclusion policy in the late 1630s, we estimate, based again on the data presented in Table 7 and on fragmentary evidence, that the annual outflow of silver to China was approximately 40,000-50,000 *kan* (150,000-187,500 kilograms), which would be a total outflow for the period of as much as 1.6-2.0 million *kan* or 6.0-7.5 million kilograms. Though we must be fully cognizant of the quality of the evidence, it perhaps is not inaccurate to conclude that the total outflow of silver from Japan to China during the eighty years from 1560 to 1640 may have been somewhere within the limits of 7.35 and 9.45 million kilograms.

As a consequence of this large outflow of silver from Japan during the 1560-1640 period, by 1640 the relative value of gold to silver in Japan had become virtually identical to that in China and in the world market, as is shown in Tables 9 and 10. This came about because the value of silver *declined*, despite the large outflow, much more than that of gold, reducing the relative value of silver to gold to the level prevailing in the world market. The principal reason for the decline in the value of silver in Japan, of course, was the sharply increased output.

On the changes in the relative price of gold and silver in Japan during this period, we will add here what we believe to be significant revisions to the long accepted Kobata estimates. Kobata's estimates indicate that the changes in the relative value of gold to silver occurred as the result of a sharp decline in the gold price (as much as 50 percent within the decade of the 1540s), accompanied by an even sharper decline in the price of silver (about 75 percent during the same period). The revised data—the price of gold and silver in terms of copper coins—which we present in Tables 11 and 12 instead show that what in fact occurred was a decline of 50 percent in the value of silver while the price of gold showed only a slight negative trend during this period.

Detailed descriptions of the procedures we adopted in revising the data are in the long note appended to Table 12, so let us here present a simple diagram (Diagram 1) to show what we believe occurred during this period. The supply of both gold and silver available in Japan rose between the mid-16th century and the early decades of the 17th century from Sg_1 to Sg_2 and from Ss_1 to Ss_2 respectively. The increase in the supply of gold was due to an increase in output and to a modest amount of imported gold while the increase in the supply of silver was due exclusively to the increase in output. (Because the new mines for both gold and silver were worked to capacity constrained only by technology, the supply schedules for both are assumed to be totally inelastic.)

Despite the increase in the domestic demand, the price of gold declined in Japan from Pg_1 to Pg_2, that is, to the price level in China and in the world gold market, because of the increased supply which exceeded the increase in demand. The price of silver also declined from Ps_1 to Ps_2, that is, to the price level in China and in the world silver market, because the increase in supply easily exceeded the increase in demand even though there was a rising

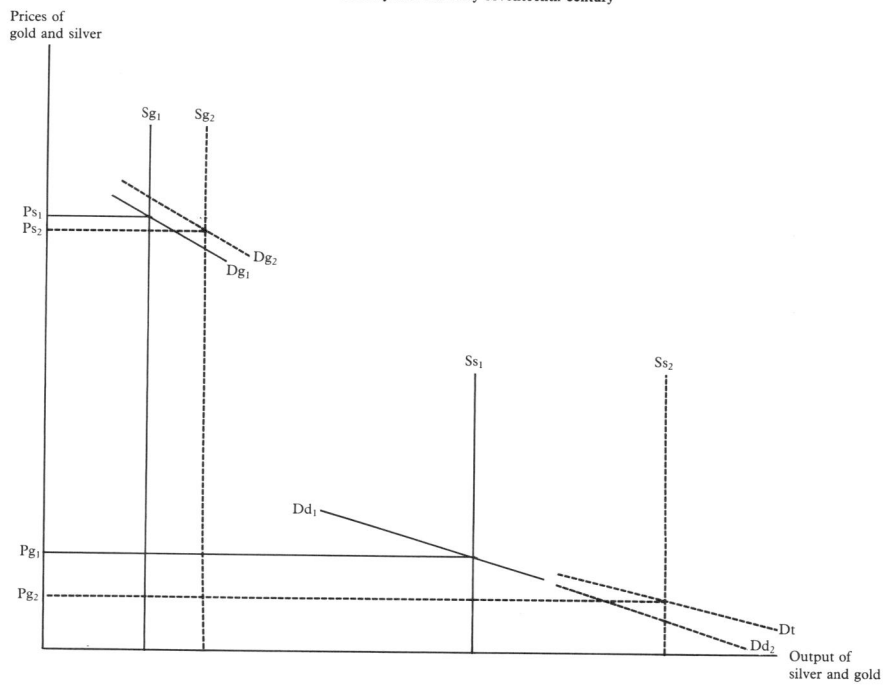

Diagram 1:
Output and demand (domestic and export) of silver and gold: The mid-sixteenth century and the early seventeenth century

total demand (export plus domestic demand Dd) because of the increased domestic demand and a large export demand. In the diagram, Dt stands for supply of silver in the world and the active trading of silver which we know accompanied it, we have drawn the Dt schedule to be more elastic than the Dd schedule. The result of these changes was that the relative price of silver to gold declined from Pg_1/Ps_1 to Pg_2/Ps_2 or from a ratio of 1:5-6 to 1:12-13.

The large outflow of silver from Japan which in the diagram represents a total magnitude of 7.35-9.45 million kilograms, undoubtedly had a significant impact on the Chinese monetary system as well as on the world markets for gold and silver. If the contemporary reports made by Spaniards to their home offices are accurate, the total amount of silver shipped from Mexico and Peru by the Spaniards (and others) to China was about 420,000 kilograms during the 1560-1600 period and 900,000 kilograms during the 1600-1640 period. This is a total of approximately 1.32 million, which is no more than one-sixth to one-seventh of the amount of silver exported from Japan to China during the same period.[59]

59. Yanai and Numata, *op. cit.*, p. 95.

Table 11. Price of gold in Japan: 1432–1622[a] (In terms of copper coins per one ryō of gold, which was 16.875 grams)

Year	Price (*mon*)	Year	Price (*mon*)
1432	2,750	1543	2,580
1450	3,000	1568	2,078
1475	3,000	1569	2,028
1477	3,000	1572	2,410
1490	3,000	1573	2,449
1493	2,600	1575	2,450
1510	2,700	1579	2,571
1518	2,469	1580	3,143
1521	3,000	1581	3,658
1524	2,900	1583	3,374
1526	3,150	1584	2,732
1527	3,050	1585	3,050
1528	3,088	1586	3,641
1529	3,150	1588	3,658
1530	3,090	1589	3,958
1531	3,061	1590	3,993
1532	3,000	1601	4,005
1533	3,000	1604	3,912
1534	3,000	1609	3,219
1535	3,003	1610	3,127
1536	2,950	1615	2,662
1537	3,120	1620	3,707
1541	2,917	1622	3,780
1542	2,700		

Notes: The price for each year is obtained by taking a weighted average of prices given for various quantities of gold. Extremely high and low prices were eliminated in taking the weighted average. In those cases in which the price of gold is available only in terms of rice, the price of rice was converted to price in terms of copper coins. Prices used in this table are for Kyoto and Nara, the two most important markets in medieval Japan. Because of the proximity of these two markets, variances in prices between these two cities were minimal.

Sources: Kyoto Daigaku Kinsei Bukka-shi Kenkyū-kai (The Kyoto University Research Group on the price history of the early modern period) ed., *15–17 seiki ni okeru bukka hendō no kenkyū* (A study on the price fluctuations in the 15th-17th centuries), (Kyoto: Tokushi-kai, 1962), pp. 33–39. Kobata, *Keitei zōho Nihon kahei ryūtsū-shi, op. cit.,* pp. 354–360 and pp. 384–405. For the price of rice in terms of copper coins, see: Tetsuo Kamiki, "Chūsei kōki ni okeru bukka hendō" (Fluctuations of prices during the late medieval period), *Shakai Keizai Shigaku,* Vol. 34, No. 1 (1968), p. 25.

Silver mines and Sung coins

Table 12. Price of silver in Japan: 1434–1647 (in terms of copper coins), (per 16.875 grams)

Year	Price (*mon*)	Year	Price (*mon*)
1434	590	1615	234
1438	523	1616	284
1447	741	1617	281
1540	828	1618	281
1571	327	1620	284
1574	279	1621	273
1575	237	1622	270
1576	278	1623	265
1577	342	1624	250
1578	322	1625	257
1579	293	1626	262
1581	410	1627	281
1583	367	1628	270
1588	400	1629	257
1589	358	1630	236
1591	410	1631	250
1594	386	1632	250
1605	356	1633	243
1608	339	1634	205
1609	264	1635	188
1610	264	1636	188
1611	231	1637	188
1612	225	1638	196
1613	220	1647	300
1614	225		

See Table 11 for Notes and Sources.

Additional Notes on Tables 11 and 12. Although accompanied by some fluctuations, the price of gold (per one *ryō*) was generally stable at the 2,700–3,000 *mon* level from the 1430s to the 1520s. Then, as seen in Table 11, beginning in the 1520s the price of gold rose to the 3,000–3,100 *mon* level, and remained at that high level into the 1540s when the price began to decline. In the 1570s the gold price remained at the low level it reached in the 1560s, only to show a rising trend again from the 1580s on. (Though we have no data for the 1550s, we can be reasonably certain of the trend, as the data for the 1560s clearly indicate the continuing low level).

In discussing the price decline of the 1540s, Kobata states, based upon his estimate, that the price of gold (per *ryō*, or 16.875 grams) fell sharply from 2,580 *mon* in 1543 to 1,300 *mon* in 1549. That is, the price of gold was halved during this short period of time. (Kobata, source 2 of Table XI, p. 400) While it is correct to say that some prices of gold, expressed in terms of copper coins, during this period remained low, that is, within the range of 1,100 *mon* to 1,500 *mon* (*Ibid.*, p. 384), we believe the uncritical acceptance of such evidence is not justified. For example, when we examine

Table 12. (*continued*)
historical sources which contain the price of gold expressed both in terms of copper coins and in rice, we discover that in 1569 one *ryō* of gold was traded for 1,200 *mon* of copper coins as well as for 2.5 *koku* of rice, and that in 1578 the same amount of gold was exchanged for 1,430 copper coins *mon* or for 6.6 *koku* of rice. (*Tamon'in nikki* contained in Tsuji Zennosuke ed., *Tamon'in nikki* (A diary of Tamon'in), (Tokyo: Sankyō Shoin, 1935), Vol. 2, p. 140 and Vol. 4, p. 100.

If we are to accept such evidence at face value, the data for 1569 mean that one *koku* of rice was valued at only 480 *mon*, despite the fact that a study by Kamiki, based on a large number of observations obtained from the records of temples in Kyoto and from other historical sources, has established that the average price of one *koku* of rice was about 998 *mon*. Similarly, the 1579 data imply that the price of one *koku* of rice was only 217 *mon* which is far too low to be accepted as a realistic price. (Kamiki, *op. cit.*, p. 25).

In explaining such large discrepancies found between the price of gold expressed in copper coins and that indicated in rice, Kobata merely states that "there is no way we can establish the price of gold in terms of copper coins because the task of determining the value of copper coin itself is extremely complex." (*Ibid.*, p. 399) However, despite this remark, he goes on to assume that the price of rice fell sharply in terms of copper coins and fails to question the reliability of such an unrealistic price as 217 *mon* per one *koku* of rice.

In preparing Tables 11 and 12, we did not use, for the 1549–1590 period, prices of gold and silver expressed directly in terms of copper coins. Instead, we used gold and silver prices which were calculated from those expressed in terms of rice, i.e., we converted the price of rice into the price expressed in copper coins. We believe our procedure is preferable to that adotped by Kobata because, among other reasons not elaborated here, the gold price which we obtained does not make the unrealistic plunge of 50 percent or more during the brief period of 1543–1549, during which no known increases in gold output or import occurred.

Even if we are to allow for some understatement of the amount of silver exported from the New World to China, the relative magnitude of the amount coming to China from Japan and from other sources clearly suggests that Japanese silver played a significant role in the changes seen in the value of silver in China. If we are correct in our assessment of the importance of the Japanese silver in China's international trade, then, given the significance of China in the world markets of silver and gold, we may conclude that Japan's silver played an active, if supporting, role in the unfolding Price Revolution which the Western world experienced during the 16th century and the early decades of the 17th. It may be justified for us to conclude, on the basis of evidence which we have presented here, that a serious reexamination of the role played by Japan in the monetary system of the world during this period is long overdue, and such a reexamination must begin with a concerted effort to obtain more reliable data.

Conclusion

The basic patterns of the flow of gold, silver, and copper coins between Japan and her trading partners are unmistakable. Broadly speaking, before the

Silver mines and Sung coins 357

mid-16th century Japan exported gold to China and imported silver and copper coins from China. Then, around the mid-16th century when the output of silver began to rise rapidly in Japan, the direction of the flows of these precious metals was reversed, and the importation of copper coins ceased altogether. Given the prevailing differentials in the relative values of gold, silver, and copper coins in China and Japan during the period between the decline of the Ritsuryō government during the 9th to 11th centuries and the adoption of the seclusion policy in the 1630s, we encountered no major problem in explaining the observed changes in the patterns of trade. However, in concluding this essay, we wish to draw the readers' attention to one important finding we made and one problem raised.

The finding that we consider most significant is the important role played by Japanese silver exported to China after the mid-16th century in China's international trade, in the gold/silver price ratio in China and in the rest of the world, and perhaps even in the course of the Price Revolution in Europe. Indeed, if our tentative assessment of the magnitude of Japanese silver exported to China—six or seven times the amount shipped from the New World by the Europeans—is correct, we must revise the earlier view and fully recognize the significant role which Japanese silver played in China's international trade.

The validity of our tentative assessment of the importance of Japanese silver can only be demonstrated or disproved by further efforts to obtain more reliable data and the analysis of such data within a correctly specified framework of analysis. As the collective efforts of monetary historians of Japan, China, and the West continue, we may soon be able to ask a tantalizing question: what would the course of the relative prices of gold to silver in China and in the rest of the world (and the price levels in the rest of the world) have been had Japan not exported the large quantity of silver it did during the 1560-1640 period?

Second, the problem we raised in this paper is: Why were such large amounts of Chinese coins imported by the Japanese and why did these coins play such a crucial and dominant role in the Japanese economy from the twelfth to the sixteenth centuries? While we are generally persuaded that the large differentials between the values of coins in China and in Japan played a dominant role in causing the large inflow of coins to Japan, we cannot ignore, in explaining the sustained inflow of the coins, such non-economic facts as the acceptance which the Chinese coins enjoyed even before the early thirteenth century when import of these coins began to increase visibly. After the decline of the Ritsuryō government, neither the Kamakura nor the Ashikaga Bakufu possessed the requisite political power to issue coins with which to replace the widely accepted Chinese coins.

Consideration of these facts—the acceptance which the Chinese coins came to enjoy in Japan and the weakness of the central governments—leads us to ask: Would the Kamakura or the Ashikaga Bakufu have minted their own coins if the differentials between the value of the coins in China and in Japan had been smaller? The Japanese clearly possessed the technological capability to mint coins, since they had the capability to produce swords highly prized in China for their quality, and privately minted illegal imitations of the Chinese coins. It is also clear that they possessed the resources to do so, since by the 15th century they were exporting copper to China.

In the face of the historical fact that the Japanese did not produce their own coins, the above counterfactual question suggests the basic difficulty of assigning relative weights to the economic factor (the large differentials between the values of coins in Japan and China) and the non-economic factors which we have presented. Thus, for now, we must be satisfied with the explanation that the differentials between the values of the coins, the absence of a strong central government, and the established position which the Chinese coins already enjoyed in Japan all contributed to the continued importation of the Chinese coins until the end of the 16th century. And perhaps we should add that there was a sustained demand for the imported coins not only because they remained the principal medium of exchange, but also because of the increasing demand created by the growth of commerce and the economy as a whole beginning in the 13th century.

We are reminded also that a obtain a truly satisfactory answer to questions relating to the huge amount of Sung coins used in Japan we must ask why the Chinese governments, from the T'ang to the Ming, adopted policies to limit the outflow of coins, despite the fact that it must have been profitable to permit, if not encourage, such an outflow. Did China suffer shortages of copper in meeting the growing needs of its own economy for a medium of exchange? Were the copper mines in the hands of the Liao, the Chin, and the Mongols, rendering the dynasties of Sung, Southern Sung, and Ming unable to obtain all the copper they required? If so, why was the relative value of copper in China seemingly, at least vis-à-vis Japan, so low (in term of gold)? Our inability to answer these questions clearly demonstrates our own limits as well as the necessity of international and collective efforts to obtain a better understanding of the interdependence of the monetary systems of the world.

Finally we conclude with two important disclaimers. One is that our study was focused on Japan's trade with China and with the Europeans and did not include an examination of Japanese trade with Korea, the Ryukyu islands, and the nations of Southeast Asia. While the volume of the trade with the latter nations was significantly smaller in comparison to that conducted with the Chinese and the Europeans, a comprehensive study of Japan's international trade, especially that for the period after the 1600s, must include an examination of the trade with her Asian neighbors.

The other disclaimer, which also explains in part our reasons for neglecting Japan's trade with Korea and the Ryukyus, is that our study of the monetary system of "closed" Japan and the rapid outflow of gold and in the inflow of Mexican and the other silver which occurred after the "opening" of Japan by Commodore Perry, are referred to the able works of Frost, Patrick, Shinbo, and other listed in the Bibliography.

Silver mines and Sung coins

Bibliography

Crawcour, Sydney. "Notes on Shipping and Trade in Japan and the Ryukyus," *The Journal of Asian Studies*, Vol. XXIII: No. 3. (May 1964).

Crawcour, Sydney and Yamamura K., "The Tokugawa Monetary System: 1787-1868," *Economic Development and Cultural Change*, Vol. 18, no. 4 (1970).

Duffy, W. and Yamamura K., "Integration of Tokugawa Monetary Markets, 1787-1867: A Spectral Analysis," *Explorations in Economic History*, Vol. 8, No. 4 (1971).

Frost, Peter, *The Bakumatsu Currency Crises* (Cambridge, Mass.; Harvard University Press, 1970).

Hall, J. W. and Toyoda, T. eds., *Japan in the Muromachi Age* (Berkeley: University of California Press, 1977).

Hanley, S. B. and Yamamura, K., *Economic and Demographic Change in Pre-industrial Japan* (Princeton N.J.; Princeton University Press, 1977).

Hino Kaisaburō, "Hokusō jidai ni okeru dō tetsu sen no chūzō gaku ni tsuite" (On the amount of copper and iron coins minted during the Northern Sung period) *Shigaku Zasshi*, vol. 46, no. 1 (1935). pp. 46-105.

Hino Kaisaburō, "Hokusō jidai ni okeru dō tetsu sen no jukyū ni tsuite," (On the demand for copper and iron coins during the Northern Sung period), *Rekishigaku kenkyū*, vol. 6 No. 5, 6, 7 (1936).

Hirano Kunio, "Nihon ni okeru kodai kōgyō to shukōgyō," (Mining and handicraft in ancient Japan), in Ishimoda Sho ed., *Kodaishi koza* (Lectures on the ancient history), vol. 9 (Tokyo: Gakuseisha, 1963).

Hosokawa Kameichi, *Jōdai kahei keizaishi* (A monetary history of the ancient period) (Tokyo: Moriyama Shoten, 1934).

Iwahashi Masaru, "Tokugawa jidai no kahei suryō—Sato chūzaburō sakusei kahei aridaka-hyō no kentō," (The supply of monies in the Tokugawa period—An examination of the tables of stocks of monies prepared by Sato Chūzaburo) in Umemura Mataji et al. ed., *Nihon Keizai no hatten* (The development of the Japanese economy) (Tokyo: Nihonkeizai Shimbunsha, 1976), pp. 241-260.

Irita Seizō, "Hakkutsu sen ni tsuite no kōsatsu" (A study of unearthed coins), *Kōkogaku Zasshi*, vol. 20, no. 12 (1930), pp. 820-829.

Itazawa Takeo, *Nihon to Oranda*, (Japan and Holland) (Tokyo: Shibundō, 1955).

Iwao Seiichi, *Shuinsen bōekishi no kenkyū* (A study of the vermilion-seal trade) (Tokyo: Kōbundō, 1958).

Iwao Seiichi, *Shuin-sen to Nihon machi* (The vermilion-seal ships and Japanese towns) (Tokyo: Shibundō, 1962).

Kamiki Tetsuo, "Chūsei kōki ni okeru bukka hendō" (Price fluctuations in the late medieval period), *Shakai-Keizai Shigaku*, vol. 34, no. 1 (1968), pp. 21-38.

Katō Shigeru, *Tō-Sō jidai ni okeru kin gin no kenkyū* (A study of gold and silver of the T'ang and Sung periods), 2 vols. (Tokyo: Tōyō Bunko, 1926).

Katō Shigeru, "Nisso no kin gin kakaku oyobi sono bōeki ni tsuite," (On the prices and international trade of gold and silver in Japan and Sung China) *Shakai-Keizai Shigaku*, vol. 3, no. 3 (1933), pp. 1-12.

Kida Shinroku, "Nara chō ni okeru senka no kachi to ryūtsū to ni tsuite" (On the values and uses of copper coins during the Nara period), *Shigaku Zasshi*, vol. 44, no. 1 (1933), pp. 1-58.

Kobata Atsushi, "Chusei kohanki ni okeru Nissen kin gin bōeki no kenkyū" (A study on the international trade of gold and silver between Japan and Korea during the late medieval period), *Shigaku Zasshi*, vol. 43, nos. 6 and 7 (1932), pp. 693-717, pp. 873-912.

"Nihon no kin gin gaikoku bōeki ni kansuru kenkyū—sakoku izen ni okeru" (A study of the Japanese international trade of gold and silver before the period of seclusion), *Shigaku Zasshi*, vol. 44, nos. 10 and 11 (1933), pp. 1280-1318, pp. 1381-1434.

Chūsei Nisshi tsūkō bōeki shi no kenkyū (A study of the international trade between Japan and China during the medieval period), (Tokyo: Tōkō Shoin, 1941).

"Chūsei no kin gin no kakaku oyobi sono Nisshi bōeki" (Prices of gold and silver and the international trade of gold and silver between Japan and China in the medieval period), *Shakai-keizai Shigaku*, vol. 3, no. 6 (1933), pp. 42-58.

Kaitei zōhō Nihon kahei ryūtsūshi (A revised and expanded history of the uses of money in Japan) (Tokyo: Tōkō Shoin, 1943).

Nihon no kahei (Monies in Japan) (Tokyo: Shibundō, 1958).

"Kinsei shotō ni okeru gin yushutsu no mondai" (The problem of silver export at the beginning of the premodern period), *Rekishi Kyōiku*, vol. 10, no. 9 (1962), pp. 17-23.

"The Production and Uses of Gold and Silver in Sixteenth and Seventeenth Century Japan," *Economic History Review*, vol. XVIII, no. 2, 1965, pp. 245-266.

Nihon kōzanshi no kenkyū (A study of the history of Japanese mines) (Tokyo: Iwanami Shoten, 1968).

Kōda Shigetomo, *Nichi-ō tsūkō shi* (A history of the international trade between Japan and Europe) (Tokyo: Iwanami Shoten, 1942).

Kyōto Daigaku Kinsei Bukkashi Kenkyūkai ed., *15-17 seiki ni okeru bukka hendō no kenkyū* (A study of the price fluctuations during the 15th-17th centuries) (Kyoto: Tokushikai, 1962).

Leeds, L. T., "Zinc Coins in Medieval China," *The Numismatic Chronicle and Journal of the Royal Numismatic Society*, 6th series, 14 (1954), pp. 177-185.

Matsuda, Kiichi, *Nippo kōshōshi* (A history of the relationship between Japan and Portugal) (Tokyo: Kyōbunkan, 1963).

Mori Katsumi, *Shintei Nisso bōeki no kenkyū* (A revised study of the international trade between Japan and Sung China) 3 vols. (Tokyo: Kokusho Kankōkai, 1975).

Mori Katsumi and Tanaka Takeo, ed., *Kaigai kōshō-shi no shiten*, vol. 1 (A study of Japan's overseas relationships) (Tokyo: Nihon Shoseki, 1975).

Maruyama Kiyoyasu, *Hōken shakai no tsūka mondai* (Problems of Monies in feudal Japan) (Tokyo; Hakuyōsha, 1933).

Nachod, Oscar *Die Beziehungen der Niderländischen Ostindischen Kampagnie zu Japan im 17 Jahrhundert*, (Leipzig, 1897).

Okamoto Yoshitomo, *16 seiki Nichi-ō kōtsūshi no kenkyū* (A historical study of the relationships between Japan and Europe during the sixteenth century) (Tokyo: Rokkō Shobō, 1942) (Reprint: Tokyo: Hara-Shobo, 1974).

Okumura Shōji, *Hinawajū kara kurofune made—Edo jidai gijutsushi*, (From the matchlock to the Black ships—A history of technology in Tokugawa Japan) (Tokyo: Iwanami Shoten, 1970).

Ōkura-shō, ed., (Ministry of finance) *Dai Nihon kaheishi* (A history of monies in Japan) 9 vols. (Tokyo: Chōyōkai, 1925).

Patrick, Hugh T., "External Equilibrium and Internal Convertability: Financial Policy in Meiji Japan," *The Journal of Economic History* Vol. XXV, No. 2 (1965).
Pham Van Kinh, "Bo mat thuong nghiep Viet Nam thoi Ly-Tran (Vietnamese trade in the Ly-Tran period, 10th-14th centuries)", *Nghien Cuu Lich Su* (Hanoi), 189 (#6, 1979), 35-42.
Sakai, Robert K., "The Satsuma-Ryuku Trade and the Tokugawa Seclusion Policy," *The Journal of Asian Studies*, Vol. XXIII: no. 3 (May 1964).
Sasaki Ginya, *Chūsei no shōgyō* (Commerce in Medieval Japan) (Tokyo: Shibundo, 1961).
Sakudō Yōtarō, *Kinsei Nihon kaheishi* (A monetary history of Tokugawa Japan) (Tokyo: Kōbundo, 1958).
Seki Akira, and Aoki Kazuo, "Heijōkyō," in Ishimoda Shō ed., *Nihon Rekishigaku kōza* (Lectures in Japanese history), vol. 2, (Tokyo: Kawade Shobo, 19540.
Shimbo Hiroshi, "Zeni-sōba no hendō, 1736-88" (Fluctuations in the price of copper coins, 1736-88), *Kokomin Keizai Zasshi*, Vol. 133, no. 6 (1976).
"Kinsei-kōki ni okeru bukka, kinsōba, kawaseuchiginsōba—1787-1867," (Price level, gold price and discounts on silver notes—1787-1867) in Umemura Mataji et al. ed., *Nihon keizai no hatten* (The development of Japanese economy) (Tokyo: Nihon Keizai Shinbunsha, 1976).
Sogabe Shizuo, *Nissō Kin kahei kōryū-shi* (A history of trade in monies between Japan and Sung and Chin China) (Tokyo: Hōbunkan, 1949).
Sun, E-tu Zen, "Ch'ing Government and the Mineral Industries Before 1800", *Journal of Asian Studies*, 27, 4 (1968), 835-845.
Tamura Hiroyuki, *Chūsei Nitchō bōeki no kenkyū* (A study of the international trade between Japan and Korea during the medieval period) (Kyoto: Sanwa Shobō, 1967).
Tanaka Takeo, *Chusei kaigai kōshō-shi no kenkyū* (A study of international relations during the medieval period) (Tokyo: Tokyo University Press, 1959).
Wakō to kangō bōeki (Wakō and the tally trade) (Tokyo: Shibundō, 1961).
Chusei taigai kankei shi (A history of medieval international relations) (Tokyo: Tokyo University Press, 1975).
Tashiro Kazui, "Tokugawa jidai ni okeru gin yushutsu to kahei aridaka—1680 nen-dai ikō Chōsen Ryukyū e no gin yushutsu o chūshin ni (Silver export and stocks of monies during the Tokugawa period on the export of silver to Korea and the Ryukyu islands after the 1680s) in Umemura Mataji et al. ed., *Nihon Keizai no hatten* (The Economic development of Japan) (Tokyo: Nihon Keizai Shinbunsha, 1976), pp. 223-239.
Taya Hirokichi, *Kinsei Ginza no kenkyū* (A study of the silver guilds of the Tokugawa period) (Tokyo: Yoshikawa Kobunkan, 1963).
Van de Polder, Leon, "Abridged History of Copper Coins of Japan," *Transactions of the Asiatic Society of Japan*, Vol. XIX, Part II (May, 1891).
Yajima Kyōsuke, "Kahei—honpō ni okeru shutsudo senka" (Money: the coins unearthed in Japan), in Gotō Shuichi, and Ishimodo Shō ed., *Nihon kōkogaku kōza* (Lectures on the study of Japanese archaeology) (Tokyo: Kawade Shobō, 1956) vol. 7, pp. 155-177.
Yamaguchi Kazuo, "Edo jidai ni okeru kin gin ka no aridaka," (Stocks of gold and silver coins in the Tokugawa period) *Keizaigaku Ronshū*, vol. 28, no. 4 (1959).
Yamamura, Kozo, "The Decline of the Ritsuryo System: Hypotheses on Economic

and Institutional Changes," *The Journal of Japanese Studies*, Vol. 1, no. 1 (Autumn, 1974).

"Returns on Unification: Economic Development in Japan 1550-1650," Paper presented at the Japan-U.S. Conference held by the Social Science Research Council in August 1977 in Maui, Hawaii.

Yamawaki Teijirō, *Nagasaki no tōjin bōeki* (Trading with the Chinese in Nagasaki) (Tokyo: Yoshikawa Kobunkan, 1964).

"Oranda higashi Indo Kaisha to Nihon'no kin" (The Dutch East India Company and Japanese gold), *Nihon Rekishi*, no. 321 (19750.

Yanai Kenji and Numata Jirō, *Kaigai kōshōshi no shiten* (A study of Japan's overseas relationships), Vol. 2 (Tokyo: Nihon Shoseki, 1976).

12

Vietnam and the monetary flow of eastern Asia, thirteenth to eighteenth centuries*

JOHN K. WHITMORE

The question of monetary flow in East and Southeast Asia involves copper as much as it does gold and silver. Whether following the example of China or utilizing the abundant supply of "copper cash" often available from China, the countries along the main trade routes of this region of the world tended to use such "copper" (generally an alloy of copper with tin, lead, and/or zinc)[1] as the basis for their own monetary systems. Vietnam, together with Korea and Japan, received the most direct Chinese influence, and all three adopted the "cash" form of currency. In addition, during the great trading boom of the Sung dynasty (960-1279) in China, that country had such a trade deficit that a great flood of cash poured out along the international route, as far as the east coast of Africa.[2] Beyond the countries like Vietnam which already used cash due to direct Chinese cultural influence, other lands on the international route found it convenient to adopt the plentiful cash for their systems of exchange. Later the widespread Chinese merchant communities would provide an additional impetus in this direction.[3]

As evidence of the use and non-use of copper cash across late medieval Asia, let us turn first to Ma Huan, Chinese translator on several of Cheng Ho's great expeditions, and then to Tomé Pires, the well-known Portuguese apothecary of Malacca. During the first quarter of the fifteenth century, almost midway through our period of study, Ma took note of the use of Chinese copper cash in various of the countries visited by the Chinese fleets. The expeditions themselves appear to have encouraged the use of cash, and Ma's comments will help us draw an approximate line around the area with

---

*Abbreviations used: *DNNTC*—*Dai Nam Nhat Thong Chi*; *DNTC*—*Dai Nam Thuc Luc*; *NTTT*—Nguyen Trai, *Nguyen Trai Toan Tap*; *TT*—Dai Viet Su Ky Toan Thu

1. Yang, p. 37; Hartwell, pp. 283-284. As Perlin, p. 241, notes, "Historians spend too little time on the problem of copper, ..."
2. Yang, pp. 26, 38; Hartwell, pp. 284-285; Wheatley, pp. 37-38, 100-101, 113, 115; Gernet, pp. 84, 110. For Japanese coinage and their use of Sung cash, see Kobata 1980, p. iii, and Kamiki & Yamamura in this volume; on Korea, see Henthorn, pp. 65, 103.
3. Nguyen 1970, pp. 165-166.

363

currency systems based on this cash. The countries noted were located in eastern Java (Majapahit) and southern and northern Sumatra (Palembang and Lambri respectively). Ayudhya in Thailand may or may not have used the cash; the Chinese texts are ambiguous. Presumably Malacca, in the beginning of its glory, also used the cash, though the fact is not mentioned here. Champa, on the southern coast of present day Vietnam, did not.[4] Chinese cash thus followed the flow of trade through Southeast Asia, being used in Java and generally around the Straits of Malacca, but not to any degree in mainland countries off the route and not along the international route beyond Southeast Asia. From Ceylon and the coasts of India through the Middle East, the local rulers used mainly gold and silver, occasionally copper, to mint their own coins, according to Ma's account.[5]

A century later, Pires' report reflects this same pattern and shows how it had changed. Champa now used Chinese cash, presumably because of the Vietnamese conquest of 1471. East Java continued to use the cash, and those areas within its trading system, Sunda on west Java and the spice islands, used cash as well. But Malacca's emergence as the major international trading center in Southeast Asia shifted the orientation of the Straits area toward the Bay of Bengal and away from the South China Sea. Like northern Sumatra, Malacca now used small tin coins and was linked to the flow of gold and silver among Bengal, Pegu in southern Burma, and Ayudhya in Thailand. West of Southeast Asia there continued to be no sign of the use of copper cash.[6]

Gold and silver played major roles in the trading system of eastern Asia not as currencies but as commodities and merchandise for the market.[7] For the purpose of this paper, "eastern Asia" is defined as that region, illustrated in the comments above, which utilized copper cash, either coins from China or coins minted locally. Vietnam stood near the center of this system, and its monetary usage had a strong relationship to what was occurring around it. I will first briefly discuss the history of the copper monetary system and the mining of precious metals in Vietnam and then describe the commercial flow of valuable metals, gold, silver, and copper, throughout East and Southeast Asia in relation to Vietnam. The result will be impressionistic rather than quantitative. Given the general lack of figures on Vietnamese metals and money, it is difficult to provide either absolute or relative numbers for Vietnam, internally or externally. Thus, the place of Vietnam within the Asian monetary scene will necessarily remain vague.

---

4. Mills, pp. 45-47, 88, 96-97, 102, 107, 123; Huang, p. 76.
5. Mills, pp. 129-130, 136, 141, 151, 153, 156, 161, 167, 177.
6. Cortesao, pp. 93, 96-97, 99-100, 104-105, 114-115, 140, 144, 170, 181, 203, 206-207: Godinho, pp. 279, 391-398, 415-417, 519-520, 523, 593: Meilink-Roelofsz, p. 40. For later uses of Chinese cash on Bali, see Geertz, pp. 91, 178, 199, 208-209.
7. Meilink-Roelofsz, p. 40, and Godinho, pp. 392-395, 453. For Vietnam, see Nguyen 1970, p. 170; Deloustal, p. 77-78; Manquin, p. 238; Malleret, pp. 47, 137.

The Vietnamese monetary system

The form of copper cash that set the pattern for eastern Asia became established in China over two thousand years ago, during the early Han dynasty. Cash would continue to be used in everyday transactions by the Chinese until the fall of their last dynasty in 1911. Much variation occurred, however, in the quality and quantity of coins cast and the areas where they circulated. Following the Han, other items, such as grain and cloth, tended to be used, and only in the seventh century did the new T'ang dynasty introduce the coin that would serve as the model for later dynasties, Chinese and foreign.[8] The Vietnamese employed gold and silver in their exchanges before the T'ang period and during the time of control by that dynasty.[9] Nevertheless, shortly after its independence in the middle of the tenth century, Vietnam developed a long lasting style of currency similar to that of its northern neighbor,[10] while using Chinese cash as well. The Vietnamese economy at large was probably not monetized, and only the urban-administrative areas undoubtedly used cash in their exchanges.

The first historical record of a Vietnamese ruler minting his own cash comes in 984 when Le Hoan established coins for his Thien-Phuc reign period, though coins are known from a decade or so earlier.[11] During the Ly dynasty (1010-1224), minting of Vietnamese cash is known to have taken place in its first decades, particularly in the early 1040s; this was probably as much for a political as an economic purpose, to help establish the power of the throne. For eighty years thereafter, into the 1120s, there is no sign of Vietnamese cash. From that decade, the Ly appear only to have minted occasional coins[12] and generally to have relied on Chinese cash. On the one hand, minting had probably not developed a strong link to royal legitimacy, while on the other the huge amount of cash moving out of China undoubtedly allowed the Vietnamese to take their share of this reliable currency, as the Japanese did.[13]

---

8. Yang, pp. 2-3, 9, 16-17, 21-26; Hartwell, pp. 280-281; Herbert, pp. 253-269, 284; Masui, p. 2.

9. Taylor 1976b, pp. 203-204; Schafer, pp. 153, 163; Herbert, p. 272, n.68.

10. Pham 1979, p. 40, and for later descriptions, see A. de Rhodes in Nguyen 1970, p. 162, from the seventeenth century; Richard, p. 739, from the eighteenth century; and White, pp. 257-258, from the nineteenth. Two Vietnamese poems on coins may be found in Huynh, pp. 150-152.

11. *TT*, I, 169; Deloustal, pp. 59, n.2; Do 1979, pp. 26-28; Pham 1979, p. 35; Toda, p. 77. Yang, p. 26, and Toda, p. 76, also note the earlier casting by Dinh Bo Linh.

12. *TT*. I. 219, 222; Taylor 1976a, pp. 176-177; Do 1979, pp. 28-30; Gutman, p. 8. Interestingly, elsewhere on the Southeast Asian mainland and in Japan, the use of coins ceased by the late tenth century and did not begin again until the twelfth century in Japan and the fourteenth century on parts of the mainland (and in some locations, for example, Burma and Thailand, not until the nineteenth century); see Gutman, pp. 9-10, and Kobata 1980, p. iii.

13. Do 1979, pp. 30-31; Kobata 1980, p. iii; Takizawa, p. 22; Kamiki and Yamamura in this volume. Toda, pp. 65-66, makes the following statement:

The Tran dynasty (1225-1400), which followed the Ly, seems to have maintained this ambiguous pattern to the end of the fourteenth century, years that undoubtedly saw a slowly increasing use of money throughout the country and the growth of agriculture and local markets in a situation similar to that of Japan. The only existing note of concern on a monetary matter occurred at the beginning of the dynasty (1226) when the Vietnamese ruler stipulated that each short string of cash or "hundred" (*tien*) would have 69 cash in private dealings and 70 in public matters.[14] The Vietnamese, like the Japanese, probably sent less valued gold to China for its more valued cash as they relied on China for their currency rather than produce their own. The Mongol period may have caused difficulties in this direction, however. During the fourteenth century, a shortage of copper and of Chinese cash relative to demand seems to have occurred since the Vietnamese in the 1320s again began to mint their own cash and had to use zinc for the coins. Vietnamese copper coins from mid-century were conspicuously small and thin.[15] Only with the brief Ho attempt in the last years of the century and the early years of the fifteenth to copy the Chinese use of paper money so as to maintain a better control of national resources did the pattern change in any major way. Under this system, holding copper cash was declared illegal; all coins had to

---

>... nearly all the coins which were in circulation up to the fifteenth century were actually cast in the provinces of Kuang-tung, Kuang-si, and Fu-kien, and brought direct to Annam [Vietnam] in Chinese junks. The coins thus imported were smaller than the ordinary Chinese cash; they bore the Nien-hao or *name of reign* of various Emperors of the Sung dynasty in China, and are still to be found in large numbers. In Annam coins were cast only under a few kings of the dynasties Dinh, former Le, Ly, and Tran, who reigned before that period.

and, pp. 79-80;

>... the use of small thin cash was first introduced, on account of the great scarcity of copper then existing in the Kingdom. The Chinese traders immediately took advantage of this circumstance, and had Chinese cash recast into smaller ones, exporting them from their own country into Annam. In consequence of the abundance of coins caused by this proceeding the manufacture of cash was suspended by the Annamese government for a period of fifty years.

No source is supplied for these statements, but they may have been drawn from a Chinese work (see p. 43). On the Vietnamese coinage up to the fifteenth century, see pp. 75-90.

Toda's initial outline of Vietnamese monetary history made no claim to being either "complete or correct" (p. 43). Later studies, presumably more solidly based, include D. Lacroix, *Numismatique annamite* (Saigon, 1900); A. Schroeder, *Annam, Etudes numismatiques* (Paris, 1905); and B. J. Permar, *Catalogue of Annam Coins, 968-1955* (Saigon, 1963). Do 1979; 1980, used Lacroix and Permar.

14. Phan, III, 61; Deloustal, p. 60; *TT*, II, 8. Gernet, p. 79, notes that in late Sung China the official short string had 77 cash, though the number actually varied depending on the kind of transaction. In Hangchou, the Sung capital during the thirteenth century, it contained only 50 cash. See also Yang, pp. 34-37; Hartwell, p. 283.

15. Deloustal, p. 60, n.3; *TT*, II, 126; Do 1979, pp. 30-33; 1980, pp. 50, 53. See Kamiki & Yamamura in this volume for "the diverging value of copper coins in China and Japan."

be turned over to the state. Only some very small coins, perhaps ceremonial in nature, were minted.[16]

Following the twenty year occupation of Vietnam by the Ming dynasty of China (1407-1427), in which both Chinese copper cash and paper money were used,[17] the new Le dynasty (1428-1787) found itself faced with the question of what monetary system to employ. As the dynastic founder Le Loi commented in an edict opening the matter for discussion,

> Money *[tien]* is the lifeblood of the people; We cannot run short of it. Our country produces much copper, yet the Ho destroyed the old copper cash, leaving only 1% of it. . . . in ages past there were those who believed that gold and silver, leather and silk, copper cash, paper money, all were equal as currency. In these circumstances, which is the best to use.[18]

The decision was, for the first time, to develop good Vietnamese copper cash, most likely because of the link to royal legitimacy in the Chinese pattern. Le Loi and his successors minted their own coins. The founder set the number of cash per short string *(tien)* at 50, 28% less than under the Tran, though a decade later his son increased the number 20% to 60. The long string or "thousand" *(quan)* was ten short strings, that is, 600 cash. This number remained standard into the nineteenth century.[19]

As Le Loi feared, there immediately appeared the problem that would last to the end of his dynasty in the 1780s—a shortage of copper cash. Already, by 1434, Le Loi's son and successor had announced that there would be no sorting through strings of cash in order to pick out only the good and leave aside worn and damaged pieces. Government officials were especially condemned in this matter. The theme appeared again in later centuries as the government continually sought to keep a shortage of cash from adversely affecting the economy. It reappears in 1486, 1658, and 1741. Sorting out cash had to be reported by any who had knowledge of it: if it were a serious offense, exile was the penalty (1658). Hoarding good coin was forbidden. By 1741, the government was willing to allow half value for a broken coin, at the same time as it was mining more copper and minting

---

16. Phan, III, 61-62; Deloustal, pp. 60-62; Do 1979, p. 31; 1980, pp. 50-54; Pham 1979, p. 40; Whitmore 1976b.

17. Whitmore 1977a, p. 71; Aurousseau, pp. 83ff.

18. Phan, III, 62; Deloustal, p. 63 (mistranslation); *NTTT*, p. 134; *TT*. III, 70-71; Pham 1979, p. 35; Do 1980, pp. 52-53.

19. Phan, III pp. 62-63; Deloustal, pp. 63-66, 74, *TT*, III, 62, 66, 128; Nguyen 1970, p. 163; Masui, p. 2; Bowyear in Lamb, p. 53; Malleret, pp. 47, 162; Richard, p. 739; White, p. 257.

Toda, p. 102, noted, ". . . from the date of the accession of the Le family there was a manifest improvement in the manufacture of coins; excellent metal was used for the casting, and the work is equal to the best specimens of coins circulating in China at that time."

The Japanese, on the other hand, continued to rely on imported Chinese cash, particularly those of the Yung-lo period (1402-1424); Takizawa, pp. 22, 26-28. Only in 1636 did a Japanese regime successfully mint its own coins.

more cash than during any prior period. In this effort to keep up with the monetary demand, to keep the economic blood flowing as it were, the government decentralized minting operations and turned to private enterprise for a time in the middle of the eighteenth century. The need for currency was so great that the government even countenanced counterfeiting, despite the official death penalty due it.[20]

With the growth of population and transactions during these centuries,[21] the Vietnamese government continued to search for the means to maintain a sufficient supply of coins. One way was to keep old coins in circulation. The notice for 1658 made reference to the good coinage of the second half of the fifteenth century (the Quang-thuan, 1460-1469, and Hong-duc, 1470-1497, periods), and a notice from 1745 strenuously barred discriminating between old coins and those newly minted.[22] The continued use of old coins is confirmed by a Japanese source from the 1790s which noted that the coins from Le Loi's reign (the Thuan-thien period, 1428-1433) were still in use.[23]

Other possibilities were the manufacture of smaller copper coins, as occurred in the 1440s and again in the late sixteenth century under the Mac, the use of one coin for 50 or 100 cash (mid-eighteenth century), the attempted reduction in the number of cash per string for certain transactions (to 36 in the eighteenth century, and the reduction of copper content in the coin.[24] The latter undoubtedly occurred, particularly in the eighteenth century, since debasement was the only way (until the switch to silver) to avoid a commercial depression. However, more detailed numismatic studies need to be undertaken before we can begin to raise questions which relate changes in coinage to the historical situation.

From the use of less copper in the coin, it was a short step to the minting of coins out of less valued metals, as was done with zinc in the 1320s. Le Thanh-tong, ruler in the second half of the fifteenth century, would not allow non-copper cash, but the Mac tried zinc and iron coins during their dynasty in the sixteenth century.[25] In the mid-seventeenth century, the government in Thang-long (Hanoi) made strong efforts to remove from circulation cash made of zinc, tin, cast iron, and iron, which villages specializing in minting coin had begun to cast, presumably to fill the need for coinage at the local

20. Phan, III, 63-65; Deloustal, pp. 65-68, 70, 72-73; *TT*, III, 296; Toda, pp. 106-108; Nguyen 1970, pp. 164-165. The Vietnamese situation was undoubtedly aggravated by shortages of cash in China itself, as during the fifteenth century; Huang, pp. 75-76. For the problem of good and bad copper coins in Japan, see Takizawa, pp. 22-30, 33-34.

21. Nguyen Thanh Nha stresses population growth and rise in monetary velocity in his study; see his conclusions, Nguyen 1970, pp. 229-233.

22. Phan, III, 63-65; Deloustal, pp. 67, 71-72. Toda, p. 103, noted the good quality of the late fifteenth century coins.

23. Muramatsu-Gaspardone, pp. 73-74. See also the poem, "Postscript to a Book on Coins", in Huynh, p. 149.

24. Toda, pp, 103, 111-112, 120; Kamiki and Yamamura in this volume; Yang, p. 37; Nguyen 1970, p. 163; Deloustal, p. 74.

25. Deloustal, pp. 66-67; *TT*, IV, 121; Toda, p. 119.

level. All such coins were to be destroyed. In 1663, the state was prepared to give two short strings of copper cash for each long string of zinc cash. Yet, in the mid-eighteenth century, with the availability of Chinese tutenag (a copper, zinc, nickel, or tin alloy) and a growing demand for coins, the Vietnamese government yielded to the temptation of officially sponsoring such coinage. In 1741-42, it allowed undamaged zinc coins to be circulated and broken ones in circulation to be turned in at a very low rate for good ones. The result appears to have been, in a situation relatively similar to that of the early Sung dynasty (960-1040) in China, a certain stability in prices as the rise in monetary demand countered the inflationary effect of debasement and higher mint output.[26]

More importantly, in 1740 the Vietnamese for the first time began to use silver as official currency, together with the copper cash. Previously, silver had been little valued as money, despite the fact that China had been using uncoined silver much more extensively since the fifteenth century. The Vietnamese state had tried to use silver to supply its troops in the seventeenth century, but merchants had driven the value of the metal down and had made the task of supply quite difficult. Now the state actively sought to control the market value of the metal. In cutting up a bar of silver, the government declared that one ounce (37.3 grams) of silver was worth two long strings or 1200 copper cash and one piece of silver brought two short strings or 120 copper cash. Silver was to be used for all transactions worth two short strings or more. As Phan Huy Chu, an early nineteenth century Vietnamese scholar, commented, "From this time, the value of silver stabilized, and the rich merchants were no longer able to reap great profits."[27] Our Japanese source of the 1790s noted that the Vietnamese used small unmarked coins of silver and gold which merchants weighed and nicked off to meet the price of purchase.[28]

Yet, in the final analysis, it was foreign cash, copper and later tutenag, mainly Chinese but also Japanese, which helped the Vietnamese government to keep up with the indigenous demand for money. Of this foreign supply of cash, a distinction was made between the standard, officially minted copper coins of China (such as those of the K'ang-hsi and Ch'ien-lung periods in the seventeenth and eighteenth centuries) and other coins, some made specifically for export to Vietnam. Generally, the former coins might be worth anywhere

26. Phan, III, 63-64; Deloustal, pp. 67-71, 75-77; Nguyen 1970, pp. 157-162, 167; Dermigny, p. 409; Toda, p. 106; Hartwell, p. 284; Leeds, 181. The date of 1752 given in Deloustal, p. 75, and followed by Nguyen 1970, p. 158, is Khanh-duc 4, actually *1652*.

Toda, pp. 67-68, commented, "We have searched in vain for any law relating to the different standards of copper, zinc, and lead coins. Their value depends altogether on the market, which in ports open to foreign trade is regulated by the price of the Mexican dollar."

27. Phan, III, 64; Deloustal, pp. 69-70; see also Nguyen 1970, pp. 160, 170; Dermigny, pp. 419-420; Huang, p. 76; and Masui, pp. 2, 5-7.

28. Muramatsu-Gaspardone, p. 74; White, p. 258, notes the use of gold and silver in the early nineteenth century.

from 40% to 70% more than the latter. For example, one ounce of silver, as we have seen officially brought 1200 regular copper cash. The imported copper cash, on the other hand, went 860 to the ounce of silver. Basically, the relative values were dependent on the market rather than any officially designated worth. Despite the rising prices of Chinese tutenag and Japanese copper from the beginning of the eighteenth century, the demand for coins in Vietnam was such that, as noted, general market prices remained relatively stable. The desirability of the official Chinese coins was so great that when the Trinh took the Nguyen capital in the south in 1774 they found, we are told, a hundred million of them from the T'ang, Sung, and Ch'ing reign periods.[29]

During the period of their independent state, the Vietnamese committed themselves to the basic Chinese pattern of copper coinage, only using silver in the eighteenth and nineteenth centuries. The extent of this commitment may be seen in the retention of cash as the monetary system in the southern Vietnamese realm even though it had no copper resources. As a result, the southern realm followed much the same monetary pattern as that of the northern government described above.[30] This chosen monetary system could not be fulfilled by the resources of Vietnam alone. The country became increasingly involved in the international exchange of metals as it exported its own gold and silver, as well as foreign silver, and imported the raw copper, copper cash, and tutenag needed for its expanding economy. By the first half of the eighteenth century, a general shortage of currency was the major result.[31]

Vietnamese metal production

Though few figures exist for metal production in Vietnam before the nineteenth century, we may use such data as are available to gain a general impression of mining there. The main source of valuable metals was the Viet-bac, the mountainous region north of the Red River delta. This area was important in supplying China with gold during the T'ang occupation. The gold was initially exploited by working the streams, and only later, after the twelfth century, does mining seem to have taken place. The amount of gold taken out in the first centuries of Vietnamese independence was moderate. Amounts recorded in the *Dai Viet Su Ky Toan Thu*, the Vietnamese historical chronicle, range from 47 ounces (1.75 kg.) to 112 ounces (4.18 kg.) in presentation to the Court by different localities at different times.[32]

---

29. Nguyen 1970, pp. 165-167; Dermigny, pp. 410, 413-414, 416. Toda, p. 69, noted an occurrence of Chinese counterfeiting for the Vietnamese market in mid-nineteenth century Hong Kong. John Munro made a good statement at the conference on the benefits of offering imitation coin for both external minters and internal fiscal policy.
30. Nguyen 1970, pp. 164, 167-169; Malleret, pp. 137, 160-161.
31. Phan, III, 65; Deloustal, p. 73; Nguyen 1970, p. 230.
32. Schafer, pp. 162-163; *TT*. I. 279; Pham 1976, pp. 48-49.

Silver, too, came out of the mountains, though we have no amounts linked to particular locations. During the T'ang period, it was, to quote Edward Schafer, "a commonplace" in Vietnam, unlike in the rest of China at that time. From the tenth century both gold and silver are recorded as part of the tribute sent by the Vietnamese to China. The only specific indications of the amount of silver available in these early centuries are the 1680 ounces (62.66 kg.) used to cast one bell and the 800 ounces (29.84 kg.) of silver and 310 ounces (11.56 kg.) of gold for three other bells in the first decades of the eleventh century. More bells and some Buddha images are also recorded, though without figures for the metals involved.[33] While the amounts of these metals were seemingly not great, their local availability in the age before the great American and Japanese production of the sixteenth century had a value in and of itself.

Copper does not figure nearly so much in the early historical records as do silver and gold. It was not exploited by the T'ang, and the first mine mentioned was opened in the northern mountains of Lang-son only in 1198. References to the use of copper for Buddhist bells and images, weapons, utensils, and the aforementioned cash occur at scattered points in the records. In 1035, 6000 *can* (3580.8 kg.) of copper went into a temple bell and six years later 7560 *can* (4561.8 kg.) were used for the image of a Buddha and two Bodhisattvas; the mid-1050s saw 12,000 *can* (7161.6 kg.) go into a huge bell for a special temple.[34] Much of the copper may have come from the cash flow leaving China.

According to a Sung dynasty source, the Vietnamese produced gold, silver, copper, tin, and lead. In the early fourteenth century, Le Tac's *An-nan Chih-lueh* noted these metals plus iron and pointed to locations in Thai-nguyen and Cao-bang provinces as producing gold and silver.[35] Our most detailed early source comes from the period of the Ming occupation in the first decades of the fifteenth century. This work records reports from the various colonial prefectures (the old Vietnamese provinces) on the production of gold and silver. In the northern mountains, Thai-nguyen had seventeen localities producing gold, and Lang-son four, while Tuyen-hoa had silver. In the western mountains, Quang-oai drew gold from fifty-nine places, Gia-hung from five, Ning-hoa from three, Quy-chau from one, Ngoc-lam chau from six, and Tra-long chau from three.[36] The 1430s give us the first Vietnamese geography, the *Du Dia Chi*. In this work, modeled on a classic Chinese text ("The Tribute of Yu"), the great scholar Nguyen Trai and his colleagues briefly described each province and its major resources. According to this work, Thai-nguyen had one location which produced silver, copper, lead, and gold; Lang-son had one locality producing gold and lead and

33. Schafer, p. 163; Pham 1976, pp. 49-50; *TT*. I. 192, 195, 209, 212, 230, 239.
34. Schafer, p. 164; Pham 1976, p. 50; *TT*, I, 212, 218, 230; II, 16.
35. Pham 1976, p. 50; Le, p. 149 (Chinese); Aurousseau, p. 66; Li, 1, p. 17a.
36. Aurousseau, p. 66; on this work, see Whitmore 1977a, p. 77, n.34.

another with copper and silver; in Cao-bang, one place produced gold; Tuyen-quang produced gold, silver, iron, and tin from one location and copper from another; and Hung-yen to the west had silver, copper, and tin.[37] Copper and iron were considered important items by the Vietnamese government in the fifteenth century, while the other metals were merely seen as taxable commodities and not as a basis for the monetary strength of the kingdom.[38]

Only with the monetary crisis of the eighteenth century does our information on metal production pick up again. We continue to see the importance of what Nguyen Thanh Nha has called "le riche quadrilatere" which included the provinces of Thai-nguyen, Tuyen-quang, Lang-son, Cao-bang, and Hung-hoa. The Vietnamese government, like its contemporary in China, began to take a much greater interest in developing mines. Records after mid-century mention eight copper mines, two gold mines, two silver mines, and one mine each of zinc and tin. The emphasis on copper reflects the problems of the period. The government gave out much of this mining in concessions, and the industry was dominated by the fluid world of Chinese entrepreneurs and laborers who applied the more advanced technology of their country. As an example of copper production, two of the best mines put out about 450,000 *can* (268,560 kg.) and 350,000 *can* (208,880 kg.) annually.[39]

A significant change that had taken place by this time lay in the expansion of the Vietnamese population southward under the aegis of the southern Vietnamese regime. While copper, lead, and tutenag came from abroad, gold and to a lesser degree silver were found and exploited in the new territories which had once belonged to Champa. A high quality gold came mainly from the provinces of Quang-nam and Thuan-hoa where both mining and panning took place in a relatively simple fashion. One of the major mines could produce 1000 *hot* (10 ounce bars) of gold annually (284 kg.), and the tax on gold production yielded over 800 ounces (30 kg.) per year.[40]

This eighteenth century distribution of valuable metals in Vietnam is confirmed by materials from the following century. The English envoy John Crawfurd, visiting the capital of Hue in 1822, commented on the gold and silver mines north and west of Hanoi which were manned entirely by Chinese

---

37. *NTTT*, pp. 205, 207, 218, 219, 221; Nguyen 1970, p. 87, n.48, except for his Thai-nguyen listing.

38. *TT*, III, 64, 282, 283, 288. The acts of 1484 dealt with gold, silver, tin, and lead.

39. Nguyen 1970, pp. 86-89, 199; Phan, III, 76-79; Deloustal, pp. 167-176; Toda, p. 64; Sun 1967, pp. 52-55; 1968, pp. 843-845; Woodside, pp. 277-278, based on articles in Japanese by Fujiwara and Wada. *NTTT*, p. 221, and Nguyen 1970, p. 87, n.48, provide a list of 26 mines based on an unknown document entitled *Ban-quoc San-xu Ky* ("A Record of the Productive Regions of the Country") which could be from the first half of the eighteenth century; of these mines, 15 produced gold, 10 silver of different kinds, 5 copper, 4 lead, 2 tin, and 2 iron (about half produced more than one metal).

40. Nguyen 1970, pp. 90-91; Malleret, pp. 46, 138, 173; Lamb, pp. 131, 136, 176; Manquin, pp. 239-241.

laborers. He gained no figures on the gold production, but heard one of 215,000-220,000 ounces (over 8,000 kg.) for silver annually.[41] The latter figure was most likely higher than that of the previous century since the new Nguyen dynasty stressed payments in silver.[42] There seem to have been a total of 124 mines of all sorts in the north during the first half of the century, 38 of them in Thai-nguyen.[43] The official geography of the second half of the century recorded three mines (gold) in Cao-bang, ten (5 gold, 5 silver) in Thai-nguyen, four (gold) in Lang-son, four (3 gold, 1 silver) in Hung-hoa, four (3 gold, 1 silver) also in Tuyen-quang, and one (gold) in Bac-ninh, all north or west of Hanoi. These were open mines and do not count ones that had been closed. In the center, Quang-nam and Binh-dinh both produced gold, though the text gave no indication of either mines or amount produced. Taxes on the total production, mainly from Thai-nguyen, annually amounted to 3 kg. (over 75 ounces) of gold and 31 kg. (820 ounces) of silver.[44]

Given the Vietnamese choice of copper as the basic metal for their currency, Vietnam's supply of valuable metals dictated involvement in foreign trade. The Vietnamese needed to import copper, whether raw or already minted, while gold and silver served as commodities on foreign markets.

The international monetary flow

In the first centuries of Vietnamese independence, there exists little direct evidence of Vietnamese involvement in foreign trade. Previously, while under the control of the T'ang, a port area in the upper Red River delta had served a role, sometimes primary, sometimes secondary, in the international trade of its day. After the fall of the T'ang and the rise of the Sung in the late tenth century, the Chinese had seen little commercial advantage in regaining their hold over Vietnam,[45] particularly because of the development of ports on their southeastern coast. Heretofore we have generally assumed that the Vietnamese cared little for trade, but recent work has begun to postulate a

41. Lamb, p. 262.
42. Woodside, p. 277.
43. *Ibid.*, p. 138.
44. *DNNTC* (Hanoi), III, 57; IV, 141, 170-171, 314, 345, 379, 408-409; (Saigon), "Cao-bang", p. 51; "Bac-ninh", p. 104; "Quang-nam", p. 123; "Binh-dinh", p. 103. Toda, pp. 60-63, provided the following figures on mines, generally similar to the Vietnamese figures:

    Cao-bang—4 gold
    Thai-nguyen—4 gold, 5 silver, 1 tin
    Tuyen-quang—4 gold, 1 silver, 1 copper/silver
    Lang-son—2 gold
    Hung-hoa—2 gold, 2 copper
    Bac-ninh—1 gold
    Son-tay—1 copper

The annual payments to the government, according to Toda, were 150 ounces of gold (double the Vietnamese figure), 820 ounces of silver, 13,000 pounds of copper, and 600 pounds of tin.

45. Hirth and Rockhill, p. 45.

role for the Vietnamese, at least from the eleventh century, in a route that went across to the Mekong River via Nghe-an province and down to Cambodia, and its flourishing civilization of Angkor.[46]

Thirteenth to fifteenth centuries

Our best starting point is Chao Ju-kua's commercial report of 1225.[47] Written to portray Chinese knowledge of the world at large, this work illustrates the commercial routes extending from east to west. The major points on the main international route were the southeast coast of China, the area of Southeast Asia around the Straits of Malacca, and the Middle East, that is, the eastern Mediterranean. These same years saw east Java increasingly become the major center of trade in Southeast Asia and a key point on the east-west route.[48] As Marco Polo wrote in the late thirteenth century,

> The island is of surpassing wealth, producing . . . all . . . kinds of spices, . . . frequented by a vast amount of shipping, and by merchants who buy and sell costly goods from which they reap great profit. Indeed the treasure of this island is so great as to be past telling.[49]

How did Vietnam fit into this trade? and what part did metals play in it? Information contained in Chao's text would suggest that gold and silver were important Vietnamese products which moved along both primary and secondary trade routes. Chao tells us that gold and silver were among the top Vietnamese products. He also notes that Cambodia was in the market for these precious metals, producing none itself, and that Kuala Berang (Kelantan) on the east coast of the Malay Peninsula, Srivijaya on Sumatra, and east Java all sought gold and silver as desired commodities.[50] In addition, the *Dai Viet Su Ky Toan Thu* refers to commercial contacts at different times between the Vietnamese port of Van-don on the edge of the Red River delta and ships from Java, the island world at large, and around the Gulf of Siam.[51]

Java gained a great amount of copper, gold, and silver primarily in the pepper, clove, and nutmeg trade which it had controlled for centuries. This exchange was a major component in the flood of cash from China, and with the increased European demand for spices from the thirteenth century, it would appear that a flow of western monies moved through the Middle East and India to this distant island. Javanese sources and various travelers' accounts all attest to the plentitude of gold and silver accumulated on that

---

46. Hall 1975, pp. 321, 325; Hall and Whitmore, p. 335, n.64; Pham 1979, pp. 38-39.
47. Hirth and Rockhill, pp. 1-39; Wheatley, pp. 5-18; Hall and Whitmore, 321-322.
48. Wisseman, pp. 206-208; Whitmore 1977b, pp. 143-145.
49. Benda and Larkin, p. 12; see also Godinho, p. 277.
50. Hirth and Rockhill, pp. 46, 53, 61, 69, 78; see also Pham 1979, p. 39, quoting Hoang Xuan Han, *Ly Thuong Kiet* (Hanoi, 1949), p. 106. Chou Ta-kuan, p. 34, affirms the Cambodian desire for gold and silver at the end of the thirteenth century.
51. *TT*, I, 281, 295; II, 152, 163; Do 1979, pp. 32-33; Pham 1979, pp. 38-40.

island.[52] It seems logical then that the Vietnamese gold and silver were exchanged for other goods (such as spices) on the international trade route, eventually reaching Java, that these metals also went to Cambodia via the postulated Mekong route, and that Vietnamese gold went to China in exchange for the needed copper cash, as noted earlier.

Additionally, Vietnamese gold and silver went north to China in tribute within the East Asian political system. From the tenth century, the Vietnamese sent these metals as part of a varied offering on their tributary missions. In 1289, an undoubtedly significant embassy to the Yuan dynasty (just two years after the Vietnamese had defeated the Mongols) carried gold and silver to offer to Kubilai Khan in addition to a number of other goods. The tribute included items with gold and silver inlay and gold plating and objects made of gold and/or silver, particularly five gold gongs weighing 100 ounces and ten silver gongs weighing 300 ounces. The total amount of gold surpassed 250 ounces (9.32 kg.) and of silver over 300 ounces (over 11 kg.).[53]

This flow of Vietnamese gold and silver north and south seems to have been typical of the thirteenth and fourteenth centuries, as the poles of exchange remained China and Java. The fifteenth century saw changes in this pattern with the rise of Malacca as the major focal point of international trade in Southeast Asia and with Ming China's government expeditions and strictures against foreign trade. Though the Ming controls eased unofficially later in the century, they led to official Chinese encouragement of Ryukyuan traders as intermediaries for the Southeast Asian trade. From the late fourteenth century into the sixteenth, ships of the Ryukyu islands were heavily involved in this regional commerce and dealt with Siam, Malacca, and the islands beyond, carrying among many other items copper coins from China and gold and silver from Japan to these ports.[54]

For Vietnam, first came the two decades of Ming occupation. In the Vietnamese folk memory, this was a time of brutal exploitation. Gold and silver were merely two of the seventy-nine different types of natural resources

---

52. Whitmore, 1977b, pp. 145-146; Wheatley, pp. 45, 100-101, 107; Ashtor, pp. 195-197, 207, 239-242, 264-266, 275-276, 297-300, 325-326; Colless, pp. 137-138; Godinho, pp. 279-280, 587-594, 614.

The papers at the conference gave a strong sense of the movement of precious metals out of the Mediterranean (and East Africa), through the Middle East, to India. In particular, see Bacharach and Curtin in this volume. Marie Martin also gave evidence on this movement at the conference. A good part of the money may not have stopped in India, but instead have accompanied Indian textiles into Southeast Asia and especially Java to gain the spices. The Javanese then sent textiles, Chinese cash, and their rice further east for the cloves and nutmegs, retaining the precious metals for themselves (Meilink-Roelofsz, pp. 83-84, 93-100, 105-115). A study is needed of the spice trade along its full length, including the question of how far Western coins (such as the Venetian gold ducat) went.

53. Pham 1976, p. 49; Phan, IV, 156-157.

54. Higgins, pp. 31-32; Wills 1974, pp. 7; Sakamaki, pp. 383-389; Crawcour, pp. 377-378; Innes 1980, pp. 33-34; Cortesao, pp. 130-131; Godinho, pp. 281-282; Toyoda, pp. 31-32; Kerr, pp. 63, 74, 76-78, 81, 88-96, 124-130.

exploited in Vietnam by the Chinese. To quote from the Vietnamese victory edict of 1427: "To extract gold ore the people were obliged to confront pestilential vapors as they dug the mountains and washed the sands."[55] The yearly take by the colonial government (presumably applicable for the years 1416 to 1423) was 573 ounces (21.37 kg.) of gold and 1072 ounces (39.99 kg.) of silver. This production came mainly from the nothern mountains (Lang-son, 29 ounces of gold and 145 of silver, Tuyen-hoa, 0 and 859, and Thai-nguyen, 144 and 65), but also from Thanh-hoa (180 gold) and Nghe-an (220 gold) in the south. This southern gold must have come from Champa down the coast or the Lao territories to the west since there is no record of gold production in either of those two provinces. Generally these metals seem to have been used for the provincial administration, though there is a record of 32 ounces (1.19 kg.) of gold being officially sent to the Chinese Court in 1410. These amounts, of course, do not include whatever peculation occurred among the colonial officials themselves, and much of the metals may have gone to China under private auspices.[56]

Beginning in the 1430s, the Ming state dissociated itself from the earlier activities that had so strongly involved it with Southeast Asia. In the process, Malacca emerged as the major trading center of the region.[57] It became strongly integrated into the older Javanese system and fed the products of Southeast Asia into the international route. Vietnam played a minor role in the overseas trade of the period, though its metals held a certain significance for it. Our evidence comes from the first major Portuguese work dealing with Asia, that written by Tomé Pires shortly after the conquest of Malacca in 1511. Champa, in what would become central Vietnam, produced "a good quality of tested gold" from mines and sent it north to Vietnam.[58] Pires' comment on Vietnam and its merchandise speaks for itself: "Especially gold and silver, much more so than in Champa."[59]

Thus both Chao Ju-kua in the thirteenth century and Tomé Pires in the sixteenth noted the special place in Vietnamese production and trade of these precious metals. Vietnam sent this bullion as its main product to Malacca, particularly in exchange for sulphur. The Vietnamese gold, judged the best of Southeast Asia, joined the larger amount coming from Sumatra to help make Malacca the main center for gold in Asia, if not in the world. Much of the gold went on to India.[60] Canton, however, was the main port of call for the Vietnamese, and they went from there to Malacca on Chinese or Ryukyuan junks. Few Vietnamese ships went directly to Malacca. Curiously, since the

55. Truong, p. 56.
56. Whitmore 1977a, p. 71; Aurousseau, pp. 83, 87-88, 94-97. There is a curious claim in *TT*. III. 14, that the Chinese did not mine silver until 1424.
57. Wolters, chapter 11.
58. Manquin, pp. 40-41.
59. *Ibid.*, p. 44.
60. Meilink-Roelofsz, pp. 80-81; Manquin, p. 240; Godinho, pp. 277-280, 283-285, 445, 450, 523.

Ryukyus produced sulphur and carried Japanese metals, there is little evidence of direct trade between these islands and Vietnam.[61]

Gold went to China from Vietnam officially, as tribute from the new Le government, as well as unofficially, in the form of trade to Canton. There are records of numerous Vietnamese tribute missions to Peking, but most of them do not state what constituted the tribute. At the beginning of the dynasty, the Vietnamese sent two "golden men" of 100 ounces (3.73 kg.) each, together with a silver incense burner, a pair of silver vases, and numerous other non-metallic items. In 1431, the Vietnamese ruler wished to discuss the amount of tribute owed, since the Chinese were demanding 50,000 ounces (1865 kg.) of gold yearly (sic). He wanted it presented once every three years, as before. The Ming ruler insisted on the yearly tribute of gold in 1433, and the next year, following Le Loi's death, the Vietnamese complied, sending gold as well as another "golden man".[62] We may assume that Vietnamese gold was a significant element in the frequent tribute missions to the Chinese capital throughout the fifteenth century.

Sixteenth to eighteenth centuries

A series of episodes in the sixteenth century led to a drastic change in the trading situation of Southeast and East Asia. First came the Portuguese penetration, displacing the trading center at Malacca, moving to the Spice Islands in the years 1511 to 1513, and as a consequence opening them to international competition in succeeding years. Eventually, the house of Malacca settled at Johor, on the southern tip of the Malay Peninsula, and handled the eastern trade, while to the west Acheh, on the northern tip of Sumatra, rose to deal with the Bay of Bengal commerce. By mid-century the diffusion of new smelting techniques to Japan from China via the Korean peninsula stimulated sharp increases in the island nation's silver output. Concurrently, a newly ascendant class of Japanese military lords was developing a taste for luxurious Chinese silks. Since the tribute trade between China and Japan, the only legal commercial channel linking the two countries, lapsed after the 1549 mission to Peking, the demands for Japanese silver on the mainland and for Chinese silk in the land of the rising sun had no legitimate suppliers. In this void, smuggling and piracy began to flourish on the Chinese coast. Finding that military measures were insufficient to quell the freebooters, the Ming court lifted its ban on private trade with Southeast Asia (but not with Japan, due to the menace of *wako* piracy) in 1567. By providing coastal merchants with an approved outlet for their goods, abrogation of the prohibition reduced the incentive to engage in smuggling. It also contributed to the decline of the Ryukyuan trade by undermining the Ryukyuan position as

61. Manquin, pp. 45; Sakamaki, p. 387; Godinho, p. 592.
62. Phan, IV, 158-159, 172; Li, 6, pp. 15b-16a; *TT*, III, 75-76, 78, 80. The Ming seem to have made a habit of seeking gold in tribute at this time; see Kobata 1965, p. 249, for demands made on Korea and the latter's import of Japanese gold to satisfy them.

commercial middlemen between China and Southeast Asia, and it encouraged Japanese to sail to Southeast Asian ports where they could swap silver for Chinese silk. Late in the century, first the Spanish, then the Dutch, entered the scene and soon had their own centers, Manila and Batavia.[63]

These events would in time have major implications for Vietnam, but in the sixteenth century itself there is little evidence to say how Vietnamese trade and its flow of metals were involved in the changes taking place. Presumably Vietnamese commerce continued with Canton and, in the second half of the century, began with the new Portuguese center of Macao. The Vietnamese may have had some contact across the South China Sea with Brunei, Johor, and Portuguese Malacca as well.[64] Two internal aspects of Vietnamese history undoubtedly affected such trade as there was: heavy fighting in the first and last quarters of the century, with an undercurrent of resistance in between, and the increasing Confucian influence, with its bias against trade, which appears to have occurred under the new Mac dynasty. A Portuguese priest, visiting the Red River delta in the third quarter of the century, noted the peaceful and prosperous existence of the people while stating that no significant international trade took place there.[65] The Mac controlled the northern mountains and continued to send gold and silver to Peking in tribute. Their first mission, in 1542, carried four sets of gold censors and vases weighing 100 ounces, one gold turtle (90 ounces), a silver crane and base (50 ounces), two sets of silver vases and censors (150 ounces), and twelve silver trays (641 ounces), a total of 190 ounces (7.09 kg.) of gold and 841 ounces (31.37 kg.) of silver. As the warfare became more intense in the final decades of the century, the Mac were unable to fulfill their tribute obligations.[66]

A key development in these years that would have great significance for later Vietnamese trade was the establishment of the Nguyen family on the then southern border of Vietnam. Participants in the anti-Mac resistance that in 1592 retook the Capital of Thang-long (Hanoi) for the Le, the Nguyen made this southern fief the base for their struggle with their rivals, the Trinh, who came to control the Capital. By 1570, the Nguyen had gained control of the provinces of Thuan-hoa and Quang-nam and in the following decades built up a regional government that was to last for two centuries. Sometime around 1600, the port of Hoi-an, or Faifo as it was known to the Europeans, a small town at the mouth of the Thu-bon River (thirty kilometers south of

---

63. Godinho, pp. 276-277, 548, 574, 783-787, 790-791, 813; Whitmore, 1977b, p. 147; Wills 1979, p. 213-215. Robert Innes supplied the information and language on East Asia in this paragraph; see also Innes 1980, pp. 10, 21-45, 51-56, 532-533, 542-543, 619-620.

64. Ch'en, p. 19; Manquin, pp. 182, 185, 227, 236; Wills 1974, pp. 7-8; Innes 1980, pp. 49-51.

65. Boxer, ed., p. 73; Lach, p. 565. On sixteenth century Vietnam, see Whitmore 1976a.

66. Phan, IV, 172-173.

Danang), developed in response to the new currents of international trade and to the needs of the southern administration for goods and funds.[67]

The conjunction that led to the increasing importance of this port on the south-central coast of Vietnam combined the Chinese private commercial thrust, the flow of American silver, the Japanese connection, Portuguese commercial efforts, and the desire of the Dutch to tap, and if possible control, the trading networks of Asia. In part, Hoi-an (Faifo) was a Chinese port away from China, existing in a freer commercial atmosphere than that of China. The central activity was the Chinese trade, and to pay for the silk, etc., of China much metal flowed through the port. Both for the Vietnamese and for the varied foreign traders, these metals, gold, silver, copper, and tutenag, as well as copper cash, were vital to the economies of the seventeenth and eighteenth centuries.

American silver began to enter Asia from the east almost as soon as the Spanish had set themselves up in the Philippines during the 1560s. The means by which this silver continued on into Asia was mainly the effort of private Chinese traders and the commercial network formed by them across East and Southeast Asia. Before the end of the sixteenth century, the Chinese community in Manila had grown quite large and come to dominate the trade there. Chinese junks, bringing silk and a host of other goods to exchange for the silver, came not only from the China coast but also from other Chinese communities of Southeast Asia as Hoi-an, Phnom Penh, and Ayudhya in Thailand. While the number of junks varied from year to year, as many as fifty arrived annually up to the 1640s. Some of the large quantities of silver taken by the Chinese junks undoubtedly made their way to Hoi-an for more silk, Vietnamese as well as Chinese, and for other goods, such as pepper.[68]

Japanese contact with Vietnam began at least from the mid-1580s and in succeeding decades exchanged Japanese and Spanish silver for Chinese and Vietnamese silk. In addition, when the Ming rescinded the ban on private trade in 1567, they maintained a prohibition against the export of copper and strengthened it in later years. This ban created a serious monetary problem for the Nguyen who lacked copper mines within their domain and a great opportunity for the Japanese merchants who flocked to Hoi-an. Spending

---

67. Ch'en, pp. 6-12; Manquin, pp. 185-187. As Robert Innes pointed out to me, "In the 1600s, Danang (or Tourane) and Faifo were distinct entities even though they appear to have been linked by an inland canal. Tourane served European vessels, while Chinese and Japanese junks that could cross the shallow bar at the river mouth went to Faifo, which had better access to produce from inland areas."

68. Schurz, pp. 26-27, 50, 63-98, 144, 146, 151, 153; Higgins, p. 32; Te Paske in this volume; Atwell, pp. 1-2; Chaunu; Iwao, p. 3; Vickery, pp. 509-522; Dermigny, pp. 100, 105-106, 194-195, 338-340; Godinho, pp. 518-519. An important study needed at this time concerns the interlocking trade of the overseas Chinese through the various ports of East and Southeast Asia during the sixteenth to eighteenth centuries. See, for example, Virpahol, pp. 40, 42-43, 47, 51-53, 58-59, 72-73, 96, 99, 122-123, 135.

copper cash imported from their homeland, Japanese residents of Hoi-an achieved dominance over the local markets for raw silk and sugar. This imported cash may have financed much of the development of new agricultural lands in Quangnam and newly annexed territories further south. In the early seventeenth century, almost two thirds of the ships officially designated by the Japanese for overseas travel were coming to the mainland ports of Southeast Asia and were increasingly focusing on Hoi-an. The total ships departing Japan in the years 1604-1616 (195) included 48 for the Nguyen domain, 11 for Vietnam itself, 5 for Champa, 24 for Cambodia, and 36 for Ayudhya. Of the 356 ships leaving Japan in the years 1604-1635, 174 of them went to the eastern mainland of Southeast Asia (the Nguyen realm—87, Vietnam—37, Champa—6, Cambodia—44), plus 56 each for Ayudhya and Luzon. Thus, Hoi-an formed a major part of a trade network that included Manila, Phnom Penh, and Ayudhya. This Vietnamese port received half of the eastern mainland ships and almost a quarter of all the ships leaving Japan in the first third of the century. By the 1630s, Hoi-an became the major port for Japanese overseas trade. As testified to by one of the first missionaries, the southern Vietnamese lord gained much from the trade stimulated by the Chinese and Japanese merchants.[69]

In the port of Hoi-an, the Chinese and Japanese each had their own enclaves, adjacent to each other. Coming with the northeast monsoon at the end of the (lunar) year, ships from Japan brought large amounts of silver and copper cash which went mainly for silk, sugar, aloeswood, deerskins, rayskins, and ceramics. Japanese traders controlled the local silk and sugar markets by prepayment with the imported cash. The Chinese merchants gathered during this four month "fair" and traded their silk, copper cash, and tutenag for the Japanese silver and the goods of Southeast Asia, pepper in particular, sandalwood, camphor, and other aromatics. The Vietnamese welcomed the silver, copper, tutenag, and Chinese books, sold their gold, sugar, and silk, as well as the imported silver, and drew revenue from the exchanges which took place on their soil. The Portuguese mingled with the Chinese traders, dealt in silver, silk, gold, and pepper, and generally tried to facilitate matters to their benefit. They brought American and Persian silver via Goa as well as American silver from Manila and Japanese silver. The Dutch, also carrying American silver, made contact with the Chinese in Hoi-an not long after their arrival on the Asian scene. The Vietnamese too were interested in the Dutch trade, emphasizing silk and the Chinese contact available there. Portuguese enmity and Japanese competition were, however, to keep the Dutch from

---

69. Innes 1980, pp. 6, 53-54, 56-62, 66, 164-165, 213, 635; Nguyen 1970, pp. 189-190; Ch'en, pp. 13-14; Lamb, p. 21: Innes 1975; Iwao, pp. 1-5, 8-10; Takase, pp. 28-29. Robert Innes supplied the information on the Japanese-Vietnamese copper trade and updated the ship figures.

establishing a permanent base in Nguyen territory, and the Dutch generally came to work the Asian metal trade out of Ayudhya and northern Vietnam.[70]

The relative importance of the various merchant groups for the Nguyen may be seen in the duties charged them at this time. The lowest rates went to those on immediately adjacent coasts (500 long strings to arrive, 50 to depart), then the Chinese merchants of Fu-chou, Ayudhya, and Manila (2000 and 200), those of Shanghai and Canton (3000 and 300), the ships from Macao and Japan (4000 and 400), and finally the other European ships (8000 and 800). This might indicate that Chinese who brought silk and Southeast Asian spices were treated better than those traders (Japanese, local Portuguese, and especially other Europeans) who, generally carrying bullion, came to buy these goods.[71]

The transformation of the international trade routes in the sixteenth and seventeenth centuries also saw a major change in the role of precious metals in that trade. Though metals had been traded earlier, there had generally been a relatively even balance among the goods being exchanged, only China in the Sung dynasty having a large trade deficit requiring copper cash to fill. Now changes in the Ming tax system—by 1600 almost all state revenues were collected in silver[72]—and the great demand for Chinese goods, particularly silk, meant that the grey metal had to fill the gap. (This would remain the case until the nineteenth century when the large sale of opium shifted the trade balance against China.) In the first years of the new economic circumstances, the relative value of gold and silver required much adjustment, being part of the general trade and its fluctuations. Almost a century elapsed before a stable ratio between the two precious metals came about.

In prior centuries, the value of gold to silver seems to have varied from 1:3-4 (Japan) to 1:12-13 (China). During the early Ming dynasty, it had decreased to 1:5-6 in China. But the middle of the sixteenth century saw the major jump in the production of Japanese silver, and its value dropped sharply to 1:10-11 by the end of the century. This relatively cheap silver led to its export in exchange for Chinese and Vietnamese silk and gold. American silver from the Philippines or via Europe compounded the flow. In China,

---

70. Ch'en, pp. 14-15, 19-21; Innes 1980, pp. 62-66, 98, 100, 150, 188, 240, n.241, 376, 505-508, 524, 585-586, 608, 648-649; Nguyen 1970, pp. 190, 201-202, 205; Manquin, pp. 188-196, 200, 237-238; Godinho, pp. 502, 508-515, 520-522, 524, 527-531, 817, 819; Schurz, pp. 131-133; Dermigny, p. 112; Kobata 1965, pp. 245-246; Atwell, pp. 2-3; Wills 1974, pp. 9-10, 20; Iwao, p. 10; Kato, pp. 41-49, 64-67; Tashiro, pp. 86, 94; Boxer 1974, p. 70; Viraphol, p. 13; Smith, p. 92. The VOC was able to trade in Hoi-an only between 1633 and 1638 and in 1652.

Cross in this volume points out how the Portuguese gained Peruvian silver through Buenos Aires and undoubtedly shipped much of it out to Goa.

71. Nguyen 1970, p. 39; Manquin, pp. 189-190; *DNTL*, I, 146-147 (10, 26b-27a). Innes notes that this information is also to be found in Le Qui Don's eighteenth century work, *Phu Bien Tap Luc* (Saigon, 1972), I, 67-69, and he suspects that the size of the duty was related to the usual value of the cargoes carried by each type of ship.

72. Yang, p. 67; Hucker, p. 292; Innes 1980, pp. 26-27.

with its great demand for silver, the ratio slowly changed to 1:8 and through the first half of the seventeenth century dropped to 1:13-14 at which point it and the Japanese rate stabilized for a time. Gold was cheaper in Southeast Asia than in China and Japan. In Vietnam, it steadily increased in value until the 1630s when the ratio reached 1:10.[73] The Vietnamese ratio seems to have remained at a somewhat lower level through the seventeenth and eighteenth centuries, probably due to the local production of gold.

In the 1630s, major changes began to occur both in the trade routes and in the metals being exchanged. The Tokugawa government in Japan banned travel by the Japanese and expelled the Portuguese in 1635 and 1639 respectively. Yet these changes had only limited impact on areas such as Central Vietnam, despite its heavy dependence on trade with Japan. Chinese junks and the Dutch were able to expand their activities in the absence of their former Japanese and Iberian competitors. After 1635, Japanese merchants, both those at home and those residing overseas, maintained their involvement in the export of copper and other monetary metals from the island nation to Vietnam by chartering Chinese junks or investing money in junk voyages. Barred from personal involvement, however, the Japanese soon lost their paramount position in the Nagasaki-Hoi-an trade to the Chinese.[74]

The Japanese community in Hoi-an atrophied, and the rising tide of private Chinese trade, based in Southeast Asia as well as in China, became the dominant theme in East Asian exchange. Ships manned by Chinese sailors traded from the varied Southeast Asian ports to China and Japan, carrying silk and sugar from Hoi-an. The flood of Chinese refugees following the Manchu conquest of China in 1644 reinforced this trend. Portuguese traders continued to deal in the Asian trade, but being cut off from Japan hurt them and they generally followed the same pattern as the Chinese, trading silk for pepper, silver, and copper. Nevertheless, they continued to flourish in Vietnam through the mid-seventeenth century, gaining excellent profits in silver by trading Chinese cash. The Dutch meanwhile made several attempts to take advantage of the need for Japanese copper in Hoi-an, but, despite their access to Japan, their efforts floundered in the political differences existing between them and the Nguyen and in the Chinese competition.[75]

---

73. Kamiki and Yamamura in this volume; Kobata 1965, pp. 247, 250-256; Boxer 1970, pp. 459-464; Huang, pp. 79-81; Rawski, pp. 75-77; Dermigny, p. 420; Lamb, p. 22; Godinho, pp. 524, 530-531; Innes 1980, pp. 25-27, 591, 608. This discussion is meant merely as a crude indication of shifting and relative values across eastern Asia and makes no pretense to describe the cultural desirability of gold and silver in the various areas; see Miskimin in this volume.

74. Robert Innes provided the information on Japanese trade after 1635; see also Innes 1980, pp. 3-5, 7, 155-164, 169-171, 378, 635-636.

75. Kobata, 1965, p. 256; Manquin, pp. 195-205, 208-209, 238; Ch'en, pp. 15-17, 21-22, 35-36; Viraphol, pp. 58, 61, 72; Crawcour, pp. 378, 380; Nguyen 1970, pp. 191-192, 197-199, 205-207; Boxer 1974, pp. 69, 73-74, 77-84; Dermigny, p. 411; Innes 1980, pp. 188-189, 208, 634.

A silver shortage, economic problems, and the change in Chinese dynasties during the 1640s led to more instability. Ming loyalists maintained themselves along the southern coast of China until the 1680s, and for twenty years from the early 1660s the new Ch'ing dynasty cut off all coastal activity. This left the diminished East Asian sea trade to the rebels as the Cheng forces on Taiwan kept control of the Southeast Asian trade with Japan, including that of Hoi-an, Phnom Penh, and Ayudhya. After the late 1660s, copper ore displaced silver as the major Japanese export, responding to Japanese worries as well as to the Southeast Asian need for cash, cannon, and ritual objects. The Chinese ships making the Nagasaki—Hoi-an run exchanged the Japanese copper ore and cash mainly for raw silk, high quality Vietnamese gold, and sugar. The Nguyen government, like the Thai monarchy in Ayudhya, placed an official monopoly on the purchase of copper ore and cash so as to use it for their money supply and other official purposes. Between 1660 and 1684, the Nagasaki city elders operated a special mint to produce copper cash for export. Central Vietnam was evidently dependent on the output of this mint. In 1688, four years after the closing of the mint, the Nguyen ruler wrote to the shogun complaining of a coin shortage and requesting the export of one million coins to his country. In the trade contacts across the South China Sea, ten ships from Hoi-an traded for American silver in Manila from 1658 to 1678, before the link between Manila and Ayudhya became the dominant one, displacing both Hoi-an and Phnom Penh.[76]

With the Ch'ing conquest of Taiwan in 1683, the sea routes opened again for the coastal Chinese and a great increase in commerce took place at Hoi-an. Junks came not only from China itself (Amoy and Canton), but also from Nagasaki, Ayudhya, Phnom Penh, Manila, and Batavia, according to the 1696 report of the Englishman Thomas Bowyear. Ten to twenty of these ships arrived a year, about half from China itself and the rest Chinese junks from the other ports. Portuguese traders continued to be effective, swapping tutenag for pepper. The Tokugawa regime limited the movement of silver, and copper was the main export from Japan. Though the Japanese authorities restricted the number of ships and amount of trade in Nagasaki, they left open an allotment for junks from Hoi-an. Junks from Nagasaki also would stop first in China and bring at great profit the cash necessary for the Vietnamese economy. Ayudhya sent sixteen ships to Manila (Hoi-an and Phnom Penh only one each) from 1690 to 1700, and it was the junks from Batavia and Manila which brought American and other kinds of silver to

---

76. Atwell, pp. 8-22; Wills 1979, pp. 216-229; Ch'en, pp. 22-23; Kobata 1965, p. 256; Hall 1949, pp. 445, 448-450; Nguyen 1970, p. 164; Manquin, pp. 202, 233, 241; Innes 1975; 1980, pp. 10, 66, 171-177, 282-284, 298-303, 306, 311-314, 407-409, 413-414, 428, 495, 498, 507, 525-526, 587-588, 593-597, 621, 636; Kobata 1980, p. v; Wills 1974, pp. 14-17, 24-28, 32-38, 47, 54, 82, 119, 151, 153, 159-160, 179, 192-193; Iwao, pp. 11-16; Kato, pp. 44, 50; Tashiro, pp. 95, 97; Chaunu; Dermigny, pp. 100, 103, 134-138, 411-412; Viraphol, pp. 13, 22, 29, 43-45, 58-59, 143.

Hoi-an, while the Vietnamese provided gold.[77] Some gold and silver, mainly of Laotian origin, also came from the Cambodian ports in exchange for copper. The Chinese were the main traders in the eastern mainland ports, but Vietnamese merchants were to be found in Ayudhya and Laos as well as Cambodia, exchanging silk and sugar for the available products.[78]

The Japanese copper trade grew to a peak in the final years of the seventeenth century. But, whereas Ming bans on the export of copper and on direct trade with Japan had encouraged the export of Japanese copper to Vietnam throughout most of the seventeenth century, a growing shortage of the red metal in China prompted the Ch'ing to encourage its import from the island nation in 1685 and thereafter. This Chinese competition for what were soon to be decreasing supplies of Japanese copper (exports reached their peak of 12,027,000 pounds in 1698) was probably the cause of the copper shortage complained of by the Nguyen in 1688. The diversion of Japanese copper to China once again made the Vietnamese dependent on South China as a source of coinage. Through the 1690s, an average of three or four Chinese ships per year arrived in Nagasaki from Hoi-an, with as many as seven in any one year. Increasingly thereafter, the amount of copper supplies from Nagasaki and the number of ships from Hoi-an dropped drastically, there being no more than two ships in any one year. By 1715, trade in general and the purchase of copper ore in particular were greatly reduced.[79]

During these same years, the increasing European trade came more and more to focus on the port of Canton where the Ch'ing dynasty was centralizing control of southern shipping. From 1717 to 1722 and occasionally thereafter, Ch'ing authorities banned mainland Chinese trade and travel to Southeast Asia. By 1730, the result for Hoi-an of this turbulence in East Asian commerce seems to have been a slackened trade with Japan (though perhaps more Southeast Asian based junks put into the Vietnamese ports because of it) and the Vietnamese dependence on a Chinese link to Canton for dealing with European merchants. The Chinese traded Vietnamese products to the Europeans and their own and European goods to the Vietnamese. Undoubtedly the latter gained Chinese cash through the trade. In general, the com-

---

77. Wills 1979, pp. 231-233; Innes 1980, pp. 1, 177-179, 209, 318-323, 341, 428-429, 435-436, 636; Ch'en, pp. 23-25; Lamb, pp. 36, 52-53; Hall 1949, pp. 445, n.4, 452-454; Manquin, pp. 210-213; Boxer 1970, pp. 470-472; 1974, pp. 82, 84, 89; Glamann, pp. 51-64, 167, 172-173, 175-178; Nguyen 1970, pp. 170-171; Iwao, p. 13; Tashiro, p. 95; Wills 1974, pp. 195-196; Chaunu; Viraphol, pp. 45-46, 54, 59-60; Dermigny, pp. 100, 137-140, 193-195, 284, 311-312. See Dermigny, p. 406, for loads of copper and tutenag taken out of China by various English ships from 1700 to 1734, and Leeds, 180, for zinc going from Canton to Sweden in 1745.

78. Nguyen 1970, pp. 193-195; Groslier, pp. 152-153, 162-163; Skinner, pp. 7-13; Smith, p. 84.

79. Robert Innes supplied the information in this paragraph, see also Innes 1980, p. 5-6, 8, 12, 330, 333-336, 344-346, 349-356, 439, 495-497, 515, 526-530, 567-569, 597-599, 608-609, 636; Hall 1949, 455-457; Viraphol, pp. 60, 64-67, 143.

mercial situation of Hoi-an appears to have suffered through these years.[80]

The next two decades, through the 1740s, however, saw a major upsurge in Hoi-an's situation, perhaps in response to non-competitive prices (as for gold) and a tightening of Ch'ing control over trade in Canton. Chinese and some European shipping frequented the Vietnamese port. By 1750, some sixty to eighty Chinese junks came annually to Hoi-an, and contemporary sources show it to have been "a Chinese town supported by flourishing foreign trade."[81] Indeed, the Chinese population of the town seems to have doubled from the beginning of the century.[82] The local gold played a part in this trade, and the moderate cost of gold (perhaps 1:9, as in the north; see below) would have drawn buyers from China in exchange for copper and especially Chinese tutenag. One source indicated that the price of gold jumped by a quarter when the junks arrived from China. Gold may also have moved into Vietnam from further west since the ratio in Laos and Cambodia was 1:7.[83] On the other hand, Vietnamese merchants used "many loaves of gold" in their trade with Cambodia.[84]

International commerce and the flow of metals was generally beneficial for the Nguyen regime established in the new southern territories of Vietnam. For the Vietnamese state itself, in the Red River delta to the north, economic exchange was much more secondary an enterprise. The main center of trade sat on the outskirts of the Capital (Thang-long, now Hanoi), though commerce also took place on the coast in Nghe-an province a bit south of the delta. This latter location may have been good for contact with the Lao territories (as earlier). The Chinese and the Dutch were the main foreign merchants from the 1630s on, while the Portuguese were able to maintain a fairly constant low-level trade. Difficulties arose time and again through the seventeenth century due to the Trinh court's generally negative attitude toward trade.[85]

---

80. Ch'en, pp. 25-26; Lamb, pp. 37; Nguyen 1970, pp. 196-197, 214-215; Manquin, pp. 215-224; Glamann, pp. 178-180; Innes 1975; Tashiro, pp. 93, 98-101; Dermigny, pp. 140-159, 170-171, 193-195, 274-287, 322-327, 406, 412-416; Chaudhuri, pp. 386, 388; Gaastra in this volume; Wills 1979, p. 233; Viraphol, pp. 50-51, 55-57, 70-73, 94, 121-123.

81. Ch'en, p. 18. On Canton in these decades, see Dermigny, pp. 194, 197-198, 279, 318-321, 335, 355, 356, 364, 431, 546.

82. Ch'en, pp. 18-19, 26; Nguyen 1970, 207, 215-216; Manquin, pp. 224-226; Malleret, pp. 158-160; Glamann, p. 180; Lamb, p. 63.

83. Holm (from Launay, I, 613, which gives the information on the rise of the price of gold, and Jesuites, XVI, 152); Malleret, 46-47, 137, 160; Hall 1949, pp. 459, n.58; Dermigny, pp. 419, 421, 432. MacCartney's statement of 1793 would seem to confirm such a flow of metals, ". . . silver came here from abroad and was exchanged for gold at a very great profit for those who import it." Recently, however, local silver had been exploited and a better balance achieved; Manquin, p. 239.

84. Nguyen 1970, p. 193, with the quoted phrase from the 1755 letter of a Catholic priest.

85. *Ibid.*, pp. 126, 190, 199-201, 205-208; Manquin, pp. 227-235; Lamb, pp. 31, 57; Innes 1980, pp. 189-191.

The Vietnamese pattern of trade in the north was similar to that in the south, though on a much smaller scale, and it followed the same vicissitudes described. This trade too was based on the exchange of silk, Chinese and Vietnamese, for metals (silver, copper cash, and copper ore).[86] Whereas 87 ships received Japanese seals for trade with Hoi-an from 1604 to 1635, only 37 went to Vietnam itself, less than the 44 which went to Cambodia.[87] Through the decades following the closure of Japan, an occasional year saw more than one ship in Nagasaki from Vietnam, but in general three ships came from the south for every one from the north. These were almost entirely Chinese junks. The Trinh too relied on imported copper coins. After the mid-seventeenth century, Dutch records frequently refer to shipments of copper cash to the north (or Tonquin as they called it). In 1671, for instance, the Dutch shipped 318 chests of Japanese copper cash to Vietnam. Two years later the shipment was worth 30,000 ounces of silver, and in 1675 the company contracted with the Nagasaki city elders for a supply of coins worth 50,000 ounces. The mint in the Japanese port city was not the only source of copper cash. Dutch merchants stationed in the Vietnamese capital during 1653 noted the arrival of a vessel from Macao which was laden primarily with copper coins minted by Chinese in Macao.[88]

A better perspective for the trade of northern Vietnam is to view it as part of the commercial network, both land and sea, of south China. The Chinese community in the northern mountains of Vietnam and along the coast provided a matrix by which goods moved to and from adjacent Chinese territory. In particular, silver began to move north out of the mines into China throughout the eighteenth century.[89] The Dutch had tapped into this network in the previous century by exchanging silver and copper cash for gold to trade in India. Another network linked to this trading pattern was the Lao which sent gold to Vietnam as well as down river to Cambodia and overland to Ayudhya.[90] The movement of Lao gold into Vietnam may have

86. Lamb, pp. 29, 34; Hall 1949, p. 445; Manquin, p. 238; Iwao, p. 16; Kato, p. 42; Yamawaki, pp. 111-112; Innes 1980, pp. 649-650, 654, 668.
87. Ch'en, p. 13. Robert Innes updated the ship figures (Innes 1980, p. 58).
88. Information supplied by Robert Innes from Van Dam, pp. 444-448, and Coolhaas, II, 653. Toda, pp. 66-67, in the late nineteenth century, noted,

> Coins are also cast in Macao for circulation in Annam [Vietnam]; and from a very recent report addressed by the Governor of that Colony to the Portuguese Government it would appear that there exist at the present moment six manufactories of Annamese coins, employing 12 furnaces and 320 workmen, and producing daily 700,000 cash.

89. Nguyen 1970, p. 208; Innes 1975; see Woodside, pp. 270-278, for the early nineteenth century.
90. Nguyen 1970, p. 195; Lamb, p. 55; Holm (from Marini, pp. 336-339); Smith, pp. 50, 58, 59, 61-62, 68, 85-86, 88-89, 149. Smith describes how the Dutch sought gold to trade in India, and Innes has commented, "Chinese gold was being shipped to India, as was Southeast Asia's Japanese gold." See Innes 1980, pp. 302, 525-526, 591; and also Chaudhuri, pp. 181-182, and Godinho, pp. 285, 392, 398, 451, 454.

been based on the Lao ratio of 1:7 being less than the Vietnamese ratio (1:9) and the Chinese 1:11-12 (ca. 1740).[91] The development of copper mines in the first decades of the eighteenth century, concurrent with a brief shortage of world copper, led both Chinese and Dutch to consider importing Vietnamese ore, though such trade did not develop to any great extent.[92]

The little interest the Trinh had in the benefits of international trade tended to fade after the end of hostilities with the Nguyen in the 1670s. In addition, the direct contact between China and Japan after 1685 undercut the demand for inferior Vietnamese silk in Japan. The Dutch were gone by 1700, and trade continued in the hands of the Chinese merchants who crisscrossed East and Southeast Asia.[93] More important for the Vietnamese rulers was the tribute relationship with Peking. Following the Restoration of the Le dynasty in 1592 the Le sent 100 *can* (59.68 kg.) of gold and 1000 ounces of silver (37.3 kg.), but the Ming insisted on a gold image, so the Vietnamese sent two, of gold and silver (weighing 10 ounces, 0.37 kg.), and the standard tribute—two pairs of silver vases and five small silver incense burners. The Ming at first refused, then accepted the Le once more in a tributary status. The Vietnamese were to send a double tribute every six years. This continued until the fall of the Ming in the 1640s. The Nguyen had no official contact with the Chinese court, and the Trinh via the Le, were the ones who sent the tribute. The latter made official contact with the new Ch'ing dynasty in 1664 and the Manchus maintained the pattern established by the Ming, declaring that the Vietnamese were to present goods from their own territories:

As to tribute objects, in each case they should send the products of the soil of the country. Things that are not locally produced are not to be presented. Korea, Annam, . . . all have as their tribute their customary objects.[94]

Among the "customary objects" of the Vietnamese, we have seen were silver and gold, which went north with embassies about every four years from the 1660s to the 1750s.[95]

The same sort of trade in the Lao-Vietnamese highlands continued through the nineteenth century, with "ticals, rupees, gold and silver bars, Mexican silver, and piastres" in use; N. S. Adams, *The Meaning of Pacification, Thanh-hoa Under French Rule, 1885-1908*, Ph.D. dissertation, Yale University, 1978, from Anonymous, "Notice sur le territoire des Houa Phans Thang Hoc", *Bulletin economique de l'Indochine francaise* (1898), 231.

91. Nguyen 1970, pp. 160, 170, notes that in the mid-eighteenth century, silver when monetized (see above) was valued at two long strings of copper cash (*quan*) to the ounce and gold was eighteen *quan* to the ounce; the ratio was therefore 1:9. Dermigny, pp. 430-433, noted that the value of silver in China rose towards the end of the seventeenth century to an official rate of 1:10 and then steadily declined through the eighteenth.

92. Glamann, p. 180; Hall 1949, pp. 457. Viraphol, p. 287, n.90, mentions that in 1718 the K'ang-hsi Emperor allowed Chinese ships to trade with Vietnam, despite the ban on trade to Southeast Asia, in order to continue to obtain copper there.

93. Innes 1980, pp. 191, 280, 311, 324; see Dermigny, pp. 193-194, for junks going from Vietnam to Batavia.

94. Translated from the *Ch'ien-lung Hui-tien* (1764) by Fairbank and Teng, p. 143.

95. *Ibid.*, pp. 193-194; Phan, IV, 173-175; *TT*, IV, 211-215.

Each triennial tribute (offered double every six years) included four gold censors and vases (totaling 209 [perhaps 290] ounces) and twenty silver trays (weighing 692 ounces). Also needed on the missions were 990 ounces of silver and 20 ounces in gold ingots for other presentations. There might additionally be a mission of thanks, condolences, congratulations, or of report (of a significant event). The mission of thanks carried one gold censor and vase set (57.5 ounces), a silver crane and stand (48.4 ounces), and one silver censor and vase set (50.4 ounces), while those of congratulations and report offered one gold turtle (18 ounces), a silver crane and stand (50 ounces), and a silver censor and vase set (49 ounces). Eventually, it would appear, the obligatory tribute objects were replaced by an equivalent value in gold and silver ingots.[96] Thus every six years approximately 500 ounces (18.65 kg.) of gold and 2500 ounces (93.25 kg.) of silver moved from Vietnam to Peking.

Yet the important interrelationship between Vietnam and China lay in the regional economic integration of Kuangtung and northern Vietnam rather than in the formal political relationship. The new Nguyen dynasty of early nineteenth century Vietnam made a much greater use of silver in its monetary system than had prior rulers. By this time the British merchants had begun to shift the balance of their China trade as they brought opium from India, and through the 1820s and 1830s silver poured out of China to pay for it. The flow became so strong that it pulled silver out of northern Vietnam via the resident Chinese community. The result was a serious deflation which shook the internal strength of the dynasty, a deflation remarkably similar to that of the end of the Ming dynasty two hundred years earlier in China.[97]

The Vietnamese choice of monetary systems, first copper and then, in the eighteenth and nineteenth centuries, silver, exposed them to the vicissitudes of international money flows. For the thirteenth and fourteenth centuries, this was the copper cash of China which was, to all appearances, generally adequate for the Vietnamese economic situation. Vietnamese gold and silver moved along the trade routes of Southeast Asia as commodities and into China as trade and tribute. Increasingly from the fifteenth to the eighteenth centuries, however, a shortage of cash plagued the growing Vietnamese economy and made it dependent to a great extent on the Japanese copper ore as well as on the flow of cash from both China and Japan. The Vietnamese seem to have been prepared "to pay well", in Frank Perlin's phrase,[98] for currency, and their gold and silver continued to leave the country as both commodity and tribute. In the process Vietnam found itself a junction for the world flow of precious metals—American silver via both Acapulco (and the Spanish in Manila) and Europe (the Portuguese in Macao and the Dutch in

---

96. Phan, IV, 175-176.
97. Woodside, pp. 276-278; Masui, p. 8; Atwell, p. 22. See also Nguyen The Anh, "Quelque aspects economiques et sociaux du probleme du riz au Vietnam dans le primiere moitie de XIXe s.", *Bulletin de la Societe des Etudes Indochinoises*, n.s., 42, 1-2 (1967), 5-22; and Bui Quang Tung, "La succession de Thieu-tri", *ibid.*, 23-175, especially 52-78.
98. Perlin, p. 239.

Batavia) and Japanese silver and copper from Nagasaki. Most important in the movement of copper ore and cash was the overseas Chinese network of trade across East and Southeast Asia. Local movement of gold and silver took place in the eastern mainland among Laos, Cambodia, and the two segments of Vietnam.

Nevertheless, it would appear that these monetary flows were generally secondary to the internal socio-economic developments of the country in their influence on Vietnamese life. Vietnam did not directly experience the massive impact of the sixteenth century silver boom (American and Japanese) as did other countries, notably China. The movement of silver and the growth in international trade during the seventeenth and eighteenth centuries seem to have speeded developments already beginning to take place. Only in the nineteenth century did the Vietnamese find themselves in a position where the international monetary situation seriously affected the internal scene.

*My thanks to Dr. Norman G. Owen for his reading of the initial drafts of this paper, to Dr. Robert L. Innes for his reworking of East Asian sections cited above and for his general comments, and to the conference members for their comments and the contributions of their papers.

Bibliography

Ashtor, E., *A Social and Economic History of the Near East in the Middle Ages*, Berkeley, Cal., 1976.
Atwell, W. S., "Notes on Silver, Foreign Trade, and the Late Ming Economy," *Ch'ing-shih Wen-t'i*, 3 (1977), 1-33.
Aurousseau, L., ed., *Ngan-nan Tche Yuan* (An-nan Chih Yuan), Hanoi, 1932.
Benda, H. J., and J. A. Larkin, eds., *The World of Southeast Asia*, New York, 1967.
Boxer, C. R., ed., *South China in the Sixteenth Century*, London, 1953.
Boxer, C. R., "*Plata es Sangre:* Sidelights on the Drain of Spanish-American Silver in the Far East, 1550-1700", *Philippine Studies*, 18, 3 (1970), 457-478.
"Macao as a Religious and Commercial Entrepot in the Sixteenth and Seventeenth Centuries", *Acta Asiatica*, 26 (1974), 64-90.
Chaudhuri, K. N., *The Trading World of Asia and the British East India Company, 1660-1760*, Cambridge, 1978.
Chaunu, P., *Le Philippines et le Pacifique des Iberiques* (16e, 17e, 18e s.), Paris, 1960, Serie 13, 14, pp. 148-219.
Ch'en, Ching-ho A., *Historical Notes on Hoi-an (Faifo)*, Carbondale, Ill., 1974.
Chou Ta-kuan, *Notes on the Customs of Cambodia*, trans, from Paul Pelliot, Bangkok, 1967.
Colless, B. F., "Majapahit Revisited: External Evidence on the Geography and Ethnology of East Java in the Majapahit Period", *Journal of the Malayan Branch of the Royal Asiatic Society*, 48, 2 (1975), 124-161.
Coolhaas, W. P., ed., *Generale Missiven van Gouverneurs- en Raden aan Heren XVII der Verenigde Oostindische Compagnie*, Vol. 2 (1637-1655) in Rijks Geschiedkundige Publicatien, no. 104, 's-Gravenhage, 1960.
Cortesao, A., ed., *The Suma Oriental of Tomé Pires*, London, 1944.
Crawcour, S., "Notes on Shipping and Trade in Japan and the Ryukyus", *Journal of Asian Studies*, 23, 3 (1964), 377-381.
Dai Nam Nhat Thong Chi, trans., vols. 3-5, Hanoi, 1971; Saigon, 1964, 1966, 1967.
Dai Nam Thuc Luc, vol. 1, Tokyo, 1961.
Dai Viet Su Ky Toan Thu, trans., second ed., four volumes, Hanoi, 1971-1973.
Deloustal, R., "Ressources financieres et economiques de l'etat dans l'ancien Annam", *Revue Indochinoise*, 1-2 (1925), 59-78; *Bulletin des Amis de Vieux Hue*, 19 (1932), 157-218.
Dermigny, L., *La Chine et l'Occident*, Le commerce a Canton au 18e s., 1719-1833, vol. 1, Paris, 1964.
Do Van Ninh, "Tien Co Thoi Ly-Tran (Money of the Ly-Tran Period, Tenth-Fourteenth Centuries)", *Nghien Cuu Lich Su* (Hanoi), 189 (#6, 1979), 26-34.
"Tien Co Thoi Ho (Money of the Ho Period, 1400-1407)," *Nghien Cuu Lich Su* (Hanoi), 191 (#2, 1980), 50-54.
Fairbank, J. K., & Teng Ssu-yu, *Ch'ing Administration*, Cambridge, Mass., 1961.
Geertz, C., *Negara, The Theatre State in Nineteenth Century Bali*, Princeton, 1980.
Gernet, J., *Daily Life In China on the Eve of the Mongol Invasion*, trans., New York, 1962.
Glamann, K., *Dutch-Asiatic Trade, 1620-1740*, The Hague, 1958.
Godinho, V. Magalhaes, *L'Economie de l'Empire Portugais aux XVe et XVIe S.*, Paris, 1969.
Groslier, B. P., *Angkor et le Cambodge au XVIe S.*, Paris, 1958.

Gutman, P., "The Ancient Coinage of Southeast Asia", *Journal of the Siam Society*, 66, 1 (1978), 8-21.
Hall, J. W., "Notes on the Early Ch'ing Copper Trade With Japan", *Harvard Journal of Asiatic Studies*, 12 (1949), 444-461.
Hall, K. R., "Khmer Commercial Development and Foreign Contacts Under Suryavarman I", *Journal of the Social and Economic History of the Orient*, 18, 3 (1975), 318-336.
Hall, K. R., and J. K. Whitmore, "Southeast Asian Trade and the Isthmian Struggle, 1000-1200", in K. R. Hall and J. K. Whitmore, eds., *Explorations in Early Southeast Asian History*, Ann Arbor, Mich., 1976, 303-340.
Hartwell, R., "The Evolution of the Early Sung Monetary System, A.D. 960-1025", *Journal of the American Oriental Society*, 87 (1967), 280-289.
Henthorn, W. E., *A History of Korea*, New York, 1971.
Herbert, P. A., "A Debate in T'ang China on the State Monopoly on Casting Coin", *T'oung Pao*, 62, 4-5 (1976), 253-292.
Higgins, R. L., "Pirates in Gowns and Caps: Gentry Law-Breaking in the Mid-Ming", *Ming Studies*, 10 (1980), 30-37.
Hirth, F., and W. W. Rockhill, *Chau Ju-kua, His Work on the Chinese and Arab Trade in the Twelfth and Thirteenth Centuries...*, St. Petersburg, 1911.
Holm, D. F., "Indochina 1650-1800 Through the Eyes of French Missionaries", Ms., Yale University, 1971.
Huang, R., *Taxation and Governmental Finance in Sixteenth Century Ming China*, Cambridge, 1974.
Hucker, C. O., *China's Imperial Past*, Stanford, Cal., 1975.
Huynh Sanh Thong, ed., *The Heritage of Vietnamese Poetry*, New Haven, Conn., 1979.
Innes, R. L., "Japanese Trade With Southeast Asia in the Seventeenth Century and Its Socio-Political Impact on Vietnam," Ms., University of Michigan, 1975.
Innes, R. L. *The Door Ajar*, Japan's Foreign Trade in the seventeenth century, Ph.D. dissertation. University of Michigan, 1980.
Iwao Seiichi, "Japanese Foreign Trade in the 16th and 17th Centuries", *Acta Asiatica*, 30 (1976), 1-18.
Jesuites, *Lettres edifiantes et curieuses*, vol. XVI, Toulouse, 1810.
Kato Eiichi, "The Japanese-Dutch Trade in the Formative Period of the Seclusion Period", *Acta Asiatica*, 30 (1976), 34-84.
Kerr, G. H., *Okinawa*, Rutland, Vt., 1958.
Kobata Atsushi, "The Production and Uses of Gold and Silver in Sixteenth and Seventeenth Century Japan", *Economic History Review*, 18, 2 (1965), 245-266.
Kobata Atsushi, "Forward: Studies in the History of Japanese Currency Systems", *Acta Asiatica*, 39 (1980), iii-vi.
Lach, D. F., *Asia in the Making of Europe*, Chicago, 1965.
Lamb, A., ed., *The Mandarin Road to Old Hue*, London, 1970.
Launay, A., ed., *Histoire de la Mission de Cochinchine*, Vol. 1, Paris, 1923.
Le Tac, *Annam Chi Luoc* (Chinese *An-nan Chih Lueh*), Hue, 1961.
Leeds, L. T., "Zinc Coins in Medieval China," *The Numismatic Chronicle and Journal of the Royal Numismatic Society*, 6th series, 14 (1954), 177-185.
Li Wen-feng, *Yueh Ch'iao Shu*, n.p., n.d. (Hong Kong? 1960s?)
Malleret, L., *Pierre Poivre*, Paris, 1974.
Manquin, P.-Y., *Les Portugais sur les cotes de Viet-nam et du Champa*, Paris, 1972.

Marini, G. F., *Relation nouvelle et curieuse des royaumes de Tonquin et de Lao*, Paris, 1666.

Masui Tsuneo, "Silver and China in the Nineteenth Century", *Acta Asiatica*, 10 (1966), 1-15.

Meilink-Roelofsz, M. A. P., *Asian Trade and European Influence, 1500-1620*, The Hague, 1962.

Mills, J. V. G., ed., *Ma Huan, Ying Yai Sheng Lan, The Overall Survey of the Ocean's Shores, 1433*, Cambridge, 1970.

Muramatsu-Gaspardone, Mme., "Nampyoki", *Bulletin de l'Ecole Francaise d'Extreme Orient*, 33, 1 (1933), 35-120.

Nguyen Thanh Nha, *Tableau economique du Vietnam aux XVIIe et XVIIIe s.*, Paris, 1970.

Nguyen Trai, *Nguyen Trai Toan Tap* (Collected Works of), trans., Hanoi, 1969.

Perlin, F., "A History of Money in Asian Perspective", *Journal of Peasant Studies*, 7, 2 (1980), 235-244.

Pham Van Kinh, "Mot so nghe thu cong hoi the ky 10-14; Nghe det—nghe gom—nghe khai khoang luyen kim (A number of trades in the 10th-14th centuries: Weaving, Ceramics, Mining and Metallurgy)", *Nghien Cuu Lich Su* (Hanoi), 168 (#3, 1976), 42-53.

Pham Van Kinh, "Bo mat thuong nghiep Viet Nam thoi Ly-Tran (Vietnamese trade in the Ly-Tran period, 10th-14th centuries)", *Nghien Cuu Lich Su* (Hanoi), 189 (#6, 1979), 35-42.

Phan Huy Chu, *Lich Trieu Hien Chuong Loai Chi*, trans., 4 vols., Hanoi, 1961.

Rawski, E. S., *Agricultural Change and the Peasant Economy of South China*, Cambridge, Mass., 1972.

Richard, J., "History of Tonquin", trans. in J. Pinkerton, ed., *A General Collection of the Best and Most Interesting Voyages and Travels in all Parts of the World* (London, 1811), IX, ca. 739-740.

Sakamaki, S., "Ryu-kyu and Southeast Asia", *Journal of Asian Studies*, 23, 3 (1964), 383-389.

Schafer, E. H., *The Vermilion Bird*, T'ang Images of the South, Berkeley, Cal., 1967.

Schurz, W. L., *The Manila Galleon*, New York, 1939.

Skinner, G. W., *Chinese Society in Thailand*, Ithaca, N.Y., 1957.

Smith, G. V., *The Dutch in Seventeenth Century Thailand*, DeKalb, Ill., 1977.

Sun, E-tu Zen, "Mining Labor in the Ch'ing Period", in A. Feuerwerker et al., eds., *Approaches to Modern Chinese History*, Berkeley, Cal., 1967, 45-67.

Sun, E-tu Zen, "Ch'ing Government and the Mineral Industries Before 1800", *Journal of Asian Studies*, 27, 4 (1968), 835-845.

Takase Koichiro, "Unauthorized Commercial Activities by Jesuit Missionaries in Japan", *Acta Asiatica*, 30 (1976), 19-33.

Takizawa Takeo, "Early Currency Policies of the Tokugawa, 1563-1608", *Acta Asiatica*, 39 (1980), 21-41.

Tashiro Kasui, "Tsushima Han's Korean Trade, 1684-1710", *Acta Asiatica*, 30 (1976), 85-105.

Taylor, K. W., "The Rise of Dai Viet and the Establishment of Thang-long", in K. R. Hall and J. K. Whitmore, eds., *Explorations in Early Southeast Asian History*, Ann Arbor, Mich., 1976a, pp. 149-191.

The Birth of Vietnam, Sino-Vietnamese Relations to the Tenth Century and the Origins of Vietnamese Nationhood, Ph.D. dissertation, University of Michigan, 1976b.

Toda, E., "Annam and its Minor Currency", *Journal of the North-China Branch of the Royal Asiatic Society*, n.s., 17, 1 (1882), 41-220.
Toyoda Takashi, *A History of Pre-Meiji Commerce in Japan*, Tokyo, 1969.
Truong Buu Lam, ed., *Patterns of Vietnamese Response to Foreign Intervention*, New Haven, Conn., 1967.
Van Dam, P., *Beschryvinge van de Oostindische Compagnie*, Vol. 2, Book 1, edited by F. W. Stapel in Rijks Geschiedkundige Publicatien no. 74, 's Gravenhage, 1931.
Vickery, M. T., *Cambodia After Angkor, The Chronicular Evidence for the Fourteenth to Sixteenth Centuries*, Ph.D. dissertation, Yale University, 1977.
Viraphol, S., *Tribute and Profit, Sino-Siamese Trade, 1652-1853*, Cambridge, Mass., 1977.
Wheatley, P., "Geographic Notes on Some Commodities involved in Sung Maritime Trade", *Journal of the Malayan Branch of the Royal Asiatic Society*, 32, 2 (1959), 5-140.
White, J., *A Voyage to Cochin China*, orig. 1824, London, 1972.
Whitmore, J. K., "Mac Dang Dung", in L. C. Goodrich and Fang Chao-ying, eds., *Dictionary of Ming Biography*, New York, 1976a, pp. 1029-1035.
"Crisis, Reform, and Defeat, Vietnam and Ho Quy Ly, 1360-1407", ms., 1976b.
"Chiao-chih and Neo-Confucianism: The Ming Attempt to Transform Vietnam", *Ming Studies*, 4 (1977a), 51-91.
"The Opening of Southeast Asia, Trading Patterns Through the Centuries", in K. Hutterer, ed., *Economic Exchange and Social Interaction in Southeast Asia*, Ann Arbor, Mich., 1977b, pp. 139-153.
Wills, J. E., Jr., *Pepper, Guns, and Parleys*, Cambridge, Mass., 1974.
Wills, J. E., Jr., "Maritime China from Wang Chih to Shih Lang: Themes in Peripheral History", in J. D. Spence and J. E. Wills, Jr., eds., *From Ming To Ch'ing, Conquest, Region, and Continuity in Seventeenth Century China*, New Haven, Conn., 1979, 201-238.
Wisseman, J., "Markets and Trade in Pre-Madjapahit Java", in K. Hutterer, ed., *Economic Exchange and Social Interaction in Southeast Asia*, Ann Arbor, Mich., 1977, pp. 197-212.
Wolters, O. W., *The Fall of Srivijaya in Malay History*, Ithaca, N.Y., 1971.
Woodside, A. B., *Vietnam and the Chinese Model*, Cambridge, Mass., 1971.
Yamawaki Teijiro, "The Great Trading Merchants, Cocksinja and His Son", *Acta Asiatica*, 30 (1976), 106-116.
Yang Lien-sheng, *Money and Credit in China*, Cambridge, Mass., 1961.

Part III

New World metals

13

South American bullion production and export 1550-1750

HARRY E. CROSS

From the mid-16th century until almost the present day, the Americas accounted for the overwhelming majority of the world's bullion production. More specifically, between 1550 and 1800, Mexico and South America contributed more than 80% of the silver and more than 70% of the gold produced in the world. In terms of money supply, the mines and placers of North and South America comprised the world's chief source of currency. Given man's proclivity to seek wealth and to increase his standard of living, it is no surprise that New World bullion greatly accelerated the process of settlement in the Americas and largely financed the expansion of European trade with the rest of the globe. The spread of Western commerce and, with it, Western culture, significantly altered the historical development of Europe, Africa, and Asia.

The chronological bounds of this volume, as they pertain to the New World, span its discovery in 1492 to the mid-18th century. During these two and a half centuries, South America's precious metal deposits dominated Latin American bullion output. Production consisted mainly of silver from the Viceroyalty of Peru (including present-day Bolivia) and gold from Colombia and Brazil. The purpose of this chapter is to trace the course of bullion production in colonial South America and to determine direction flows and magnitude of export. The available basic data series are presented in a comparable format after being analyzed for reliability. Because of widespread fraud and effective contrabanding, official production and export figures must be viewed with great care. Where possible, therefore, official statistical series are amended to take into account bullion not registered with government treasuries, and bullion which evaded customs agents. Finally, the chapter examines the immediate destinations of exported South American gold and silver and attempts to place in perspective the proportional contributions of different Latin American regions to the world's monetary supply. Before tackling the questions of production and export, however, it will be useful to develop a sense of Spanish coinage classification and of the changes in the bimetallic ratio.

New World coinage

The Spanish Crown required that all New World silver be brought to either a mint or a royal treasury office immediately after refining. Royal officials used this law to control bullion production, taxation, and metal fineness. The Crown established mints in Mexico City in 1535, and in Lima and Potosí in Peru between 1568 and 1572.[1] Before these mints initiated operations, miners and merchants exported silver bars which had been marked by treasury officers as having paid the royal fifth *(quinto)*. Until coins became available for local transactions, Spanish colonists utilized crude pieces and bits of adulterated silver.

Once established, the mints turned out great quantities of silver specie; gold was not coined in Spanish America until the end of the 17th century, and then only on a limited scale. The principal unit of currency circulated by the Mexico City and Peruvian mints was the peso of eight *reales*, sometimes called the *peso fuerte*, or the *duro real de a ocho*.[2] Although several other kinds of pesos appeared in the 16th and 17th centuries, the peso of eight and its fractions generally dominated Hispanic American silver mintage. After 1683, the Crown ruled that all silver mined in Spanish America had to be registered at a mint and be stamped into pesos and reales.[3] For the purpose of comparability, all production, mintage, and export data appearing in this study are presented in Spanish pesos of eight. In addition, when it is convenient, bullion and specie figures are expressed in kilograms. (Appendix I depicts Spanish measurements of weight as they pertained to bullion, and measurements of coinage as they applied to silver.)

The Spanish peso underwent one debasement in 1728, when the fineness of the coin was reduced from 0.931 to 0.917. This event ended a remarkable 231-year period, stretching from the reforms of Isabel and Ferdinand in 1497 to the reign of Philip V, in which silver pesos were required to be 0.931 percent pure.[4]

The Spanish Crown continually set out detailed regulations regarding minting procedures, coin weights and finenesses: reality often deviated from the regulations. Pesos were crudely milled; rarely did the mints achieve a uniformly circular shape. Round coins generally lost their original configuration, thanks to the widespread practice of clipping.[5] When dealing with Spanish

---

1. Clarence H. Haring, *Trade and Navigation Between Spain and the Indies in the Time of the Hapsburgs* (Gloucester, 1964), pp. 287-290.
2. Humberto F. Burzio, *La ceca de la villa imperial de Potosí y la moneda colonial* (Buenos Aires, 1945), pp. 112-113.
3. D. A. Brading and Harry E. Cross, "Colonial Silver Mining: Mexico and Peru," *The Hispanic American Historical Review*, volume 52 (November, 1972), number 4, p. 565.
4. Burzio, *La ceca de la villa imperial de Potosí*, pp. 35-36.
5. See, C. R. Boxer, "Plata es Sangre: Sidelights on the Drain of Spanish-American Silver in the Far East, 1550-1700," *Philippine Studies*, volume 18 (July, 1970), number 3, pp. 476-477; Burizo, *La ceca de la villa imperial de Potosí*, pp. 54-55; Burzio, "La moneda primitiva del Peru

American pesos, therefore, a variance of plus or minus 5.0 percent in actual silver content should be allowed to account for the irregularities of mintage and clipping.[6]

For Brazil, it is not necessary to apply the same qualification. The Portuguese Crown did not require its subjects to present gold bullion for mintage. Miners and merchants could exchange gold bullion for coin in Brazil after the establishment of mints in the early 18th century, but most preferred their gold in bar or dust form. Whether in coin or bar, Brazilian gold was relatively pure, since on the one hand, it was difficult to counterfeit or adulterate, and on the other, the Portuguese Crown prided itself in producing only the highest quality gold coins. As a consequence, from the time that Brazilian gold became a significant factor in the world's monetary development in the late seventeenth century until the end of the reign of Dom João V in 1750, there occurred no debasement in the content of Portuguese and Brazilian gold coins.[7]

Unlike the relative constancy of fineness of Spanish American silver coin and Brazilian gold specie, the exchange or price ratio of silver to gold changed markedly during the period covered by this study. At the opening of the Age of Discovery in the late 15th century, about ten units of silver in Spain purchased the same amount of goods as one unit of gold. The monetary reforms of 1497 formalized this relationship which was approximately maintained until the mid-16th century. By the decade of the 1560s, however, the value of New World silver output for the first time surpassed the value of its gold production.[8] As the skyrocketing silver production of the Americas, especially that of Potosí, dramatically increased the supply of silver specie, the value of silver fell 60 percent relative to gold. By the mid-17th century, the bimetallic ratio in the Spanish Empire hovered between 15:1 and 16:1. It remained at this level until the late 19th century. (See Table 1.)

In China, the ratio of silver to gold in the years 1580-1630 fluctuated between 5.5:1 and 8:1.[9] The relative price of silver to gold in China, there-

---

en el siglo XVI," *Boletin de la Academia Nacional de la Historia* (Argentina), volumes 22-23 (1947-1948), pp. 408-418.

6. Lee K. N. Chaudhuri, *The Trading World of Asia and the English East India Company* (Cambridge, 1978), Table A.8. p. 185 for example of variations in the fineness of four Spanish-American coins assayed in 1738. The weights of new coin from the mints of Potosí and Lima were remarkably consistent and varied no more than a grain or two from the 416 grains per peso prescribed by the Crown. See, Robert I. Nesmith, "A Hoard of Lima and Potosí 'Cobs', 1654-1689," *The American Numismatic Society Museum Notes*, volume 1 (1946), pp. 81-99.

7. Boxer, *The Golden Age of Brazil, 1695-1750* (Berkeley and Los Angeles, 1969), pp. 56-57, 324. For equivalences of Brazilian gold coins see pp. 354-356. Also, Boxer, "Brazilian Gold and British Traders in the First Half of the Eighteenth Century," *The Hispanic American Historical Review*, volume 49 (August, 1969), number 3, pp. 470-471.

8. Calculated by determining the respective *values* of imported American silver and gold into Spain from E. J. Hamilton's table on the weight of treasure imports. Earl J. Hamilton, *American Treasure and the Price Revolution in Spain, 1501-1650* (Cambridge, Mass., 1934), p. 42.

9. Boxer, "Plata es Sangre," p. 461.

Table 1. Bimetallic ratio for Spanish empire, 1497–1750

Period	Ratio of silver to gold
1497–1536	10.11 to 1
1537–1565	10.61 to 1
1566–1608	12.12 to 1
1609–1642	13.13 to 1
1643–1686	15.45 to 1
1687–1728	16.60 to 1
1729–1750	16.00 to 1

Sources: 1497–1650: Earl J. Hamilton, *American Treasure and the Price Revolution in Spain, 1501–1650* (Cambridge, 1934), p. 71; Burzio, *La ceca de la villa Imperial*, p. 43; F. P. Braudel and Frank Spooner, "Prices in Europe from 1450 to 1750," in *The Cambridge Economic History of Europe. Volume IV: The Economy of Expanding Europe in the Sixteenth and Seventeenth Centuries* (Cambridge, 1967), p. 459.

fore, was about twice what it was in Europe and the Americas. This discrepancy in silver prices meant that Far Eastern goods could be exchanged for silver in the New World or in Europe at perhaps double (or more) their purchase price in Asia. High transportation costs naturally lessened the discrepancy in silver prices, but excessive demand for spices by Westerners and the relative cheapness of Chinese fine cloth seem to have balanced the expense of shipping. To illustrate this, one need only cite the observation of the Viceroy of Peru, who in the late 16th century penned the following in Lima: ". . . a man can clothe his wife in Chinese silks for two hundred reales, whereas he could not provide her clothing of Spanish silks with two hundred pesos."[10] Thus, a vigorous trade developed between Europeans and East Asians in the 16th century. Differences in the bimetallic ratios between the West and the East, then, encouraged long distance commerce and promoted European expansion. American silver production helped lower the real value of silver to a price which made trade with Asia more profitable.

Discrepancies in the bimetallic ratio also generated commercial activities between neighboring countries and regions. In the late 17th century, for example, the ratio in Spain and the New World was 16.60 to 1, while in Hamburg it averaged only 15 to 1.[11] The 10 percent difference in values stimulated trade between Spain and the rest of Europe and helped drain silver away from Spain. In a similar manner, commercial activities between Japan and China grew in the 16th century as a partial consequence of a 10 to

---

10. Woodrow Borah, *Early Colonial Trade and Navigation Between Mexico and Peru* (Berkeley and Los Angeles, 1954). p. 123.

11. J. Laurence Laughlin, *The History of Bimetallism in the United States* (New York, 1896), Appendix II. These figures are actually from A. Soetbeer who collected his bimetallic ratios from the records of a Hamburg merchant house.

20 percent higher silver valuation in China.[12] Even in the New World, the difference in the real price of silver fostered intercolonial trading activities. The early Mexico-Peru trade is perhaps the best example of this process. In the second half of the 16th century and the first part of the 17th, Peru was silver rich and Mexico was relatively silver poor. Additional Mexican demand for silver occurred with the opening of the Philippines trade in the 1580s. It became highly profitable, therefore, to send Mexican products to Peru, and later to forward Chinese silks from the Manila Galleon, in exchange for silver pesos. Although the official bimetallic ratio within the Spanish Empire did not take into account the difference in market values, there nevertheless occurred a flow of specie from a region abundant in silver to one with a demand for it. This imbalance in the demand for silver gave rise to a rather active intercolonial trade and partly fostered the early economic development of Mexico.[13]

New World production

Contrary to the expectations of the early explorers of the Caribbean, Mexico, and Peru, New World bullion did not flow in great abundance into Spain's public and private coffers. Most of the initial returns consisted of gold gleaned by native slaves from the placers of Hispanola. In all, Spaniards reported recovering only 14,118 kilograms of gold in the first two decades of the 16th century.[14] As a comparison, the amount collected in these twenty years amounted to about a third of *annual* New World bullion production a century later. The conquests of Mexico (1521) and Peru (1535) resulted in the capture of the relatively small treasures of the Aztecs and Incas, but more significantly, they opened the door to the major strikes the Spaniards had been awaiting.

An important series of discoveries began in the 1540s and included Antioquia in Colombia, Zacatecas and Guanajuato in Mexico, and Potosí in Peru. Although these strikes bolstered silver production in the New World, the widespread application of a new refining technique (the mercury almagamation process) was required to send silver output soaring. By the 1560s and 1570s, the New World silver boom was well under way. Supplemented with gold from Colombia, the mines of Mexico and Peru yielded quantities of bullion greater than the world had ever experienced. As we have suggested, for the two centuries, 1550-1750, South American mines provided the bulk of this extraordinary bullion production.

12. Evelyn Sakakida Rawski, *Agricultural Change and the Peasant Economy of South China* (Cambridge, Mass., 1972), pp. 75-76; See also, the chapter in this volume by Professors Kamiki and Yamamura.
13. Borah, *Early Colonial Trade and Navigation*, passim.
14. Hamilton, *American Treasure*, p. 42; also, see, Alfred G. Humpherys, "The Monetary History of the Caribbean During the Sixteenth Century," Ph.D. dissertation, University of New Mexico, 1973.

This output originated from two political units: the vast Spanish Viceroyalty of Peru, which extended in a wide belt down the western coast of the continent from Panama to Chile, and the equally vast Portuguese colony of Brazil. Within the Viceroyalty of Peru, bullion production centered in a few argentiferous mountains in Upper Peru, now called Bolivia, and in the streambeds of central and lowland Colombia. In terms of contributions to global production, Upper Peru flourished from the 1570s until perhaps the 1690s, while Colombian placer fields enjoyed fairly consistent prosperity from the mid-16th until the 18th century.[15] Despite a century of searching the expanses of the interior by Paulistas, gold was not discovered in Brazil until the last decade of the 17th century. Production, which came from alluvial streams far inland, increased throughout the first half of the 18th century, probably peaking in the decades of the 1740s and 1750s.[16]

How important was New World silver and gold output in the worldwide scheme of bullion production and what were the proportional contributions of Spanish and Portuguese America? The answers to these questions depend upon the reliability of production data presently available, and as we shall soon see, there are few, if any, accurate data series for the globe or for Latin America during this period. For the moment, we must rely upon the best published compilations and estimates with the assumption that if they do not provide the most trustworthy figures, at least they afford an idea of relative proportion. One of the most comprehensive collections of bullion production data ever assembled was researched in the 1920s by the U.S. Bureau of Mines. Those who worked on these studies (gold and silver) reviewed and analyzed the available literature: as a result, the study represents in the broadest terms the trends of gold and silver production for the New World.

Several consistency checks of the Bureau's data with later estimates reveal an acceptable tolerance for our purposes. For 17th century Mexico, for example, the Bureau's figures are within 10 percent of the most recent, and hopefully the most accurate estimate.[17] In the case of Colombia, the Bureau's researchers correctly dismissed Adolf Soetbeer's underestimation of gold production in the sixteenth and seventeenth centuries in favor of Vicente Restrepo's carefully reasoned and documented figures.[18] It is worth stressing that the massive data developed by the Bureau of Mines may be subject to criticism on the grounds that some series may not precisely reflect certain regional production. The data appear more reliable, however, if one's goals are to depict the proportions and the broad sweep of bullion output. (See Table 2.)

---

15. Vicente Restrepo, *A Study of the Gold and Silver Mines of Colombia* (New York, 1886). Table on p. 201 and accompanying discussion.
16. Boxer, *The Golden Age of Brazil*, chapters II and III, pp. 30-83.
17. Here I compare the Bureau's figures with estimates obtained by utilizing the techniques outlined in Appendix III of this paper.
18. Restrepo, *A Study of the Gold and Silver Mines of Colombia*, passim.

South American bullion production and export 1550-1750

Table 2. New World regional production as a percent of total world output, 1500–1800, by centuries

	Percent silver			Percent gold		
Region	16th	17th	18th	16th	17th	18th
Viceroyalty of Peru	57.1	61.0	32.5	35.7	60.1	36.0
Brazil	–	–	–	–	1.7	44.1
Mexico	11.4	23.4	57.0	3.4	4.3	4.8
Total New World as a percent of total world output	68.5	84.4	89.5	39.1	66.1	84.9

Sources and notes: Charles White Merrill, *Summarized Data of Silver production* (U.S. Bureau of Mines, Economic Paper No. 8, Washington, D.C., 1930), table facing p. 56; Robert H. Ridgeway, *Summarized Data of Gold Production* (U.S. Bureau of Mines, Economic Paper No. 6, Washington, D.C., 1929), table facing p. 64. Note that the Viceroyalty of Peru included all Spanish possessions in South America until it was divided into three viceroyalties between 1739 and 1776. I have adjusted the figures of the U.S. Bureau of Mines to take into account the estimates for Japanese silver exports in the 16th and 17th centuries as advanced by Professors Kamiki and Yamamura elsewhere in this volume. As a result, Japan's share of total world silver output increases from 3.4 percent and 5.9 percent in the 16th and 17th centuries respectively to 9.1 percent and 8.8 percent. The shares of Mexico and Peru are reduced correspondingly by several percentage points. I have extrapolated an overall average for Japan of 25,000 kgs annually for the entire 16th century (this is estimated from Kamiki and Yamamura's figures of from 33,750 to 48,750 kgs annually for the period 1560–1600). For the 17th century, I have used their estimate of 39,893 kgs yearly for 1601–1694.

Looking first at the bottom line of Table 2, we find that in each successive century the New World increased its share of total world output. In the 16th century, when Mexico and Peru actively produced bullion for only 60 years, the new Spanish colonies accounted for about three-quarters of the world's silver output and more than a third of its gold. By the 18th century, the combined figures are indeed impressive, with the New World furnishing 89.5 percent of silver and 84.9 percent of gold production. That a new region's precious metal production could completely dwarf the world's existing output indeed signals a quantum increase in available specie. And for those who subscribe to the quantity theory of money and prices, it also implies a considerable impact upon the economies of those countries which were "processing" the new specie. At a minimum, the first two hundred years of sustained bullion production by the Spanish American colonies doubled the existing European stock of silver. Estimates for Europe's silver supply in 1500 range from a low of around 3,000 tons to F. Braudel's admittedly speculative calculation of 37,000 tons.[19] Between 1540 and 1700, the New World produced at least 40,000 tons and possibly as much as 60,000 or 70,000 tons of silver.[20] It is fair to estimate that 80 percent of this production directly passed to or through the European economy. The consequences of doubling or even tripling Europe's stock of money and the resultant increase in the velocity of specie (i.e., the increase in commercial activity it caused, the re-minting and re-issuing of Spanish and Portuguese coins, etc.) in a relatively short span of time accelerated the development of the European economy in the 16th and 17th centuries.

Table 2 also confirms our earlier assertion that South American bullion provided the bulk of world supply during the period 1550-1750. Despite the decline of Potosí, beginning in the 1640's, the Viceroyalty of Peru accounted to 60 percent of the world's silver production in the 16th and 17th centuries. For the same period, the Peruvian mines yielded about 70 percent of all New World silver. The nadir of Peru's output, roughly 1690-1760, coincided with the gold boom in Brazil; in terms of value, then, Peru's losses in silver production were more than offset by Brazil's new gold discoveries. As a result, South America retained its position as the world's number one bullion producer until surpassed by Mexico's unparalleled silver boom of 1740-1810.

Production and export: the Viceroyalty of Peru

Although Indians continued to mine and refine silver after the Spanish conquest, the silver boom of Peru did not begin until the discovery and exploitation of Potosí in the late 1540s. Within a few years, the town of Potosí had grown to be the largest city in the Viceroyalty, yet the technology of its mining industry remained remarkably primitive. Thousands of Indian refiners utilized small clay smelters which could be positioned around the mountains of Potosí to best catch the fanning winds necessary for ore reduction.[21] As long as the grades of argentiferous ores continued to be exceptionally high (as they often were near the surface), native refining techniques proved adequate to the task. As the richer ores dwindled, however, silver yields began to decline. Registered output fell from 2,599,720 pesos (68,759 Kg pure silver) to 1,082,585 pesos (28,633 Kg) in the seven years between 1565 and 1572.[22] The successful introduction in 1573 of the mercury

19. Braudel and Spooner, "Prices in Europe from 1450 to 1750," pp. 442-446.
20. Estimate of 40,000 tons calculated from Brading and Cross, "Colonial Silver Mining," p. 579, and Hamilton, *American Treasure*, p. 42; the higher estimate from Charles W. Merrill, *Summarized Data of Silver Production*, table facing p. 56.
21. Bernabé Cobo, *Historia del Nuevo Mundo* (2 volumes; Seville, 1890-1891), first published 1653, see vol I, p. 308; Juan López de Velasco, *Geografía y descripción universal de las Indias (1574)* (Madrid, 1894), p. 503; Pedro Vicente Cañete y Domínquez, *Guía historica, geográfica, física, política, civil y legal del gobierno e intendencia de la provincia de Potosí (1789)* (Potosi, 1952), pp. 54, 111.
22. Calculated from tax records of Lambertó de Sierra, "Reales quinto pagados á S.M. desde 1 de enero de 1556 hasta 19 de julio de 1736," *Colección de documentos inéditos para la historia de España* (volume V; Madrid, 1844), pp. 172-173.

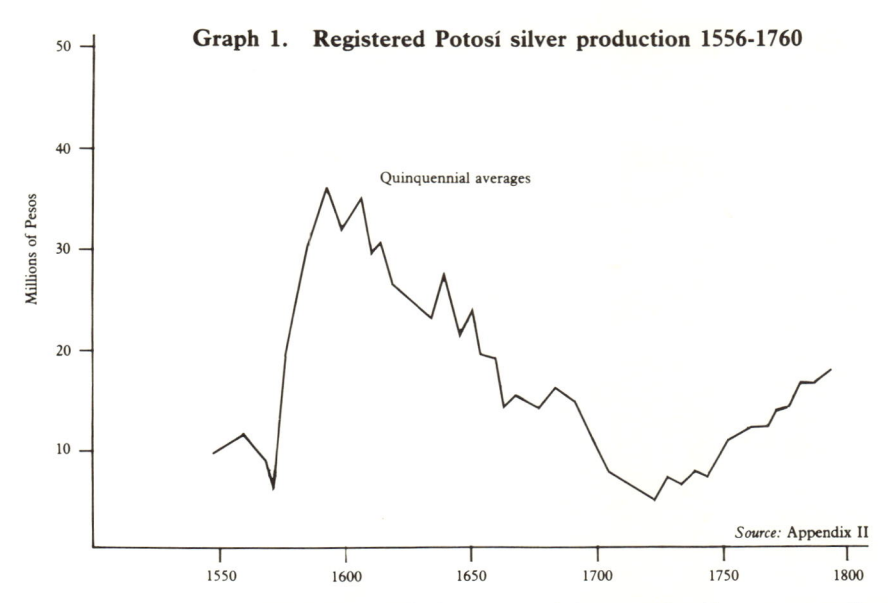

Graph 1. Registered Potosí silver production 1556-1760

amalgamation process, which profitably treated low-grade silver ore, radically transformed the industry. The new refining process required large-scale plants and greater organization of labor. Independent Indian refiners soon became employees of Spaniards, and the viceroy implemented a system of forced Indian labor (the *mita*) which brought natives from as far away as six hundred miles to work in and around the mines.[23] The innovative refining method coupled with the reorganization of the industry sent the curve of silver production soaring upwards. From slightly more than one million pesos in 1572, registered Potosí production rose to the then astronomical figure of 7,632,275 pesos (201,864 Kg pure silver) in the year 1585. Graph 1 depicts the magnitude of this rise and describes the course of subsequent output.

From the peak years of 1580-1650, in which Potosí silver output never fell below 4.2 million pesos annually, production slipped steadily until it bottomed out in the 1720s at about 1.1 million pesos. The causes of this decline were: (1) the gradual exhaustion of silver ores and the deepening of the mines; (2) a general population decline which restricted the availability of forced labor; and (3) a contraction of the supply of essential mercury from the mines of Huancavelica.[24] Although Potosí recovered somewhat in the latter half of the 18th century, it never regained the startling levels of production it had attained during its florescence. Instead of dominating

---

23. See sources for the history of the Potosí labor system in Brading and Cross, "Colonial Silver Mining," p. 558, note 42.

24. See Günter Vollmer, *Revölkerungspolitik und Revölkerungsstruktur im Vizekönigreich Peru zu Ende der Kolonializeit, 1741-1821* (Berlin, 1967); Brading and Cross, "Colonial Silver Mining," p. 563, graph I.

Table 3. Registered silver production in the District of Charcas, Peru and registered silver production at Potosí 1611–1700 (Pesos of eight reales and by quinquennia)

Quinquennium	Total registered production (millions)	Total Potosí production (millions)	Percent Potosí of total
1611–15	37.7	32.3	85.7%
1616–20	39.3	27.8	70.7
1621–25	32.6	26.9	82.5
1626–30	34.2	26.0	76.0
1631–35	30.6	24.8	81.0
1636–40	33.8	29.5	87.3
1641–45	28.3	22.8	80.6
1646–50	28.7	24.2	84.3
1651–55	22.0	20.5	93.2
1656–60	21.6	20.3	94.0
1661–65	15.3	15.5	–
1666–70	18.1	16.3	90.1
1671–75	16.5	13.0	78.8
1676–80	14.9	14.8	99.3
1681–85	17.7	17.3	97.7
1686–90	16.5	16.0	97.0
1691–95	15.7	13.5	86.0
1696–00	14.9	11.2	75.2
Totals	438.4	372.7	85.0%

Sources and methods: Total registered silver production for the District of Charcas (Potosí and Oruro) is derived from, P. J. Bakewell, "Silver mining in the economy of seventeenth century Spanish America," a paper presented at the 1976 meeting of the American Historical Association Convention in the session entitled *Latin America, The Seventeenth Century Crisis,* appended graph. Bakewell's figures are presented in marcs (8 oz.) of silver; these have been converted to pesos of eight reales at the rate of 8.5 pesos per marc. Potosí production from, Lamberto de Sierra, "Reales quintos pagados á S. M. desde 1 de enero de 1556 hasta 19 de julio de 1736," in *Colección de documentos inéditos para la historia de España* (volume V; Madrid 1844), pp. 173–184. Data converted from royal tax register.

South American silver production as it had for nearly two centuries, by 1750 Potosí was only one of several important contributors.

The present historical literature does not contain any serial data on the total silver production for the Viceroyalty of Peru during the period under consideration. As we have just seen, there is a significant run of figures for the mines of Potosí. In addition, thanks to the recent research of P. J. Bakewell, there is data for the principal mining district of Peru called Charcas. (Charcas included some of the major silver-producing mines in the Viceroyalty, namely

Porco and Oruro, besides Potosí.) These two series, coupled with the proportional distribution of silver production according to the U.S. Bureau of Mines and recent estimating techniques based upon mercury consumption, enable us to construct an estimated silver production series for South America.

The first step in this exercise is to establish a relationship between Potosí production and total output for the district of Charcas. Since Bakewell's data does not begin until 1611, determining Potosí's share of Charcas' total production will yield a basis for estimating Charcas output before 1611. Table 3 compares total registered production for Charcas to similar data for Potosí for the period 1611-1700.

According to these series, Potosí silver accounted for 85 percent of Charcas' overall production. Assuming that this relationship held true for the years 1571-1610, a period which included some of Potosí's most prolific years, then we can safely extrapolate the output of Charcas for that period. The application of this ratio to Potosí production results in the following series of figures for the district of Charcas: 1571-80, 34.1 million; 1581-90, 80.7 million; 1591-00, 85.1 million; and 1601-10, 80.4 million. These additional figures when combined with those of Bakewell provide a complete series of silver production for Charcas for the years 1571-1700. Now, in the same manner that Potosí output accounted for a particular average percentage of the district, so too, did the district account for a standard percent of the Viceroyalty's production. The proportional data furnished by the U.S. Bureau of Mines study indicates that for the 16th and 17th centuries, Charcas (roughly today's Bolivia) mined 70 to 75 percent of South America's silver. The other 25 to 30 percent originated from Lower Peru (today's Peru), Chile, and Colombia. If these proportions approximate reality, we need only add 25 percent to the figures for Charcas to obtain total minimum registered silver production for the Viceroyalty of Peru.[25] Up to this point, the derived data should yield a series which reflects real registered production with an accepted tolerance of plus or minus 10 percent. A substantial stumbling block remains, however, in our quest to determine total output, and that problem concerns unregistered silver.

Unregistered and/or contraband bullion has long been the bane of modern historians seeking to unravel the history of colonial mining. The simple problem is that, from our point of view, fraud and smuggling by their nature leave little or no historical record. Volumes of recurring complaints and plaintive discussions and petitions make us only certain such crimes existed and were widespread. Recently, however, a method has been devised by which silver outuput may be reliably estimated from serial data of known

---

25. It would be mathematically correct to add 33 percent to the Charcas figures since the rest of the Viceroyalty accounted for 25 percent of total silver output (or one-third of Charcas' total). I have chosen the lower multiplier of 25 percent with the intent of biasing the estimates downward.

Table 4. Registered and estimated silver production, District of Charcas, 1611–1650 (by decades)

Decade	Registered production (millions of pesos)	Estimated production (millions of pesos)
1611–20	77.0	77.1
1621–30	66.8	72.7
1631–40	64.4	105.9
1641–50	57.0	77.6
Totals	265.2	333.3 (+ 25.7%)

Sources: Registered production from Table III; estimated production from Appendix III.

mercury consumption,[26] since mercury was used up at a relatively constant and predictable rate during the refining process. In the district of Charcas, a *quintal* of mercury (a hundredweight) produced on the average between 120 and 130 marks of pure silver.[27] Since we have a good idea of mercury consumption in the district of Charcas, based upon the works of the Chaunu and Guillermo Lohmann Villena, we can calculate the quantity of silver produced by the amalgamation process.[28] To achieve a reasonable estimate of silver output, it is also necessary to add 20 percent to the total to account for the smelted metal. The results of these computations appear in Appendix III.

We now have a set of estimated decennial figures during 1611-1650 for the district of Charcas which should reflect the realities of silver production. Table 4 compares this deduced output with silver officially registered in Charcas by royal treasury agents. We can take the difference between the two series to be an indication of unregistered silver bullion.

In the most important mining district of the Spanish Empire, where bureaucratic regulations and government control would presumably be most efficient, unregistered silver exceeded registered silver by 25 percent in roughly the first half of the 17th century. Assuming that Spanish miners, merchants, and officials defrauded the royal treasury to the same extent in the

---

26. See, Brading and Cross, "Colonial Silver Mining," pp. 568-579.

27. Cañete y Domínguez, *Guía historica*, p. 70. This ratio is most likely too low since Bakewell's data for Potosí during 1593-1685 reveal a return of 138 marks of silver for each quintal of mercury consumed. P. J. Bakewell, "Registered Silver Production in the Potosí District, 1550-1735," Table 2, p. 98.

28. In the early seventeenth century, at least 95 percent of all mercury consumption in the Viceroyalty of Peru occurred in the district of Charcas. See, Carmen Báncora Cañero, "Las remesas de metales preciosos desde el Callao a España en la primera mitad del siglo XVII," *Revista de Indias*, volume 19 (January-March, 1959), number 75, p. 36; Carta, Marqués de Montesclaros to King, Los Reyes, 16 October 1611, in *Relaciones de los Vireyes que han gobernado Perú* (volume II; Madrid, 1871), pp. 411-412.

South American bullion production and export 1550-1750

Table 5. Estimated silver production, Viceroyalty of Peru, 1571–1700 (by decade, millions of pesos, and kilograms)

Decade	Total registered silver production District of Charcas	Estimated silver production Viceroyalty of Peru	Millions of kilograms pure silver
1571–80	34.1*	49.4	1.3
1581–90	80.7*	117.0	3.1
1591–00	85.1*	123.4	3.3
1601–10	80.4*	116.6	3.1
1611–20	77.0	111.7	3.0
1621–30	66.8	96.9	2.5
1631–40	64.4	93.4	2.5
1641–50	57.0	82.6	2.2
1651–60	53.6	63.2	1.7
1661–70	33.4	48.4	1.3
1671–80	31.4	45.5	1.2
1681–90	34.3	49.6	1.3
1691–00	30.6	44.4	1.2
Totals	718.7	1,042.2	27.7

(*) estimated

Sources and methods: Estimated of Charcas production equals known Potosí production divided by 0.85. Potosí production from Appendix II. Estimated silver production, Viceroyalty, is registered Charcas production plus 25 percent for the output of Lower Peru, Chile, and Colombia, and 20 percent more to account for unregistered silver.

forty years before 1611 and in the fifty years after 1650, we have a fair estimate of the proportions of "un-taxed" silver for the period 1571-1700. To insure our estimate against minor errors, and to bias our general figures downward, we can accept a figure of 20 percent of registered production as representative of illegal silver bullion. This information, along with the previously established production relationship between Charcas and the Viceroyalty, permits the calculation of total estimated silver output for South America. To the Charcas series, therefore, we add 25 percent for the production of Lower Peru (including the mines of Castrovirreina), Chile, and Colombia, and 20 percent for unregistered silver. The resulting data is presented in Table 5.

As a ratification of the estimated series in the above table, we can compare the calculated figures for registered silver production with a thirty-year run of data obtained by Alvaro Jara from royal treasury accounts. His figures for the whole of the Viceroyalty cover the years 1571-1600; they are contrasted to ours in the following tabulation.

Table 6. Official registered and estimated registered silver production Viceroyalty of Peru, 1571–1600 (millions of pesos)

Decade	Registered Viceroyalty output after Jara	Estimated registered output after Table 5
1571–80	35.6	42.6
1581–90	98.1	100.9
1591–00	104.1	110.6
Totals	237.8	249.9 (+ 5.1%)

Sources: Alvara Jara, *Tres ensayos sobre economia minera Hispanoamericana* (Santiago, 1969), pp. 115–116; Estimated registered output equals total registered silver production from the district of Charcas (see Table 5) plus 25 percent.

Table 7. Colombian gold production, 1537–1800 (annual averages; pesos of eight)

Period	Annual output in silver pesos	Kilograms of gold (annual average)
1537–1600	793,642 pesos	1,749
1601–1700	1,700,000	2,810
1701–1800	1,940,000	3,159

Sources: Vicente Restrepo, *A study of the gold and silver mines of Colombia* (New York, 1886), p. 201. Note that the pesos of eight are converted, after obtaining their weight in pure silver, to gold kilograms at the ratios of 12:1 for the 16th century, at 16:1 for the 17th and 18th centuries.

The comparison reveals that our estimated registered silver production figures for the three decades in question are within 5.1 percent of the official record. We consider this an acceptable tolerance and conclude that the estimated production series for the Viceroyalty in Table 5 is a reasonable representation of the magnitude and trend of silver output.

If we do not have much solid data about silver production in the Viceroyalty of Peru, we have even less about gold output. Most gold originated in Colombia with the remainder equally scattered among Upper and Lower Peru and Chile. Judging from the most reliable estimates, Colombian gold mining flourished and generated a goodly share of the world's output. In the 16th, 17th, and 18th centuries, this region accounted for 18%, 39%, and 25%, respectively, of total global production.[29] Vicente Restrepo, who published in 1886, is still considered to be the authority on the volume of

29. Ridgeway, *Summarized Data of Gold production*, Table facing p. 64.

Colombian precious metal output for the colonial years. Although meticulously derived, Restrepo's data, unfortunately, are lump sums for each of the three colonial centuries. As a result, we have little or no sense of the rhythm or growth of the industry. The secular trend reveals an apparent increase in production. The figures are shown above in Table 7.

Data on gold production for the rest of the Viceroyalty of Peru is scanty. There were active gold workings in Lower Peru and Chile throughout the colonial era, but the estimates of their output seem highly inflated. One scholar, however, tells us that in the 17th century the Crown rarely exported more than 10,000 pesos per year in gold from the main Peruvian port of Callao—a fact which suggests limited gold mining.[30] Nevertheless, if we accept only the Colombian figures as approximations, then the Viceroyalty's gold production in the 17th century equalled one-third of the value of silver output. Such a finding suggests that gold production in Colombia deserves more attention by Andeanist researchers.

Most of the silver and gold produced in the Viceroyalty of Peru found its way into export channels with little delay. Miners, merchants, and the Crown exported much of it directly from the mints to Europe and Asia. Since the monetized populations were largely European and Hispanized Indians and mestizos, another substantial portion of bullion output went to satisfy the demand for imported goods. Even money expended in local or regional economies was soon shipped out of the Viceroyalty. Indian laborers pressed into service at the Potosí mines, for example, used their peso wages to pay the royal tributes of their home villages.[31] Pesos sent to the region of Ecuador to pay for cloth were quickly exchanged by Spanish manufacturers and merchants for imported goods.[32] Pesos paid to large landowners in the region of Argentina for work animals were soon spent on European wines, cloth, hardware, and even slaves.[33] Finally, the increasing portion of Crown income in the Viceroyalty which was "retained" in Peru for royal expenditures nevertheless became an export. The costs of defense (i.e., ships, shipbuilding, canon, hardware, etc.) involved spending much money on imports from Europe; and salary disbursements to Spanish functionaries, which accounted for a great proportion of royal expenditures in the Indies, supported lifestyles

---

30. María Encarnación Rodríguez Vicente, "Los caudales remitidos desde el Peru a España por cuenta de la Real Hacienda. Series estadisticas (1651-1739)," *Anuario de Estudios Americanos*, volume 21 (1964), p. 5. The eighteenth century was apparently a much different story according to Viceroy Amát y Juniet who reported that during the 1760s and 1770s, Peru exported a million pesos in gold annually. See, Amát y Juniet, *Memoria de gobierno*, pp. 228-229.

31. Same as note 23.

32. Robson B. Tyrer, "The Demographic and Economic History of the Audiencia of Quito: Indian Population and the Textile Industry, 1600-1800," Ph.D. dissertation, University of California, Berkeley, 1976.

33. See, Marie Helmer, "Commerce et Industrie au Pérou à la fin du XVIIIe siècle," *Revista de Indias*, volume 10 (1950), pp. 519-526.

imbued with European goods. All the trends, therefore, favored the movement of bullion out of the New World.[34]

In the long run, the annual amounts of bullion exports should have roughly paralleled the curve of bullion production. Since we do not yet possess serial data on total silver and gold shipments from the Viceroyalty of Peru, it is virtually impossible to test this assertion. If an official series for bullion exports existed, it would reveal trends similar to the curve of precious metal output, but it would far understate the real quantities exported. The reason for this discrepancy lies in the tremendous amount of bullion clandestinely shipped out of the Indies.

The principal trading route connecting the Viceroyalty of Peru with Spain passed through Callao on the Peruvian coast, the isthmus of Panama, and on to Seville. In theory, nearly all communication and commerce between the Iberian peninsula and South America was restricted to this channel, according to the monopolistic system designed by the Crown. As long as the demand for imports was relatively low, the Seville-Lima monopoly largely succeeded in supplying Peru with European goods. The silver boom at Potosí in the 1570s and 1580s, however, resulted in a seven-fold increase of available silver. This rapid expansion of export capacity caused significant growth in the demand for imports, which the merchant monopolists of Seville and Lima could not or would not meet. Prices rose dramatically, as did the possibilities of profits.[35] As a consequence, Spanish and especially Portuguese investors and merchants were quick to take advantage of the favorable circumstances; they infiltrated Spain's transatlantic monopoly and developed an effective and widespread contraband network.[36]

From the mid-16th century until nearly a century later, trade continued between Mexico and Peru. As we have seen, the principal exchange in this commerce involved Chinese and Mexican goods for Peruvian silver. At the height of its success (1580-1610), an average 2-3 million pesos (53,000 kg to 79,000 kg pure silver) was sent annually from Peru to Mexico with official

---

34. The continual draining of bullion from the New World is also evidenced by the general lack of specie except in a few large urban centers. See, Borah, *Early Colonial Trade and Navigation*, pp. 83-84; Brading and Cross, "Colonial Silver Mining," p. 576; Tyrer, "The Demographic and Economic History of the Audiencia of Quito;" Ricardo Levene, *La moneda colonial del Plata* (Buenos Aires, 1916), passim. Work by numismatists also indicates that South American silver coins were largely uncirculated. See, Nesmith, "A Hoard of Lima and Potosí 'Cobs', 1654-1689," pp. 95-96.

35. Licenciado Cepeda, President of the Audiencia of Charcas, to King, 9 December 1586, La Plata; Licenciado Matienzo to King, 3 February 1578, Potosí; Audiencia de Charcas to King, 12 February 1579, La Plata, all in Roberto Levillier, ed., *La Audiencia de Charcas: correspondencia de presidentes y oidores* (3 volumes; Madrid, 1918-1922), vol I, p. 470, vol II, p. 263.

36. See H. E. Cross, "Commerce and Orthodoxy: A Spanish Response to Portuguese Commercial Penetration in the Viceroyalty of Peru, 1580-1640," *The Americas*, volume 35 (October 1978), No. 2, pp. 151-167.

sanction. According to Woodrow Borah, the trade also involved "large amounts" of contraband silver, so the actual average exported during this period probably exceeded 100,000 kg per year.[37] A slight amount of this silver remained in Mexico or was remitted to Spain, while the rest was sent to the Philippines via the Manila Galleon. Relentless pressure by Seville monopolists, who saw their commercial opportunities limited by Mexican and Chinese goods, forced the Crown to stifle this trade completely by the 1630s.

A highly lucrative contraband trade developed in the region of the Rio de la Plata (Argentina) after the founding of Buenos Aires in 1580. The route through Buenos Aires, Cordoba, Tucumán, and on to Potosí, often called the "back door" of the Viceroyalty, provided merchants the easiest access to major Peruvian markets; it was, however, an illegal conduit for the importation of European goods. Until the mid-17th century, however, Spanish royal authority in the region was minimal; and, in 1613, the King himself acknowledged the widespread corruption in the port's administration.[38] Under these conditions, enterprising merchants stood to gain significant advantages over competitors restricted to the legal convoys: they avoided the bureaucratic delays and expense of Seville and other entrepôts; they circumvented many of the taxes which inflated final destination prices; and they saved considerably on the costs of transportation since the Rio de la Plata route was much shorter and involved fewer transshipments than that through Panama. The benefits of shipping via Buenos Aires afforded Spanish and Portuguese businessmen the leverage of offering lower market prices in Peru without diminished rates of return.[39] The Spanish Crown facilitated the movement of contraband goods to Peru through the "back door" of the colony by granting two important concessions: in 1595, several Portuguese merchants negotiated an *asiento* (contract) to import slaves through Buenos Aires; and seven years later, the port of Buenos Aires secured trading rights with Brazil. In effect, these concessions opened Peru's back door by serving as a front for large-scale contraband operations.[40]

Contraband traffic up the Rio de la Plata elicited unremitting protests from Spaniards dependent upon the transatlantic monopoly. The crown could not fail to acknowledge the complaints: a royal decree issued in 1619, for example, stated that, "A great quantity of merchandise and iron has

37. Borah, *Early Colonial Trade and Navigation Between Mexico and Peru*, pp. 88, 123.
38. Real Cédula to Audiencia de Charcas, 20 October 1613, Madrid, Biblioteca Nacional, Lisbon, Pombahna Mss. 249, ff. 11-12.
39. See letter of Portuguese merchant on profits of Rio de la Plata trade in, Raul Molina, *Las primeras experiencias comerciales del Plata: El comercio maritimo, 1580-1700* (Buenos Aires, 1966), pp. 82-83, 127; also, Alice P. Canabrava, *O Comércio portugues no Rio da Prata (1580-1640)* (São Paulo, 1944), p. 26; C. R. Boxer, *Salvador de Sá and the Struggle for Brazil and Angola, 1602-1686* (London, 1952), pp. 77-78.
40. Molina, *Las primeras experiencias comerciales*, pp. 65, 88-122; Boxer, *Salvador de Sá*. p. 77; Antonio Dominguez Ortiz, *Los Judeo-conversos en España y América* (Madrid, 1971), pp. 135-136.

entered Potosí through the port of Buenos Aires. It was carried from the city of Lisbon in ships which were dispatched from Brazil. The trade has been so excessive that goods are brought from Potosí to sell in Lima."[41] As noted, the very nature of contrabanding precludes the collection of statistics, but some scattered data is available. During the decade 1616-1625, recorded imports for Buenos Aires (the sum of legal and confiscated goods) were 7,957,579 pesos, while exports for the same period amounted to only 360,904 pesos.[42] The annual trade deficit, which was met with smuggled silver from Upper Peru, amounted to at least three-quarters of a million pesos per year. Between 1619 and 1623, port officials seized a total of 3,656 slaves from illegally landing vessels; their market value in Lima would have approached two million pesos.[43] These numbers do not include those slaves legally imported and those which evaded port officials. The size of these figures clearly indicates a flourishing and considerable illicit trade up the Rio Plata from the 1580s until probably the 1640s. During its most successful years, no less than 1-2 million pesos (roughly 25,000 to 50,000 kg) flowed illegally from the mines of Peru out throught the port of Buenos Aires. These totals equalled from 15 percent to 30 percent of the silver output of Potosí.

As evidenced by the decree above, the bulk of silver exported legally and clandestinely from the port of Buenos Aires ended up in Portuguese hands. Portuguese shippers, who dominated maritime traffic in the Rio de la Plata estuary, carried Peruvian silver to Brazil and Portugal.[44] Besides Portuguese vessels, ships from Holland and England occasionally landed in Buenos Aires and successfully traded their goods for specie.[45] Thus, hundreds of thousands of kilograms of silver sent from Peru to the port of Buenos Aires and on to Europe completely avoided registration by Spanish officials and was transported to other European countries.

Such an illicit and prosperous trade was not destined to endure, since the Seville-Lima monopoly, a powerful force in Spain, actively lobbied for the closure of Buenos Aires after the turn of the 17th century. By 1625, the

41. Real Cédula, 18 September 1610, Madrid, Archivo General de Indias, Buenos Aires, legajo 5, ff. 45-46; also, Real Cédula to Audiencia de Charcas, 20 October 1613.

42. For trade figures, Juan Agustín García, *La ciudad indiana (Buenos Aires desde 1600 hasta mediados del siglo XVIII)* (Buenos Aires, 1900), p. 246; for Potosí production see Appendix II of this paper.

43. Numbers of slaves derived from ship inventories in, Molina, *Las primeras experiencias comerciales*, pp. 192-195, 232-233; slave prices for Peru from Frederick P. Bowser, *The African Slave in Colonial Peru, 1524-1650* (Stanford, 1974), Appendix B.

44. Rozendo Sampaío García, "Contribuicão ao Estudio de Aprovisionamento de Escravos Negros da América Española (1580-1640)," *Anais do Museu Paulista*, volume 16 (1962), pp. 7-196; Marie Helmer, "Comércio y Contrabando entre Bahia e Potosí no Século XVI," *Revista de Historica* (São Paulo), volume 4 (1953), pp. 195-212; Canabrava, *O Comercio Portugues*, passim.

45. See for example, "Corto y verídico relato de la desgraciada navigación de un buque de Amsterdam, 1598-1601," and, "Informe del Gobernador Valdés, Buenos Aires, 1599," *Anales de la Biblioteca* (Buenos Aires), volume 4 (1905), pp. 371-496.

Crown had established a customs house between Buenos Aires and Potosí, and officially registered trade began to decline thanks to a 50% charge on all goods. Despite this severe measure, the trade continued through the 17th century, although much reduced.[46] By the 1680's, the Portuguese showed a renewed interested in the Rio de la Plata trade with Peru, and by 1700, Portuguese ships again plied the estuary. The key entrepôt for this rejuvenated commerce was the newly established Portuguese colony of Sacramento, located across the estuary from Buenos Aires.[47] The success of this second phase of the Rio Plata trade has been documented by Vitorino Magalhães Godinho, who comments that by the beginning of the eighteenth century, all fleets arriving in Lisbon from Brazil carried Peruvian silver. Another scholar, H. E. S. Fisher, states that by the years 1760 and 1761, the silver trade through the colony of Sacramento reached the sums of 1.0 million and 2.5 million pesos, respectively (26,000 kg and 65,000 kg).[48]

If contraband activities flourished outside the Seville-Lima monopoly, fraudulent practices also prospered within it. The *Carrera de Indias* (the annual Atlantic convoy) carried unknown quantities of illegal goods to the New World and returned fortunes in smuggled silver. Shipmasters and captains of accompanying galleons took advantage of their opportunities for easy profits. Several Spanish admirals became wealthy men after a few voyages to the Indies. Even merchant houses, approved by the monopoly for the Indies trade, made a practice of altering and forging bills of lading and bribing customs officials in order to smuggle silver into Spain.[49] Finally, the numerous slave asientos, usually granted to Portuguese, offered opportunities to import far more Africans and merchandise than the concessions permitted. In this manner, Portuguese merchants routinely shuttled illegal goods and slaves to the Viceroyalty of Peru and extracted contraband silver.[50]

Brazilian gold production and export

A government faced with regulating and taxing a silver mining industry during the colonial era had a far easier task than the one required to oversee

---

46. Molina, *Las primeras experiencias comerciales*, pp. 219-222; Canabrava, *O Comércio Portuguese*, passim.

47. Mario Rodriguez, "Colonia do Sacramento: Focus of Spanish American Rivalry in the Plata, 1640-1683," Ph.D. dissertation, University of California, Berkeley, 1952.

48. Vitorino Magalhães Godinho, "Le Portugal, les flottes du sucre et les flottes de l'or (1670-1770)," *Annales-Economies-Societes-Civilisations*, volume 5 (April-June, 1950), number 2, p. 191; H. E. S. Fisher, "Anglo-Portuguese Trade, 1700-1770," *The Economic History Review*, second series, volume 16 (1963), number 2, p. 225.

49. John H. Parry, *The Spanish Seaborne Empire* (New York, 1970), p. 135; Haring, *Trade and Navigation*, pp. 111-115.

50. See, Bowser, *The African Slave*, pp. 32-33, 55-59, 61-62; Rodriguez Vicente, *El Tribunal del Consulado de Lima en la primera mitad del siglo XVII* (Madrid, 1960), pp. 71-73; Boxer, *Salvador de Sá*, p. 80; also, Robert Ricard, "Los Portugueses en las Indias Españolas," *Revista de Historia de America*, number 34 (1952).

gold mining. The silver industry was, by its nature, a fairly concentrated activity, and as we have seen, at any given time a few prosperous mining centers accounted for the majority of output. Colonial gold mining, however, had a different character, since ores were superficial and widely scattered. This natural dispersion kept mining operations small-scale and transitory. Moreover, because gold had a much higher value than silver, concealment was less of a problem for gold smugglers. As a result of these circumstances, Brazilian gold production, which flourished for the first two-thirds of the 18th century, was riddled with fraud and contraband. The significance of this fact for modern historians is that it is impossible to document Brazil's gold output or export with any certainty. Estimating gold production is also a problematical exercise since its refining process (generally simple panning and sluicing) required no precisely measurable input from which to determine output.

The only conceivable method of estimating Brazilian gold output is locating some official tax or export records and adding to this a percentage for unregistered bullion. The difficulty with such a procedure is that determining the percentage of unregistered gold is pure guesswork. Contemporary estimates of "un-taxed" gold range from two-thirds to nine-tenths of total production. C. R. Boxer, who has studied this problem extensively, implies that the two-thirds estimate is probably closer to the truth.[51]

Although there are a variety of ranging estimates of Brazilian gold production in the 18th century, none seems to have a solid documentary basis. For our purposes, the best series has been assembled by Magalhães Godinho, whose data is based upon fleet registration records in Lisbon. An arrangment of Magalhães Godinho's figures appears in Table VIII.

As with the case of Peruvian silver, the majority of Brazilian gold was exported to Portugal, so the magnitude of exports should closely resemble real production. Keeping in mind Boxer's suggested percentage of unregistered bullion, we can be confident that if the figures in Table VIII are doubled, they will represent at a *minimum* the actual gold production of Brazil. When this calculation is applied to the data, we find that between 1712 and 1755 (and most likely into the 1770s), annual gold output averaged over 21,000 kg. The equivalent of 21,000 kg of gold in Spanish pesos of eight computes to more than 13 million per year. The conversion of Brazilian gold into the Spanish monetary unit affords a mean of comparison by which we may contrast the contributions of the two colonies. The most productive decade in the colonial history of the Viceroyalty of Peru occurred in 1591-1600 when, according to Table V, its mines yielded an annual average of 12.3 million pesos. Brazil's best decade, 1741-1750, produced the equivalent of 16.2 million pesos per year. From a longer perspective, the most prosperous forty-year period of Brazilian gold mining, 1716-1755, reveals an average annual production of 13.5 million pesos, while Peru's outstanding four

---

51. Boxer, *The Golden Age of Brazil*, pp. 59-60.

Table 8. Registered gold exports from Brazil to Lisbon, 1699–1755 (in kilograms and in pesos of eight)

Period	Kilograms of gold (annual averages)	Equivalent in Spanish pesos of eight reales
1699	725 Kg	439,054
1701	1,785	1,080,981
1703	4,350	2,634,323
1712–15	10,788	6,533,120
1716–20	8,900	5,389,763
1721–25	10,400	6,298,151
1726–30	7,000	4,299,237
1731–35	10,200	6,264,602
1736–40	10,800	6,633,108
1741–45	12,400+	7,615,791
1746–50	14,000+	8,598,473
1751–55	13,300+	8,168,550
Annual average 1712–1755	10,867	6,580,962

Sources and methods: Yearly and period averages have been calculated from annual figures given in, Vitorino Magalhães Godinho, "Le Portugal, les flottes du sucre et les flottes de l'or (1670–1770)," *Annales-Économies-Sociétés-Civilisations*, volume 5 (April-June, 1950), number 2, pp. 192–193. For two years, 1732 and 1735, Godinho lacks the data. I have estimated these at 7,000 kilograms each, which equals the lowest annual amount exported during the period 1725–1755.*
Kilograms of gold have been converted to Spanish pesos of eight reales by the following method. Kilos converted to pounds and ounces (Kg × 2.2 × 16 = oz.). Ounces of gold to silver at the prevailing rate of 1:16 (oz. × 16). Ounces of pure silver converted to Spanish pesos of eight which were of an approximate fineness of 0.931 percent (oz. silver ÷ 0.931). After 1728, the fineness was reduced to 0.917 percent.
*Note that the data for 1741–1755 represent minimum exports.

decade period (1581-1620) yielded 11.7 million pesos yearly. If we take into account inflation and the fact that there was a larger world monetary stock by the eighteenth century, the magnitude of Brazil's success as compared to Peru's is somewhat mitigated. However, in terms of contributing to world economic development and monetary supply, Brazil's gold production during the first half of the 18th century at least equalled the more famous Peruvian boom of a century earlier.

Much as Seville acted as a simple entrepôt for American silver in the 17th century, so too, did Lisbon function as an entrepôt for Brazilian gold in the 18th. England became Portugal's principal trading partner in the second half of the 17th century, and as Portugal increasingly relied upon imported goods, its trade deficit with England grew. For the five year period 1702-1706,

Portugal's deficit with England ranged between £266,000 and £596,000 (1.3 million and 3.0 million pesos). As a result of Brazil's gold boom, merchants in Lisbon met this deficit with payments in bullion. Later figures of registered gold imports into England from Lisbon for the 1750s and 1760s demonstrate that the equivalent of about 5 million pesos (8,141 kg pure gold) was sent annually.[52] If we take into account contraband, as rampant between Falmouth and Lisbon as between Brazil and Portugal, possibly as much as 10 million pesos per year (16,281 kg gold) passed into England from Portugal in the first half of the 18th century.[53] Thus, 80 percent of Brazil's gold output eventually found its way to England; and it may be no coincidence that the Industrial Revolution followed on the heels of the expansion of British money supply resulting from the importation of Brazilian gold.

Conclusion

Our main interests in this chapter have been bullion production and flows. Since the greater portion of American silver went to Europe, it will be instructive to compare estimates of Peruvian output with registered imports into Spain. The standard series for Spanish bullion imports is that of E. J. Hamilton, whose figures are often cited in the literature. Hamilton's data include all treasure (silver and gold) from the New World, including Mexico. Our data pertain only to silver production in the Viceroyalty of Peru.

With the exception of the decades 1571-80 and 1591-1600, the estimated silver production of Peru is greater than overall treasure imports as given by Hamilton. If we added Mexican bullion and Colombian gold to the Peruvian estimates, all the figures for New World production would far exceed registered imports. Assuming that at least three-quarters of New World bullion was exported to Europe and that our estimating techniques are sound, it is clear that Hamilton's imports do not accurately reflect the levels of American production and that they far understate the amounts of bullion flowing to Europe.[54] There exists additional documentary evidence which casts doubt on Hamilton's series.

Two excellent studies based upon royal treasury records, by Carmen Báncora Cañero and María Rodriguez Vicente, document public bullion shipments from Lima to Spain in the 17th century. Their figures pertain only to that money sent to Spain on account of the royal treasury from Lima. They do not include public remissions from other treasury offices in the New

---

52. Fisher, "Anglo-Portuguese Trade," pp. 223-224; Boxer, "Brazilian Gold and British Traders," p. 471.

53. This estimate is similar to Boxer's. See, Boxer, "Brazilian Gold and British Traders," p. 470.

54. Of interest are the figures of Michel Morineau, "D'Amsterdam a Séville: De quelle réalité l'histoire des prix est-elle le miroir?", *Annales—Économies—Sociétes—Civilisations*, volume 22 (January-February, 1968), number 1, p. 196, for precious metal imports into Spain between 1660 and 1700. These data reveal that for the four decades, Spain never received less than 100 million pesos in precious metal imports. This information meshes particularly well with

South American bullion production and export 1550-1750

Table 9. Registered bullion imports into Seville from the New World and estimated silver production for the Viceroyalty of Peru, 1571–1700 (million of pesos of eight)

Decade	Treasure imports after Hamilton	Estimated Peruvian silver output
1571–80	48.2	49.4
1581–90	88.0	117.0
1591–00	115.2	123.4
1601–10	92.3	116.6
1611–20	90.4	111.7
1621–30	86.0	96.9
1631–40	55.3	93.4
1641–50	42.2	82.6
1651–60	17.6	63.2
1661–70		48.4
1671–80	data not available	45.5
1681–90		49.6
1691–00		44.4

Sources: Peruvian production from Table V; Seville imports from Hamilton, *American Treasure*, p. 34. Hamilton's figures converted from pesos of 13-1/4 reales to pesos of 8 reales.

World, nor do they include the much larger amounts of silver and gold sent on private account. Further, the monies remitted by the Lima treasury represented only that portion of Crown income remaining after the operating costs of the Viceroyalty had been met. Our expectation, therefore, is that public remissions from Lima would comprise but a fraction of total registered bullion imports into Spain. For the two decades of the 1640s and 1650s, Hamilton shows *total* treasure imports of 42.2 and 17.6 million pesos. Corresponding figures of Lima's public remittances by Báncora Cañero and Rodriguez Vicente are 20.3 and 13.9 million, respectively, or 48.1% and 79.0% of Hamilton's totals.[55] The conclusion deduced from this comparison is the same conclusion evidenced in Table IX: the import data series of Hamilton must be viewed with great reservations.

The divergence of Hamilton's series for imports into Spain and ours for silver production in Peru can be traced to (1) alternate trade routes employed

---

our estimated silver production for the Viceroyalty of Peru for it shows that approximately a third to a half of all bullion imports during the last half of the seventeenth century originated with Peruvian silver. It also helps explain the discrepancy between our estimates and Hamilton's figures. Unfortunately, Morineau's data are undocumented although he extracted them from the same archive as Hamilton; we must therefore accept them with caution until they are fully substantiated.

55. Báncora Cañero, "Las remesas de metales preciosos," p. 86; Rodriguez Vicente, "Los caudales remitidos desde el Peru," p. 19.

by merchants exporting precious metals and (2) widespread corruption and contraband. Quantities of silver left the New World through the ports of Buenos Aires and Sacramento and through the Manila galleons. At the peak of these activities, perhaps as much as 6 million pesos per year (159,000 kg), or half the silver output of Peru, was diverted to these channels from the Seville trade. Moreover, untold amounts of unregistered bullion routinely accompanied the annual fleets from the New World to Spain, and additional thousands of kilograms went to independent merchants operating under slave asientos. In large part, Portuguese were the beneficiaries of this substantial "unrecorded" commerce; consequently, they were responsible for dispersing much American bullion around the world.

When Peruvian output began to fade in the second half of the 17th century, Portugal's fortune as a commercial empire also began to fade. Portuguese access to bullion once again increased with the exploitation of Brazilian gold fields after 1700. However, a severe trading deficit with England forced Portugal to send the bulk of its gold to England, thus stifling Portugal's chances to repeat its previous commercial successes. Spain suffered much the same fate in the 17th century as its bullion—registered and contraband—was continually drained away to other European countries.

We can draw several final conclusions from the estimated silver and gold output presented in the previous pages. First, the New World sent more bullion and specie to Europe than commonly believed by those who rely on Hamilton's figures. In terms of the sheer value of production, the Brazilian gold boom of 1700-1760 equalled if not surpassed for a few decades the brilliant output of Potosí and Peru during 1580-1640. Second, the curve of estimated Peruvian production indicates that there occurred no drop as precipitous as Hamilton's figures suggest. Indeed, Hamilton reveals an 85% decline in imported treasure between 1600 and 1660, while the estimated series shows only a 49% diminution in output. Consequently, monetary flows from the New World did not diminish to the disastrous extent implied by registered bullion imports into Spain. Third, the curve of silver output in the Viceroyalty of Peru established that from a peak in the 1590s, production only slowly drifted downwards for the first fifty years of the 17th century. A sharper slump occurred in the 1650s and 1660s, with a leveling trend for the remainder of the century. Whereas Peruvian mining production in the second half of the 17th century was only 50% that of the first half of the century, the overall decline cannot be called catastrophic. Paced adjustments in the Peruvian economy resulting from the diminished capacity to import, may have sufficiently stimulated local industrial production to avoid the depression so often attributed to the 17th century New World.[56]

56. For examples of the depression thesis as applied to Peru or to the New World in general, see, Parry, *The Spanish Seaborne Empire*, p. 227; Charles Gibson, *Spain in America* (New York, 1966), p. 160; J. H. Elliot, *Imperial Spain, 1469-1716* (New York, 1963), p. 288.

APPENDIX I. **Spanish American weight and coinage equivalents, sixteenth and seventeenth centuries**

Weight
    Spanish pound = 2 marcs = 460 grams
      (*libra*)   (*marcos*)
    1 marc = 8 ounces = 230 grams
              (*onzas*)
    1 ounce = 8 eighths = 28.75 grams
              (*ochavas*)

Silver coinage (1565–1728)
    1 marc = 8.4 pesos = 67 reales
    1 peso = 8 reales = 28.75 grams silver @ 0.931 fineness

Silver coinage (1728–1771)
    1 marc = 8.5 pesos = 68 reales
    1 peso = 8 reales = 28.75 grams silver @ 0.917 fineness

Note. This appendix contains information which directly relates to the peso of eight units of currency. There existed a variety of other monetary units and subdivisions which, taken as a whole, comprised a rather complex metrology. For our purposes, however, knowledge of the basics of the peso is sufficient to understand the data presented in this paper.

Sources. Humbero F. Burzio, *La ceca de la Villa Imperial de Potosí y la moneda colonial* (Buenos Aires, 1945), pp. 39–42; Manuel Moreyra Páz-Soldan, *Apuntes sobre la historia de la moneda colonial en el Perú* (Lima, 1938), *passim;* Ricardo Levene, *La moneda colonial del Plata* (Buenos Aires, 1916), *passim.*

APPENDIX II. Silver production at Potosí, Peru, 1556–1760 (pesos of eight reales)

Quinquennium	Total production	Average annual production	Average annual production in Kilograms
1556–1560	10,328,795	2,065,759	58,686
1561–1565	10,992,525	2,198,505	62,458
1566–1570	10,034,380	2,006,876	57,014
1571–1575	7,224,530	1,444,906	41,048
1576–1580	21,832,775	4,366,555	124,050
1581–1585	33,016,000	6,603,200	187,591
1586–1590	35,631,715	7,126,343	202,453
1591–1595	38,457,050	7,691,410	218,506
1596–1600	33,833,290	6,766,658	192,235
1601–1605	36,671,100	7,334,220	208,359
1606–1610	31,612,695	6,322,539	179,618
1611–1615	32,269,035	6,453,807	183,347
1616–1620	27,845,695	5,569,139	158,214
1621–1625	26,939,405	5,387,881	153,065
1626–1630	26,049,450	5,209,890	148,008
1631–1635	24,831,775	4,966,355	141,090
1636–1640	29,519,725	5,903,945	167,726
1641–1645	22,752,060	4,550,412	129,273
1646–1650	24,207,115	4,841,423	137,540
1651–1655	20,517,305	4,103,461	116,576
1656–1660	20,338,565	4,067,713	115,560
1661–1665	15,497,760	3,099,552	88,055
1666–1670	16,272,595	3,254,519	92,458
1671–1675	13,006,820	2,601,364	73,902
1676–1680	14,849,780	2,969,956	84,338
1681–1685	17,255,350	3,451,070	98,042
1686–1690	15,992,585	3,198,517	90,867
1691–1695	13,468,405	2,693,681	76,525
1696–1700	11,236,010	2,247,202	63,841
1701–1705	8,620,510	1,724,102	48,980
1706–1710	8,681,440	1,736,288	49,326
1711–1715	6,121,515	1,224,303	34,781
1716–1720	7,165,980	1,433,196	40,716
1721–1725	5,704,135	1,140,827	32,409
1726–1730	7,726,090	1,545,218	43,898
1731–1735	7,244,440	1,451,110	41,224
1736–1740	8,566,766	1,713,353	48,675
1741–1745	8,267,747	1,653,494	46,976
1746–1750	10,082,220	2,016,444	61,549
1751–1755	11,765,544	2,353,108	66,850
1756–1760	12,928,272	2,585,746	73,460

Appendix II. (continued)

Sources and methods. The figures from 1556 until 19 July 1736 are converted from the royal tax of 20 percent (the *quinto*). Figures from 20 July 1736 to 1760 are converted from royal tax returns of 10 percent of total production (the *diezmo*). Lamberto de Sierra, "Reales quintos pagados á S. M. desde 1 de enero de 1556 hasta 19 de julio de 1736," and, "Reales diezmos pagados á S. M. desde 20 de julio le 1736 hasta 31 de diciembre de 1783," in *Colección de documentos inéditos para la historia de España* (volume V; Madrid, 1844), pp. 173-184. These figures closely parallel the most recent data as given by P. J. Bakewell, "Registered silver production in the Potosí district, 1550-1735," *Jahrbuch für Geschichte von Staat, Wirtschaft und Gesellschaft Lateinamerikas*, volume 12 (1975), Table I.

APPENDIX III. Estimated annual silver production, Charcas, 1611-1650

Decade	Average annual mercury consumption in quintals (qq)	Estimated annual production, pesos (qqX1020X1.20)
1611-20	6,300	7,711,200
1621-30	5,937	7,266,888
1631-40	8,649	10,586,376
1641-50	6,340	7,760,160

Sources and methods. Total mercury consumption equals Huancavelica production plus periodic imports from Spain. Pierre and Hughette Chaunu, *Seville et l'Atlantique, 1504-1650* (8 vols; Paris, 1955-1959), vol 8, part 2:2, pp. 1974-1975; Guillermo Lohmann Villena, *Las minas de Huancavelica en los siglos XVI y XVII* (Seville, 1949), pp. 452-455. Estimated production calculated in the following manner: mercury, the key ingredient in the amalgamation refining process, was consumed in a relatively constant proportion to silver produced. At Potosí, a quintal of mercury yielded on the average 120 to 130 marks of silver. See, Pedro Vicente Cañete y Dominguez, *Guía histórica, geográfica, física, política, civil y legal del gobierno e intendencia de la provincia de Potosí* (1789) (Potosí, 1952), p. 70. We have chosen a multiplier of 120 marks which yields a production of 1,020 pesos (at .931 fineness) per quintial. To this we add 20 percent to account for smelted silver. Despite the successful introduction of the amalgamation process at Potosí and elsewhere, Indians and small-scale refiners continued to reduce silver ores by smelting. See, José de Acosta, *The Natural and Moral History of the Indies* (Madrid, 1608), p. 211, and Alvaro Alonso Barba, *El arte de los metales* (London, 1923), pp. 197-198; Manuel de Amát y Junient, *Memoria de gobierno* (Seville, 1947), p. 261.

14

New World silver, Castile and the Philippines 1590-1800

JOHN J. TEPASKE

The Spanish discovery and conquest of the New World had a profound effect, not only on America and Spain but also on the rest of Europe and the world at large. Spanish exploitation of large deposits of gold and silver, in particular, played a large role in molding the economic patterns of the early modern period. Among other things American bullion influenced the creation of what one scholar has named "the modern-world system,"[1] the decline of Spain and the supposed seventeenth-century depression in the Spanish Indies, and the economic crises of Europe and the economic growth of certain areas of Asia during the same period. In fact, American silver was so ubiquitous that merchants from Boston to Havana, Seville to Antwerp, Murmansk to Alexandria, Constantinople to Coromandel, Macao to Canton, and Nagasaki to Manila all used the Spanish peso or piece of eight (*real*) as a standard medium of exchange; these same merchants even knew the relative fineness of the silver coins minted at Potosí, Lima, Mexico, and other sites in the Indies thousands of miles away. Yet despite the worldwide importance of American treasure, little is known concerning either the quantities produced or the flow of these precious metals once they left the Indies, crucial factors for a better understanding of economic developments in America, Europe, and the world at large during the early modern period.

Several decades ago Earl J. Hamilton's classic work on American treasure and the price revolution in Spain provided at least some basis for establishing the silver flow out of the Indies to Spain.[2] That book both demonstrated the effect of the importation of American bullion on Spanish and European prices—a subject still hotly debated—and provided estimates on the public and private specie remitted to Castile for the period 1503-1660. This essay intends to build on Hamilton's earlier investigations in order to provide new insights into the quantity and flow of American bullion to Castile and Asia. More specifically, it will (1) provide new data for the period 1581-1660 to

---

1. Immanuel Wallerstein, *The Modern World-System: Capitalist Agriculture and the Origins of the European World-Economy* (New York, 1974).
2. Earl J. Hamilton, *American Treasure and the Price Revolution in Spain, 1501-1650* (Cambridge, Mass., 1934).

juxtapose against Hamilton's figures for remissions of public silver or crown revenues from the Indies to Castile; (2) put forward new estimates on the shipment of royal silver from the Indies to Spain, 1661-1800; (3) measure over time the proportions of crown income being retained in the Indies; (4) compare the flow of treasure out of the two viceroyalties of Peru and New Spain (Mexico) and assess the importance of each for the metropolis; and finally (5) take still another look at the significance of the Philippines as the receptor for New World bullion and as the entrepôt or conduit for siphoning American bullion into Asia.

The flow of crown silver from the Indies to Castile, 1590-1800

As noted previously Earl J. Hamilton has delineated general patterns in the flow of American treasure to Spain for the period 1503 to 1660. Arranged by decade in Table 1, his figures demonstrate that during the first half of the sixteenth century treasure imports into Castile from the Indies rose gradually from 2 to 17 million pesos (51,000 kilograms to 435,000 kilograms); after 1550 remissions increased dramatically until they reached 115 million pesos (2,940,000 kilograms) by the last decade of the sixteenth century, the high point in the flow of American bullion to Castile during the colonial period. After 1600 silver remissions began to decline, slowly at first, and after 1630 more precipitously. In fact by the decade 1651-1660 the total amount of American silver registered in Seville was less than 18 million pesos (460,000 kilograms) or an average annual influx of less than 2 million pesos (51,000 kilograms), almost seven times less than the yearly average for the last decade of the sixteenth century.

Hamilton's estimates are based on the registry lists recorded in Seville by agents of the House of Trade *(Casa de Contratación)*, who supervised the unloading of the treasure fleet, but these figures can be complemented and compared to another source as well: the records *(cartas cuentas)* kept by royal accountants in the Indies. Found in all major centers of the Spanish empire in America, these accounts detailed the major sources of income and expenditures.[3] For shipments of crown revenue to Spain, the accounts for the treasuries *(cajas)* of Lima, Mexico, and Veracruz documented the flow of public bullion out of Peru and Mexico bound for the king's coffers in

---

3. Virtually every important area of the Spanish empire in America had a treasury district *(caja)*. By the end of the colonial period New Spain had twenty-three treasuries and Peru, including Upper Peru (present-day Bolivia), had twenty-five; the Río de la Plata ultimately had fourteen *cajas* and Chile five. These accounts are being compiled and put into machine-readable form by Professors Herbert S. Klein and John J. TePaske under a grant from The Tinker Foundation and the National Endowment for the Humanities. For a more detailed discussion of the accounts as a source, see John J. TePaske, "Quantification in Latin American Colonial History," in V. Lorwin and J. Price, eds., *Dimensions of the Past: Materials, Problems, and Opportunities for Quantitative Work in History* (New Haven, Conn., 1972), 431-501.

Castile.[4] Table 2 records the flow of crown silver to Spain for the period 1591 to 1800 as listed in these accounts. To 1660 data from these records generally confirm Hamilton's figures gleaned from the lists kept by the House of Trade, although with some discrepancies. They show, for example, that the last decade of the sixteenth century marked the period of the greatest outflow to Spain for bullion consigned to the crown, that after 1600 remissions of specie began declining, and that the steepest drop occurred during the middle of the seventeenth century. Hamilton's estimates, however, show that the precipitous decline began in the 1640's when registered crown revenue sent from the Indies to Castile fell by 5 million pesos (128,000 kilograms). The records of the royal accountants in Lima and Mexico reveal that the most severe drop came a decade later.[5]

Hamilton's data end in 1660, but the colonial accounts document the flow of crown silver to Spain after that date. These figures show a continued decline to the end of the seventeenth century, when, in the last decade—the nadir—only little more than 3 million pesos (77,000 kilograms) reached Castile from the New World. The first half of the eighteenth century saw a slight increase in silver flows to the metropolis, but never more than 10 million pesos (256,000 kilograms) for any decade. In the 1750's, however, despite an end to all remissions of crown revenue from Peru, the influx from Mexico increased substantially to produce more than 16 million pesos (413,000 kilograms) for the coffers of Ferdinand VI, and except for a sharp drop in the 1760's occasioned by the war with England and the temporary loss of Havana, remissions remained at that level until the 1790's, when the volume jumped to almost 40 million pesos (969,800 kilograms). In summary, therefore, data from the accounts demonstrate that remissions of crown revenues to Castile

---

4. The materials for this analysis were compiled from the Contaduría Sections and Audiencia Sections of the Archivo General de Indias in Sevilla. For Lima the accounts may be found in Contaduría, Legajos 1693-1772 and Lima 1137-1152 and various accounts in Sección de la Real Hacienda of the Archivo General de la Nación in Lima; for Mexico, Contaduría, Legajos 677-844 and Mexico, Legajos 2029, 2045-2070; for Veracruz, Contaduría, Legajos 878-891 and Mexico 2920, 2921, 2924-2940; and for Acapulco, Contaduría, Legajos 897-90 and Mexico 2031, 2110-2115.

5. The discrepancies are obvious but might be explained in a number of ways. Wherever the amounts taken from the accounts are less than those listed by Hamilton, one can surmise that Hamilton's figures are larger because they included remissions from Cartagena in New Granada and Amatique and Trujillo in Honduras. When Hamilton's figures are smaller, one can surmise that some of the monies originally intended for Seville and listed that way in the accounts were (1) diverted to imperial defense; (2) siphoned off by royal officials at transfer points such as Veracruz, Panama, Portobelo, or Havana; or (3) lost to enemy corsairs and pirates. Also differences might be explained by the delays in shipping the silver—the delays between the time the amounts were listed in the accounts in the Indies and the time they were registered in Seville. For the period 1591-1660 Hamilton's figures indicate that 125,266,541 pesos went from the Indies to Castile in public remissions of treasure. Data from the accounts of Lima, Veracruz, and Mexico indicate that 150,299,255 pesos were actually registered for remission to Castile, a difference of over 25 million pesos. This may be close to the amount actually lost in shipwrecks or to enemy pirates during that period.

from the Indies had the following pattern from 1591 to 1800: significant silver remissions 1591-1610 with a slow decline to 1660, a steeper drop after that to the end of the century, except for the decade 1671-1680. A modest rise in the flow of crown silver occurred during the first half of the eighteenth century, followed by a sharper increase after 1750. After 1780 the influx from Mexico jumped rapidly, doubling in the decade 1781-1790 and reaching almost 40 million pesos (969,800 kilograms) by the end of the century.

How can these patterns be explained, particularly the gradual and then severe drop in remissions in the seventeenth century? Hamilton has ascribed this decline to a number of factors—the increased cost of extracting precious metals in the Indies, depletion or exhaustion of the mines, oppressive taxes on silver shipments or their sequestration by the king once they reached Spain, decimation of the labor supply available to work the mines, an increase in the Far Eastern trade, and retention of larger proportions of treasure in the Indies.[6] On balance all of these factors played a role in the drop in the bullion flow to Spain, but for Peru two appear preeminent—the exhaustion of the mines and the retention of larger proportions of crown revenues in the Indies. Potosí, for example, was the richest mine in the Spanish empire at the end of the sixteenth century. In the 1570's its annual production was 10,000 marks (87,500 pesos or 2,200 kilograms), rising spectacularly to 90,000 marks (787,500 pesos or 20,100 kilograms) annually in the 1580's, and remaining at that level until 1610 when a secular trend downward set in.[7] In part, a shortage of mercury, oppressive tax policies, decline in the labor supply, and increasing costs of production may all have played a role in causing a downturn in mining production, but the depletion of the deposits at the Cerro de Potosí seems primarily responsible.

A second important factor in the decrease in remissions was the retention of greater proportions of crown revenue within the viceroyalty (Table 3). In the last decade of the sixteenth century, for example, 64% of all revenue flowing into the matrix treasury of Lima went to Castile. That sum dropped to 45% in the decade 1601-1610 and to about one-third for the next two decades. In the 1630's the program inaugurated by the Conde Duque de Olivares to garner more income for Castile from Spain's overseas kingdoms saw remissions rise to 46% of all crown revenues coming into the Lima exchequer. A slight decline occurred in the next two decades from 34% to 27%, but after 1660 remissions plummeted sharply. Although revenues into

---

6. Hamilton, *American Treasure*, pp. 36-37.

7. See Peter J. Bakewell, "Registered Silver Production in the Potosí District, 1550-1735," *Jahrbuch Für Geschichte von Staat, Wirtschaft, und Gesellschaft Lateinamerickas*, 12 (1975), 67-103, and David A. Brading and Harry E. Cross, "Colonial Silver Mining: Mexico and Peru," *Hispanic American Historical Review*, 52 (November, 1972), 569. Also a check on a list of annual tax revenues imposed on mining production in Potosí indicates the same trend: "'Manifesto' de la plata extraida del cerro de Potosí (1556-1800)," *Academia Nacional de Historia* (Buenos Aires, 1971), 25-40.

the Lima treasury remained fairly steady—maintained artificially by the resort to forced loans and donations, exactions on officeholders, and the sale of land titles[8]—remissions to Castile amounted to 12% of total income in the 1660's, 8% in the 1670's, and a mere 1% in the 1680's. From 1690 to 1750 the percentage sent to Castile shrunk to 10% or less except for the decade 1731-40 when 11% of the paltry sums flowing into the Lima treasury went back to Spain. But even more important is that the Lima treasury generated scant revenue after 1690—less than 2 million pesos annually until the 1760's—diminishing the volume of pesos shipped to the metropolis to miniscule amounts.

In Peru retention of increasing proportions of dropping crown revenues occurred at the same time the viceroy and his aides in Lima had to meet fixed or increased expenditures for defense, administration, education, and charity. This meant a smaller surplus to ship to Seville. Within the viceroyalty more monies went for subsidies to the garrisons at Concepción, Chiloé, and Valdivia in Chile and for the soldiers and defenses at Panamá; for fitting out galleons and frigates at Callao and Guayaquil; for financing the war against the Araucanians in Chile; for strengthening fortifications at harbors like Callao along the Pacific coast; and for paying the salaries of militia soldiers needed to meet the increasing number of enemy pirates and rebellious Indians threatening the peace of the community. As the Spanish population of Lima and Peru grew, the viceroy also had to spend more for churches, missions, hospitals, poorhouses, education, and the University of San Marcos. All this occurred in the face of decreasing royal revenues caused primarily by the decline in mining production. The drop in remissions may also have been rooted in still another factor: the conscious or unconscious decision of authorities in Spain to use crown revenues in the Indies for defense of their overseas kingdoms. Losses in the Caribbean in the first half of the seventeenth century to the Dutch, French, and English and the increasing threat of foreign attack may have motivated the move toward increased expenditures for imperial defense.

In Mexico the general pattern was the same as in Peru (Table 4)—retention of increasingly larger proportions of crown revenue within the viceroyalty. For the fifty years 1590-1640 roughly half of all crown income produced in Mexico was remitted abroad, either to Castile or to the Philippines. For the next forty years, approximately one-third of this revenue flowed out of the viceroyalty to Castile or the Far East, diminishing in the last two decades of the century to less than one-quarter of all crown revenues generated in Mexico. Unlike Peru, however, the reduced flow of Mexican bullion into Castilian coffers could not be attributed to a decline in mining production. Mining output did not drop, and for the seventeenth century the

---

8. Kenneth J. Andrien, "The Sale of Fiscal Offices and the Decline of Royal Authority in the Viceroyalty of Peru, 1633-1700," *Hispanic American Historical Review*, 62 (February, 1982), 49-71.

secular trend for most mining areas of Mexico was generally upward, moderately so to be sure, but upward nevertheless.[9]

If mining output increased slightly and produced revenues for the crown, why did lesser proportions of crown revenue reach Castile from Mexico? Two factors seems responsible. First, like Peru, a greater share of crown income remained within the viceroyalty; and second, there was a drain of Mexican tax revenues to the Philippines. (The latter point will be discussed in more detail in the next section of this paper.) In New Spain as in Peru increasing amounts of tax revenue went for support of imperial defense—outfitting and supplying the Caribbean fleet (Armada de Barlovento); supporting presidios and garrisons in Campeche, Cuba, Florida, Puerto Rico, Santo Domingo, and other Spanish outposts on the Gulf of Mexico or Caribbean; and maintaining presidios on the troublesome northern frontiers of the viceroyalty to prevent intrusions of the Chichimecas. Ecclesiastical and administrative salaries, missions, charity, pensions, church building, drainage of Lake Texcoco, flood control, and public works also took their share of crown revenues. Unlike Peru, however, public revenues did not decrease as expenditures increased. Greater proportions of crown tax revenue simply were set aside for imperial defense and other purposes within New Spain.

The English historian John Lynch has argued that the retention of a larger sum in the Indies stimulated local industry, and a good case might be made for his contention.[10] If crown revenues which ordinarily might have gone to Spain remained in the colonies for shipbuilding, church construction, clerical maintenance, and public works, this surely must have acted as a stimulus or support of the colonial economy. At the same time the increased need for such things as cannon, powder, naval stores, wine, biscuits, and firearms surely helped to develop new colonial economic enterprises and gave the overseas kingdoms more economic independence from the mother country.

How important was the flow of silver from the Indies for the support of the Spanish crown? In view of the scanty evidence available on the Spanish financial structure in the sixteenth and seventeenth centuries, this is difficult to determine; but clearly a lesser amount of bullion flowing into Castile from the Indies meant less money available for warmaking on the continent, the

---

9. John J. Tepaske and Herbert S. Klein, "The Seventeenth-Century Crisis in New Spain: Myth or Reality?" *Past & Present*, 90 (February, 1981), 116-135. For this article we ran regression analyses on the mining revenues for all eight mining regions of Mexico during the seventeenth century. For Durango, Guadalajara, Guanajuato, Pachuca, and Zacatecas silver production rose moderately. For San Luis Potosí production dropped slightly, probably because of a mercury shortage; at the mines of Sombrerete new veins of silver dried up quickly; and in Mexico the creation of new treasury districts previously reporting mining tax income at the matrix treasury accounted for the decline there. Overall, however, silver production in New Spain rose slightly during the seventeenth century.

10. John Lynch, *Spain under the Habsburgs*, II, *Spain and America, 1598-1700* (New York, 1969), p. 195.

support of the king and his court, administrative expenses in Castile, or repayment of Spanish loans to European bankers. And in the late sixteenth century a goodly proportion of royal income came from the Indies. In 1598, for example the total income flowing into the royal exchequer amounted to approximately 13,500,000 pesos (345,000 kilograms). Of this total, remissions from the Indies constituted a bit over 3 million pesos (77,000 kilograms) or 22% of all royal income. Of the various broad categories within the royal accounting system (*servicios, maestrazgos y hierbas, arbitrios, bulas de cruzada*, etc.) only the revenue from peninsular sales taxes (*alcabalas y tercias*) surpassed that of the income from the Indies, constituting 29% of total crown revenue for 1598.[11]

But 1598 was in the halcyon days when remissions from the Indies to Castile were at their peak, and as the seventeenth century wore on, the significance of bullion shipped from the Indies for support of the crown diminished greatly. One scholar familiar with imperial finance during the first half of the seventeenth century estimates roughly that only one-sixth or less of all royal revenues came from the Indies, but this proportion was probably closer to one-tenth or less. During the first twenty years of the reign of Philip IV (1621-1640), for example, income from the Indies amounted to only 9% of all royal revenues collected during that period.[12] And in view of the shrinking amount of silver flowing to Castile after 1640 that percentage may have dropped still more, except perhaps for the 1670's when Mexican tax revenues produced almost 10 million pesos (256,000 kilograms) and Peru another 2 million (51,000 kilograms) for the exchequer of Charles II.

Viewed in a different light, however, diminishing revenue from the Indies for support of royal enterprises on the continent may not have been as serious for the crown as is generally supposed because royal revenues generated in the New World supported and maintained Spain's overseas dominions. The early part of the seventeenth century saw increasing intrusions by the English, Dutch, and French in the Caribbean and the loss of Spanish islands to these interlopers. Foreign corsairs, pirates, and privateers had also become more numerous, not only on the Spanish Main but also in the Pacific along the coasts of Peru, Central America, and Mexico. In 1670 England founded Carolina which posed a real threat to Spanish control of Saint Augustine and the Bahama Channel on the route of the treasure galleons going home to Spain. To meet these threats, therefore, the crown began to pour ever larger sums into imperial defense. This meant larger subsidies for key outposts such as Cuba, Santo Domingo, Puerto Rico, Florida, Panamá, Puertobelo, Cartagena, Callao, Veracruz, Concepción, Valdivia, and Buenos Aires and additional funds for shipbuilding to build more effective Spanish naval squadrons in the

---

11. Modesto Ulloa, *La hacienda real de Castilla en el reinado de Felipe II* (Rome, n.d.), p. 534.
12. Antonio Domínguez Ortiz, *Política y hacienda de Felipe IV* (Madrid, 1960), pp. 333-342.

Gulf of Mexico, Caribbean, Atlantic, and Pacific. Unlike English America, however, these costs were not borne by the crown but by the colonials and indigenous peoples from the taxes and tribute contributed to the royal treasury. Unlike English America, Spanish colonists and their Indian vassals in the Indies paid the costs of imperial defense, and the Spanish empire in the New World never became a liability for the mother country in the sense that it drained revenues away from the metropolis. Thus, it might be argued that despite the drop in remissions to Spain and ever larger proportions of royal tax revenues being retained in the Indies, the crown was pursuing a wise course that assured continued Spanish dominance and expansion of its far-flung American empire in the eighteenth century.

The importance of Mexico and Peru as the major providers of silver for the metropolis also shifted during the colonial period. Peru was clearly more significant as a source of revenue for the crown during the first half of the colonial period to 1660. Between 1580 and 1660 Peru sent over twice as much bullion to Castile as did New Spain. Abruptly, however, after 1660 Peruvian shipments of silver to Spain declined drastically. To 1660 Peru had provided 60% or more of the revenues remitted from the Indies to the metropolis. In the decade 1661-1670 that share dropped below 50% for the first time and plummeted to less than 20% in the following ten-year period. After 1680 both the proportion of total income remitted and the amount of bullion actually sent from Peru was miniscule. From 1680 to 1750, when remissions stopped entirely, the average was only 8% of the total income being produced in the viceroyalty, and bullion shipped out of Peru to Spain never averaged more than 170,000 pesos (4,346 kilograms) a year for any decade.

In the 1660's, therefore, Mexico replaced Peru as the prime source of imperial revenue in the Indies and never lost that position for the remainder of the colonial epoch. Still, Mexico never produced the large sums of bullion that Peru had furnished Philip II and Philip III at the end of the sixteenth and beginning of the seventeenth century. In fact to 1750 Peru still outstripped Mexico in the *total* amount of gold and silver provided to royal authorities in Spain, 113 million pesos (2,900,000 kilograms) or 52% from Peru to 103 million pesos (2,600,000 kilograms) or 48% from New Spain. In the last half of the eighteenth century, however, when Peru stopped remissions altogether, Mexico forged ahead, largely because of the 40 million pesos (969,800 kilograms) sent back to Spain in the 1790's. Overall, for the period 1591 to 1800, Mexico accounted for approximately 65% of all bullion sent to Castile from the New World consigned to the king; Peru remitted 35%, most of that during the late sixteenth and early seventeenth centuries.

The foregoing discussion has focused solely on the quantity, flow, and significance of *public* treasure—specie and bullion from tax revenues collected for the king and his officials in Spain or the Indies. The flow of privately owned silver in the hands of private individuals or corporate groups has been ignored, but for good reason. Except for Hamilton's figures to 1660 no *comparable* data are available to document the remissions of private consignments of silver after 1660. Although the Spanish historian, Antonio García-Baquero González, establishes some patterns for the period 1717 to 1778, his data do not come from the same sources as those of Earl J. Hamilton and cannot be used for analytical purposes.[13] Thus, one can only speculate on whether the flow of private silver to the metropolis followed the same pattern as that of crown silver. Probably, it did. For the earlier period to 1660 Hamilton demonstrates that roughly three out of four pesos shipped to Spain were privately owned and that this ratio persisted throughout the epoch 1503-1660. He also shows that remissions of private consignments of silver dropped when remissions of public treasure declined. Probably these patterns did not change appreciably in the late seventeenth and eighteenth centuries.

One additional problem—among many—concerns the flow of these private consignments of treasure once they reached Seville or Cádiz. Where did the silver go after it was turned over to agents in these Spanish entrepôts? Precise measurement of this flow is difficult to document since the silver was shipped—literally—to the ends of the earth. The bullion flowed out of Spain to England, France, and the Low Countries for purchase of manufactured goods unavailable in Castile. From English, French, Flemish, or Dutch ports Spanish pesos were transshipped through the Baltic or Murmansk into Scandinavia and Russia and traded for furs. In Russia, whatever silver was not hoarded or fashioned into religious icons went southeastward along the Volga into the Caspian Sea to Persia, where it was sent overland or by sea into Asia. Spanish-American bullion also flowed out of Spain through the Mediterranean and eastward by land and water routes to the Levant. India procured its American silver by means of the traffic from Suez through the Red Sea and into the Indian Ocean, overland from the eastern end of the Mediterranean through Turkey and Persia to the Black Sea, and finally into the Indian Ocean, or directly from Europe on ships rounding the Cape of Good Hope following the route discovered by Vasco da Gama. The latter way was also used by Portuguese, Dutch, and English ships to carry Spanish American treasure directly to Asian ports to exchange for Asian goods. Lastly—and long ignored—American silver found its way to the Orient by way of the Pacific route from Acapulco to Manila.

The Philippine subsidy, Philippine trade, and the flow of American silver to Asia, 1591-1800

Earl J. Hamilton and others have argued that the Philippines were an important outlet for both the public and private specie flowing out of the Indies. On the one hand crown revenues went from Mexico to Manila for support of the Spanish military, administrative, and religious establishments

---

13. Antonio García-Baquero González, *Cádiz y el Atlántico (1717-1778)*, 2 vols. (Sevilla, 1976). See particularly Vol. II, pp. 250-252.

in the islands and for maintenance of the two Manila galleons carrying on the legal trade between the Philippines and Mexico. On the other hand American silver consigned to merchants, factors, or fortune hunters flowed to the Philippines both legally and illegally to purchase a wide variety of oriental goods available there to be resold later at high profits in the Indies. This trade was apparently crucial in reducing the Atlantic trade which fell off so drastically after 1620 and sharply reduced the profits of the Seville merchants who controlled the Atlantic commerce with the New World.[14]

Like remissions to Castile, it is far easier to document the flow of *public* monies to the Philippines. In the public sector remissions of Mexican silver to Manila began soon after the discovery of the islands in 1565 and continued throughout the colonial period where it supported the administrators, soldiers, and priests laboring in the islands and generally maintained the Spanish colonial enterprise in the Philippines. How significant was this flow of treasure to the islands (Table 4)? On balance, compared to the flow of bullion to Castile, remissions to the islands were relatively modest. In fact for the 220-year period running from 1581 to 1800 remissions of silver for the support of the Philippines constituted only 17% of all crown revenues remitted outside the Indies from Mexico; the other 83% went to Castile. In all, the islands drained off 44 million pesos (1,123,000 kilograms) from Mexico during this epoch. Although this amounted to an average of only a bit over 200,000 pesos (5,100 kilograms) annually, during crucial periods the Philippines took significant sums which might have gone to Castile. During the first half of the seventeenth century, for example, a critical epoch which saw the sharp decline of Spain, the Philippines siphoned off over a quarter of all bullion shipped out of New Spain, and in the 1640's, perhaps in response to the Dutch threat in the islands, over 40% of Mexican surplus income.

That 44 million pesos (1,123,000 kilograms) went to the Philippines seems an enormous sum, yet from a royal perspective this was a small expense for maintaining the Spanish hold on the Philippines. Representing an average annual cost of 200,000 pesos (5,100 kilograms), this Mexican silver paid for administrative, military, naval, and clerical stipends; fortifications; fitting out galleons and frigates; missionary endeavors; church construction; arms and munitions; expeditions to insure Spanish control of the islands; and a host of other things. And for the crown, Mexico, not Spain, bore the burden of these expenses. Moreover, 200,000 pesos was really a small sum, hardly crucial in surmounting the myriad of financial problems afflicting Spain, even in the seventeenth century.

Much more important, and much more difficult to document than this flow of crown silver to the Philippines was the drain of bullion in the private sector from the trade in Asian goods in Manila. The development of this Asia-Philippines-Indies trade has been well described by William Lytle Schurz

---

14. Pierre Chaunu, *Séville et l'Atlantique (1504-1650) La Conjuncture (1593-1650)*, Tome VIIbis (Paris, 1959).

and Pierre Chaunu,[15] but briefly, commercial activity in the islands began almost at the moment the Spanish occupied the Philippines in 1565. Ships from China primarily but also from Indochina, Indonesia, India, the Moluccas and other areas of the Orient called at Manila to trade their silks, tapestries, feathers, brocade, pepper, fans, ivory, jade, and other fine goods and spices for Spanish silver. Spanish captains also engaged in the trade, and where they were able, they bought goods in Chinese and other ports to sell in Manila or the Indies. Manila, however, was the principal entrepôt for the exchange of New World specie and Asian commodities. Initially, this port was open to all Spanish captains from the Indies who had the ships, money, and courage to make the long voyage to the Philippines; and vessels from Mexico, Peru, Panama, and Guatemala all called at the islands with their cargoes of silver to exchange for the *"ropas de China"* available in the Manila marketplace. Loaded with Asian goods, these ships returned to the Indies to sell their wares for high profits, sometimes enabling merchants to quadruple or quintuple their original investments.

For a few years royal policy toward this early trade was one of salutary neglect. As long as Spanish ships in the Philippines paid duties on imports and exports in Manila and in their port of entry in the Indies, there were no restrictions placed on those who could engage in the trade. In fact, it appeared as if this commerce was beneficial for all—the crown, merchants, and the Spanish residents of both the Philippines and the Indies. The influx of large numbers of Asian and Spanish ships into the Philippines and the collection of import-export taxes *(almojarifazgos)* helped pay the costs of supporting and maintaining the islands, solidifying Spanish occupation, providing the funds for further expansion, and generally insuring the security of the new colony. Traders could also make high profits from the sale of Asian goods in the Indies, and in the New World the colonials could enjoy the silks and spices brought in by these merchants, higher quality luxury goods at cheaper prices than those imported from Europe on the *flota* and the *galeones*.

Philip II, however, did not see these as advantages. Responding to the complaints of the merchants in Seville, who saw their profits from the sale of goods in the Indies shrinking, the king quickly took steps to regulate and restrict the rapidly growing commerce between the Philippines and the Indies. In 1582, for example, he issued a royal order prohibiting the direct trade between the Philippines and Peru and also the import or sale of Philippine goods in Peru, even those transshipped from other ports in the Indies.[16] Nine years later in 1591 he ordered an end to all trade between the Philippines and Guatemala and Tierra Firme (Panamá) and reiterated his prohibition concerning Peru, evidence perhaps that his earlier *cédula* had

---

15. William Lytle Schurz, *The Manila Galleon* (New York, 1939) and Pierre Chaunu, *Les Philippines et le Pacifique des Ibériques (XVIe, XVIIe, XVIIIe siècles). Introduction méthodologique et indices d'activité, Ports-Routes-Trafics, XI* (Paris, 1960).

16. Woodrow Borah, *Early Trade and Navigation between Mexico and Peru*, Ibero-Americana 38 (Berkeley and Los Angeles, 1954), pp. 118-119.

been ignored.[17] Two years later Philip II restricted the trade even more: commerce between the Indies and the Philippines was to be carried on only by two galleons of 300 tons each, solely to keep the islanders supplied. These Manila galleons, as they came to be called, could carry annually no more than 250,000 pesos worth of oriental goods from the Philippines to Acapulco and no more than 500,000 pesos (12,800 kilograms) of silver to Manila.[18] These strict limitations remained in effect for over a century until 1702 when the galleons were allowed to carry 300,000 pesos worth of goods from Manila and 600,000 pesos (15,300 kilograms) in silver from Acapulco. In 1734, probably in response to rising prices, the ratio rose to 500,000/1,000,000 pesos (12,400/24,800 kilograms).[19]

Apparently these prohibitions were ignored, and reports of *Peruleros* (ships from Peru) calling at the Philippines and returning to Callao with their cargoes of Asian goods were common.[20] Mexican silver also flowed out to the islands in large sums, far exceeding the 500,000-peso limitation. In fact at the opening of the seventeenth century the drain of pesos from Mexico to the Orient through the Philippines was estimated at 5 million pesos (128,000 kilograms) annually, with a reported 12 million pesos (307,000 kilograms) being smuggled out in 1597.[21] Moreover until 1620 the number of Asian vessels calling at the Philippines also kept increasing as did the tax revenues on the goods they carried, all evidence that the Philippine-Indies commerce was flourishing despite royal restrictions to the contrary.

After 1620, however, *official* records show a sharp drop in the Pacific trade. Fewer Asian vessels evidently called at Manila, and import-export taxes collected at both Manila and Acapulco dropped off. Several factors seem responsible. First, royal officials in the islands began to seize the *Peruleros* and other illegal ships engaged in the Asian trade in order to profit from the seizures of silver or to keep the islands maintained when the subsidy did not arrive. Second, ships from other parts of the Indies and from Peru probably found it less costly and less dangerous to purchase Asian goods at or near Acapulco and eliminate the risks and costs of the illicit voyage across the Pacific. Third, the Dutch began to intrude on the trade in Asian commodities and diverted these goods away from Manila. Fourth, internal problems in China, particularly along the South China coast, reduced the trade from that area, which provided the bulk of the oriental silks and spices available in the Philippines. Lastly, and by far the most important factor, was the increase in

---

17. Real cédula, Madrid, 6 Febrero 1591. *Recopilación de leyes de los reynos de las Indias* (Madrid, 1681), Libro VIII, Título XXXXV, Ley V.

18. Real cédula, Madrid, 11 Enero 1593. *Recopilación de leyes de los reynos de las Indias* (Madrid, 1681), Libro VIIII, Título XXXXV, Ley VI.

19. Schurz, *Manila Galleon*, p. 155.

20. Borah, *Early Trade*, pp. 116-127.

21. C. R. Boxer, "*Plata es Sangre:* Sidelights on the Drain of Spanish-American Silver in the Far East, 1550-1700," *Philippine Studies,* 18 (July, 1970), 464.

smuggling. Although this cannot be documented except by reports of occasional seizures of contraband goods by some ardent royal official in Manila or Acapulco, perusal of the Acapulco accounts after 1620—sloppily kept and inconsistent in their entries, particularly compared with those of Veracruz for the same period—and the continuous flood of royal orders repeating the restrictions and prohibitions on the Philippine-Indies trade testify to the widespread growth of smuggling, which continued unabated throughout the colonial period. The contraband trade was simply too advantageous for the people of the Indies to ignore, and Manila and Acapulco were too far removed from Spain and the centers of empire in the Indies for the crown to exercise effective control. That the Philippines siphoned off large sums of silver from the New World cannot be denied, but measuring that flow is virtually impossible.[22]

The restrictions imposed on the Philippine trade also graphically illustrate Spanish mercantile policy toward her overseas kingdoms. After a brief period of neglect Philip II sharply limited the Philippine-Indies trade and might have prohibited the commerce altogether except for the need to maintain the Spanish, administrative, military, and religious structure in the Philippines. Two fundamental bases of Spanish colony policy emerge from Philip's limitations: (1) the absolute necessity of maintaining royal control over all commerce and (2) protection of the monopoly of the Seville merchants who dominated the Atlantic trade in European goods. No amount of exhortation to the contrary could convince Philip II or his successors that Spain would be well served by pursuing another course, at least until the eighteenth century. In 1637, for example, a procurator representing Philippine and Mexican interests at the court of Philip IV called for the lifting of restrictions on New World trade with the islands. He argued that by increasing American commerce with the Philippines additional revenues from import-export taxes could accrue to the crown and lessen the drain of silver out of Mexico for maintenance of royal officials and clerics in the islands. Moreover, although the Seville merchants would suffer from the competition coming from an increased flow of oriental goods into the Indies, the crown was better served because profits from the Atlantic trade now went to the hated enemies of Spain—the Dutch, Genoese, French, or English. Was not the crown far better off having that silver flow to China? China, after all, was a friendly

---

22. Table 5 may give some clue to the remissions of private silver to the Philippines, but in my view the data are unreliable and do not reveal what actually went to the islands. Legally registered private silver sent to the islands was taxed at the rate of 5/6 of 2% or 1.67%. Table 5 shows how much was remitted to the Philippines according to the figures laid down in the Acapulco accounts, the total remitted from the Indies to Castile taken from Earl J. Hamilton, the grand total of all privately owned silver sent out of the Indies to both the Philippines and Castile, and the percentage of that silver remitted to the Philippines for the period 1591-1660. That 13% of all private silver shipped out of the Indies between 1651 and 1660 went to Manila may be significant, but probably a great deal more slipped out of the Philippines than these figures reveal.

power and would not use American silver to bring down Spain.[23] The crown, however, ignored these representations.

Profit, either for the crown or private individuals not a part of the Seville monopoly, was not a motive behind royal trade policy. The rationale for the Columbus voyage in 1492 had been to get to the riches of the East by sailing west. In 1565, three-quarters of a century later, Spain finally got its opportunity to procure the spices and fine cloth of Asia through its new possessions in the Philippines. Like the Portuguese in Goa and Macao and the Dutch later in Batavia and Nagasaki, the Spaniards had the chance to open a lucrative trade in oriental commodities not only on the Acapulco-Manila axis but directly with Europe as the Portuguese had done. Philip II rejected that possibility in favor of a policy of severe limitation and control. The result: a continued flow of crown silver out of Mexico for support of the Philippines and a sharp increase in smuggling from which the Spanish government could obtain no benefit whatsoever. Interestingly, even in the private sectors in the Philippines, Spaniards rejected the opportunity for profits to be made in commerce. After 1571 Chinese immigrants flooded into Manila to control most mercantile activities, and like the Genoese and Flemish in Seville, dominated the trade and took the profits to be made from it.[24]

Conclusions

Conclusions on the flow of American silver to Europe and the Far East can be classified three ways: highly probable, tentative, and highly speculative. And of those propositions characterized as highly probable, most are based on analysis of the flow of bullion in the *public* sector: (1) Remission of crown revenues from the Indies to Castile reached a high point at the end of the sixteenth century and dropped off steadily in the seventeenth century. Never, even with the boom in the Mexican economy and the advent of Bourbon reforms, did remissions of crown revenues to Castile from the New World reach the apogée at the end of the sixteenth century. (2) In both Peru and Mexico larger and larger proportions of crown revenues remained in the Indies to finance imperial defense, costs of administration, welfare, charity, and religious endeavors. (3) To 1660 Peru was more important than Mexico in providing funds for the royal exchequer in Spain; after 1660 Mexico became more significant and retained that position until the end of the colonial period. Overall, Mexico provided 65% of all revenues remitted to Spain, Peru 35%. (4) The middle of the seventeenth century (c. 1660) marks a significant shift in royal policy toward the Indies. Consciously or unconsciously royal authorities began allocating far greater proportions of royal

---

23. *Memorial Informativa al Rey Nuestro Señor en Su Real y Supremo Consejo de' las Indies*, Dn. *Juan Grau y Monfalcon, Procurador Gen en esta Corte* (Madrid, 1637). British Museum, Additional Manuscript 13,992.

24. Schurz, *Manila Galleon*, p. 81. He estimates 30,000 Chinese inhabitants living in the segregated Parián district of Manila in 1650.

income being produced by its overseas dominions to their own defense. (5) The Philippines drained off some revenue being produced in Mexico (44 million pesos or 1,122,000 kilograms), but overall this was a cheap price to pay for maintenance of the Spanish hold on the Philippines and the support of the Manila galleons.

The more tentative conclusions are these: (1) During the reign of Philip II (1556-1598) remissions of treasure from the Indies to Castile constituted a significant source of income for the metropolis, as much as one-quarter of the revenues accruing to the crown from all sources. (2) In the seventeenth century beginning c. 1620 crown revenues generated in the Indies and remitted to Spain grew far less significant and represented miniscule proportions of royal income after 1660. (3) After 1660 as lesser amounts of silver flowed from the New World to Spain in the public sector, lesser amounts also flowed to Castile in the private sector. (4) Spain's need to defend her overseas dominions against foreign intrusions led to greater proportions of revenue being retained in the Indies. (5) The Philippines were the major conduit for the flow of American silver westward into the Orient; this westward flow was as significant as the flow of silver from the east, particularly in the late sixteenth and early seventeenth centuries.

More speculative are the conclusions that (1) the legal and illegal trade in oriental goods brought from the Philippines to the New World drained off significant amounts of American silver and was one cause of the steep drop in the Atlantic trade after 1620. (2) American silver was primarily responsible for the economic development and prosperity of certain areas of India and China during the seventeenth century. And lastly (3) smuggling of Asian goods was rampant in Mexico and all over the Spanish New World, and this smuggling was deleterious to royal interest in Spain and the New World throughout the colonial period.

EXPLANATION OF TABLES AND CONVERSION FACTORS

All tables have been put into both pesos of 272 maravedís and kilograms of silver. From 1535 to 1728 pesos de ocho, as they were called, had 25.561 grams of fine silver. From 1728 to 1772 the silver content dropped to 24.809 grams. In the period 1772-1786 the peso contained 24.433 grams and from 1786 to 1825, 24.245 grams of pure silver. For convenience in calculating kilograms of silver this conversion factor of .025561 has been used to 1730; .024809 from 1731 to 1770; .024433 from 1771 to 1790; and .024245 from 1791-1800.[1] The conversion of marks to pesos is taken from Peter Bakewell's article on Potosí silver production in which he notes that 2380 maravedís made 1 mark of fine silver; or, 1 mark of fine silver equalled 8.75 pesos of 272 maravedís.[2] In Table IV remissions of public silver to the Philippines includes the costs of fitting out the Manila galleons and paying the salaries of those serving on these vessels, an entry in the Acapulco accounts which read SALARIOS DE GENTE DEL MAR DE FILIPINAS. Also included in this sum are entries for GASTOS EXTRAORDINARIOS DE FILIPINAS, REGISTRADO A FILIPINAS, and similar listings for public monies remitted to the islands.

1. Humberto Burzio, "El 'peso de plata' hispanomericano," *Historia* (Buenos Aires), 3 (1958), 42.
2. Peter J. Bakewell, "Registered Silver Production in the Potosí District, 1550-1735," *Jahrbuch Für Geschichte von Staat, Wirtschaft, und Gessellschaft Lateinamerikas*, 12 (1975), 77-78.

Table 1. Treasure imports into Castile from the Indies, 1503–1660*

	Public			Private			Totals	
	Pesos of 272 Maravedis	Equivalent in kilograms of silver	% of Total	Pesos of 272 Maravedis	Equivalent in kilograms of silver	% of Total	Pesos of 272 Maravedis	Equivalent in kilograms of silver
1503–1510	514,636	13,155	26	1,449,620	37,054	74	1,964,256	50,209
1511–1520	948,721	24,250	26	2,672,347	68,308	74	3,621,068	92,558
1521–1530	508,270	12,992	26	1,431,691	36,595	74	1,939,961	49,587
1531–1540	2,950,201	75,410	32	6,294,790	160,901	68	9,244,991	236,311
1541–1550	3,888,601	99,397	22	13,420,916	343,052	78	17,309,517	442,449
1551–1560	8,597,920	219,771	29	20,957,158	535,686	71	29,555,078	755,457
1561–1570	9,271,714	236,994	22	32,665,259	834,957	78	41,936,973	1,071,951
1571–1580	16,458,532	420,697	34	31,781,373	812,364	66	48,239,905	1,233,061
1581–1590	25,798,410	659,433	29	62,227,652	1,590,601	71	88,026,062	2,250,034
1591–1600	34,738,540	887,952	30	80,429,809	2,055,866	70	115,168,349	2,943,818
1601–1610	24,931,089	637,264	27	67,398,553	1,722,774	73	92,329,640	2,360,038
1611–1620	19,126,038	448,881	21	71,271,339	1,821,767	79	90,397,377	2,310,648
1621–1630	15,733,273	402,158	18	70,237,962	1,795,353	82	85,971,235	2,197,511
1631–1640	15,592,931	398,571	28	39,706,145	1,014,929	72	55,299,074	1,413,500
1641–1650	10,437,237	266,786	25	31,806,791	813,013	75	42,244,028	1,079,799
1651–1660	4,707,433	120,327	27	12,920,005	330,248	73	17,627,438	450,575
Totals	194,203,546	4,964,038	26	564,671,410	13,973,468	74	740,874,946	18,937,506

*From Earl J. Hamilton, *American Treasure and the Price Revolution in Spain, 1501–1650* (Cambridge, Mass., 1934). p. 34.

Table 2. Public revenues remitted to Castile from Mexico and Peru, 1591–1800

	Remitted from Peru		Remitted from Mexico		Total remitted to Castile		Remissions of public treasure to Castile (E. J. Hamilton)		% from Peru	% from Mexico
	In Pesos of 272 Maravedis	Equivalent in Kilograms of Silver	In Pesos of 272 Maravedis	Equivalent in Kilograms of Silver	In Pesos of 272 Maravedis	Equivalent in Kilograms of Silver	In Pesos of 272 Maravedis	Equivalent in Kilograms of Silver		
1591–1600	19,957,476	510,133	9,333,073	238,563	29,290,549	748,696	34,738,540	887,952	68	32
1601–1610	17,249,406	440,912	10,711,341	273,793	27,960,747	714,705	24,931,089	637,264	62	38
1611–1620	11,711,677	299,362	6,104,678	156,042	17,816,355	455,404	19,126,038	488,881	66	34
1621–1630	11,553,339	295,315	6,606,659	168,873	18,159,998	464,188	15,733,273	402,158	64	36
1631–1640	17,484,705	446,927	8,732,471	223,211	26,217,176	670,138	15,592,931	398,571	67	33
1641–1650	14,956,483	382,302	2,981,421	76,208	17,937,904	458,511	10,437,237	266,786	83	17
1651–1660	8,595,357	219,706	4,317,139	110,350	12,912,496	330,056	4,707,433	120,327	67	33
1661–1670	3,568,493	91,214	3,991,220	102,020	7,559,713	193,234	Figures not available after 1660		47	53
1671–1680	2,089,103	53,400	9,967,125	254,770	12,056,228	308,170			17	83
1681–1690	307,387	7,857	4,770,990	121,951	5,078,377	129,808			6	94
1691–1700	842,091	21,525	2,580,425	65,958	3,422,516	87,483			25	75
1701–1710	1,658,007	42,381	5,240,459	133,951	6,898,466	176,332			24	76
1711–1720	77,411	1,978	7,186,736	183,700	7,264,147	185,678			1	99
1721–1730	1,035,681	26,473	6,098,799	155,891	7,134,480	182,364			15	85
1731–1740	1,427,272	35,409	8,510,710	211,142	9,937,982	246,551			14	86
1741–1750	545,000	13,522	6,436,368	159,680	6,981,368	173,201			8	92
1751–1760	–0–		16,123,480	400,007	16,124,480	400,007				100
1761–1770	–0–		7,482,801	185,641	7,482,801	185,641				100
1771–1780	–0–		15,787,230	385,729	15,787,230	385,729				100
1781–1790	–0–		29,593,184	723,050	29,593,184	723,050				100
1791–1800	–0–		39,845,843	966,062	39,845,843	966,062				100
Totals	113,058,888	2,888,415	212,402,152	5,296,592	325,461,040	8,185,008			35	65

Table 3. Public revenues from the matrix treasury of Lima remitted to Castile and retained in the Indies, 1591–1750*

	Total revenues into the Lima treasury		Remitted to Castile			
	In Pesos of 272 Maravedís	Equivalent in kilograms of silver	In Pesos of 272 Maravedís	Equivalent in kilograms of silver	% remitted to Castile	% retained in the Indies
1591–1600	33,683,302	860,979	19,957,476	510,133	59	41
1601–1610	38,006,115	971,474	17,249,406	440,912	45	55
1611–1620	34,664,082	886,049	11,711,677	299,362	34	66
1621–1630	33,319,795	851,687	11,533,339	295,315	35	65
1631–1640	37,686,150	963,296	17,484,705	446,927	46	54
1641–1650	44,162,913	1,128,848	14,956,483	382,302	34	66
1651–1660	32,077,933	819,944	8,595,357	219,706	27	73
1661–1670	30,655,309	783,580	3,568,493	91,214	12	88
1671–1680	27,249,881	696,534	2,089,103	53,400	8	92
1681–1690	27,166,234	694,396	307,387	7,857	1	99
1691–1700	19,606,978	501,174	842,091	21,525	4	96
1701–1710	17,338,287	443,184	1,658,007	42,381	10	90
1711–1720	9,564,178	244,470	77,411	1,978	1	99
1721–1730	14,063,286	359,472	1,035,681	26,473	7	93
1731–1740	12,454,339	308,980	1,427,272	35,409	11	89
1741–1750	13,394,174	332,296	545,000	13,522	4	96
Totals	425,092,956	10,846,363	113,058,888	2,888,416	27	73

*After 1750 there were no remissions from Peru to Castile listed in the accounts.

443

Table 4. Remissions of public revenues from Mexico to Castile and the Philippines, 1581-1800

	Remitted to Castile		Remitted to Philippines		Total remitted outside Mexico		% to Castile	% to Philippines	% of total revenue remitted
	In Pesos of 272 Maravedis	Equivalent in kilograms of silver	In Pesos of 272 Maravedis	Equivalent in kilograms of silver	In Pesos of 272 Maravedis	Equivalent in kilograms of silver			
1581–1590	9,040,136	231,075	1,259,651	32,198	10,299,787	263,273	88	12	
1591–1600	9,333,073	238,563	466,016	11,912	9,799,089	250,475	95	5	44
1601–1610	10,711,341	273,793	1,174,782	30,030	11,886,123	303,823	90	10	50
1611–1620	6,104,678	156,042	2,541,652	64,967	8,646,330	221,009	71	29	55
1621–1630	6,606,659	168,873	3,620,573	92,545	10,227,232	261,418	65	35	54
1631–1640	8,732,471	223,211	3,672,874	93,882	12,405,345	317,093	70	30	48
1641–1650	2,981,421	76,208	2,206,810	56,408	5,188,231	132,616	57	43	36
1651–1660	4,317,139	110,350	1,508,388	38,556	5,825,527	148,906	74	26	36
1661–1670	3,991,220	102,020	1,379,509	35,262	5,370,729	137,282	74	26	34
1671–1680	9,967,125	254,770	1,628,439	41,625	11,595,564	296,395	86	14	34
1681–1690	4,770,990	121,951	1,952,190	49,900	6,723,180	171,851	71	29	24
1691–1700	2,580,425	65,958	1,661,385	42,467	4,241,810	108,425	61	39	21
1701–1710	5,240,459	133,951	1,248,873	31,922	6,489,332	165,873	81	19	
1711–1720	7,186,736	183,700	1,010,868	25,839	8,197,604	209,539	88	12	
1721–1730	6,098,799	155,891	1,889,403	48,295	7,988,202	204,186	76	24	
1731–1740	8,510,710	211,142	1,510,826	37,482	10,021,536	248,624	85	15	
1741–1750	6,436,368	159,680	1,761,649	43,705	8,198,017	203,385	79	21	
1751–1760	16,123,480	400,007	2,106,972	52,272	18,230,452	452,279	88	12	
1761–1770	7,482,801	185,641	1,948,564	48,342	9,431,365	233,983	79	21	
1771–1780	15,787,230	385,729	2,900,749	70,874	18,687,979	456,603	84	16	
1781–1790	29,593,184	723,050	2,239,929	54,728	31,833,113	777,778	93	7	
1791–1800	39,845,843	966,062	4,936,031	119,674	44,781,874	1,085,736	89	11	
Totals	221,442,288	5,527,667	44,626,133	1,122,885	266,068,422	6,650,552	83	17	

Table 5. Remittances of privately owned silver to Castile and the Philippines, 1591-1660

	To Philippines from Mexico		To Castile from the Indies[1]		Total private remittances from the Indies		
	Pesos of 272 Maravedís	Equivalent in kilograms of silver	Pesos of 272 Maravedís	Equivalent in kilograms of silver	Pesos of 272 Maravedís	Equivalent in kilograms of silver	Percentage remitted to Philippines
1591–1600	578,170	14,779	80,429,809	2,005,866	81,007,979	2,070,645	1%
1601–1610	3,516,513	89,886	67,398,533	1,722,774	70,915,066	1,812,660	5%
1611–1620	5,048,118	129,035	71,271,339	1,821,767	76,319,457	1,950,802	7%
1621–1630	5,423,822	138,638	70,237,962	1,795,353	75,661,784	1,933,991	7%
1631–1640	3,509,871	89,716	39,706,145	1,014,929	43,216,016	1,104,645	8%
1641–1650	1,759,706	44,980	31,806,791	813,013	33,566,497	857,993	6%
1651–1660	2,015,681	51,523	12,920,005	330,248	14,935,687	381,771	13%
Totals	21,851,881	558,557	373,770,604	9,553,950	395,622,486	10,112,507	6%

[1]From Earl S. Hamilton, *American Treasure and the Price Revolution in Spain, 1501–1650* (Cambridge, Mass., 1934), p. 34.

15

The exports of precious metal from Europe to Asia by the Dutch East India Company, 1602-1795

F. S. GAASTRA

On the 27th of October 1758 Reinier Jan Elsevier, master of the Chinaman *De Vrouwe Petronella Maria*, signed a bill of loading covering an unspecified number of Mexican reals to an amount of 280,860 guilders, that were handed over to him by the directors of the Dutch East India Company. These reals, unsorted coins of various valuations, were packed in thirty chests, each chest containing four bags weighing 100 marks. The chests, double-locked and covered with canvas, were carefully stowed in the cabins of the ship's officers.[1] It was a usual cargo for outgoing China- or East Indiamen in the 18th century, when trade between Europe and Asia was based on a steady flow of precious metal to the east. As one of the largest East India Companies, the *Verenigde Oostindische Compagnie* (VOC) was responsible for a considerable part of the exports of silver and gold to Asia.

The first aim of this essay is to provide insights into the amount and composition of these exports of the VOC, that possessed a strictly maintained monopoly of trade from the Netherlands to Asia from 1602 till 1795. Secondly, the factors within the tradesystem of the Company, that influenced the size and changes in the flow of *contanten*, as precious metals was described in the Company's books, will be analysed. I do not intend to go into the main question, why there was such an imbalance in the trade between Europe and Asia nor shall I be concerned with the impact of the great import of this silver and gold on the Asian economy.[2]

---

1. The bill of loading in *Algemeen Rijksarchief* (A.R.A.), The Hague, *Koloniaal Archief* (K.A.) 3878. About the *mark realen* see appendix I. Much of the information in this paper has been assembled in the course of a research project entitled "Dutch-Asiatic Shipping in the 17th and 18th centuries," directed by J. R. Bruijn, I. Schöffer, E. S. van Eyck van Heslinga and the present author. As a result of this project, tables concerning *Outward-bound voyages from the Netherlands to Asia and the Cape (1595-1794)* and the *Homeward-bound voyages from Asia and the Cape to the Netherlands 1597-1795*) have been published (*Rijks Geschiedkundige Publicatiën* 166 and 167, The Hague 1979). See also F. S. Gaastra, "De Verenigde Oostindische Compagnie in de zeventiende en achttiende eeuw; de groei van een bedrijf. Geld tegen goederen. Een structurele verandering in het Nederlands-Aziatische handelsverkeer," *Bijdragen en Mededelingen betreffende de Geschiedenis de Nederlanden*, 91 (1976), pp. 249-272.

2. See for the factors creating the flow of precious metal to Asia K. N. Chaudhuri, *The*

Every year the highest council of the company servants in Asia, the *Hoge Regering* in Batavia, sent its "orders" *(eisen)* to the directors in the Netherlands, in which the amounts of goods and precious metal needed for Asiatic trade were given. At some periods the government of Ceylon also sent in an independent order. The *Heren XVII* then decided how far they were prepared to meet the orders and turned the requirements for shipping these over to six chambers.[3] The division of the sum was according to the norms laid down in the charter; Amsterdam had to pay half, Zeeland a quarter and the four smaller chambers, Rotterdam, Delft, Hoorn and Enkhuizen, were each charged with a sixteenth. In the large chambers of Amsterdam and Zeeland there were separate committees of the boards of directors charged with the buying in of precious metal, namely the departments of "receipts" or of the "treasury". Unfortunately, little is known of the activities of these various committees, so that we do not know precisely how the gold and silver was bought. The large number of reports of the *Generaalmeesters* of the Mint show that the *wisselbanken* (banks of exchange), above all those in Amsterdam and, to a lesser extent, Middelburg and Rotterdam, played an important role.[4]

Although dealing in precious metal by the *wisselbanken* was forbidden, this prohibition was largely ignored; thus in 1628 the bank of Amsterdam bought gold from the West Indian Company and sold it to the VOC.[5] On the other hand the authorities realised that, as regards the observation of the

---

Trading World of Asia and the English East India Company 1660-1760 (Cambridge, 1978), chapter 8, "The Export of Treasure and the monetary system." For a criticism on Chaudhuri's view: K. Glamann, "Aedelmetalstrømme og verdenshandel" i 16.-18. Arhundrede, *Historisk Tidsskrift* (Copenhagen), 80 (1980), 69-81 (with English summary). For the impact on the Asian economy: Om Prakash, "Bullion for Goods: International Trade and the Economy of Early Eighteenth Century Bengal" *Indian Economic and Social History Review*, vol. 13, nr. 2 (1976), pp. 1-29, and Dietmar Rothermund, *Europa und Asien im Zeitalter des Merkantilismus*, (Darmstadt 1978), pp. 147-165.

3. The organization of the VOC was very complex. The central governing body, the *Heren XVII*, consisted of seventeen delegates chosen from the board of directors *(bewindhebbers)* of each of the six chambers. Amsterdam, the most important Chamber, was represented by eight delegates.

The *Heren XVII* established the general policy; in Asia, the *Gouverneur-Generaal en Raden* (also called the *Hoge Regering*) were subordinate to its decrees and in patria the *bewindhebbers* of the Chambers had to execute its orders.

For the organisation see: Pieter van Dam, *Beschryvinge van de Oostindische Compagnie* (Ed. F. W. Stapel). Four books in seven volumes. *Rijks Geschiedkundige Publicatiën, Grote Serie* LXIII, LXVIII, LXXIV, LXXVI, LXXXIII, LXXXVII and XLVI; (The Hague, 1927-1954). F. S. Gaastra, "De VOC in Azië tot 1680" and "De VOC in Azië, 1680-1800", in *Algemene Geschiedenis der Nederlanden*, vol. VII and vol. IX, (Bussum, 1979 and 1980).

4. The *Generaalmeesters* were charged with the supervision of the minting houses and watched over the observation of the official monetary regulations. H. Enno van Gelder, *De Nederlandse Munten*, (Utrecht, 1972), 91-92.

5. J. G. van Dillen (ed.), *Bronnen tot de Geschiedenis der Wisselbanken* (Amsterdam, Middelburg, Delft, Rotterdam), 2 parts, *Rijks Geschiedkundige Publicatiën* 49 and 60, (The Hague, 1925); Part I, 67.

regulations concerning the price of precious metal, they had more control over transactions with the banks than with private merchants. Therefore, at the end of the 17th century, the *Generaalmeesters* requested that the VOC should be required to purchase gold and silver exclusively from the *wisselbanken*. A resolution to this effect had already been taken by the *Heren XVII* in 1668, but this decision had no general validity.[6]

In the 18th century the bank of Amsterdam ceased trading in specie, but maintained its importance in a different way. After 1683 it was possible to set specie—Dutch and foreign coins, but also bullion—in the bank at "pawn." Merchants could thus deposit their surplus in the bank and their accounts were credited with almost the total sum. Thus they had their capital immediately at their disposal. A six month receipt, which could be extended, was given while the bank charged a low rate of interest for this, 1/4% for silver and 1/2% for gold. The system was introduced because it was expected that, with the arrival of the annual fleets from Spain, an "incredible" amount of precious metal would be brought to the bank. The measure did indeed greatly benefit the trade in precious metal and, according to Van Dillen, the bank was able to fulfil a useful reservoir function. In this respect as well Amsterdam functioned as a staple market, thus ironing out irregularities in supply by forming reserves, so that prices were stabilised. Undoubtedly the VOC was able to profit from this.[7]

The export of precious metal from the Netherlands was in general free, and in so far as limitations existed, the Company, with the powerful support of the city of Amsterdam, was normally able to avoid them. The limiting regulations were generally concerned to strengthen the position of the mints and the improvement of the mint system in the Republic. The *Generaalmeesters* of the mints attempted to have as much of the precious metal as possible minted in the Netherlands. However, after 1647 the export of bar silver was allowed and when, in 1690, the regulation was introduced that anyone who wanted to export silver had to offer a similar quantity to the bank of exchange or the mint master, the directors asked for dispensation and meanwhile disobeyed the injunction.[8] When, in 1701, the export of silver was for a short time forbidden at the request of some of the Dutch mints, the VOC was exempted to the extent that it could send out the cash it had already bought to Asia.[9]

Other equally fruitless complaints that the *Generaalmeesters* continually

---

6. Pieter van Dam, *Beschryvinge*, I, i, 634.

J. G. van Dillen, *Bronnen Wisselbanken*, I, 220, 303-304.

7. J. G. van Dillen, "Een boek van Phoonsen over de Amsterdamse Wisselbank", in *Economisch-historisch Jaarboek VII* (1921), 16-17; id. *Van Rijkdom en Regenten*, (Den Haag, 1970), 446.

8. *Algemeen Rijksarchief* (A.R.A.), Den Haag; *Koloniaal Archief* (K.A.) no. 192, resolutions *Heren XVII* 13-XI, 1691.

J. G. van Dillen, "Amsterdam als wereldmarkt der edele metalen in de 17e en 18e eeuw," *Economisch-Historische Herdrukken*, Den Haag 1964, 235-270.

9. J. G. van Dillen, *Bronnen Wisselbanken*, I, 315.

made against the export were related to its driving up the price. The Company was blamed for buying the metal at a price higher than that officially determined. Particularly when the directors increased the dispatch of money to Asia, as around 1680, 1700 and after 1720 this reproach came to the surface; in 1722 it became impossible for the mints to compete with the VOC and therefore could not strike any coins of reasonable quality—except on the account of the VOC.[10]

The resolutions of the *Heren XVII* are the only source which give a uniform survey of the amount of precious metal exported every year throughout the Company period from 1602 to 1795. Moreover, from 1678 it is possible to investigate with the help of the same resolution books how far the chambers actually implemented the decisions. Thus the Seventeen recorded to the stuiver precisely by how much the chambers were above or below their quotas. If too little was sent out, the chamber was required to make up the deficit, but if too much had been delivered, then it was counted against the next year. In rough terms, the chambers abided by the decisions.[11] However, in the years after 1780 it became steadily more difficult to meet obligations, above all for the Zeeland chamber, which on 14 December 1786 was ƒ449,471 in arrears—and two years later still ƒ348,968. Finally the day of reckoning came and a sum of ƒ1,300,000, destined for Bengal in 1788, was never sent out.[12] For the period before 1678, however, this check is not available. Comparisons with reports from Batavia, in the *Generale Missiven* for instance, suggest that in general the resolutions were complied with; but these figures must be treated with caution.[13] The survey in Table 1 is derived from the resolutions, with one important exception: for the period after 1756 the Chinese data have been taken from the correspondence of the "Chinasche Commissie." In that year the China trade was more or less divorced from the

---

10. J. G. van Dillen, *Bronnen Wisselbanken*, I, 358-360. Similar cases are to be found in England, K. N. Chaudhuri, *The Trading World of Asia and the English East India Company 1660-1760*, (Cambridge, 1978), p. 160.

11. See also K. Glamann, *Dutch-Asiatic Trade 1620-1740*, (Kopenhagen-Den Haag, 1958), 288.

12. A.R.A., K.A. 232, res. *Heren XVII* 19-XII 1786; A.R.A., K.A. 234 res. *Heren XVII* 28-XI 1788.

13. As shown below it is difficult and dangerous to compare the figures of the Dutch sources with the Asiatic papers, because of the difference in the accounting year (for Batavia from September 1 to the end of August). According to lists made up at Batavia over a number of years between 1658/59 and 1671/72 (the "*Bevindingen op de eisen*"), A.R.A., K.A. 10076-10084 in six years the chambers sent more than the resolutions of the *Heren XVII* stated (in total ƒ628,240) and in five years sent less (ƒ1,806,468). The greatest gap occurred in 1665/1666, when from the 2 million fl. in the resolutions only an amount of fl. 819,866 was fulfilled, which might be due to the war with the English. The *Generale Missiven* (III, 616, 669 and 736) however provide a somewhat better picture for the chambers: according to that source, an amount of fl.4,769,637 was sent in 1667-1669, while the *Bevindingen* only stated fl. 3,615,462. (The resolutions: fl. 3,800,000).

The exports of precious metal from Europe to Asia

Table 1. The exports of precious metal by the VOC to Asia, according the resolutions of the *Heren XVII*[15] (in guilders)

1602–10	5,179,000	1700–10	39,275,000
1610–20	9,658,000	1710–20	38,827,000
1620–30	12,479,000	1720–30	66,779,000
1630–40	8,900,000	1730–40	42,540,000
1640–50	8,800,000	1740–50	39,940,000
1650–60	8,400,000	1750–60	55,020,000
1660–70	11,900,000	1760–70	54,588,000
1670–80	10,980,000	1770–80	47,726,000
1680–90	19,720,000	1780–90	48,042,000
1690–1700	29,005,000	1790–95	16,168,000

rest of the Company's organisation and the amounts were determined by this committee, which was concerned with sailings to Canton.[14]

When the major trends in the amount and changes of the exports of money are considered, then it appears that, after an initial rise till about 1630, the exports remain fairly constant, the sums vary from a half to one million guilders a year, with a rise around 1642, when maybe the two million guilder mark was passed, and in 1664/65, when the two million guilder point was again reached.[16] After a gap in 1672-1673, exports increased, to a peak in 1700-01 of four million guilders. The decade of the 1720s saw the largest exports but throughout the whole 18th century the company—whatever its financial position may have been—was able to maintain its exports of precious metal at a high level.

Thus it is possible to use the resolutions of the *Heren XVII* to sketch a rough trend in the export of precious metal from the Netherlands to Asia. It is much more difficult, however, to be more precise about the amounts of

14. See below page 459.
15. The years are counted according to the season of the equipment from the Netherlands: between September and June normally three fleets were sent out. This means that a ten year period runs from about June 1 1610 up to the end of May 1620.
Source: The Resolutions of the *Heren XVII*. The amounts mentioned there differ only in detail from the lists in Pieter van Dam, *Beschryvinge*, I, i, 364 (for the years 1640-1702) and A.R.A., K.A. 8403, coll. Hope (for the years 1640-1751).
See also K. Glamann, *Dutch-Asiatic Trade*, 287-288 (1714-1733). Up till 1623 the value is given in Reals of eight, in the Table converted into guilders according to a rate of 48 stuivers a Real. A minor distortion is possible because the value given in 1628, 1632, 1634 and 1635 may also have covered the value of the exported goods.
16. According to the resolution of the *Heren XVII* an amount of 1,6 mill. fl. was to be sent, but in reality 2,2 mill. fl. were sent out (Pieter dan Dam, *Beschryvinge*, I, 1, 633 and W. Ph. Coolhaas (ed.), *Generale Missiven van Gouverneur-Generaal en Raden aan Heren XVII der Verenigde Oostindische Compagnie*, II, Rijks Geschiedkundige Publicatiën, Grote Serie, LXII. The amount for the season 1664/65 was sent in 1667 (see note 13).

gold and silver which was shipped each year to Batavia or the other factories, both because little is known about the prices which the various chambers paid for gold and silver, and because it is difficult to be certain in what form —gold or silver, minted or not minted—the *contanten* were sent out.

From reports of the *Generaalmeesters* of the Mint it is clear that the prices paid by the VOC were higher than the officially determined prices for which the banks had to deliver to the mints. Rather more certainty can be gained from the resolution of the *Heren XVII* of March 1683, which laid down that "all specie," in so far as it was in ready money, had to be reckoned at the normal rate of exchange. This decision was considered necessary to impose more uniformity on the divergent practices of the chambers, and it was maintained in the eighteenth century.[17] A greater part of the *contanten* however, consisted not of ready money, but of bullion or reals, the prices of which fluctuated according the situation of the market.[18]

Frequently the resolutions give indications of the composition of the annual exports of precious metal, but it is certain that the chambers did not abide by them and were evidently influenced by the circumstances of the silver and gold market. It is however possible to give a rough survey on the basis of data from Batavia and the resolutions, which would serve more or less as a supplement to that supplied by Glamann in his *Dutch Asiatic Trade*.[19] In 1602, the Seventeen laid down that only Spanish pieces of eight should be despatched, and that, when these were not available, they should be replaced by gold. In the first twenty years these Spanish coins, well known in Asia, were predominantly exported. However, reals became gradually scarcer and thus more expensive in the Netherlands—in 1621 war with Spain was resumed and in these years their price rose from 46 stuivers (in 1602) to about 50. The alternative was found in the *negotiepenningen*, Dutch coins which were minted mainly for export, while at the same time small silver coins such as *schellingen*, single and double stuivers (so called *payement*) and, from 1646, bar silver was sent out. Sometimes, thanks to reports from Batavia, it is known what the composition of the cargo was. Thus, in 1634 ƒ872,707 was received, which consisted very largely of *rijksdaalders* and, for the rest, of reals, *leeuwendaalders* or *kroonen* and *payement*. The *Hoge Regering* wrote about this that the *rijksdaalders* and *kroonen* were increasingly well received in Asia, although reals remained necessary for the Indonesian archipelago. The *kroonen* were above all advantageous in Surat.[20]

---

17. A.R.A., K.A. 4195/8345 (Chamber Zeeland); Pieter van Dam, *Beschryvinge*, I, i, 634; K. Glamann, *Dutch-Asiatic Trade*, 293.

18. For the eighteenth century the "Cours van Koopmanschappen tot Amsterdam" provide information about prices of silver and gold. The data in Appendix 2, taken from N. W. Posthumus, *Nederlandsche Prijsgeschiedenis*, (Leiden, 1943), are based on that source.

19. K. Glamann, *Dutch-Asiatic Trade*, 51-72. See also Appendix I and V.

20. W. Ph. Coolhaas (ed.), *Generale Missiven*, I, 453 (Missive 27-XII 1634). Fl. 206,708 was sent in reals of eight, fl. 490,905 in *Rijksdaalders*, fl. 142,519 in *Leeuwendaalders* and fl. 32,575 in *payement*.

In 1634, no gold was sent out. However, already in 1602, with the first despatches under the auspices of the VOC, gold was added to the cargo, including ƒ247,500 in *rosenobels*.[21] In later years too, in 1615 and 1619, for example, *rosenobels* were included, destined for Ambon. In 1619 there was also a substantial sum in gold sent to Coromandel, while after 1620 gold was also delivered to Surat. This gold was i.a. despatched in bars—in 1619 the alloy was the same fineness as the pagoda—and as Hungarian ducats.[22] It would seem from the orders from Batavia, the resolutions of the *Heren XVII* and the information about receipts in Asia that from 1632 there was a period without gold which lasted till the end of the 1650s. In 1658 Batavia began once again to ask for gold and in 1662 the directors decided to despatch ƒ300,000 worth of gold in ducats or bars. At the end of the 1670s the gold export slumped again for a short period because of high prices for silver in Coromandel, normally the main gold consuming area, which, according to an order from Batavia, made the export of silver coins to that region more profitable.[23]

In the last decades of the 17th century most of the larger coins minted in the Dutch Republic and the Southern Netherlands disappeared, only the ducaton or "silvery knight" was sent in great quantities, together with small change. The Mexican reals were preferred to the Sevilian or Peruvian; the reals were valued by weight and called *"markrealen"*. The bar silver was of the fineness of the Bengal Rupee, the bar gold of the fineness of the South Indian pagoda. Thus, according to the normal pattern of the 18th century, silver was exported in reals, ducatons and bar, gold in ducats and bar. Only the proportions varied. It is apparent that in the beginning of the century there was a shortage of reals, so that the chambers could not fulfil the resolutions in this respect.[24] Instead of reals, ducatons were sent out, while for about ten years after 1718 bar silver made up the largest part of the

21. Pieter van Dam, *Beschryvinge*, I, i. 16.
22. A.R.A., K.A. 249, res. *Heren XVII* 25-I-1619; id. 184, res. *Heren XVII* March 1626, no. 6; H. T. Colenbrander en W. Ph. Coolhaas (ed.), *Jan Pietersz. Coen, Bescheiden omtrent zijn bedrijf in Indië*. 7 parts, 's-Gravenhage 1919-1953; Part IV, 328 (letter from the *Heren XVII*, 30-XI-1615).
23. K.A. 10062 (Orders from Batavia, 1676-1680) and 10063 (id., 1681).
24. K. Glamann, *Dutch-Asiatic Trade*, 291.

According to the *Generale Journalen* of 1700/01, 1701/02 and 1702/03 (K.A. 10751, 10752 and 10753) the actual composition of the shipments received in Batavia and Ceylon in those years was as follows: (in fl. 1000):

	1700/01	1701/02	1702/03
gold: ducat	192	456	315
bar	167	110	
silver: mark real	708	639	548
ducatons	2,138	1,877	1,831
bar	920	567	491
payment	1,460	1,413	1,588
	5,585	5,063	4,773

consignments. During the 1730s both the last mentioned categories were roughly of the same amount. In 1740 and 1741 the *bewindhebbers* complained about the "excessive" rise of the silver price and fresh minted ducatons—the minthouses charged the VOC with 5% extra costs—were considered too expensive. The chamber in Amsterdam preferred ducatons of older dates, only Hoorn and Enkhuizen were allowed to acquire ducatons from the mint. This resolution in fact reflects a general tendency in the division of the shipments over the chambers; Amsterdam and, to a lesser degree, Zeeland were ordered to provide those treasures that could be obtained from the *Wisselbanken*, while the smaller chambers were ordered to ship Dutch coin, ducatons as well as small change. The reason for this was probably to give the regional minthouses a proper proportion in the Company's orders. In the course of the 18th century, reals regained their importance, because the China trade was almost exclusively financed with *mark realen*. The quantities of *payement*, sometimes reaching one million a year before 1730, were declining later in the age.

The export of gold was even more irregular. The first twenty years of the 18th century saw a regular annual shipping of gold to Asia (ƒ4,475,000 in 1710/11-1719/20 and ƒ4,954,313 in 1720/21-1729/30). From 1720 till 1724 export was negligible, then a sharp rise occurred but in 1728 there was a sudden drop and during the next two decades nearly no gold was exported. In 1751 the export of gold was resumed and soon reached unprecedented heights (in 1760/61-1769/70 for instance ƒ14,397,941). According to the resolutions, the gold shipments were in the beginning equally divided between ducats and bars, but later during the 1760s, mainly bar gold was sent. This gold was consigned to Coromandel, ducats were sent to Ceylon, Batavia and Bengal (Venetian ones preferred) as well.

The archeological findings confirm this picture, although Peter Marsden demonstrates in his detailed analysis of the treasure of the VOC ships *Hollandia* and *Amsterdam* that the cargo of a single ship cannot be considered representative for the consignments of a whole season. But it is clear that the Company many times had fresh-minted Spanish-American coins at its disposal. The *Vergulde Draeck*, that had sailed from the Netherlands on the 4th of October 1655 and wrecked on the West coast of Australia, carried reals dated 1654; the *Hollandia*, an East Indiaman that sailed out in 1743 and was wrecked the same year near the Scilly Islands, carried some eleven thousand reals; the bulk of these were minted in 1740, 1741 and 1742. In the light of the above mentioned resolution of 1740 it is no surprise that the *Hollandia*, equipped by the Amsterdam chamber, had only a small number of newly minted ducatons on board. As for the ducatons, these coins were partly directly acquired from the minthouses, the other ones, sometimes of a far older date, are preserved so well, that they evidently originated from the reserves of the Wisselbank. After 1680 the provincial mints in the republic were forbidden to strike small coins for internal circulation; they got special licenses to coin the *payementen* ordered by the VOC.[25]

The major share of the precious metal from the Netherlands was in the first instance delivered to the Company's exchequer in Batavia. From there, that which was not needed for the activities in the capital itself was distributed to other offices. During the long period that the VOC existed this naturally was not always done to the same extent. Some factories exhibited a continually expanding demand for silver; others diminished over the course of time and similar shifts occurred in those establishments which exported precious metal. Although a few important regional studies have appeared it is still impossible to give a reasonably complete picture of the inter-Asian trading pattern. In general terms, silver went to the archipelago, Ceylon, and Bengal and gold to Coromandel coast. The pepper districts in Sumatra and Borneo asked for reals, while in other parts of the archipelago, Dutch coins became current money. The rise of the Bengal trade was one of the most significant changes in the trading system of the VOC. Bengal commodities accounted for 36 percent of the total returns to the Netherlands at the end of the seventeenth century and between 1710-1720 this percentage had risen to 39.[26] Most of the goods from Bengal had to be bought with silver and therefore bar silver as well as reals and ducatons found their way to the melting-pots in Bengal to be reminted in rupees. Trade with Coromandel required gold and silver, although not in such quantities as in Bengal. Surat was in general a silver importing area, but sometimes the factory in Surat provided silver for Bengal. Persia, which, before 1640, was sent a lot of money, itself became an exporter of gold and silver.[27]

Nevertheless, not all the money from the Netherlands arrived at its final destination via Batavia: in a number of cases ships were sent directly to such stations as Surat, Persia, Ceylon, Coromandel, Bengal and, in the eighteenth century, Canton.[28] An important motive for this so-called "direct-shipping" was the gain in time and thus—with the avoiding of an extra shipment in Batavia—an improvement in the quality of the return goods.

As against this, full profit could not be taken in this way of the staple market for Asian products which the Company possessed at Batavia and could use within the inter-Asian trade. This meant that the absence of spices

---

25. C. R. Boxer, Plata es Sangre: "Sidelights on the drain of Spanish-American Silver in the Far East, 1550-1700," *Philippine Studies* 18 (1970), no. 3, pp. 457-478. Peter Marsden, "A reconstruction of the treasure of the Amsterdam and the Hollandia, and their significance," *The International Journal of Nautical Archaeology*, 7 (1978), no. 2, pp. 133-148; H. Enno van Gelder, "Munten," in *Prijs der zee. Vondsten uit wrakken van Oostindiëvaarders*. Catalogus Rijksmuseum, Amsterdam (1980), ed. by J. B. Kist and J. H. G. Gawronski. In this catalogue is an exhaustive bibliography concerning wrecks of Dutch East Indiamen.

26. Om Prakash, "Bullion for Goods."

27. See below, page 465, and also Appendix III.

28. For "direct shipping" in the seventeenth century, see Pieter van Dam, *Beschryvinge*, III, 499-512.

Table 2. Direct trade from the Netherlands to Surat and Gombroon[29] between 1622–1634

	Number of ships	Precious metal	% of the total exports of precious metal in those years
1622–23	2	ƒ 200,000	± 18%
1623–24	1	ƒ 235,000	± 40%
1624–25	3	ƒ 600,000	± 33%
1633–34	2	ƒ 152,000	± 16%

and pepper in the cargoes sent to these harbours directly from the Netherlands had to be compensated for with precious metal. Thus the ships which sailed directly to Surat and Gombroon (in Persia) carried substantial capital outlays. After 1634 this traffic was stopped. The demand for Persian silk, which had given rise to it, decreased and the directors preferred pepper to be delivered to Persia in place of cash.

Between 1634 and 1662/63 the position of Batavia as central rendezvous was not threatened, but then deep rivalry between the *Hoge Regering* and the ambitious governor of Ceylon, Ryklof van Goens developed. Van Goens pressed to have Ceylon given a position equivalent to that of Batavia, and wished that the island should become a rendezvous for the Indian stations. Direct sailings from and to Ceylon began in 1662, and in 1682 the directors took the important step of extending this direct route to Bengal and Coromandel. The motives which played a role in this decision included the desire to thwart the English. The directors considered that during the years round 1672, when the VOC had been unable to send out much cash because of war conditions, English competitors had taken the chance to capture the Indian textile markets.[30]

In 1682, the ship *Purmer*, under merchant Abraham Lense, was sent out as an experiment. Lense did not succeed. As a result of long delays in Ceylon and Coromandel he arrived in Bengal too late. No less than six English ships were ahead of him. While the mint worked his competitors' silver, Lense had either to wait or to buy requirements at a very unfavourable price with the reals brought in the *Purmer*.[31] Since he was moreover considered by the local

---

29. A.R.A., K.A. 184, res. *Heren XVII* 16-X-1623; Pieter van Dam, *Beschryvinge*, III, 499-512; H. Dunlop (ed.), *Bronnen tot de geschiedenis der Oostindische Compagnie in Perzië*. Rijks Geschiedkundige Publicatiën, Grote Serie LXXII, (Den Haag, 1930), 122, 124, 130 and 135. H. Terpstra, *De opkomst der Westerkwartieren van de Oost-Indische Compagnie (Suratte, Arabië, Perzië)*, (Den Haag, 1918), 96-97.

30. Pieter van Dam, *Beschryvinge*, II, ii, 41.

31. Pieter van Dam, (*Beschryvinge*, II, ii, 46) states that the English, equally unable to have their silver minted in a short time, used to borrow local money, but thereafter could pay off their debts and buy commodities for the next season.

The exports of precious metal from Europe to Asia

Table 3. Precious metal, sent from the Netherlands to Bengal and Coromandel 1760-1780[36]

	Total sent to Asia	To Bengal	%	Coromandel	%
1760–70	ƒ 54,588,000	ƒ 9,294,000	19%	ƒ 7,190.000	13%
1770–80	ƒ 47,726,000	ƒ 10,950,000	23%	ƒ 8,790,000	19%

VOC servants as a snooper—as the directors indeed intended!—and received too little assistance, he advised against continuing direct sailings.

Nevertheless, the route was continued and every year ships were fitted out and provided with cash for Coromandel and Bengal, travelling on to one or other of these areas after putting in at Ceylon. The return fleet was generally rather larger, since ships from Batavia sailed back to the Republic via these factories on the east coast of India, where return goods were taken on board.[32] After 1682 the resolutions generally tell how much money was provided for the ships on the new route. In the period 1682-1692 a total of ƒ17,770,000 was despatched, of which ƒ970,000 (5%) was for Bengal and ƒ1,530,000 (9%) for Coromandel. This last sum consisted almost entirely of gold and maybe served as compensation for the declining supply from Japan at this time.[33]

In the course of the 18th century the position of Bengal in the Asiatic trading system underwent a great change. The substantial delivery of sugar from Java and of sugar and silk from China to the west coast of India meant that the market there for Bengal products stagnated. An alternative was sought, and found, for the disappearance of this flourishing trade: the trade from Bengal to Canton replaced it, though this activity was generally controlled by the English.[34] For the VOC Bengal took a larger place in the direct traffic with the Netherlands. From 1736 four ships returned annually from the Ganges delta, and after 1750 one or two were sent there each year. These ships carried no goods or money for Ceylon.[35] Part of the large sums that were taken on the trip out were destined for the servicing of debts, which gave the Bengal office a heavy interest burden. Between 1750 to 1753 22% of the total *contanten* sent out was destined for Bengal, in total ƒ5,312,744. In the period 1760-1780, the division between Bengal and Coromandel coast is set out in Table 3. The figures show that Bengal, but also the establishment

---

32. A.R.A., K.A. 191 res. *Heren XVII* 7-XII-1684.
33. There were no shipments to Bengal or Coromandel in 1683. For the season 1688-1689 the proportion destined for these regions is not clearly stated.
34. Holden Furber, *John Company at Work. A Study of European Expansion in India in the late Eighteenth Century*. (Cambridge, (Mass), 1948), 162-164.
35. A.R.A, K.A. res. *Heren XVII* 12-XI-1736; A.R.A., K.A. 207, res. *Heren XVII* 11-III-1750; A.R.A., K.A. 210 res. *Heren XVII* 22-X-1757.
36. The amounts mentioned for Coromandel sometimes include a modest export to Ceylon.

Table 4. Precious metal sent from the Netherlands to Canton, 1718-1734[39]

	Ships to Canton	Silver to Canton	Total precious metal to Asia	% (2 from 3)
1728–29	1	ƒ 300,000	ƒ 5,356,000	5.3%
1729–30	1	ƒ 250,000	ƒ 4,725,000	5.2%
1730–31	3	ƒ 725,000	ƒ 4,825,000	15%
1731–32	3	ƒ 900,000	ƒ 3,862,000	19%
1732–33	3	ƒ 660,000	ƒ 4,250,000	13,6%
1733–34	1	ƒ 300,000	ƒ 2,375,000	12,5%

on the Coromandel coast, were to a large degree responsible for the enormous requirements for money of the VOC in Asia in the 18th century. The pressure was already to be noted by the end of the 17th century, as Bengal received steadily more silver from Batavia.

As regards China, the divorce from the normal Company pattern went even further. In 1690, a stop was put to the sending of Company ships to China from Batavia, as the profits were too small.[37] However, the traffic between Batavia and China continued by Chinese junks, which had travelled this route for centuries and arrived at Batavia either directly or via Macao, Manilla or Macassar. As well as tea these ships brought unminted gold, Mexican reals, copper and silk. The most important trade article of the Company was pepper, while large quantities of cinnamon were sold to the Manilla traders.[38] What with the growing demand for tea in Europe, which became manifest in the second decade of the 18th century, it seemed necessary to have recourse to the silver supplies of the company. Displeasure with the inability of Batavia to supply enough tea and with the quality of this tea, plus the desire to defeat the competition, in particular of the Ostend Company, decided the *Heren XVII* to begin direct traffic to Canton from the Netherlands. Between 1728 and 1734, twelve ships travelled this route, with cargoes almost exclusively consisting of silver. The amounts shipped are shown in Table 4. After 1734, the organisation of the China trade was, for

---

37. John E. Wills jr., *Pepper, Guns & Parleys. The Dutch East India Company and China, 1662-1681*. (Cambridge (Mass.) 1974), 195.

38. J. de Hullu, "Over den Chinaschen handel der Oost-Indische Compagnie in de eerste dertig jaar van de 18e eeuw." *Bijdragen tot de Taal-, Land- en Volkenkunde Nederlandsch-Indië*, LXXIII, (Den Haag, 1917), 32-71.

39. According to the Resolutions and K. Glamann, *Dutch-Asiatic Trade* 289. If the ships from Zeeland carried as much silver as stated in the resolutions of this chamber, then the amounts exported to Canton are in 1730-31 fl. 850,000 and in 1732-33 fl. 680,000.

For more detailed information about the cargo of these ships, see C. Jörg, *Porselein als handelswaar. De porselein als onderdeel van de Chinahandel van de V.O.C. 1729-1794*, (Groningen, 1978). An English edition of this work is in preparation.

The exports of precious metal from Europe to Asia

Table 5. Precious metal sent from the Netherlands to Canton, 1756–1793[42]

	Silver to Canton	Total precious metal to Asia	% silver
1756–60	ƒ 3,354,000	ƒ 23,554,000	14%
1760–70	ƒ 11,352,000	ƒ 54,588,000	19%
1770–80	ƒ 9,876,000	ƒ 47,726,000	20%
1780–90	ƒ 12,826,000	ƒ 48,042,000	25%
1790–95	ƒ 4,620,000	ƒ 16,168,000	29%

various reasons, returned to Batavia. Thereafter two ships went annually from Batavia to Canton and from there one ship returned directly to the Netherlands, as they did later in larger numbers. The sum of precious metal for Canton in these expeditions was reckoned at ƒ600,000 a year in 1734 and this was thus sent to Batavia. During the period before 1756, in which the Canton trade was organised in this way, the resolutions do not allow the sums for China to be distinguished. In his great work on the Canton trade in the eighteenth century, Dermigny does give figures on the import of silver by the VOC to Canton, but these are not trustworthy.[40]

The organisation of this trade chosen in 1734 was equally of short duration. In the belief that silver export to Canton in fact brought more advantage and that pepper and spices could be unloaded elsewhere at higher prices, the direct sailings were reintroduced in 1756. A separate committee of the Seventeen, the *Chinasche Commissie* received the task of managing the Canton trade, and in this manner the traffic was continued until the end of the VOC's existence.[41]

It would appear that the percentage of the money sent out to Canton by the VOC, in relation to the total precious metal despatched to Asia, for a long time remained below, or scarcely above, 20% and that this share only became larger after 1780. In this, the development of the China trade of the Dutch company differed from that of its competitors. Although Dermigny considers that the directors did not fully profit from the possession of Batavia as a favourably situated staple market for Asian products, the proportion of silver in the cargoes delivered was smaller for the VOC than with its competitors.[43]

---

40. Louis Dermigny, *La Chine et l'occident. Le commerce à Canton au XVIIIe siècle 1719-1833*, (Paris 1964), p. 735. According to information of Dr. C. Jörg (University of Leiden) the figures mentioned there differ from those in the "*Kasjournalen*" of the V.O.C. at Canton.

41. J. de Hullu, "De instelling van de Commissie voor den handel der O.I.C. op China in 1756" in *Bijdragen tot de Taal-, Land- en Volkenkunde van Nederlandsch-Indië* 79 (1923), 523-545.

42. A.R.A., K.A. 3878-3887 *(Chinasche Commissie)*.

43. Louis Dermigny, *La Chine et l'occident*, 688, 690.

In explaining the development of VOC-exports of precious metal we will have to take into consideration two different but in many ways related sets of factors: namely those connected with the European Asian trade and those connected with the intra-Asian trade of the Company. For a discussion of these factors Glamann's work about the Dutch-Asiatic trade serves as an important guide.

It would seem logical first to compare the export of silver and gold with the flow of trade in the opposite direction, that is the import of Asian commodities. It is a well-known adagium that the first aim of the East Indian companies was to provide the European market with Asian products. The orders for the *retouren* from Asia made by the directors were based on their knowledge of the market in Europe, and these orders determined for a great deal the amount of the investments for the trade. Chaudhuri could prove such a correlation between export and import by the English East India company and we can expect the same situation for the VOC.[44] A comparison of the two streams of trade show significant differences in the long term development.

In rough terms, the "returns" increased until the end of the 1680s and, after a short break, this growth continued and reached a peak between 1720 and 1730 as did the export of cash. The value of the returns was continually higher than that of the exports, but the relationship between the two altered considerably: for the period before 1700 the ratio was about 1:2 (ƒ125 million in cash as against ƒ242 in returns), but between 1700 and 1795 it was nearly 5:7 (448 million as against 626 million).[45] Glamann's analysis on the Dutch-Asiatic trade shows clearly that there was a link between the European demand for Asiatic products and the rise in the export of gold and silver.[46] Demand was increasingly for those products, such as cotton, silk and coffee, which came from areas where gold and silver was needed to keep trade flowing, namely Coromandel, Bengal and, for coffee, Mocha. A short time later there was another new product, namely tea, which had to be paid for with silver in Canton.

It is highly probable, that the amount of the European demand had a direct influence on the value of the exported *contanten* on the short term, but a positive proof needs a more detailed analysis than can be carried out within the scope of this paper. But on the long term the structural change in the composition of the returns from Asia gave rise to an enormous expansion of the export of treasure to Asia.

It is often assumed that precious metals were the only commodity that the Europeans could offer to Asia, but there was certainly some demand for European products in Asia, above all manufactures and the annual export of cloths of various sorts was not negligible. Over a long period Japan was a

44. K. N. Chaudhuri, *The Trading World of Asia*, 79-95.
45. F. S. Gaastra, "De VOC in Azië, 1680-1800", table 5.
46. K. Glamann, *Dutch-Asiatic Trade*, 62-69, 263.
47. F. S. Gaastra, *Geld tegen Goederen*, 251.

good market and in the second half of the 18th century Leiden *polemieten* and higher quality cloth were very attractive in China.[47] Another article was lead, which could serve as ballast on the outgoing ships and sold in Asia wherever possible. In addition ivory (presumably bought from the West Indian Company) was sent to Asia while mercury and vermillion formed part of the cargo every year.

According to a report made up by Joan van Hoorn in Batavia (in his function as *directeur-generaal*, the second in rank in the Batavian government) in 1694, the Company had exported commodities to an amount of ƒ38,686,195 to Batavia and Ceylon, while the value of the *contanten* dispatched from patria was some ƒ50 million. Between 1700 and 1750 no less than ƒ100,660,131 was received in goods, as against ƒ228,265,232 in precious metal. As often in the VOC books, no distinction was made between goods for the Company's own use and products for sale. Perhaps the latter category was smaller in the 18th century than before. Figures in the *Generale Missiven* show, that over a number of years between 1650 and 1695 an average of ƒ500,000 a year in *koopmanschappen* (merchandise) were sent to Asia. In some years the value of the merchandise was equal to the value of the contanten. But after 1700 it would seem that only ƒ250,000 of trade goods were delivered.[48] It can be safely concluded that the enormous rise of the Dutch-Asiatic trade in the 18th century did not lead to a corresponding rise in the export of trade commodities.

In a certain way the traffic bills of exchange could serve as an outlet for the capital needs of the Company's factories in Asia. In Asia that money deposited in the Company's treasury to be repaid on a bill of exchange in the Netherlands was also reckoned under *secours uit de lieve vaderland*. In the first instance this was a service for company servants returning home. They were forbidden to take any money back with them. The Asian operation thus received money, which the chambers in the Netherlands had to pay several months later, and it is understandable that this money was entered in the Batavia books with the Dutch *secours*. Especially in 18th century much wider use was made of this way of making money over, both by the VOC employees and by other Europeans in Asia.

The level of the sums offered thus depended on the possession of money by private individuals and on their desire to transfer this money to Europe. The Company could only influence it by making the conditions more or less favourable. For long period, from 1664 to 1735, it generally gave 4% interest. Large profits could thus be made as a consequence of the difference in rate between the ducaton in Asia and in Europe—at the beginning of the 18th century twenty *per cent* could be made. The ducaton, worth 63 stuivers in the Netherlands, was reckoned at 78 in Asia. The consequence was that ducatons were increasingly smuggled on the outgoing ships to be sold directly to the Company in Batavia, reckoned on the books at 78 *stuivers* and paid out at that

---

48. A.R.A., Coll. Hudde, no. 10; K.A. 4464, 214. and the *Generale Journalen* 1700/1701, 1730/1731 and 1734/1735.

Table 6. Amounts paid on the bills of exchange from Asia by the Dutch East India Company, 1640-1795[50] (in guilders)

		1700-10	6,387,000
		1710-20	11,219,000
		1720-30	7,956,000
		1730-40	16,814,000
1640-50	3,765,000	1740-50	13,892,000
1650-60	4,506,000	1750-60	23,619,000
1660-70	2,492,000	1760-70	37,900,000
1670-80	4,304,000	1770-80	35,878,000
1680-90	8,024,000	1780-90	40,015,000
1690-1700	7,555,000	1790-95	13,375,000

rate in the Netherlands plus 4% interest. In 1734-5, some four million guilders were transferred. At the same time attempts to keep returns to the Netherlands within a limit failed. Some stabilisation was reached when, in 1738, it was laid down that the ducaton would be reckoned at 72 stuivers without interest. Nevertheless, it remained attractive to smuggle money to Asia—there is the example of Captain Klump of the *Amsterdam*, wrecked in 1749, who took money out for various individuals with the promise to repay them at the Cape or on arrival in Batavia.[49]

Originally bills of exchange could only be made out by the Governor-General and Council in Batavia and by the governments of Ceylon and the Cape of Good Hope. In the middle of the 18th century the possibility was created of using the other offices, such as Bengal, Coromandel and Surat. Bengal in particular was important, because many English made use of this office—to the great profit of the director who first put the money so received out at high interest on his own account! The division of the payment between the chambers was dealt with at the autumn meeting of the *Heren XVII*. As the amounts became larger, the payment posed greater problems for the VOC. In 1760, the directors considered that the "disbursal of so much ready money" at one moment put too much pressure on the rate for bank-money and decided henceforth to pay two-thirds after the autumn sale of Asian products and one-third after the spring sale. Clearly with payment at one time the VOC was forced to make too heavy a call on the *wisselbank*, which would diminish public confidence in bank-money, that was at a premium because the backing of precious metal was thought so solid.

---

49. Peter Marsden, *The Wreck of the Amsterdam*, (London, 1974), p. 34.
50. Pieter van Dam, *Beschryvinge*, I. 1. 364 (for the period 1640-1660); A.R.A., K.A. 8304 (coll. Hope) (for the years 1660-1757) and the Resolutions of the *Heren XVII*. See also F. S. Gaastra, *Geld tegen Goederen*, 259.

The sums were of such an extent that the Company was undoubtedly thus provided with a large part of its cash. Without information on the motives behind the *Indische geldeis* (the orders from Batavia) or on the *Heren XVII's* satisfaction with it, it is difficult to establish whether these money transactions directly influenced the fluctuations in the sending out of gold and silver. In one case, in the period 1733-37, when ƒ13,181,000 was made over to private individuals, this seems probable: as against this high sum, for the time, ƒ14,075,000, lower than in the years before or after, was sent out as *contanten*. In any case, with regard to its policy on the order of money, the *Hoge Regering* had to take account of the money offered against a bill of exchange. In 1748 it restricted the bills on the Netherlands in order to allow the *Heren XVII* to meet its large 1749 order. And in 1760 the government at Batavia complained against the introduction of restrictions on the money traffic, but promised to attune its orders to the amount received against bills of exchange. Was the promise kept? In the period immediately afterwards both the quantities of money exported and the amounts made over from Asia were particularly large.

During the second half of the 18th century, the chambers paid almost as much on bills of exchange as on precious metal shipped out. While the Asian side of the business became more and more dependent on money offered by individuals, the chambers in the Republic found the growing burden of payment steadily more difficult. It is clear that the money requirements became continually larger through the eighteenth century and the Asian side of the business weighed continually harder on the Dutch side.

In contrast to its English competitor, the Dutch East India Company became heavily involved in the intra-Asian trade. In fact, trade and shipping in Asian waters became a major activity for the VOC, one that laid the very foundation of its prosperity in the 17th century. The financial figures about the Company's trade in Asia reflect this development. In the first thirty years of the VOC's existence, the trade capital in Asia was built up with the cash sent from the Netherlands. Jan Pietersz Coen's (*gouverneur-generaal* in 1619-1623 and 1627-1629) insatiable need for money and ships was the result of his intention to give the Company a place within the Asian commerce such that enough profit could be made to provide the directors with the returns.[51] This was never fully achieved, but from 1636 onwards the activities in Asia bore enough fruit to put Batavia in a position to increase its Indian capital itself, while at the same time part of the *retouren* for *patria* could be paid for. In this way the chambers in *patria* were subsidized to an amount of half a million guilders each year until the 1680s. Thereafter, the favourable situation came to an end. From 1696 onwards, the financial position deteriorated; profits decreased and after 1725 became losses. According to the historian Mansvelt, who made an analysis of the Company's bookkeeping system, a total disap-

51. N. Steensgaard, *The Asian Trade Revolution of the Seventeenth Century. The East Indian Companies and the decline of the caravan trade*, (Chicago-London, 1973), p. 140.

pearance of the business capital overseas was averted by the receipt of money against the bills of exchange in Asia and by ever-growing exports of treasure from the Netherlands. The directors could originally cope with this easily because of the profits that they made as a consequence of the high prices in Europe for Asian products, but after 1744 it became steadily more difficult to keep up the level of the Indian capital and in the 1760s the VOC was forced to safeguard its position by abandoning a great part of that capital.

This reversal of fortunes did not escape the attention of the Company's representatives in Asia. In 1747 they ordered no less than 9 million guilders in *contanten*, arguing that the Indian capital was rapidly diminishing and that Batavia was unable to fulfil the requests of the various Asian factories. Most of all Bengal was in urgent need of silver; the VOC factory at Hoogli had been forced to take loans and requested 5.25 million guilders in silver.[52] As far as the *Heren XVII* sent support—in 1748/49 this was not 9 but only 4.8 million—this was not meant to strengthen the trading position of the VOC in Asian trade—as figures about the merchant fleet of the Company show, this trade decreased—but to allow Batavia to procure return goods required. As Mansvelt put it, the whole concern became an "import house" for colonial products.[53]

During the 17th century the Asian commerce of the VOC did not only yield enormous profits, it gave the Company also the opportunity to acquire the silver and gold that it needed to negotiate in Asia itself. The Dutch succeeded in tapping the flows of treasure that entered into Asia via the Levant and by the famous silver-galleons sailing from Acapulco to Manila, and it could lay hands on precious metal that was mined in various Asian countries as well.

In the middle of the century Japan had become the most important supplier of silver. Its exclusive trade with Japan (besides the Dutch only the Chinese were admitted to this country) gave the Company an important lead over its competitors. Exports of treasure by the VOC-factory in Japan became substantial after 1630; during the 1650s and 1660s these exports ran at a rate of a million guilders a year. Part of this Japanese silver was exchanged for gold at Taiwan, and sometimes silver received in Batavia from the Netherlands was also sent to Taiwan to be changed.[54] From there the "treasure-fleets" sailed straight to India via Street Malacca, thus bypassing the headquarters in Batavia. When the Dutch were driven off the island in 1662, this possibility disappeared and the resumption of gold exports from Europe around 1660 seems to be connected with these events. Moreover, as Chaudhuri already could demonstrate on the base of the English experience, gold

---

52. K.A. 10067. (*Eis van Batavia*).
53. W. M. F. Mansvelt, *Rechtsvorm en geldelijk beheer bij de Oost-Indische Compagnie*, (Amsterdam, 1922), pp. 1-17, 78-112; F. S. Gaastra, "De VOC in Azie tot 1680".
54. W. Ph. Coolhaas (ed.) *Generale Missiven* II, 268 (9-VII-1645). For figures about the export of treasure from Japan and Persia see Appendix IV.

became a far more profitable commodity in Asia during these years than silver.[55] Against this background the ban on the export of silver by the Japanese authorities in 1668 might have been not such a serious set-back for the Dutch trade: the factory at Deshima easily switched over to gold. Japanese gold kobans proved very profitable in Bengal where these coins could be sold for 19 or 20 rupees. Perhaps these imports of gold contributed to the depreciation of gold in these regions after 1675. It is remarkable however, that the Dutch company servants gave exactly the same reason for this downfall of the gold price as their English colleagues: it was "principally caused by the fact that the nawab did not pay his soldiers with silver rupees, but with gold muhrs (valued at 15 rupees)". The Batavia government added to this, that it "did not matter for the Company, as long as the factory Surat could sent four or five hundred thousand rupees each year to Bengal."[56] In any case it became soon necessary for the VOC to look for alternative sources of treasure, because the Japan trade was put under severe restrictions after 1685. The decrease of the Japan trade clearly influenced the rise in the exports from the Netherlands; the *bewindhebbers* used this argument to increase the shipments to Batavia in 1687 to three million guilders.[57]

Along the ancient trade route from Europe to Asia silver and gold came to Mocha, Persia and Gujarat. The Dutch factory in Gombroon in Persia became the most important exporter for the Company in this region. Apparently the exports from Gombroon started in 1643, when *capitaal* worth ƒ235,000 was sent to Surat. Over the years the export rose from ƒ383,000 in 1646 to more than one million guilder in 1650. But after 1660 the export slackened down and became more irregular. It was reported that in the year 1671 and 1672 and in 1686 that trade had come to a temporary standstill. In the last decade of the 17th century the exports were resumed and around 1700 Persia was the greatest source of treasure for the VOC in Asia. In these years the Gombroon factory exported predominantly European (Venetian, Hungarian) gold ducats to Malabar, Ceylon and South-India and to a lesser extent some silver *abassies* to several destinations including Bengal.[58]

Mocha was of far less significance for the Dutch; the Company did not possess a factory at this port and left the trade and shipping to Asian merchants. Surat, which heavily depended on the commerce with Mocha, became of interest as a source of silver for the Company in 1660, when Persian exports slowed down. Data about the export of treasure from Surat can be found in the *Generale Missiven*, and the letters of the *Hoge Regering* to the *Heren XVII*, containing a survey of the commercial and political situation of the Company's affairs in Asia. These data are not complete but it is possible to get an impression of this trade. It was mainly silver rupees that

55. K. N. Chaudhuri, The Trading World of Asia, 176-178.
56. W. Ph. Coolhaas (ed.), *Generale Missiven* IV, 163 (13-II-1677).
57. A.R.A. coll. Hudde, no. 5.
58. See appendix IV.

were exported. In 1662 rupees to an amount of some ƒ400,000 (a rupee was equal to ƒ2.10, after 1665 it was valued at ƒ2.20) were sent to Bengal. In 1667 it was reported that 152,000 rupees and 11,500 gold ducats (a ducat was worth 6.26) were shipped to Madurai in South India. In these years the exports became regular, varying between ƒ250,000 and ƒ500,000 a year; the destinations were the VOC factories in Malabar, Coromandel and Bengal as well as Ceylon. As a result of this trade policy, the Dutch in Surat were sometimes forced to raise funds at the local market and thus the factory in this place had to bear the burden of interest for investments elsewhere in India. In 1671 the *directeur* in Surat, Adriaan Bogaart, protested against this and proposed to spend more money in buying on the markets in Gujarat region. The Batavia government rejected his proposal and preferred to employ the funds available at Surat in other Indian factories.[59]

The Dutch Company exploited still other possibilities in Asia in order to obtain silver and gold, but these were less important. For a long time the VOC tried to develop goldmining at the West coast of Sumatra, but the costs and certainly the enormous loss of lives of the slave workers were never outweighed by the results and the exploitation was finally given up in 1737. Furthermore gold was bought in Macassar. Attempts to open trade relations with Manila in order to direct part of the American silverstream to Batavia or other VOC establishments failed. According to the regulations of the Westfalen peace-treaty between the Dutch Republic and Spain (1648), trade between the Company and Manila was forbidden. The Dutch at Batavia had to see how, from 1660 onwards, a flourishing trade between Bantam and Manila began to develop, a trade in which the Sultan of Bantam as well as Indian (mostly Armenian) merchants and English, Danish and French company servants participated. When the Company in 1684 gained control over Bantam this trade came to an end, but the VOC was unable to revive the trade for its own interest. It was tried to engage Indian merchants, for instance the wealthy Surat merchant Adbul Ghafur, in the trade between Manila and Batavia, but apparently the Company left too little room for private traders and thus could not integrate them in its own rigid trade system. In the 18th century things became different. Then only a minimal cover was sufficient for ships sailing from Batavia to Manila. Although the actual size of this trade is not known, it must have been of importance. When in the years 1745 and 1746 the Manila galleons did not sail because of fear of English war ships, the Dutch governor-general decided to equip two ships directly to the American west coast in order to obtain the silver—an expedition which failed because the Spanish American authorities refused any negotiations.[60]

---

59. W. Ph. Coolhaas (ed.), *Generale Missiven*, III, 426, 435 (26-XII-1662), 504 (30-I-1666), 591 (5-X-1667); 775 (19-XII-1671).

60. Dutch Manila trade: M. P. H. Roessingh, "Nederlandse betrekkingen met de Philippijnen, 1600-1800," *Bijdragen tot de Taal-, Land- en Vokenkunde*, 124 (1968), 482-504; F. S. Gaastra, "Merchants, Middlemen and Money. Aspects of the trade between the Indonesian Archipelago

Whatever the size of this trade with Manila might have been, it could not compensate the loss of Asian sources for treasure elsewhere in Asia. The factory at Deshima continued to export gold kobans until 1747, but with interruptions and in very small amounts. Persia's role in the Company's trade came to an end with the fall of the Safavids in 1722. In 1765 the VOC withdrew completely from the Persian Gulf. Surat exports also came to an end, in fact the VOC needed cash there in order to pay its bills of exchange that were used to buy coffee in Mocha. While in the 17th century the Asian trade was the primary source for precious metal for the Dutch East India Company, in the 18th century the factories had completely to rely on the export from the Netherlands.

In conclusion it can be restated that the export of precious metal was of great extent. Relatively, the shipments in the 17th century were fairly small and the exports of European products not so insignificant as sometimes has been claimed. The massive flow of treasure to the East is, judging from the evidence of the VOC trade, really a 18th century phenomenon. The reasons for this are twofold. In the 18th century the trade between Europe and Asia increased enormously, while moreover because of changes in European demand new products were introduced that could only be paid by silver and gold. At the same time the Company lost its advantages in Asia, where during the 17th century Batavia was able to furnish itself with ample cash, and showed a remarkable flexibility in responding to changing conditions in the various Asian markets. From 1680 onwards the Dutch East India Company had to adjust itself to more unfavourable conditions and saw its lead over the European competitors disappear completely. The large sums in silver and gold which the *Heren XVII* sent out in the 18th century to Batavia, Canton and Bengal are perhaps impressive, but are an expression of the changed structure of trade rather than implying that the Company was still in a strong position.

---

and Manila in the 17th century." (Paper presented to the third Indonesian-Dutch conference, Bilthoven 1980).

Appendix 1: Coins exported by the VOC

Coin	Date minted	Weight grams	Fineness	Ratio in stuiver Dutch Republic				Ratio in stuiver Asia			
				1586	*1621*	*later* *1659*	*1615*	*1640*	*1652*	*1656*[c]	
				68		100					
Dutch gold[a,b]											
Gouden dukaat	1586–1808	3.51	0.986							132	
Dutch silver[a,b]				*1606*							
Leeuwendaalder[d]	1606–1713	27.60	0.750	38	40	42	40	48	42	48	
Nederlandse Rijksdaalder[e]	1606–1700	29.03	0.885	47	50	52	60	52	60		
Zilveren Rijder[f]	1659–1798	32.78	0.941	–	–	63	–	–	–		
	1619–1664	1.31	0.333	1	1	1	1	1	1	1¼	
Stuiver	1680–1791	0.81	0.583								
Foreign				*1621*		*1615*		*1640*	*1652*	*1656*	
Patagon[g]	–	28.10	0.875	47		–		–	50	60	
Ducaton[h]	–	32.48	0.944	60		–		–	63	63	
Reals of 8[i]	–	27.08	0.950	48		48		60	50	60	

[a]Monetary system: 16 penningen = 1 stuiver; 20 stuivers = 1 gulden.
[b]Among other coins exported to Asia were: The *Halve Zilveren Rijder* or *Halve Ducaton*; *Schelling* (a coin of 6 stuiver) and *Dubbele Stuiver* or *Dubbeltje* (2 stuiver); *Driegulden* (31.82 gr., fineness 0.920); *Gulden* (10.61 gr., 0.920) and *Halve Gulden* (5.30 gr., 0.920). These coins were minted since 1694 (With a V.O.C.-mark between 1786–1791).
[c]A long struggle between the *Heren XVII* and the *Hoge Regering* about the rate of the silver coins ended in 1656 with a success for the Batavian government: according to its wishes, the value of the most coins was raised, as shown in the table. But two years later, the *Heren XVII* decided that the value of the small change should be raised too. Thus the *Stuiver* became 1¼ stuiver, the *Dubbele Stuiver* 2½ stuiver and

468

Appendix 1: (*continued*)

the *Schelling* 7½ stuiver. This decision created many difficulties and was the origin of the "light" and "heavy" money. The smallest unit of account of the V.O.C., the "light" Stuiver was no longer represented by the "heavy" Stuiver coin. The *Heren XVII* intended that the Rijksdaalder of 60 light stuiver should be equal to 48 "heavy" stuiver. But it proved difficult to maintain the stability of the monetary system in this way, the large and better silver coins went permanently out of circulation.

[d]Also called: *Kroon, Hollandse Daalder*.

[e]Also called: *Provinciedaalder* and after 1659: *bankrijksdaalder*. The Rijksdaalder à 60 stuiver became the unit of account in the Company books.

[f]Ducaton; between 1728 and 1741 with the VOC mark. The case of the *Zilveren Rijder* illustrates the instability in the money rates. In 1676, the Zilveren Rijder was put at 90 "light" stuiver, in 1682 brought back to 75 light stuiver., in 1686 it was 90 light stuiver again and in 1700 97½ or 78 heavy stuiver. In 1710 Batavia was ordered by the Seventeen to accept ducatons for bills of exchange at a lower rate: 82½ light or 66 heavy stuiver and in 1738 the rate, in this case, was 90 light or 72 heavy stuiver.

[g]Minted in the Southern Netherlands; also called *Kruisrijksdaalder* or *Kruisdaalder*.

[h]Southern Netherlands.

[i]Minted in Spain as well as the Spanish Americas. The V.O.C. exported also "*Mark realen*," a number of Spanish coins which were valued by the weight of 1 mark. In 1686 1 Mark Sevilla real was equal to 22 fl. 8 stuiver and 1 Mark Mexican real was equal to 22 fl. 4 stuiver. In 1755 the V.O.C. paid 24 fl. 3 stuiver for 1 Mark Mexican real.

Sources: K. A. 4195/8345; P. van Dam, *Beschryvinge*; H. Enno van Gelder, *De Nederlandse Munten*; K. Glamann, *Dutch-Asiatic Trade*; C. Scholten, *De Munten van de Nederlandsche Gebiedsdelen overzee 1601–1948*, (Amsterdam, 1951).

Appendix 2: Gold-silver ratio and prices of silver and gold in Amsterdam

Table 1. Pure silver per guilder, in grams

1597–1603	11.17
1604–1606	10.94
1607–1610	10.89
1611–1619	10.71
1620–1659	10.28
1660–1681	9.74
1682–1844	9.61

Source: N. W. Posthumus, *Nederlandsche Prijsgeschiedenis* (Leiden, 1943), Part I, CXIII.

Table 2. Gold-silver ratio in the Netherlands, 1600–1800

In five years	Fine gold per guilder	Gold-silver ratio (mint)	Gold silver ratio (market price in Amsterdam)
1600–04	0.992	1:11.21	
1605–09	0.913	1:11.95	
1610–14	0.867	1:12.40	
1615–19	0.858	1:12.48	1:12
1620–24	0.845	1:12.17	
1625–29	0.845	1:12.17	
1630–34	0.845	1:12.17	
1635–39	0.839	1:12.25	
1640–44	0.77	1:13.35	
1645–49	0.738	1:13.93	
1650–54	0.738	1:13.93	
1655–59	0.738	1:13.93	
1660–64	0.738	1:13.20	
1665–69	0.738	1:13.20	
1670–74	0.738	1:13.20	
1675–79	0.738	1:13.20	
1680–84	0.738	1:13.09	
1685–89	0.738	1:13.02	
1690–94	0.738	1:13.02	
1695–99	0.738	1:13.02	
1700–04	0.738	1:13.02	
1705–09	0.738	1:13.02	
1710–14	0.738	1:13.02	
1715–19	0.738	1:13.02	1:14.75
1720–24	0.738	1:13.02	1:15.71
1725–29	0.738	1:13.02	1:15.16
1730–34	0.738	1:13.02	1:14.66

Table 2. (continued)

In five years	Fine gold per guilder	Gold-silver ratio (mint)	Gold silver ratio (market price in Amsterdam)
1735–39	0.738	1:13.02	1:14.71
1740–44	0.738	1:13.02	1:14.42
1745–49	0.738	1:13.02	1:14.64
1750–54	0.66	1:14.56	1:14.26
1755–59	0.66	1:14.56	1:14.31
1760–64	0.66	1:14.56	1:14.61
1765–69	0.66	1:14.56	1:14.63
1770–74	0.66	1:14.56	1:14.62
1775–79	0.66	1:14.56	1:14.58
1780–84	0.66	1:14.56	1:14.61
1785–89	0.66	1:14.56	1:14.79
1790–94	0.66	1:14.56	1:15.02
1795–99	0.66	1:14.56	1:15.45

Source: N. W. Posthumus, *Nederlandsche Prijsgeschiedenis*, I, CXXII.

Table 3. Prices of fine gold and fine silver at Amsterdam, 1719–1795 (per mark in guilders; 1 mark = 246.067 grams)

Year	Gold (annual average)	Silver (annual average)
1719	375.30	25.45
1722	380.29	24.20
1728	368.95	24.30
1731	369.20	25.40
1732	369.42	25.55
1733	373.95	25.39
1734	378.37	25.39
1735	381.02	25.34
1736	371.68	25.42
1737	370.56	25.42
1738	373.52	25.53
1739	375.95	25.61
1740		26.00
1741	373.33	26.02
1742	372.81	25.89
1743		25.90
1744	376.46	25.93
1745	376.30	25.75
1746	377.62	25.81
1747	377.74	25.83
1748	379.19	25.88
1750	370.13	25.68

Table 3. (continued)

Year	Gold (annual average)	Silver (annual average)
1751	374.34	26.01
1752	372.91	26.15
1753	367.80	26.28
1754	374.77	26.35
1755	374.94	26.23
1756	374.40	26.05
1757	372.51	25.84
1758	369.68	26.11
1760	371.27	25.71
1761	373.35	25.98
1762	375.42	25.70
1763	372.59	25.25
1764	375.10	25.23
1765	376.46	25.50
1766	375.74	25.86
1767	375.46	25.75
1768	375.76	25.68
1769	376.06	25.65
1770	375.42	25.59
1771	375.89	25.72
1772	377.89	25.96
1773	374.09	25.63
1774	376.07	25.66
1775	376.06	25.75
1776	375.08	25.75
1777	374.45	25.83
1778	375.09	25.80
1779	375.12	25.53
1780	374.53	25.49
1781	374.22	25.50
1782	374.73	25.65
1783	376.96	26.20
1784	378.64	25.81
1785	383.04	25.75
1786	378.62	25.45
1787	378.58	25.66
1788	383.69	26.07
1789	383.77	26.03
1790	387.04	25.90
1791	380.27	25.53
1792	378.07	25.41
1793	387.33	25.78
1794	395.94	25.82
1795	380.74	24.90

Source: N. W. Posthumus, *Nederlandsche Prijsgeschiedenis*, I, 394–397.

The exports of precious metal from Europe to Asia 473

Appendix 3

In their order for 1749 (K. A. 10067), the *Hoge Regering* gave a summary of the shipments of precious metal from Batavia to the various factories in Asia between September 1, 1746 and August 31, 1747. The distribution was as follows:

Ambon	fl.120,345	40% ducatons, 60% payement
Banda	48,955	24% ducatons, 74% payement, 2% copper coins (duiten)
Bantam	445,105	92% reals of eight, 3% ducatons, 5% payement and copper coins
Bengal	2,999,043	46% ducatons, 10% reals of eight, 11% rupees, 11% mark realen, 21% bar silver and 1% various other coins
Banjermassin	180,000	in reals of eight
Coromandel	1,117,666	37% bar silver, 18% bar gold, 13% mark realen, 23% reals of eight, 5% Austrian thalers, 3% coubangs and 1% various other coins
Cheribon	94,598	ducatons
Ceylon	4,320	ducats
China	203,147	51% reals of eight, 49% bar silver
Jambi	25,600	reals of eight
Tulangbawang	6,105	81% reals of eight, 19% payement
Macassar	43,125	payement
Malacca	45,000	reals of eight
Palembang	96,000	reals of eight
Semarang	133,025	89% ducatons, 11% payement and copper duiten
Siam	83,400	72% reals of eight, 18% ducatons
Ternate	10,035	49% payement, 51% copper duiten
Tandjungpura	8,010	78% ducatons, 12% payement
Surat	177,065	bar silver
Total	5,840,544	

The valuation was according to the Batavian standard: the ducaton was valued at 78 stuiver, the real at 60 stuiver, the mark realen were fl.29.06 per mark and 1 mark of bar silver was equal to fl.31.82.

Appendix 4

Table 1. The company's exports of silver and gold from Europe (I), Persia (II) and Japan (III), 1640–1660 (in guilders)

	I	II	III
1639/40	fl.800,000		3,720,313
1640	800,000		
1641	1,200,000		769,500
1642	1,600,000	235,000	1,063,050
1643	1,000,000	250,000	1,949,400
1644	1,000,000	400,000	2,089,050
1645	1,000,000	383,000	1,222,650
1646	600,000	531,000	1,041,105
1647	400,000	715,605	1,083,000
1648	800,000	660,000	1,054,500
1649/50	400,000	1,094,451	1,003,200
1650	400,000	424,666	1,311,000
1651	500,000		1,368,000
1652	400,000	549,195	1,655,850
1653	800,000		980,400
1654	1,000,000	583,000	1,154,250
1655	1,000,000	768,238	1,672,950
1656	1,000,000	589,174	949,050
1657	1,200,000	1,050,000	1,435,716
1658	1,700,000		1,620,795
1659/60	400,000		1,202,700
1660	1,200,000	772,560	1,756,876

N. B.: Although the information is not complete, it would seem from the orders of Batavia, from the resolutions of the *Heren XVII* and from the information about the receipts in Asia, that the exports from Europe in these years were almost completely in silver.

The exports from Persia were abbassies (a Persian silver coin), reals of eight and gold ducats; they were sent to Surat, Coromandel and Batavia.

The exports from Japan were in silver.

Sources: I (export from Europe): the resolutions of the *Heren XVII*. II (export from Persia): Coolhaas (ed.), *Generale Missiven*, II, 205, 247, 274, 294, 317, 377, 418, 550–551, 735; III, 39, 173, 228, 334. III (export from Japan): Oskar Nachod, *Die Beziehungen der Niederländischen Ostindischen Kompagnie zu Japan im Siebzehnten Jahrhundert* (Leipzig, 1897), Appendix E, ccvii–ccviii

Table 2. The company's export of precious metal from Persia, 1700–1704 (in guilders)

1700/01	ƒ 776,682
1701/02	1,162,738
1702/03	762,779
1703/04	792,041

These exports were for the greater part in gold ducats, and for the rest in abbassies. The ducat was valued according to the Batavian standard at ƒ 6.38, the abbassie at ƒ 0.89.

Appendix 5. Quantity of silver and gold exported by the Dutch East India Company. *Estimation* based on the resolutions of the *Heren XVII*

	Silver	Gold
1602/10	53,726 kg	248 kg
1610/20	102,816	49
1620/30	123,360	40
1630/40	89,436	147
1640/50	90,464	–
1650/60	86,352	–
1660/70	91,556	1,845
1670/80	107,524	240
1680/90	172,980	1,269
1690/1700	259,518	1,476
1700/1710	334,423	3,303
1710/20	325,517	3,656
1720/30	579,425	4,701
1730/40	390,636	926
1740/50	377,240	–
1750/60	487,966	4,458
1760/70	377,631	9,503
1770/80	363,508	7,305
1780/90	459,233	1,882
1790/95	144,140	862

N. B.: The conversion into kilograms is based on the intrinsic value of the guilder according to appendix 2, tables 1 and 2. Since 1723 the resolutions about the export of *contanten* sometimes included copper coins, mostly *duiten* (1/8 stuiver or ƒ0.00625, 3.84 gram) only seldon penningen or halve duiten (1/16 stuiver or ƒ0.003125). Millions of these coins had been shipped to Batavia and Ceylon: 1720/30 ƒ 115, 000; 1730/40 ƒ 695,000; 1740/50 ƒ 685,000; 1750/60 ƒ 841,000; 1760/70 ƒ 845,000; 1770/80 ƒ 726,000; 1780/90 ƒ 645,000; 1790/95 ƒ 400,000.

16

Silver in seventeenth-century Surat: Monetary circulation and the price revolution in Mughal India

JOSEPH J. BRENNIG

European penetration of Asian seas and markets in the 16th and 17th centuries marks a turning-point in the integration of world history into a common stream. But to recognize the importance of this development is not the same as to explain in what manner and at what pace it occurred. Moreover, the distinction between symbolic events in this history of European expansion, such as the Portuguese rounding of the Cape and the entrance of the Dutch and English East India Companies into Asian trade, and the real process of change needs to be maintained. Thus in the case of the Portuguese, J. C. van Leur and M. A. P. Meilink-Roelofsz have shown that attempts to make the formation of the Portuguese Empire in Asia the crucial development in the history of sixteenth century Asian trade neglect both the vigor of indigenous Asian commercial institutions and the limitations of Portuguese power.[1] For all their real achievements in conquest and seamanship, the Portuguese brought little substantive change to Asia.

With the role of the Portuguese now reduced within the context of a better understood Asian trading world, Niels Steensgaard has recently suggested a model of change in Asian trade and market organization which gives center stage to the Dutch and English.[2] The Companies clearly brought the resources, technology and organization which the Portuguese lacked. Moreover, they represented a major new economic force, merchant capital organized into joint-stock companies. But while these strengths of the Companies might predispose one to grant them a revolutionizing role in world trade, such a judgement must nevertheless be solidly based on empirical evidence.

This study proposes to examine the movement of precious metal from Europe to Mughal India's leading seaport, Surat, during the 17th century. It is intended as a contribution both to an understanding of world monetary

---

1. J. C. van Leur, *Indonesian Trade and Society: Essays in Asian Social and Economic History* (The Hague, 1955); and M. A. P. Meilink-Roelofsz, *Asian Trade and European Influence in the Indonesian Archipelago between 1500 and about 1630* (The Hague, 1962).

2. For example see Niels Steensgaard, *The Asian Trade Revolution of the Seventeenth Century: The East India Companies and the Decline of the Caravan Trade* (Chicago, 1973).

circulation in this period and as an evaluation of the role of the Companies in this circulation. In addition it shall examine in a preliminary way the consequences of the increased volume of precious metal imports for the economies of western and northern India. The argument is twofold: first, that the Companies, for all their growing importance in East-West trade, moved only a minor portion of the precious metal passing from Europe to India before the 18th century; and second, that the flow of silver into western India contributed to the expansion of the silver based Mughal monetary system but had only a limited impact on the price structure of northern India.

Precious metal imports into Surat

Imports of metal intended, at least in part, for monetary purposes varied in character over time in the trade of western India. Thus, although the evidence is incomplete, copper rather than silver seems to have been the important coinage metal imported during much of the 16th century. The Portuguese were active in the copper trade. Their copper shipments from the Far East supported not only their coinage in Goa and the copper coinages of neighboring states, but also the increased use of artillery, only recently introduced, by these states.[3]

Shipments of silver and gold to western India in the 16th century undoubtedly occurred, but a history of this trade is still to be written. Imports of American silver into Spain increased sharply after 1550; most probably the effects of the opening of American sources of silver did not reach the Indian Ocean before this date. The Ottoman Empire was situated to experience the impact of the increased flow of precious metal before India. The tide of American silver probably began to affect the Ottoman economy by 1580.[4] If the passage of American silver across the Mediterranean took on the order of thirty years, the more difficult passage across the caravan routes of the Ottoman and Safavid Empires was probably no more rapid.[5] Cheap American silver thus began reaching India in quantity probably at the end of the

---

3. V. M. Godhinho, *L'economie de l'empire portugais aux XVe et XVIe siecles* (Paris, 1969), p. 408. For Asian shipments of copper to Gujarat from the Red Sea see Jean Aubin, "Albuquerque et les negociations de Cambaye", *Mare Luso-indicum*, I (Paris, 1971), p. 44.

4. While the first effects of New World silver were experienced in Ottoman Turkey by 1580, the most significant response of Ottoman prices occurred between 1585 and 1606. O. L. Barkan, "The Price Revolution of the Sixteenth Century; A Turning Point in the Economic History of the Near East", Translated by Justin McCarthy, *International Journal of Middle East Studies*, 6 (1975), pp. 8ff.

5. R. W. Olson, "The Sixteenth Century 'Price Revolution' and its effect on the Ottoman Empire and on Ottoman-Safavid Relations", *Acta Orientalia*, XXXVII (1976), pp. 54-55. Information on the monetary history of Persia is meager. The Ottoman Sultans Selim I and Suleyman the Magnificent both sought to impose an embargo on precious metals entering Safavid Persia. Hostilities between the Ottomans and Safavids brought such an embargo in 1576, but its effectiveness is questionable. A reported Safavid devaluation of silver coinage in 1585 may be significant and bears further study.

16th century and possibly no earlier than the first decade of the 17th century.

Statistics suitable for a benchmark figure for precious metal imports into Surat, however, are available only thirty years later. By the 1640s the Portuguese were no longer major participants in Gujarat's international trade; Surat had acquired the factories of the English and Dutch East India Companies. The trade of Surat had recovered from the disastrous Gujarat famine of 1630-31 and had reached the level attained in the 1620s. These first statistics, as organized in Table 1, are all taken from accounts of the Dutch factory and date from the year 1643-44.

This information permits us to determine, for a single but not untypical year, the source, route, quantity and agency in the movement of gold and silver toward Surat. Much the greater part of these shipments were silver, most of which came from the West, Europe and ultimately the Americas. Japanese silver, in the form of bullion shipped by the Dutch from their factory in Taiwan was 19% of the total volume by weight. Direct shipments of silver around the Cape of Good Hope were 28% of the total, most arriving via overland routes and the markets of the Red Sea and Persian Gulf. The strength of the Companies in Surat's trade, however, is demonstrated in their import of 50% of the silver arriving, directly and indirectly, from the West. Adding Dutch imports from Taiwan, Company shipments of silver to Surat rise to 61% of the total. However, the quality of Dutch information on the cargoes of Asian ships is necessarily suspect; Asian shipments of silver may well have been higher than recorded.

If the Dutch noted the quantity of precious metal being imported into Surat after 1643-44, this information, together with the detailed financial accounts of the Surat Factory, has not survived. Such statistical information as is available in the Company archives is usually the result of the Factory's need to reply to a specific question put by the Governor General in Batavia or the "Seventeen" in Amsterdam. In the absence of a document as comprehensive as that cited for 1643-44 figures for silver imports at a later point in time must be taken from various sources.

The first details of Dutch imports of specie into Surat appear only in records from 1703-04. In this year six ships arrived from Amsterdam carrying 214,512 rials (5,749 Kgs) in silver, a substantial increase over the 53,000 rials imported from Europe in 1644.[6] Increased Dutch dependence on European sources reflects their loss of access to Japanese sources in 1668. Notably absent from this record are imports from the Gulf or Red Sea. If these figures were typical for the period, certainly an important qualification, the volume of imports into Surat had declined significantly, this despite the clear increase in exports from Europe.[7]

English figures for the same period are less certain. K. N. Chaudhuri has supplied data for silver sent from London to the E.I.C. factories in Bombay,

---

6. K.A. (*Kolonial Archief*, Den Hague), 1606, p. 188.
7. See above.

Table 1.[1] Precious metal imports into Surat, 1643–44

Ships	Description	Rials[2]	Silver[3] (kilograms)
3	English East India Company: from London with 235,240 rials	235,240	6,304
1	English East India Company: from Mocha with 36,000 rials	36,000	964
2	Dutch East India Company: from Gombroon with 250,000 *abassies*	83,000	2,224
1	Dutch East India Company: from Taiwan with 158,000 *teylen* in silver bars	187,718	5,031
1	Dutch East India Company: from Batavia with 8,000 *ryxdaalder*, 27,000 *crysdaalder*, and 20,000 *leevandaalder*	53,117	1,424
2	English East India Company: from Mocha and Basra with 39,000 abassies, 48,965 rials and 2,771 gold ducats as freight for Asian merchants of Surat	85,800	2,016
6	Asian: from Mocha (3), Gombroon (1) and Heyduda (2) with 216,135 rials and 2,813 ducats	228,793	5,792
1	Asian, the ship of Shah Jahan, Mughal Emperor: from Jedda with 113,428 rials and 10,700 ducats	161,578	3,040
Total 17		1,071,246	26,795

1. Exchange rates with rials are from *K. A.* 1062, fol. 1023b; and from K. Glamann, *Dutch-Asiatic Trade, 1620–1740* (Copenhagen-The Hague, 1958), pp. 53–55. Thus, 100 abassies = 62 rupees; the gold ducat = 8 rupees; and 100 rials = 210 rupees. The *tael* is the Chinese unit of weight. The Dutch give the value of this shipment in *florins*. The equivalent in rials is calculated at the rial = 48 *stuivers*. Gombroon, or Bandar Abbas, and Basra are located on the Persian Gulf; Jedda, Mocha and Heyduda are located on the Red Sea.
2. These figures include both gold and silver.
3. The Spanish reale of eight: 28.75 gms of silver @ 0.931 fineness or 26.8 gms of pure silver. The metric quantities of silver shipped to Surat in this year are calculated on the basis of this figure. Coins other than the rial (reale) are estimated in weights of pure silver from their rial equivalents.

Source: *K. A.*, 1056, fols. 99–100; and *K. A.* 1056, fols. 858f.

Surat and Gombroon.[8] These shipments do not seem to have changed greatly during the last decades of the seventeenth century, as the following quinquennial averages from 1685-1704 show:

1685-1688, 1692	87,000
1693-1696, 1698	70,000
1699-1704	87,000

Unfortunately, separate figures are not available for each factory. On the basis of a rough estimate, however, Surat probably consumed 75% of the total or 264,000 rials (5s = 1 rial) or 7,000 Kgs of pure silver per year. If slightly above the 235,000 rials which the English shipped from London in 1643, the difference is inconsequential. The evidence suggests that English specie imports from London had not greatly increased over the course of the century.

If Dutch and English imports of specie failed to grow, Surat's overall trade grew dramatically in the second half of the seventeenth century. Company exports of cloth increased sharply to meet new demand in Europe during the "India Craze" of the 1680s and 1690s. The Dutch paid for these exports by developing a market for Asian goods other than specie, especially spices. Surat's Asian merchants reexported these spices to markets in the Gulf and Red Sea. The Dutch it may be said contributed significantly to the increasing vitality of non-European trade links with West Asia.

Asian trade with Surat may have grown even more than European trade. The expansion of Asian trade was noted in Surat as early as the 1650s.[9] By the end of the century, Asian merchants were bringing six million rupees each year in silver coin from Mocha, the equivalent of 2,856,000 rials (76,541 Kgs of pure silver),[10] a far greater quantity than the combined imports of the Europeans. So great was the quantity of silver being received in Surat in this period (*ca.* 1690-1720) that the Dutch occasionally found it possible to export silver from Surat to its other factories.[11] Asian imports of silver in the early 18th century also reduced the role of the Europeans in the shipment of silver to Surat from a possible 50% in the 1640s to about 17% in the early 1700s.[12]

Uncertainties exist in any measurement of changes in the volume of silver imports into Surat in the 17th century, but the importance of this port in the economy of northern India makes an attempt worthwhile. New World silver probably began to arrive in western India between 1585 and 1600, but the first quantitative data available dates from the mid-1640s. At this time annual

   8. K. N. Chaudhuri, unpublished paper.
   9. An English merchant in 1660 reported that Asian ships in Surat had increased from 15 to 80 over the previous decade. W. H. Moreland, *From Akbar to Aurangzeb: A Study in Economic History* (London, 1923), p. 85.
   10. *K.A.* 1689, "Memoire ter order van de Hoog Edele gesorenge Heer Joan van Hoorn . . .," p. 164.
   11. See above.
   12. Omitted from consideration, and thus adding to the uncertainty of this final figure, are the private traders, Europeans trading in Surat in violation of Company monopolies.

imports of silver amounted to about 27,000 kilograms, some from Japan but most from Western sources. By the end of the century silver imports had grown, primarily because of the expansion of imports by non-Europeans, to roughly 100,000 kilograms per year, all now from the West. Despite the expansion of Company trade between India and Europe, silver continued to move through West Asia's traditional trade, the caravan routes of Ottoman Egypt and Arabia and the sea lanes of the Red Sea and Persian Gulf.

Surat silver and the expansion of the Mughal monetary system

After arriving in Surat specie was taken to the imperial Mughal mint in Surat and struck into Mughal silver rupees and gold *muhrs*.[13] Some of these were reshipped to other markets of the subcontinent, but most entered circulation locally. In order to understand what effect this flood of silver coin had on the internal economy of Mughal India, it is first necessary to trace the movement of the Surat rupee.

The history of the rupee in Gujarat begins with the conquest of Akbar, the Mughal Emperor, in 1572-1573. On entering Ahmadabad, the capital of Gujarat, Akbar struck both the silver rupee and the gold *muhr* to symbolize his sovereignty over the region. Akbar probably intended that the rupee should replace the local *mahmudi*, the smaller and baser coin struck by the Sultans of Gujarat, but the process was initially quite gradual. One reason may have been the size of the *mahmudi;* it could not be easily treated as a subdivision of the rupee.[14] For the remainder of Akbar's reign and well into the reign of his successor, Jahangir, only the mint in Ahmadabad struck rupees. Most probably these coins did not circulate much beyond the city limits.[15]

13. The rupee was virtually pure silver. In contrast to the currencies of the Ottomans and Safavids, Mughal currency was highly stable, debasement in metallic purity only occurring with the decline of central authority in the mid-18th century. The *muhr*, slightly heavier than the rupee, was issued primarily for hoarding and prestige purposes and it had only a limited circulation. The copper *dam* served for petty transactions. Since all three coins were struck of nearly pure metal, exchange rates between them were close to bullion exchange rates. Seigniorage and minting charges amounted to about 5.6 percent. See I. Habib, "The Currency System of the Mughal Empire (1556-1707)," *Medieval India Quarterly*, IV (1961), pp. 1-21.

14. C. R. Singhal ("Some New Coins in the Prince of Wales Museum, Bombay", *Journal of the Numismatic Society of India*, XII, 1 (June, 1950), p. 57) suggests that the so-called "coins of Gujarat Fabric," coins bearing Akbar's name but lighter than the standard rupee, were struck to facilitate a transition from the lighter local coinage to the Mughal rupee. However Akbar was apparently willing to permit local mints to continue issuing coins of the Gujarat standard. His general policy in Gujarat suggests no urgency in bringing the region into the Mughal monetary system.

15. The extent to which the rupee became the accepted coin in Ahmadabad is illustrated by the last coins of Muzaffar, the Sultan deposed by Akbar in 1572. A few years after the conquest Muzaffar escaped confinement and reestablished himself briefly in Ahmadabad. But instead of striking the *mahmudi*, the coin he issued a few years before in his own name, he struck rupees of the Mughal standard. A. Master, "The Gujarat Mahmudi", *Journal of the Royal Asiatic Society of Bengal* (1914), Num. Supp. XXIV, p. 463.

Resistance to the rupee was most marked in southern Gujarat in the valley of the Tapti River. Early in the 16th century southern Gujarat and its port of Surat were important in the production and export of textiles. Portuguese attacks on Surat and their subsequent patronage of its competitor, the port of Cambay, led to Surat's decline in the second half of the century. Whether Akbar felt it necessary to issue rupees from Surat in 1573 is uncertain. R. B. Whitehead claims to have identified a unique Surat rupee of Akbar's reign, but this identification is doubtful.[16] Even should Whitehead's reading be accepted, Surat did not issue rupees in any number. The *mahmudi* remained the coin in general use in late 16th and early 17th centuries.

Cambay was still the most important port in Gujarat at the beginning of the 17th century, but Cambay, heavily dependent on Portuguese trade, could not escape the consequences of the erosion of Portuguese dominance with the arrival of the Dutch and English. Merchants of the Dutch and English East India Companies chose to establish factories in Surat in order to avoid the Portuguese and in so doing revived Surat's trade. The English established a factory in Surat in 1610 and were engaged in regular trade by 1613. The Dutch arrived in 1617 and were fully established by 1620. Surat's commercial inactivity in preceding years is indicated by the English report that the mint had been closed for some years prior to their arrival.[17] Trade revived in the second decade of the century and the mint reopened in 1620, issuing rupees in the name of the Mughal Emperor Jahangir.

The mere striking of rupees in southern Gujarat, however, did not mean a transformation of the regional monetary system. The *mahmudi* remained important in the bazaar and as a money of account. Mandelslo, visiting Surat in 1638, found current Gujarat coin to be

... two sorts of money, to wit the *Mamoudies* and the *Ropias*. The Mamoudies are made at Surat, of silver of a very base alloy, and are worth about 12 pense ster, and they go only at Surat, Broda, Broitchia, Cambaya and those parts. Over all the kingdom besides as at Amadabath and elsewhere, they have *Ropias*. . . .[18]

Mahmudis thus circulated along the Gujarat coast (Surat, Broach and Cambay) and in the interior of South Gujarat (Baroda), while rupees circulated in Ahmadabad and, possibly, in eastern Gujarat toward the center of the Mughal Empire. Mandelslo also documents the Surat mint's continued striking of *mahmudis* long after the mint had come under imperial authority and indeed some years after it had begun to issue rupees.

Mughal control over the government of Surat remained lax until the 1640s. While the commander of the fort in Surat received his appointment

---

16. *Catalogue of Coins in the Panjab Museum, Lahore: Volume II, Coins of the Mughal Emperors* (Oxford, 1914), p. lxxxiii. For a caution on Whitehead's reading see H. Nelson Wright, *Coins of the Mughal Emperors of India* (New Delhi, 1975), p. lxxvii.

17. Moreland, p. 330.

18. *The Voyages and Travels of J. Albert de Mandelslo . . . into the East Indies*, translated by John Davies, 2nd edition (London, 1669), I, 67-68.

direct from the Emperor, the governor of the port as well as the various customs officers and the mint master held their positions on the payment of a sum to the imperial treasury. In other words, the administration of Surat was largely farmed out to the highest bidder. An exception in the Mughal mint system, the Surat mint probably struck *mahmudis* because of the relative independence of the mint-master. In 1641, after the Surat mint-master failed to make his expected payment to the treasury, Shah Jahan ended the practice of farming the Surat mint and the other offices of the port.[19]

European trade and increasingly as the century progressed the trade of Asians brought new stocks of precious metal, especially silver, into the Surat mint. In the second half of the century the flow of silver thorugh Surat made the Surat rupee the commonest issue in the Mughal Empire. With their greater abundance, rupees became the dominant coin in southern Gujarat and had begun to spread to the adjacent Mughal province to the south, Khandesh. Table 2 analyzes rupee coin hoards deposited during the reign of Aurangzeb (1659-1707) and Muhammad Shah (1719-1748) in these regions.

Circumstances governing the deposit of a hoard undoubtedly varied, but the nine hoards in Table 2 probably provide a composite picture of the coinage in circulation in the late 17th century. The two hoards from the reign of Muhammad Shah provide a picture for the early 18th century. Surat coins were clearly the most common of any mint represented; the figures suggest that half the coins in circulation in western India came from the Surat mint. Western Indian mints taken as a group provided for the bulk of those coins in circulation; a mint from central or eastern India is rarely represented in these hoards.

Most of the hoards are from Khandesh, the region to the south of the Tapti River. Burhanpur, the chief city of Khandesh, had a Mughal mint from the date of Akbar's conquest in 1599, but the few Burhanpur coins in these hoards suggest that this mint was relatively inactive. The Surat mint supplied Khandesh with coin. Silver from Surat was moving south, as evidence from the hoards found in the southern Mughal province of Ahmadnagar shows clearly.

Ahmadnagar, the northernmost province of the Mughal Deccan, was brought into the Empire by Akbar in 1600. But a revival of localist loyalties proved to be too strong for Jahangir and for a time the region regained its independence. It was firmly integrated into the Empire only with Shah Jahan's conquest of 1633. Only limited numbers of rupees were issued from the city of Ahmadnagar. Their use was restricted not only by the tentativeness of Mughal control but also by a monetary tradition more akin to the gold based coinage of southern India than to the silver based coinage of northern India.

The monetary integration of Ahmadnagar into the rupee system of North India occurred only in the second half of the century. In this period not only

---

19. Moreland, p. 177.

Table 2. Rupee Hoards of Gujarat and Khandesh

	Location	Size	Reign	Total coins*	Surat mint	%	Local mints**	%
1.	Surat	50	Aurangzeb	50	50	100	–	–
2.	Ahmadabad	194	Aurangzeb	19	12	63	16	84
			Shah Alam	5	3	60	4	80
			Jahandar Shah	1	1	100	–	–
			Farrukh Siyar	18	14	78	15	83
			Shah Jahan II	3	2	66	2	66
			Muhammad Shah	148	97	66	126	85
3.	Ahmadabad	39	Shah Jahan	2	0	0	0	0
			Aurangzeb	37 (6?)	24	77	28	90
4.	Khandesh	14	Shah Jahan	1?	–	–	–	–
			Aurangzeb	13 (1?)	5	42	6	50
5.	Khandesh	144	Shah Jahan	4	3	75	3	75
			Aurangzeb	38 (9?)	0	0	1	3
			Shah Alam	5 (3?)	0	0	1	50?
			Jahandar Shah	5	1 (1?)	25	2	50
			Farrukh Siyar	12	0	0	0	0
			Shah Jahan II	1	1	100	–	–
			Muhammad Shah	7	1	14	1	14
6.	Khandesh	16	Shah Jahan	2	2	100	–	–
			Aurangzeb	14 (1?)	7	54	9	69
7.	Khandesh	49	Shah Jahan	8 (4?)	3	75?	3	75?
			Aurangzeb	41 (4?)	21	57	28	76
8.	Khandesh	57	Shah Jahan	3	1	33	2	66
			Aurangzeb	53 (1?)	30	58	30	58

485

Table 2. (continued)

Location	Size	Reign	Total coins*	Surat mint	%	Local mints**	%
9. Khandesh	174	Jahangir	2	1	50	1	50
		Shah Jahan	78 (15?)	56	89	59	94
		Murad Bakhsh	4	3	75	4	100
		Aurangzeb	90	69	77	78	87

Source: P. L. Gupta, *Coin-Hoards from Gujarat State*, Numismatic Notes and Monographs, No. 15 (Patna, 1969), Nos. 122, 24, 18; and *Coin-Hoards from Mahrastra*, Numismatic Notes and Monographs, No. 16 (Patna, 1970), 59, 52, 65, 73, 75, 76.

Notes:
*Numbers in brackets indicate the number of unattributable coins in the hoard. Where these are as much as half the total for any reign, a question mark is placed against the final percentage.
**Local mints include Surat, Burhanpur (Khandesh), Ahmadabad, Kambayat (Cambay), and Junagarh. Rupees struck in the reign of Akbar have been omitted from hoards 5 and 8.

Table 3. Rupee hoards from Ahmadnagar

Size	Reign	Total coins	Surat mint	%	Local mints	%
1. 84	Shah Jahan	13 (3?)	4	40	5	50
	Aurangzeb	70 (3?)	24	36	32	48
2. 180	Shah Jahan	3 (1?)	1	33	1	33
	Aurangzeb	177 (8?)	87	51	105	62
3. 128	Shah Jahan	2	1	50	1	50
	Aurangzeb	126 (1?)	53	42	66	53

Source: Gupta, *Coin-Hoards of Maharastra*, Nos. 2, 25, 9.
Notes: Coins of Ahmadnagar are included under the heading of Local Mints.

were silver rupees available in unprecedented numbers from Surat, demand for rupees increased sharply. In 1681 the Mughal Emperor Aurangzeb launched a major invasion of the Deccan, moving his court and large numbers of men into Ahmadnagar and further south where they remained for the next quarter century. Ahmadnagar's mint expanded its output of rupees during this period, but as Table 3 shows, Surat rupees dominated the hoards of Aurangzeb's reign.

Coins from the reign of Shah Jahan are too few for a reliable analysis, but coins of the reign of Aurangzeb are sufficient. The pattern is similar to that found in the case of Khandesh. Written documents show that the local petty coinage of Ahmadnagar continued to enjoy popularity into the 18th century, but these hoards suggest strongly that the leading coin of important transactions became the silver rupee during the period of Aurangzeb's residence in the region.[20]

Hoard data for other regions of southern India is as yet unavailable. But these findings for western India suggest the direction taken by much of the silver struck into coin in Surat. The increased flow of silver into Surat made possible a monetary integration of the subcontinent never before achieved and destined to be more lasting than the political integration sought by Aurangzeb.

Silver imports and the price revolution in India

Possibly no impact of the importation of large quantities of coin metal into a basically stable agrarian economy is more significant than the impact on prices. A discussion of the complex questions involved with 17th century prices in India in only a few pages can be no more than suggestive and is undertaken only because some of the previous writing on the subject has been misleading.

Prices of key commodities did increase in Mughal India. Irfan Habib has demonstrated this in his pioneering work on the Mughal Indian economy.[21] What is at issue is not the fact of a price increase, but the timing of that increase and its causes. Habib attributes inflation in 17th century India to massive imports of silver. In other words inflation in 17th century India followed the same pattern as inflation in 16th century Europe; India too experienced the "Price Revolution" due to the introduction of New World

---

20. Petty coinage in this region continued to be known by local, non-Mughal terminology at the end of the 17th century, but sources from this period also indicate that the rupee had become current by this time. See G. H. Khare, "A Note on the Names of Smaller Denominations of a Rupee Current in Maharastra in the 17th Century", *Journal of the Numismatic Society of India*, XXXVI (1974), pp. 150-151.

21. *The Agrarian System of Mughal India (1556-1707)* (Bombay, 1963), pp. 384-394. See Aziza Hasan, "The Silver Currency Output of the Mughal Empire and Prices in India During the 16th and 17th Centuries", *The Indian Economic and Social History Review*, VI, 1 (March, 1969), p. 86. Hasan's faulty methodology necessitates cautious interpretation of her conclusions.

silver into the world monetary system. This discussion is intended to question this argument and to suggest an alternative explanation.

Price data for Mughal India is scattered and fragmentary and a series of prices can be constructed for only three commodities: copper, gold, and indigo. While the series for gold and copper are the most complete, indigo prices have the special value of being linked to India's agrarian system.

The changing bimetallic ratio or silver price of gold in Mughal India begins with the rupee price of the gold *muhr* given in the *Ain-i-Akbari*, the compilation of Akbar's administrative regulations, market prices and general information completed in *ca.* 1595. Since the rupee and *muhr* were nearly pure metal, this ratio of 1:9 may be taken as the ratio between the two metals.[22] Fifteen years later the ratio had risen to 1:10 in Gujarat and it remained at this level for the next decade. The first decisive increase occurred in the mid-1620s, being recorded as 1:14 in 1626. Figure 1 shows that at this time the bimetallic ratio in India rose to about the average ratio prevailing in contemporary Europe. The ratio dropped a few years later to 1:13 where it again remained stable for a long period. In the 1650s an abrupt increase occurred, peaking at 1:16 in the early 1660s and once again exceeding levels in Europe. This increase, however, proved shortlived and in the 1670s the ratio dropped sharply back to 1:13.

K. N. Chaudhuri has cited differences in the bimetallic ratio in Europe and India in explanation of the movement of silver from West to East.[23] Undoubtedly Company merchants were aware of profits to be made in arbitrage transactions: cheap European silver exported to India effectively lowered prices of Indian goods being purchased for export to Europe. But frequently fluctuations in the silver price of gold in Indian markets made certainty in arbitrage impossible. Moreover, conscious profiteering from price differentials between Europe and India, however much it might explain Company shipments of specie, has considerably less relevance for the shipment of silver by Asian merchants from West Asian markets, generally closely tied to Indian markets and therefore unlikely to have sharply differing bimetallic ratios. The flow of silver toward Surat throughout this period can better be explained by real trade factors rather than by purely monetary considerations.

The silver price of copper appears in European records as the number of *dams*—the Mughal copper coin—to the rupee and, after 1645 in the Dutch records, as the rupee price of copper bullion. These appear as Figure 2, Curve A and Figure 2 Curve B respectively. In general the price series of copper resembles that of gold. From 1595 to 1615 the rupee price of the *dam* remained stable. By 1619, however, the price of the *dam* began to rise,

---

22. For this figure and those following see Habib, *The Agrarian System*, pp. 384-387.
23. "The Economic and Monetary Problems of European Trade with Asia during the Seventeenth and Eighteenth Centuries", *The Journal of European Economic History*, IV, 2 (Fall, 1975), pp. 327ff.

peaking in the 1620s.[24] This peak was followed by a decline. Switching to Curve B and the bullion price of copper, copper showed a slight recovery in the 1640s. Both curves indicate that the price of copper, like that of gold, rose sharply in the 1650s and 1660s only to fall in the 1670s. The difference between the rupee price of the *dam* and copper bullion is peculiar, since the *dam* circulated by weight and its value virtually reflected its metallic content. Further study of the Mughal monetary system and the copper market might clarify this point. After 1665 copper bullion prices show a return to the level of the 1640s. The few *dam* prices available for the 1690s fit this pattern.[25]

The Europeans exported two varieties of indigo in the 17th century, Sarkhej indigo cultivated in Gujarat and Biana indigo cultivated in the interior in the vicinity of the Mughal capital at Agra. W. H. Moreland, the first to examine indigo as an indicator of price levels, chose Sarkhej indigo and concluded that indigo prices underwent no secular trend in the first half of the century.[26] In his critical reexamination of Moreland's work, however, Habib suggests that Sarkhej indigo, being a commodity primarily cultivated for export by the Companies and therefore subject to extraneous price fluctuations in the international market, was a poor choice for the study of prices in Mughal India. Biana indigo, cultivated inland and, at least early in the century less dominated by the Companies, better revealed local price trends. Figure 3, based largely on Habib's figures, shows that Biania indigo prices rose in a pattern similar to that observed for copper and gold.

I disagree with Habib's price history of Biana indigo in one important respect: prices did not increase in the period between 1595 and 1609 as Habib's figures suggest. Habib begins his series with the price in the *Ain* of between 10 and 16 rupees per Akbari *man* (551b) in 1595, the price in the Agra market. The next price comes from Surat market in 1609 where English merchants reported it to be between 16 and 25 rupees per Akbari *man*. Habib calculates the cost of transport from Agra to Surat to be 2.5 rupees per *man*, and on this basis the ex-Agra price in 1609 to be significantly higher than the price in 1595.

But Habib neglects the cost of handling. A breakdown of shipping costs dating from late in the 17th century shows handling costs to have been double

---

24. For the *dam* price of the rupee see Habib, *The Agrarian System*, pp. 387ff. For bullion prices see *K.A.* 1297, fol. 87 (1645-84); *K.A.* 1314, fol. 160vo; *K.A.* 1328, fol. 1556; *K.A.* 1333, fol. 1499; *K.A.* 1438, fol. 606vo; *K.A.* 1432, fol. 179; *K.A.* 1472, fol. 138vo; *K.A.* 1475, fol. 265vo; *K.A.* 1517, p. 158; *K.A.* 1541, p. 140.

25. Figure 2, Curve A is a modified version of Hasan's curve of the price of the *dam* drawn from Habib's figures. Hasan omits, however, Habib's prices for the 1690s as ambiguous. I believe they should be included but not as Habib states them. Habib assumes that these prices refer to the light *dam* which Aurangzeb struck after 1663. I would argue, however, that the older, heavy *dam* continued to circulate in great number. I have drawn Curve A, on the basis of this assumption, as representing the same weight *dam* through the century.

26. For Sarkhej indigo prices see Moreland, p. 161; for prices of Biana indigo see Habib, *The Agrarian System*, p. 86, n. 27.

simple transport costs. At this same ratio, handling costs in 1609 would have been 5 rupees per *man;* total costs of shipment 7.5 rupees per *man.* Thus adjusted, the 1609 price of Biana indigo ex-Agra was not significantly higher than the 1595 price. Significant change only occurred in the second decade of the century and paralleled changes in the price of the *dam* discussed earlier.

All three of the price series presented show similar patterns of change. Price increases occurred but the periods of increase were short and were followed by longer periods of relative stability. The most decisive increase occurred in the 1620s, for the price level achieved the end of this period, even if occasionally surpassed, was generally maintained throughout the remainder of the century. The dramatic increase of the mid-1650s and 1660s, although remarkable, was too brief to be significant. These common features in the price history of different commodities clearly suggests that the cause of inflation in Mughal India as elsewhere in the world in this period was a fall in the value of money, that is of silver. But price changes in 17th century India were unlike the long secular increase which characterized 16th century Europe; and they were certainly more moderate than the increases which occurred in late 16th and 17th century Ottoman Turkey. If silver imports were responsible for inflation in these regions, were they also responsible for the quite different pattern of price changes in India?

Unfortunately trade in western India before the arrival of the Companies remains obscure. New World silver may well have arrived on the shores of India in the last years of the 16th century, but large quantities seem doubtful. The Portuguese were more known for their shipments of copper, though they may have imported some silver as well. Possibly Asian merchants in Cambay were carrying greater quantities of silver from Mocha and Red Sea ports to India, but again evidence is lacking. But it seems unlikely that the volume of silver being imported into Cambay in the first decade of the century exceeded the volume being imported into Surat in the 1640s. Surat not only inherited the trade of Cambay, Company trade permitted Surat to exceed Cambay. Almost certainly, therefore, annual silver imports into Cambay did not exceed 20,000 kilograms; most probably they were less.

The possible impact of this amount on the monetary system and price structure of northern India must be evaluated in terms of the existing stock of silver then in circulation. The best we can say in this regard is that Akbar issued a large and diverse silver coinage. From the surviving sample of his coins, they were abundant at the time New World silver began to arrive in India. Imports of 20,000 kilograms per year may have been enough to supply the imperial mint in Ahmadabad, but it seems inadequate to account for a sudden jump in prices in the 1620s. Braudel and Spooner have cautioned against too hasty an attribution of causation to external sources of silver where existing stocks of silver in circulation represented a mass of great

27. "Prices in Europe from 1450 to 1750", in E. E. Rich and C. H. Wilson (eds.), *The Cambridge Economic History of Europe* (Cambridge, 1967), IV, p. 446.

inertia. In India, where the quantities of silver being imported were far below the quantities being imported into Europe, such a consideration should be taken seriously.

Internal factors affecting prices in Mughal India are worth considering. Their role in the Price Revolution of 16th century Europe may have been underestimated by those stressing the importance of imported New World silver. Thus I. Hammarstrom has argued that "monetary fuel for the expansion [of economic activity] is secured by the activation of money which has been lying idle, or by the creation of additional money;" and "we must assume that in earlier centuries, just as in our own, an upward economic trend would not only reduce idle savings, but also create new money by . . . the coining of existing bullion or by causing an increase of credit-money."[28] E. J. Hamilton's counter that the return of idle savings to circulation could not account for an upward trend lasting a century, valid for Europe though it may be, applies less well to Mughal India. Inflation in 17th century India, as has been argued here, was not a long term upward movement, but a short term jump from one plateau to another.

In general conditions in Mughal India during the late 16th and early 17th centuries met the description of economic expansion given by Hammarstrom. While Akbar did accumulate a large hoard of precious metals during his reign, successful conquest and an efficient revenue system permitted him to issue a large silver coinage. In addition, his long reign of nearly fifty years and the peaceful succession of his son, Jahangir, in 1605 created a strong basis for public confidence in the Mughal regime at the beginning of the 17th century. Indeed this increased confidence and a quickening of trade may have encouraged voluntary private dishoarding. Finally, it was in this period that the indigenous monetary instrument of the subcontinent, the *hundi*, became a common means of payment;[30] credit-money supplemented the coin being struck in the imperial mints.

This study has sought to provide a clearer understanding of the quantities, mechanisms and consequences of the flow of silver entering Mughal India through the western Indian port of Surat. Import figures from the 1640s and the early 18th century suggest that this flow increased on the order of three to four times. Although Japanese silver was prominent in the first half of the century, subsequently virtually all the silver arriving in Surat was of European or New World origin. Dutch and English East India Company imports of silver in the 1640s were a major share of the total, but this Company share declined relative to the Asian contribution during the second

28. "The 'Price Revolution of the Sixteenth Century'; Some Swedish Evidence", *Scandinavian Economic History Review*, V (1957), pp. 127-128.

29. "The History of Prices Before 1750," *International Congress of Historical Sciences, 1960; Rapports I*, p. 157.

30. Habib, "Banking in Mughal India", in T. Raychaudhuri (ed.), *Contributions to Indian Economic History* (Calcutta, 1960).

half of the century. Throughout the period examined, however, western silver arriving in Surat came primarily not around the Cape of Good Hope but by the caravan routes linking ports on the Red Sea and Persian Gulf with markets in Egypt, the Levant, Ottoman Turkey and Russia. Mechanisms for moving silver from West to East at the beginning of the 18th century remained deeply imbedded in the traditional structures of West Asian trade.

This study has also examined the consequences of silver imports for the internal economy of Mughal India. The clearest consequence was an increase in the size of the Mughal silver coinage in the second half of the century. Silver specie and bullion were struck into rupees in the imperial mint in Surat—increasingly active from 1620 on—and then passed into circulation. Hoard evidence from the provinces of Gujarat, Khandesh and Ahmadnagar shows that a high percentage of the rupees in circulation in western India as far south as the Deccan were issued from the Surat mint. Imported silver was following Aurangzeb's invasion of the south and helping to transform this region, at one time possessing a currency based on gold, into an extension of the monetary system of Mughal North India. This expansion of the Mughal monetary system at the very time silver imports were reaching their peak may have dampened their inflationary impact; demand for silver in southern India was expanding to meet increasing supplies.

This study also questions a consequence of silver imports previously suggested: a "Price Revolution" in 17th century Mughal India. New World silver probably made its appearance in India at the end of the 16th century, but it is unclear how the quantities of silver arriving in this period could have had more than the most marginal impact on North Indian prices by the first important price increase in the 1620s. A summary of the evidence for price movements in 17th century India fails to show a gradual upward trend paralleling international movements of silver; indeed prices stabilize after 1670 just when silver imports are beginning to accelerate. Internal developments affecting the supply of money in Mughal India offer a suggestive avenue for further study, for even a brief review of general conditions affecting the economy of Mughal India at the beginning of the 17th century shows they were appropriate for monetary expansion and an upward movement of prices. Recognition of the international repercussions of additions to the world silver supply in the 16th century should not lead to the neglect of powerful local developments affecting regional monetary history.

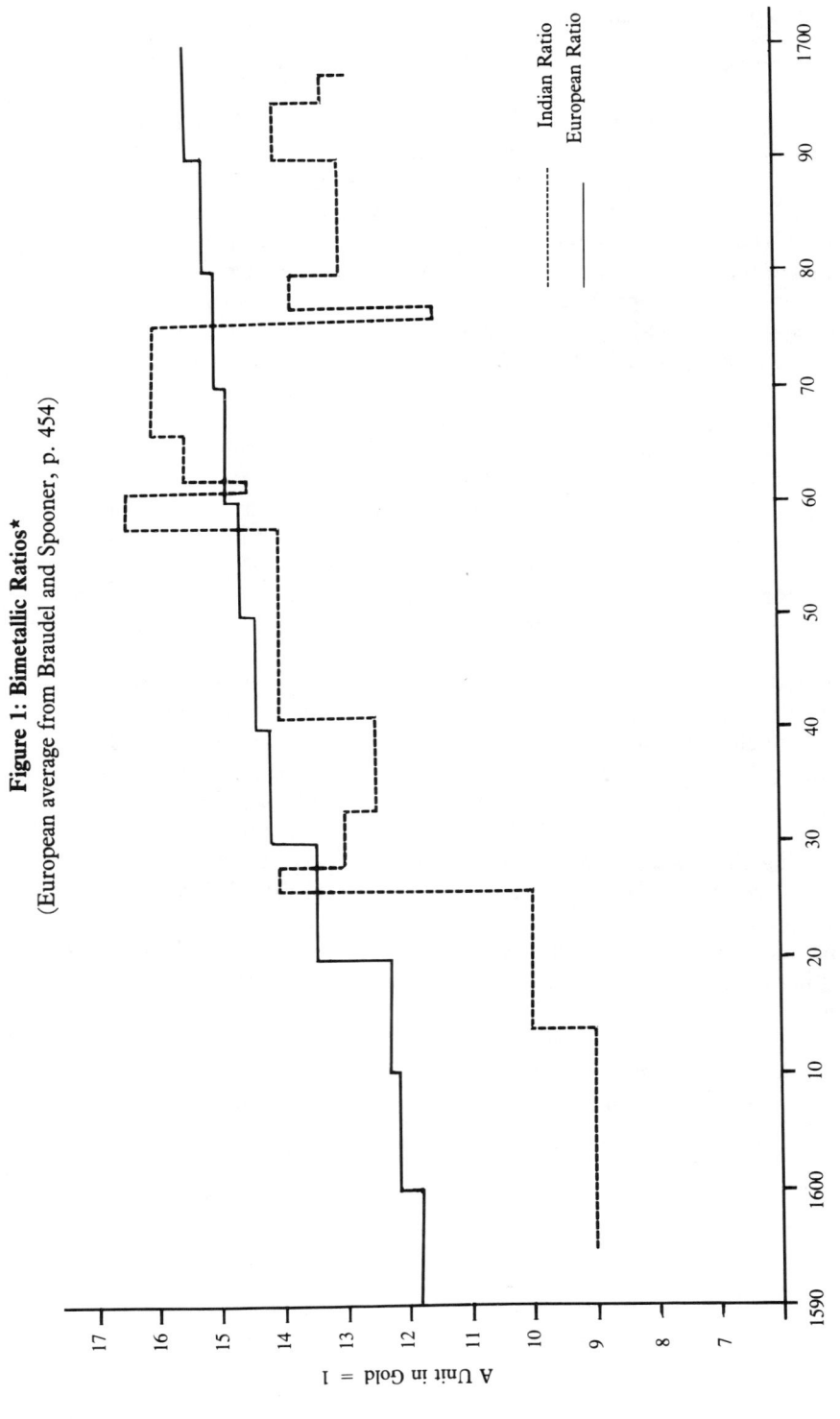

Figure 1: Bimetallic Ratios*
(European average from Braudel and Spooner, p. 454)

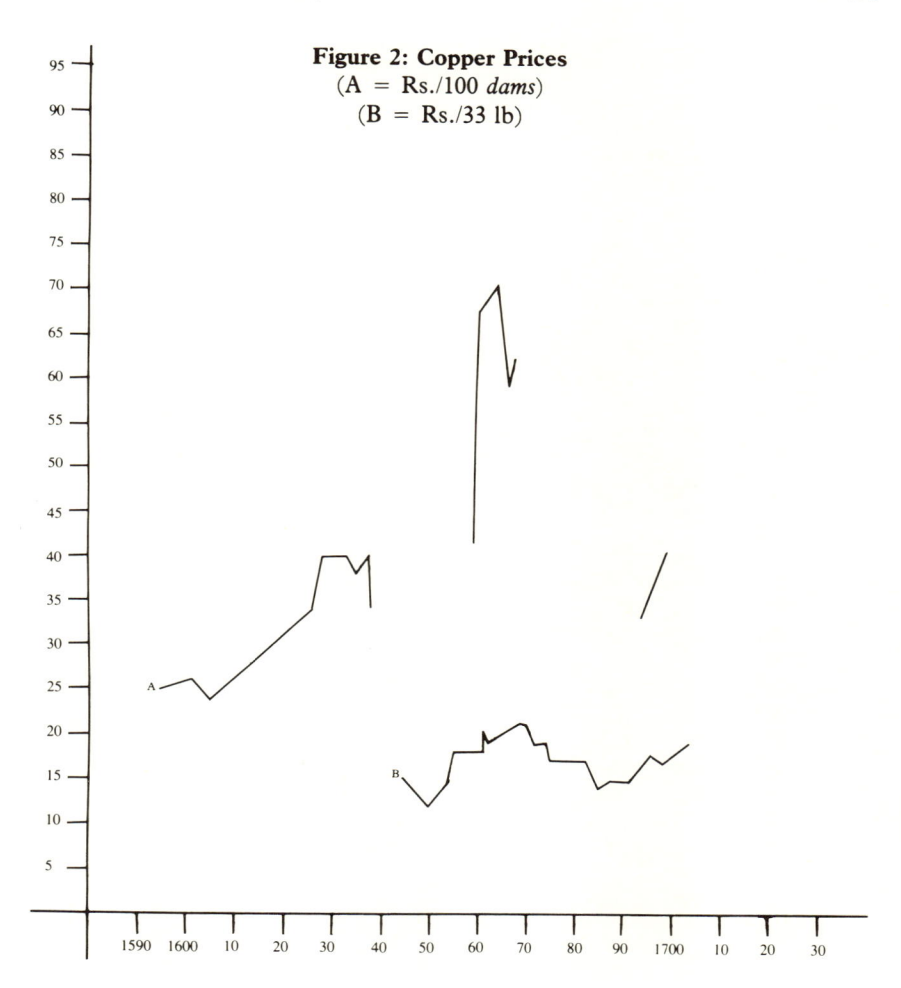

Silver in seventeenth-century Surat 495

Figure 2: Copper Prices
(A = Rs./100 *dams*)
(B = Rs./33 lb)

Figure 3: Biana Indigo Prices

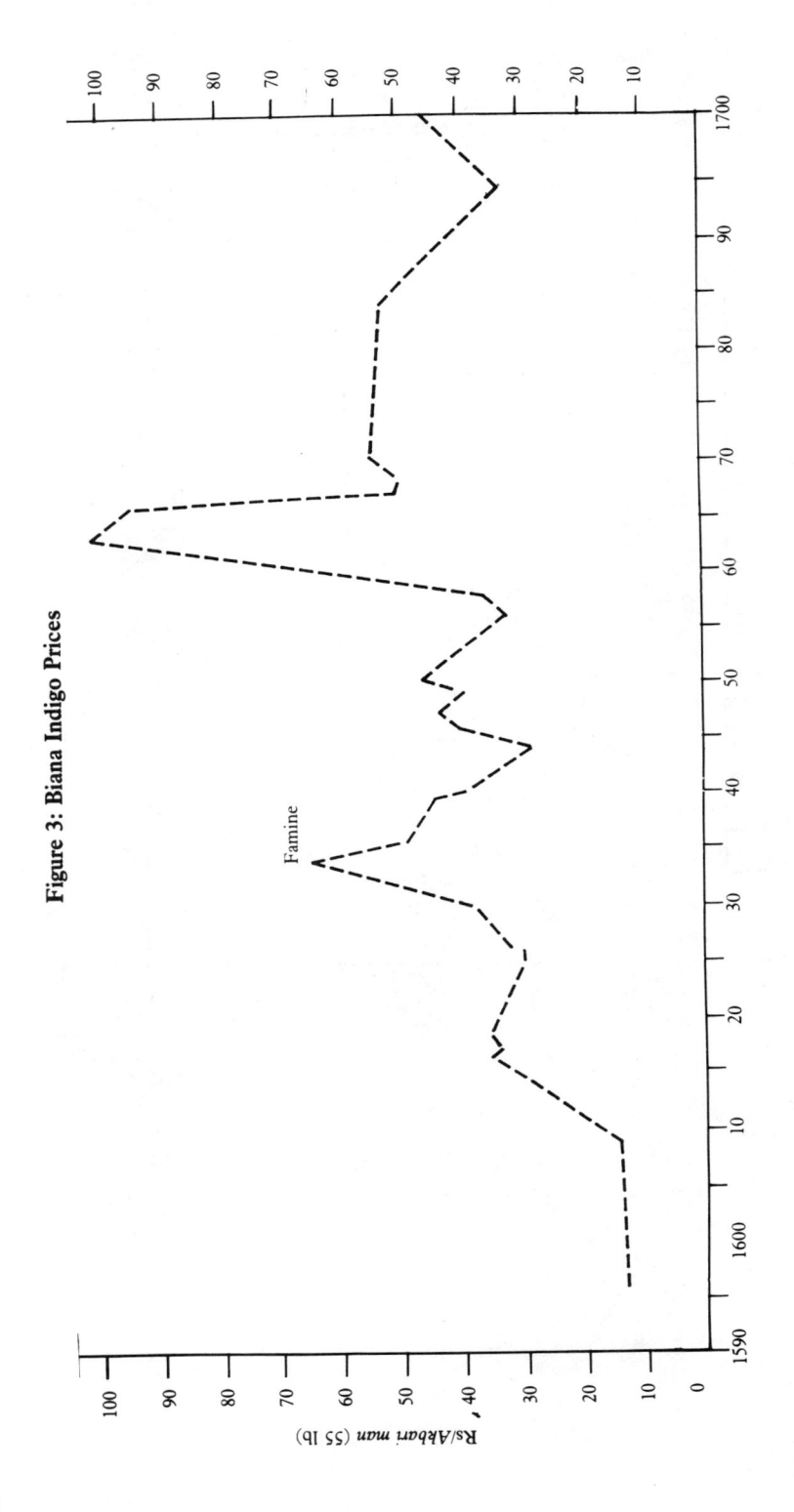

Index

Africa: 7, 8, 10, 11, 13, 15-17, 22; East Africa, 15, 16, 231, 235, 238; North Africa, 16, 17, 37, 38; Sub-Saharan, 10, 15, 16, 20, 231-33; West Africa, 37, 44, 167; *see also*, Egypt; Mediterranean; Mali; Mansa Musa; Takkrūr; Venice
Ahmadabad: 483; *see Mughal Empire*; *see also* India
Alexandria: 17, 41, 75: *see* Egypt
al-Fazari: 33
Algeria: 67; *see also*, Africa; Fatimids; Maghrib; Mediterranean
al-Maqrīzī: 159, 162-64
Almohad Dynasty: 35, 36, 38, 47, 50, 51, 162, 240; *see* Africa; *see also* Egypt
Almoravid Dynasty: 35, 36, 39, 46, 240; *see* Africa; *see also* Egypt
America: *see* New World; *see also*, New Spain; Argentina; Brazil; Chile; New Spain; Peru
Amsterdam: 24-26, 448-49; *see* East India Company; *see also*, Holland; traders, Dutch
Anatolia: 11, 15-17, 269; *see* Ottoman Empire
Arbitrage: 82, 84, 86, 111, 112, 489
Archibald, Marion: 100
Ashtor, Eliahyu: 8, 79-80, 86, 101-03, 160-64 *passim*, 310
Atwell, William: 8
Ayyubid Dynasty: 40, 160, 163-67, 169, 269; *see* Egypt; *see also* Africa
Argentina: 411-14, 420; contraband trade, 413-15. *see* New World; *see also* Europe; New Spain; Peru; Spain

Bacharach, Jere: 102
Bahmani Sultanate: 184, 210-11; *see* India
Bakewell, P.J.: 406-07
Balkans: 15-17, 41-42
Balog, Paul: 160, 165
Báncora, Cañero: 418-19
Barkan, Omer Lufti: 7
Baroda: 483; *see* Mughal Empire; *see also*, India; Surat

Batavia: 23-25, 448-67 *passim*, 479; *see* East India Company; *see also* Holland; India; traders, Dutch
Bautier, R.H.: 44
Bengal: 24, 184; and China, 207-28 *passim*; and intra Asian trade; 454-67 *passim*; *see* India; *see also*, Delhi Sultanate; Shan States; Yunnan
Berber traders: 32, 36; *see* Africa
Black Death: 12, 86-121 *passim*; *see* Europe; *see also*, England; France; Low Countries; monetary contractions
Borah, Woodrow: 413: 413
Bosman, William: 244
Bosworth, C.E.: 191
Boxer, C.R.: 414
Braudel, Fernand: 6-7, 11-12, 16, 18, 26, 31-33, 403, 491; French Annales School 6
Brazil: 16, 21, 240, 397, 399, 413-18; and Africa trade, 251-52; *see* New World; *see also* New Spain; Peru; Portugal; Spain
Brennig, Joseph: 25, 26
Broach: 25, 483; *see* Mughal Empire;*see also*, India; Surat
Bullion and Metal ratios: Byzantine, 68; China, 343-45, 352-53, 381-87 *passim*, 399-401; Egypt, 172; Europe, 118, 250, 259-61, 400; France, 125; Hungary, 68; India, 489; Japan, 343-44, 346, 352-53, 381-82, 400-01; Levant, 60; Low Countries, 117, 124-25; New Spain, 401; New World, 400-01, 436, Ottoman Empire, 285-86; Peru, 401; Spain 400; Venice, 60, 67-68; Vietnam, 381-82, 385-87; gold to copper, 259-60; gold to cowrie shell, 259-60; silver to cowrie shell, 259
Bullion and monetary flows: Alexandria, 17, 41, 45; Amsterdam, 24, Anatolia, 11, 15, 17; Arab, 11; Balkans, 15-17; Baltic, 8; Bengal, 24, 207-28 *passim*; Black Sea, 11; Bohemia, 11, 19; Bosnia,11, 16-17; Brazil, 16, 21, 240, 413-18 *passim*; Byzantium, 10, 11, 16, 270; Canton, 20,

497

457-67 *passim*; Chile, 21, 429; China, 4, 6, 8, 14, 19-20; Columbia, 21, 397; Constanople, 16-18, 42; Dahomey, 16, 233, 251, 308; Dalmatia, 11; Dutch, 15, 16, 24, 26, 447-75 *passim*; East Indies, 6, 447-75 *passim*; Egypt, 8, 11, 13, 16, 100, 103, 159-82 *passim*; 231-32, 269, 305-28 *passim*; England, 12, 15, 21, 79-96 *passim*, 97-158 *passim*; Ethiopia, 15, 18, 231-32, 319-21; Europe, 4, 7-8, 13, 18, 21, 24, 102, 425-26, 433, 447-75 *passim*; Florence, 10, 11, 30, 51-67 *passim*; France, 12, 36, 74, 79-96; Genoa, 10, 30, 41, 43-45, 49-52, 67; German, 11, 16, 53-54, 66; Ghana, 34, 36-37; Goa, 16, 232, 380; Gold Coast, 16, 232, 238-41, 247, 250-52; Hijaz, 75, 167-69, 171, 309, 312-15; Iberia, 19; India, 4, 6, 13-16, 171, 183-205 *passim*, 208, 464, 482-92; Italy, 16, 11-13, 44-49, 102, 118; Japan, 15, 18, 20, 329-62 *passim*; Levant, 6, 12, 17, 43, 49, 103, 166; Libya, 18, 317-19; Low Countries, 12, 17, 97-158 *passim*; Malacca, 20, 464; Mali, 37, 103, 309-14; Mediterranean, 4, 6-7, 11-18, 26, 31-33, 35-39, 45-53; Mozambique, 16, 232-35; Near East, 102, 110, 118; New Spain, 21, 22, 24, 397-98, 412-13, 425-29, 432; Ottoman Empire, 10, 15-18, 269-304 *passim*; Persian Gulf, 14, 15, 26, 232; Peru, 21-22, 24, 397-445 *passim*; Phillipines, 20-22, 379; 420, 425-45 *passim*; Ragusa, 11, 16, 65, 68, 75; Red Sea, 15, 18, 26, 232; Rumelia, 17, 284; Russia, 8, 11, 103; Saloniki, 59; Saxony, 11, 99; Seljuqs, 269-70; Serbia, 11, 16-17; Sicily, 10, 38, 44, 47-51; Silesia, 11; Spain, 4, 7, 17, 21, 412-17, 425-45 *passim*; Surat, 25-26, 477-96 *passim*; Syria, 13, 16-17, 39-43, 49, 74-75, 166-67; Takkrūr, 18, 168, 171, 305-28 *passim*; Tunisia, 10-11, 37-38, 47-49, 67, 241; Venice 11, 16, 53-77 *passim*; Vietnam 19, 373-89; Zambezi, 15, 235
Burma: 20, 216-24 *passim*; *see* Bengal; *see also* China, India, Shan States, Yunnan
Byzantium: 10, 11, 16, 30, 57, 269-70; *see* Mediterranean; *see also*, Egypt, Ottoman Empire; Venice

Cambay: 25, 483, 491; *see* Mughal Empire; *see also* India; Surat
Cameroon: 232; *see* Africa; *see also* Takkrūr
Castile: New World bullion, 425-445 passim, *see* Spain; *see also* New World
Ceylon: 455-57, 461-62, 465-66; *see* East India Company
Charcas mining district: 406-09, 420; *see also* Brazil; New Spain; New World; Phillipines; Potosi; Spain
Chaudhuri, K. N.: 24, 460, 464, 479-80, 489
Chaunu, Pierre: 7, 408, 435
Chile: 42, 429; bullion production in, 407-11 *passim*; *see* New World
Cipolla, Carlo: 31
Cloth: in Africa trade, 256-258
Columbia: 21, 397; bullion production, 401-11 *passim*; *see*, New World
Constantinople: 16, 17, 18, 42
Coromandel Coast: 453-66 *passim*; *see also*, Mughal Empire; *see also*, India; Surat
Cowrie shells: 232, 251; *see* Africa; *see also* Takkrūr
Craig, John: 101
Crusades: 3rd, 64; 4th, 64-65; Crusader states, 269; *see* Egypt; *see also*, Saladin
Currency: Almohad 44-45; Almoravid 35, 44; Byzantine 30, 41-44; 57-60, 66; Castile 44; Crusader States 39; Dutch 23; Egypt 13, 17-18; 39-40; 43; 161-169; 172; 270, 274; Florence 56, 59, 272; Genoa 56, 59; India 207-211 *passim*; Ottoman 16-17, 270, 274, 277-278, 281-291; 305-307; Portugal 16; Prague 69; Sicily 48, 59; Spain 17; Surat 26; Venice 10, 30; 53-62 *passim*, 290-291; 306
Curtin, Philip D.: 310

Dahomey: 16, 233, 251, 308; *see* Africa; *see also* Takkrūr
Dai Viet Su Ky Toaz Thu: 370, 374; *see also* Vietnam
Darrag, Ahmad: 310
Davies, K.G.: 261
Day, John: 99, 101-03, 107, 113, 118, 121
Delhi: 13, 188, 199; *see* India
Delhi Sultanate: 192, 193, 198, 204-15 *passim*; Alauddin Khilji, 189, 194, 197, 209, 218; *see also* India
Dermegny, Louis: 459
De Roover, Raymond: 106
Deyell, John: 24
Digby, Simon: 198

East India Company: British, 22, 24, 247, 460, 463, 477, 479, 492: Dutch (V.O.C.) 22, 23-25, 447-75 *passim*; *see* India; *see also* China; Ceylon; England; Europe; Holland; Japan; Surat
Ecuador: 411; *see* New World
Egypt: 8, 11, 13, 16, 100, 103, 159-82, 231, 269, 305-28 *passim*, Coinage in, 305-09; Copper, 168-73 *passim*, European trade, 164-68 *passim*; Fatimids, 34, 38, 39, 43, 45, 73, Indian Trade, 164-73 *passim*; Sudanic gold flow interrupted, 172; *see*

Index

Africa; Mediterranean, Ottoman Empire; Syria; Takkrūr
Ehrenkreutz,: Andrew S: 159, 162
England: 12, 15, 21, 79-158 *passim*; 414, 417-18, 420, 428; *see* monetary contractions; *see also* East India Company; France; Low Countries; India
Ethiopia: 15, 18, 231-32, 305; gold from, 172, 319-21; *see* Egypt; *see also* Africa
Europe: 4, 7, 13, 18, 21, 24, 30, 53; bullion and metal exports to Asia, 447-75 *passim*

Fatimids: 34-39 *passim*, 160-69, *passim*, 231, 241; *See* Egypt, *see also*, Africa; Mamluks; Takkrūr
Fèbvre, Lucien: 6
Fisher, H.E.S.: 415
Fitch, Ralph: 351
Florence: 10, 11, 31, 51-52, 60, 67; *see* Italy; *see also*, Mediterranean; Venice
Fondaco dei Tedeshi: 64-67; *see* Venice
Foucault, Michael: 4-5
France: 12, 36, 74, 79-96 *passim*; *see* monetary contractions; *see also*, England, Europe; Low Countries

Gaastra, F.S.: 22-24
Gautier, E.F.: 32
Genoa: 10, 30, 41-45, 50-52, 67; in Morocco, 44; traders, 45; *see*, Italy; *see also*, Venice
Germany: 11, 16; silver merchants from, 53-54, 66; *see* Venice; *see also*, Europe
Ghana: 34, 36, 37, 238; *see* Gold Coast; *see also*, Africa
Ghaznavid Dynasty: 187-92, 208; and Amir Subuktigin, 189, 195; Mahmud of Ghazni, 189-91; Masud, 190; and the Seljuqs, 190-92; *see also*, India
Ghorid Dynasty: 207; Muhammad of Ghur, 189-93; *see also*, India
Glamann, K: 452, 460
Goa: 16, 232, 380; *see also*, India; Portugal
Gold Coast: 16, 232, 238-41, 247, 250-52; *see also*, Africa; Ghana
Goltein, S.D.: 38
Gombroon (Persia): 25, 456, 465, 481; *see also*, India
Ganzáles, Antonio García-Gaquero: 433
Gopal, Lallanjii: 196
Grierson, Philip: 31-32, 76
Gresham's Law: 40, 83, 110, 169, 171, 252, 258, 292
Gujarat: 25, 465, 479, 482-83, 493; *see* Mughul Empire, *see also*, India; Surat
Gujarat Sultanate: 214: *see also*, India

Habib, Irfan: 488
Hamilton E.J.: 5, 7, 418-20, 425-27, 432-33, 492
Hammarstrom, I: 492
Hasfid Dynasty: 10, 38, 45, 50; *see also* Africa, Egypt
Hendy, Michael: 42
Herlihy, David: 100
Hijaz: 75, 167-69, 171; pilgrimages from Takrūr 309, 312-15; *see also* Africa; Egypt; India; Mansa Musa; Takkrūr
Holland: 23, 449; *see also*, Amsterdam; Bullion and monetary flows; East India Company; traders, Dutch
Hundred Years War: 115-20 *passim*; *see* Europe

Ibn Batuta: 200, 202-03, 219, 270
Ibn Fadullah al Omari: 271
India: 4, 6, 13, 14, 16, 171, 183, 247; Ahamadabad, 483; Bahmani Sultanate, 184, 210-11; Baltic 208; Baroda 483; Bengal, 24, 184, 207-28 *passim*; Broach, 25, 483; Cambay, 25, 483, 491; Central Asia, 187-89; Coromandel, 453-66 passim; Delhi, 13, 188, 199; Delhi Sultanate, 192-92, 198, 204-15 *passim*; Gujarat Sultanate, 214; horse trading, 198-201, 205; intra-Asia trade, 455-91 *passim*; Islamic, 11, 18, 183-205 *passim*; Japanese silver, 25; kingdom Mewar, 214; Lahore, 13, 188, 190, 193; Lakhnavauti, 188, 199; Malabar Sultanate, 210; Marathas, 184, 197; Malwa Sultan, 214; Mongols: 189-200 *passim*; Mughal Empire, 18, 102, 477-96 *passim*; Muslim conquest; 185-93, 207-08; New World silver, 25, 488-91; pilgrimages, 202-05; Rajputs, 185-96, 207-08; Surat, 477-96 *passim*; Taiwanese gold, 25, 464; *see above*; *see also*: China; East India Company, Europe; Ghaznavid Dynasty; Ghovid Dynasty; Goa; Gombroom; Hijaz; Japan; Khandesh; Malabar; Mines; Mocha; Persia; Phillipines; Portugal; Shan States; Vietnam
Iran: 8, 17, 18; Mongols in, 269-70; *see also*, Persia
Iraq: 8, 17; *see also* Persia
Italy: 6, 11-13, 30, 44-47, 49, 52, 102, 118, 269; *see also*, Florence; Genoa; Pisa, Venice
Ives, Herbert E: 76
Iwahashi, Masaru: 330, 351

Japan: 15, 18-20, 329-62 *passim*; Ashigaga, 337, 346, 357; Dutch in, 460-67, 479; international trade, 347-51, 464, 479, 482, 492; Kamakura, 334, 336, 339;

mines in, 330-35, 346-48; Ritsuryō, 330-37, 339, 357; Sung China 331, 336-37; Tokugawa, 330, 346-48, 382-83; Vietnamese trade, 379-80; *see also*, China; India; Phillipines; Portugal; traders, Dutch; Vietnam
Jara, Alvara: 409
Java: 20, 457; and Vietnam, 375; *see also* Batavia; East India Company; Holland; India; Sumatra; traders, Dutch; Vietnam
Johnson, Marion: 232, 253, 260-61

Kamiki, Tetsuo: 19
Khandesh: 484, 488, 492; *see* Mughal Empire; *see also*, India; Surat
Kaun-nāme-i Sultânê: 272,274; *see* Ottoman Empire
Kharijites: 34, 38; *see also* Africa; Egypt; New East
Kingdom of Mewar: 214; *see also* India
Kobata, Atpushi: 330-52
Korea: 19, *see also* China; Japan

Labouret, Henri: 240
Lahore: 13, 188, 190, 191; *see also* India, Delhi Sultanate
Lane, Frederick: 55, 60-61
La Practica della Mercatura: 71-72: *see* Florence
Levant: 6, 12, 17, 4, 46, 49, 74, 76, 103, 166, 269; *see also* Africa; Egypt; Florence; Genoa; Italy; Mediterranean; Near East; Ottoman Empire; Pisa; Venice
Libya: 18; and trans-saharan gold trade, 317-19; *see also* Africa; Egypt; Fatimids; Mamluks; Mediterranean; Near East; Normans
Lima: 398-400; 412-19, 425-29; *see also* Brazil; Europe; New Spain; New World; Peru; Potosi; Phillipines; Spain
Lisbon: 241; *see also*, Brazil; New Spain; New World; Peru; Portugal
Lloyd, Terence: 105
Lombard, Maurice: 32-33
Lopez, Roberto: 44, 79, 80 86
Low Countries: 12, 17, 97-158; *see* monetary contractions; *see also* England; Europe; France
Lynch, John: 430

Magalhães - Godinho, Vitorino: 7, 240-41, 258-59, 261, 415-16
Macao: 458; *see also*, China; Japan; Portugal
Maghrib: 34-35; 44, 47, 232, 240; *see also* Africa; Algeria; Almohad Dynasty; Almoravid Dynasty; Egypt; Fatimids; Mamluks; Mediterranean; Venice
Malabar: 465-66; *see* Mughal Empire; *see also* India; Surat
Malabar Sultanate: 210; *see also* India
Malacca: 19, 20, 22, 375-77, 464; *see also*, Batavia; China; East India Company; Holland; India; Japan; Phillipines; traders, Dutch; Vietnam
Mali: 37, 103, 309-14; *see also* Africa; Egypt; Mansa Musai; Takkrūr
Malwa Sultanate: 214; *see also* India
Mamluks: 13, 15, 34, 75, 102, 269 Bahri; 160-170 *passim*; Circassian, 160-73 *passim*; *see also*, Africa; Ayyubid Dynasty, Egypt; Ethiopia; Mediterranean; Mongols; Near East; Ottoman Empire; Takkrūr
Mansa Musa: 37, 167, 309-14; *see also* Africa, Egypt, Hijaz, Mali, Takkrūr
Mansvelt, W. M. F.: 463-64
Marathas: 184, 197; *see also* India
Marco Polo: 200, 223, 374
Mayhew, N. J.: 86, 100-01, 119
Mediterranean: 4, 6-7, 11-18 *passim*, 26, 31-33, 35-39, 45-53; money zone, 232, 243; *see also*, Africa; England; Europe; Fatimids; Florence; Italy; Maghrib; Normans; Ottoman Empire; Venice
Meilink-Roelofsz, M. A. P.: 477
Messier, Ronald: 240
Mexico: *see* New Spain
Mines: Afghanistan 208; Africa 100; Anatolia 41, 291; Arab 33; Assam 186; Balkan 41; Black Sea 67, 69; Bohemia 65-69 *passim*, 99; Bolivia 401-406; Bosnia 68-69, 99; Brazil 404; Burma 222; Corinthia 64; China 358; Colombia 401-404 *passim*; East Africa 253-254; Egypt, (Nubia), 39; Ethiopia 161, 233; Hunan 216; Hungary 64-69 *passim*, 99; India 208; Japan 218, 331-335; Katang 2; Maghrib 64-69 *passim*, 99; Mali 238; Manchuria 216; Mexico 401, 403; New World 401-404; Peru 218; Rumelia 17, 272, 291; Salonika 272; Sardinia 99; Saxony 65; Silesia 65; Senegal 33, 238-240; Serbia 43, 54, 66-69, 99; Spain 16; Sudan 67, 162, 171; Tibet 186; Tyrol 218; Vietnam 370-376 *passim*; West Africa 161; Yunnan 216; Zimbabwe 235-238
Mining technology: Japan, 347-48; Medieval 99; Roman, 99; Shan States, 216-17; South German, 99, 117; Yunnan, 216-17
Misbach, Henry L.: 41
Miskimin, Harry A.: 12, 103-06, 108-21 *passim*
Mocha: 460, 465-67, 481, 491; *see* East India Company
Monetary contractions and bullion shortages: China, 383; England, 80-81, 97-158 *passim*; Europe, 120; France, 80-81; Italian, 79; Low Countries, 97-158 *passim*; New East, 79

Index

Mongols: Golden Horde, 201; India, 189-200 *passim*; Near East, 166-67, 185, 197, 200
Morocco: 49, 51, 67, 231-32, 243-44, 308; *see also* Africa; Eygpt; Mediterranean
Moreland, W. H.: 490
Mozambique: 16, 232-35; *see also* Africa; Ethiopia; Portugal; Takkrūr
Mughal Empire: 18, 102, 477-96; Akbar, 218, 482, 492; American Silver, 488-91; Aurangzeb, 484, 488, 493, expanding monetary system, 482-88; Jahangir, 482, 492; price revolution, 488-93; Shah Jahan, 484, 488; *see* India; *see also* Surat
Munro, John: 12

Near East: 44, 53, 102, 110, 118; *see also* Africa; England; Egypt; Europe; Florence; Genoa; Maghrib; Mediterranean; Ottoman Empire; Syria
Nef, John: 99, 117
New Spain: 21-24, 297-98; 412-13, 418, 425-29, 432; bullion production in, 397-423 *passim*, 425-45 *passim*; *see* New World; *see also*, Europe; India; Japan; Mediterranean; Ottoman Empire, Peru; Phillipines; Spain
New World: 4-8, 15-24; Asian trade, 464-67; bullion production in, 397-423 *passim*, 425-45 *passim*; contraband trade, 407-09, 415, 420, 436-39; government expenses in, 411, 428-32, 438; Phillipines, 425-45 *passim*; *see also* Africa; Egypt; Europe; India; Italy; Mediterranean; New Spain; Ottoman Empire; Peru; Phillipines; Portugal; Spain
Nigeria: 232, 251; *see also*, Africa; Takkrūr
Normans: 38, 46-47

Ottoman Empire: 10, 15-18, 160, 172, 269-304 *passim*; currency, 273-89; New World silver, 272, 279-84 *passim*; trimetalism in, 288-29; *see also*, Africa; Europe; Egypt; Italy; Mediterranean; New World; Syria; Venice

Panama: 402, 412-13, 429, 435; *see also*, Europe; New World; Peru; Phillipines
Pankhurst, Richard: 23
Patterson, C. C.: 100-01, 218
Pereira, Pacheco: 240
Perlin, F.: 388
Persia: 455, 465-66; *see also*, Eygpt; Iran; Iraq; Ottoman Empire
Persian Gulf: 14, 15, 16, 232; *see also* Egypt; India; Nediterranean; Near East
Peru, Viceroyalty of: 21, 22, 24, 397-445 *passim*; bullion production, 397-445 *passim*; Charcas mining district, 406-09, 420; contraband trade, 407-09, Potosi mines, 21, 398-406 *passim*, 411-15, 420, 425, 428; *see also* Europe; Mediterranean; New Spain; New World; Phillipines; Spain
Phillipines: 20-22; government expenses, 433-34; and intra-Asian trade, 434; and Vietnam, 379; *see also*, China; Europe; Japan; New Spain; New World; Spain; Vietnam
Pires, Tomé: 363, 376
Pisa: 10, 35, 43, 50, 52; *see* Italy; *see also*, England; Europe; Florence; France; Genoa; Mediterranean; Near East; Ottoman Empire
Polanyi, Karl: 3
Portugal: 7, 8, 15, 19, 21, 24, 416-20; *see also*, Africa; Brazil; China; Europe; India; Japan; Mediterranean; New World; Spain; traders, Dutch; traders, Portuguese; Vietnam
Postan, M. M.: 81, 99-00, 104-05, 108
Postma, Johannes: 261
Potosi mining: 21, 398-406 *passim*, 411-15, 420, 425, 428; *see* Peru; *see also*; Brazil; Europe; New Spain; New World; Phillipines; Spain

Ragusa: 11, 16, 65, 68, 75; *see* Italy; *see also*, Florence; Genoa; Mediterranean; Venice
Raymond, André: 306, 308, 315
Red Sea: 15, 18, 26, 232; *see* Africa, *see also* Eygpt; Europe; Florence; Genoa; Hijaz; India; Italy; Mediterranean; Near East; Takkrūr
Restrepo, Vincente: 402, 410
Robbert, Louise Buenger: 11
Robie, Hassanein: 163, 167
Robinson, W. C.: 104, 119
Royal Africa Company, British: 244-257 *passim*; *see* Africa; *see also* England; Europe; Takkrūr
Rumelia: 17, 284; Ottoman conquest, 272, 277; *see also* Europe; Florence; Mediterranean; Ottoman Empire; Venice

Sahillioğlu Halil: 16
Saladin: 160-64, 172; *see* Eygpt; *see also* Ayyubid Dynasty; Crusades; Fatimids; Mamlukes, Mediterranean; Near East; Syria
Schurty, William Lytle: 434
Seljuqs: 269-70; *see also*, Mongols; Near East; Ottoman Empire
Senegal: 16, 35, 238-39, 250-51; *see* Africa; *see also* Takkrūr
Shafer, E. H.: 371
Shan States: 216-17, 280-24; *see* Bengal; *see also* Burma; China; India; Yunnan
Shimbo, Itiroshi: 330

Sicily: 10, 38, 44, 47-51; *See* Africa; *see also,* Florence; Genoa: Italy, Normans; Venice
Smith, Adam 5
Soetbeer, Adolf: 402
Spain: 4, 7, 17, 21, 35, 37, 44-45, 378-79, 412-17, 425-45 *passim, see* New World; *see also,* Europe; New Spain, Peru; Phillipines; Portugal
Spufford, Peter: 121
Steensgaard, Niels: 477
Sudan: *see* Takkrūr; *see also* Africa; Egypt
Sumatra: 455, 466; *see also* East India Company; Holland; India; Java; traders, Dutch
Summers, Roger: 238
Surat: 25, 26, 477-96 *passim*; intra-Asian trade, 455, 462, 466, 477; and New World silver, 479, 481; *see* Mughal Empire; *see also* East India Company; India; traders, Dutch; traders, Portuguese
Syria: 8, 13, 16-17, 39-43, 49, 74-75, 165-67, 269; *see* Egypt; *see also* Ottoman Empire
Tadic, J.: 27
Takkrūr: 18, 168, 171, 232-33, 241, 305-28 *passim*; gold and silver exchanges with Egypt, 305-28 *passim*; gold exchanges with Hijaz, 309, 312-15, with Italy, 309; gold in private trade, 315-17; monetary developments, 321-24; *see* Africa; *see also* Egypt; Mansa Musa; Mediterranean; Ottoman Empire
Te Paske, John J.: 22, 24
Thompson, J. D. A.: 107
Timur-i-lang (Amir Timur Gurgan): 189, 194-95, 211, 214, 272; *see* India
Traders, Dutch: 15, 16, 24, 26, 243-44, 251, 255, 378, 380-88 *passim*, 414, 436, 447-75 *passim*; *see* East India Company; *see also,* Ceylon; Holland; India; Japan
Traders, Portuguese: 25, 214, 232-255 *passim*; 305, 310-11, 347-48, 414-151, 451, 477, 480-88, 477, 483; *see* Portugal; *see also,* Africa; Brazil, China; Goa; India; Japan; New World; Phillipines
Tunisia: 10, 11, 37-38, 47-49, 67, 241; *see* Africa; *see also* Egypt
Udovitch, Abraham: 102, 310
Ummayyads: 34, 37; *see* Africa; *see also* Spain
Van der Wee, Herman: 117
Van Leur, J. C.: 477
Venice: 11-12, 16, 42, 50, 53-77 *passim*; monetary zone, 69-77; *see* Italy; *see also* Africa; Europe; Mediterranean; Ottoman Empire
Vietnam: 15, 19-20, 363-93 *passim*; metal productions, 370-73, 76; monetary system, 365-70, 388; New World Silver, 381, 383, 388; Nguyen Family, 20, 378-85, 388; *see* China; *see also* Japan, Phillipines
Vilar, Pierre 7, 8
Villena, Guillermo Lohmann 408
Vincente, Maria Rodrequez: 418-19
von Heyd, Wilhelm: 47
Watson, Andrew: 12, 32, 34, 59, 81, 83, 110-11, 160
West India Company, Dutch: 244, 247, 488; *see* Holland; *see also* East India Company
Whitehead, R. B.: 483
Yamamura, Kozo: 19
Zilbaldone da Canal: 66-67, 71: *see* Venice
Zirid Dynasty: 37, 240; *see* Egypt; *see also* Africa